Cover crops for sustainable soil management

It is widely recognised that agriculture is a significant contributor to global warming and climate change. Agriculture needs to reduce its environmental impact and adapt to current climate change whilst still feeding a growing population, i.e. become more 'climate-smart'. Burleigh Dodds Science Publishing is playing its part in achieving this by bringing together key research on making the production of the world's most important crops and livestock products more sustainable.

Based on extensive research, our publications specifically target the challenge of climate-smart agriculture. In this way we are using 'smart publishing' to help achieve climate-smart agriculture.

Burleigh Dodds Science Publishing is an independent and innovative publisher delivering high quality customer-focused agricultural science content in both print and online formats for the academic and research communities. Our aim is to build a foundation of knowledge on which researchers can build to meet the challenge of climate-smart agriculture.

For more information about Burleigh Dodds Science Publishing simply call us on +44 (0) 1223 839365, email info@bdspublishing.com or alternatively please visit our website at www.bdspublishing.com.

Related titles:

Advances in Conservation Agriculture Volume 1
Print (ISBN 978-1-78676-264-1); Online (ISBN 978-1-78676-266-5, 978-1-78676-267-2)

Advances in Conservation Agriculture Volume 2
Print (ISBN 978-1-78676-268-9); Online (ISBN 978-1-78676-270-2, 978-1-78676-271-9)

Improving soil health
Print (ISBN 978-1-78676-670-0); Online (ISBN 978-1-78676-672-4, 978-1-78676-673-1)

Managing soil health for sustainable agriculture Volume 1
Print (ISBN 978-1-78676-188-0); Online (ISBN 978-1-78676-190-3, 978-1-78676-191-0)

Managing soil health for sustainable agriculture Volume 2
Print (ISBN 978-1-78676-192-7); Online (ISBN 978-1-78676-194-1, 978-1-78676-195-8)

Understanding and preventing soil erosion
Print (ISBN 978-1-80146-379-9); Online (ISBN 978-1-80146-380-5, 978-1-80146-381-2)

Chapters are available individually from our online bookshop:
https://www.bdspublishing.com/webshop/

climate•SMART•publishing
IN AGRICULTURAL SCIENCE

BURLEIGH DODDS SERIES IN AGRICULTURAL SCIENCE
NUMBER 176

Cover crops for sustainable soil management

Edited by Professor Sjoerd W. Duiker, The Pennsylvania State University, USA and Dr Andy Clark, formerly University of Maryland and USDA Sustainable Agriculture Research and Education (SARE), USA

burleigh dodds
SCIENCE PUBLISHING

Published by Burleigh Dodds Science Publishing Limited
82 High Street, Sawston, Cambridge CB22 3HJ, UK
www.bdspublishing.com

Burleigh Dodds Science Publishing, 1518 Walnut Street, Suite 900, Philadelphia, PA 19102-3406, USA

First published 2026 by Burleigh Dodds Science Publishing Limited

Library of Congress Control Number: 2026935171

British Library Cataloguing in Publication Data
A catalogue record for this book is available from the British Library

ISBN 978-1-83545-120-5 (print)
ISBN 978-1-83545-121-2 (ePub)
ISBN 978-1-83545-122-9 (PDF)
ISSN 2059-6936 (print)
ISSN 2059-6944 (online)

DOI: 10.19103/AS.2025.0162

Typeset by Deanta Global Publishing Services, Dublin, Ireland

Contents

Part 2 Managing cover crops

Series list

Title	Series number
Achieving sustainable cultivation of maize - Vol 1 From improved varieties to local applications Edited by: Dr Dave Watson, CGIAR Maize Research Program Manager, CIMMYT, Mexico	001
Achieving sustainable cultivation of maize - Vol 2 Cultivation techniques, pest and disease control Edited by: Dr Dave Watson, CGIAR Maize Research Program Manager, CIMMYT, Mexico	002
Achieving sustainable cultivation of rice - Vol 1 Breeding for higher yield and quality Edited by: Prof. Takuji Sasaki, Tokyo University of Agriculture, Japan	003
Achieving sustainable cultivation of rice - Vol 2 Cultivation, pest and disease management Edited by: Prof. Takuji Sasaki, Tokyo University of Agriculture, Japan	004
Achieving sustainable cultivation of wheat - Vol 1 Breeding, quality traits, pests and diseases Edited by: Prof. Peter Langridge, The University of Adelaide, Australia	005
Achieving sustainable cultivation of wheat - Vol 2 Cultivation techniques Edited by: Prof. Peter Langridge, The University of Adelaide, Australia	006
Achieving sustainable cultivation of tomatoes Edited by: Dr Autar Mattoo, USDA-ARS, USA and Prof. Avtar Handa, Purdue University, USA	007
Achieving sustainable production of milk - Vol 1 Milk composition, genetics and breeding Edited by: Dr Nico van Belzen, International Dairy Federation (IDF), Belgium	008
Achieving sustainable production of milk - Vol 2 Safety, quality and sustainability Edited by: Dr Nico van Belzen, International Dairy Federation (IDF), Belgium	009
Achieving sustainable production of milk - Vol 3 Dairy herd management and welfare Edited by: Prof. John Webster, University of Bristol, UK	010
Ensuring safety and quality in the production of beef - Vol 1 Safety Edited by: Prof. Gary Acuff, Texas A&M University, USA and Prof. James Dickson, Iowa State University, USA	011
Ensuring safety and quality in the production of beef - Vol 2 Quality Edited by: Prof. Michael Dikeman, Kansas State University, USA	012
Achieving sustainable production of poultry meat - Vol 1 Safety, quality and sustainability Edited by: Prof. Steven C. Ricke, University of Arkansas, USA	013
Achieving sustainable production of poultry meat - Vol 2 Breeding and nutrition Edited by: Prof. Todd Applegate, University of Georgia, USA	014

Introduction

Soil is the foundation of agricultural production and a critical natural resource underpinning food security, environmental quality and human well-being. Soil health is commonly defined as the continued capacity of soil to function as a vital living ecosystem that sustains plants, animals, and humans. Central to this concept is soil life. A vast majority of soil biological activity occurs in the rhizosphere, the thin zone surrounding living roots, where complex interactions among plant roots, microorganisms, and soil fauna regulate nutrient cycling, aggregation, water dynamics and disease suppression. Consequently, one of the most widely recognised principles of soil health management is maintaining the continuous presence of living roots in the soil.

Modern agricultural systems, however, often violate this principle. Annual cropping systems in particular may leave soils bare and biologically inactive for extended periods between harvest and planting of cash crops. These fallow intervals increase vulnerability to erosion, nutrient losses, declines in soil organic carbon, and reduced biological functioning. Cover crops offer a practical and increasingly important solution to this challenge.

Cover crops are grown primarily to benefit the soil and agroecosystem rather than for direct harvest. They provide soil protection, seedbed improvement and soil enhancement between periods of normal crop production. They are known by many names that reflect their diverse functions and management contexts, including green manures, living mulches, undersown or companion crops, smother crops and catch crops. Regardless of terminology, their shared role is to occupy ecological niches left vacant by cash crops, thereby restoring biological activity, protecting soil resources and enhancing system function and resilience.

Over the past century, agricultural intensification, supported by advances in science, technology, and mechanisation, has dramatically increased productivity and improved food availability worldwide. At the same time, these gains have often been accompanied by unintended environmental and social consequences, including degradation of soil structure, wind and water erosion, eutrophication of surface waters from nutrient losses, depletion of soil organic carbon, greenhouse gas emissions, loss of biodiversity and natural habitat, and socio-economic pressures on rural communities. Addressing these challenges requires management practices that reconcile productivity with environmental stewardship. Cover crops are widely recognised as a Best Management Practice capable of contributing to such outcomes.

A growing body of research now documents the multifunctional benefits of cover crops across a wide range of agroecosystems. These benefits include improved nutrient cycling and retention, enhanced soil biological activity and biodiversity, increased soil carbon sequestration, greater water infiltration and water-holding capacity, suppression of weeds, pests and diseases, and expanded opportunities for integrated crop-livestock systems. Cover crops can also buffer cropping systems against climate variability by increasing infiltration, reducing erosion during extreme rainfall events and improving soil resilience during drought. At the same time, cover crop management involves trade-offs and challenges, including costs of establishment, risks to cash crop yields under certain conditions, and difficulties in termination. Understanding how to optimise species selection, mixtures, timing, and management is therefore critical for successful adoption.

This edited volume brings together leading researchers and practitioners to synthesise current knowledge on the role of cover crops in sustainable soil management. The chapters draw on advances in soil science, agronomy, ecology and farming systems research to provide a comprehensive assessment of how cover crops function, how they can be effectively managed, and how their use may evolve in the future.

The book is organised into four parts. Part 1 focuses on understanding the roles of cover crops, including their effects on nutrient cycling, soil biological health, biodiversity, pest and disease dynamics, weed suppression, forage production and cash crop yields. Part 2 addresses practical aspects of managing cover crops, including decision-support tools, establishment strategies, and challenges associated with termination. Part 3 provides detailed descriptions of major cover crop groups (leguminous, gramineous, and broadleaf species) across cool and warm season contexts. Part 4 looks ahead, examining strategies to increase adoption and exploring future opportunities and challenges for cover crops in a changing agricultural and climatic landscape.

Together, these chapters provide an integrated, science-based resource for researchers, extension professionals, students, policymakers, and farmers seeking to better understand and implement cover crops as a cornerstone of sustainable soil management. By synthesising current research and identifying knowledge gaps, this volume aims to support the wider adoption and continued improvement of cover cropping systems that enhance both agricultural productivity and environmental outcomes.

Part 1 Understanding the role of cover crops

Chapter 1 examines how cover crops enhance macronutrient cycling within agricultural systems. The chapter frames improved nutrient cycling as central to the development of sustainable cropping systems and organises the discussion around four primary mechanisms: (i) biological nitrogen fixation, (ii) reduction of nutrient losses from agroecosystems, (iii) improvement of soil health processes that regulate nutrient transformations, and (iv) nutrient supply to subsequent crops. Processes such as nutrient scavenging, reduced leaching and runoff, enhanced aggregation, and mitigation of nitrous oxide (N_2O) emissions are also explored.

The next chapter explores how cover crops influence soil microbial communities, soil fauna and overall biological functioning. Chapter 2 emphasises the role of root exudates, residue inputs, and rhizosphere processes in shaping microbial diversity and activity. The chapter examines how increased biological activity contributes to improved soil structure, enhanced nutrient mineralisation, disease suppression and greater resilience to environmental stress.

Building on themes in the preceding chapter, Chapter 3 examines how cover crops enhance both above- and below-ground biodiversity in agricultural landscapes. Biodiversity is treated as a functional attribute influencing productivity, ecosystem stability and long-term sustainability. Particular attention is given to arthropods which play pivotal roles in food webs, nutrient cycling, pest suppression and pollination. Evidence from perennial and annual systems shows that cover crops can increase arthropod abundance and diversity, though responses depend on species selection, management practices, landscape context and assessment metrics.

Following on from the preceding chapter, Chapter 4 evaluates the ecological mechanisms through which cover crops regulate pest and pathogen dynamics. Cover cropping is presented as a biodiversity-driven pest management strategy that supports natural enemies, disrupts pest life cycles and, in some cases, exerts bio-fumigation effects. Management practices such as mowing regimes, flowering insectary species, irrigation and strategic termination are discussed in the context of integrated pest management (IPM). The chapter shows that successful pest suppression depends on ecological context, implementation strategy and long-term system design.

Building on the previous discussion of pest management, Chapter 5 analyses how cover crops can serve as foundational components of integrated weed management (IWM) systems, particularly in the face of increasing herbicide-resistant weed populations. The chapter explains how cover crops impose ecological stressors that reduce weed fitness and population growth. It further explores how these ecological tactics can be strategically integrated with herbicide programs to slow resistance evolution and reduce reliance on chemical inputs. Management variables influencing weed suppression outcomes are reviewed alongside associated agronomic trade-offs.

Chapter 6 explores the integration of cover crops into livestock systems and mixed farming operations. Beyond ecological services, cover crops can provide high-yield, high-quality forage for grazing or stored feed. When harvested for feed, however, they must be managed more like cash crops, with attention to species selection, planting date, fertility and harvest timing.

The chapter reviews both grazing and stored-feed systems, highlighting how integrated crop-livestock approaches can enhance nutrient cycling, improve whole-farm profitability, and extend grazing seasons while maintaining soil conservation goals.

Concluding Part 1, Chapter 7 synthesises global research on yield responses following cover crop use. The chapter recognises that yield effects are variable and highly context dependent. It evaluates how agronomic variables (species, biomass, termination timing), pedoclimatic conditions and main crop type interact to determine yield outcomes. Potential mechanisms influencing yield (including improved soil moisture retention, enhanced nutrient availability, pest suppression and temporary nitrogen immobilisation) are examined.

Part 2 Managing cover crops

The first chapter of Part 2 reviews the development of tools and frameworks designed to improve species selection and management decisions. Chapter 8 highlights advances emerging from the Precision Sustainable Agriculture (PSA) network, including site-specific species selectors and nitrogen calculators applicable across diverse U.S. agroecosystems. It discusses limitations of earlier tools and outlines future needs, including improved modelling of interactions between genetics, environment and management. The chapter emphasises that decision tools must evolve alongside expanding datasets and farmer experience.

Chapter 9 addresses one of the primary barriers to successful adoption: timely and effective establishment. Because cover crop performance is sensitive to growing degree days, rainfall and local climate, establishment strategies must be site-specific. The chapter compares different establishment techniques: drilling, broadcasting with incorporation, surface broadcasting, interseeding, overseeding and frost seeding, outlining advantages and limitations of each. Drilling after harvest is generally the most consistent method when sufficient growing season remains. Interseeding and overseeding allow earlier establishment but introduce competition with the cash crop. The chapter also discusses the effects of burndown and residual herbicides on establishment success, noting particular sensitivity of legumes and brassicas to certain residual chemistries.

Complementing the previous chapter, Chapter 10 examines natural, chemical and mechanical termination strategies. Winterkill species offer natural termination options, while herbicides remain widely used for predictable control. Mechanical approaches–including mowing, tillage, roller-crimping, and undercutting–are increasingly important in reduced-tillage and organic systems. The chapter discusses termination timing relative to biomass accumulation, soil moisture and subsequent crop planting, emphasising the risk of excessive competition if termination is delayed. Case studies compare effectiveness of different methods and highlight implications for herbicide resistance management, soil cover retention and operational logistics.

Part 3 Cover crop species

Part 3 reviews the broad range of available cover crop species, divided broadly into cool and warm season species. Chapter 11 reviews temperate legume species that contribute biologically-fixed nitrogen through symbiosis with rhizobia. The chapter discusses species selection, inoculation requirements, seeding rates and termination timing, and compares monoculture and mixed-species systems. It also explores opportunities for harvesting select forage or grain legumes while maintaining soil health benefits.

Chapter 12 examines species such as Sunn hemp, cowpea and late-maturing soybean that produce large quantities of biomass and fixed nitrogen within short growing windows. These species can be planted after main crops, intercropped, or grown ahead of cash crops when sufficient growing degree days are available. Their aggressive growth contributes to weed suppression, erosion control, soil organic matter accumulation and improved microbial activity, though potential drawbacks that need managing include soil moisture competition and variable nitrogen release dynamics.

Chapter 13 focuses on grasses such as winter rye, wheat, triticale, barley, oats, black oats and annual ryegrass. These species excel at nutrient scavenging, erosion control and biomass production. Winter rye, widely used in corn-soybean systems, is highlighted for its rapid growth and winter hardiness. Differences between grasses and legumes in decomposition rate, nitrogen dynamics and functional outcomes are examined.

Complementing the previous chapter, Chapter 14 reviews sorghums, millets, teff and related species. Their rapid biomass production and extensive fibrous root systems improve soil aggregation, porosity, water infiltration and nutrient capture. Dense canopies and allelopathic effects enhance weed suppression. The chapter also addresses potential limitations such as nitrogen immobilisation and soil moisture depletion, alongside breeding advances and emerging management practices to optimise benefits.

Finally, Chapter 15 examines brassicas and other broadleaf plants that provide benefits such as nutrient capture, bio-fumigation potential, soil structural improvement, pollinator support and weed suppression. Growth characteristics, environmental adaptation, and management considerations are reviewed, along with the roles of these species in enhancing soil physical, chemical, and biological health.

Part 4 The future of cover crops

Part 4 concludes the book by looking at future challenges and how they might be addressed. Chapter 16 traces the historical evolution of cover crop use, from erosion control and nitrogen provision to water quality protection, soil health restoration and improving climate resilience. The chapter examines socio-economic, technical, and policy factors influencing adoption, including cost-share programs, risk management tools, farmer learning networks and extension programs. Achieving cumulative benefits will depend on scaling implementation across diverse production systems.

The final chapter of the book synthesises where research is going and knowledge gaps. Chapter 17 considers how climate change, technological innovation, improved modelling tools, and evolving policy frameworks may shape future adoption. The chapter challenges researchers and practitioners to refine species selection, improve system integration, and strengthen economic and ecological performance to ensure that cover crops remain central to sustainable agricultural intensification.

Part 1

Understanding the role of cover crops

Chapter 1

The role of cover crops in optimizing nutrient cycles

Yajun Peng, Claudia Wagner-Riddle and Laura L. Van Eerd, University of Guelph, Canada

1 Introduction

During the last few decades, research has resulted in significant advances in understanding the role of cover crops in sustainable agroecosystems, particularly in improving nutrient cycling. As addressed in other chapters, using cover crops contributes to soil structure improvement, soil erosion reduction, soil carbon (C) sequestration, and soil biological health (e.g. Blanco-Canqui and Ruis, 2020; Jian et al., 2020; Kim et al., 2020), which directly and indirectly influence nutrient cycling and losses. Fertilizer inputs to provide nutrients are a significant cost of production; thus, optimizing nutrient availability to crops via cover crops is a key component of sustainable agriculture. Nutrient losses that are from agroecosystems, especially the losses of nitrogen (N) and phosphorus (P), cause eutrophication, biodiversity loss, and in the case of N-related climate change (Tilman, 1999). Therefore, we need to comprehensively understand the role of cover crops in reducing nutrient losses and improving nutrient availability to subsequent crops.

Despite many studies to evaluate cover crop effects on crop agronomy and environmental impacts (e.g. Scavo et al., 2022; Crespo et al., 2021; Nitu et al., 2021; Hallama et al., 2019; Blanco-Canqui et al., 2015; Dabney et al.,

http://dx.doi.org/10.19103/AS.2025.0162.01

2010; Eriksen, 2009), several challenges remain in improving nutrient cycling with cover crops. The key challenges include how to select and manage legume cover crops to improve biological N fixation (BNF); how to improve the synchrony between soil supply (i.e. nutrient release from cover crops or organic matter mineralization) and crop demand (Fontaine et al., 2024; Hansen et al., 2021; Nevins et al., 2020); and how long-term cover cropping (≥10 years) affects soil health, soil microbial community, and nutrient cycling. Moreover, nutrient cycles have some common aspects, such as mobilization-immobilization, and only a small fraction of nutrients (i.e. inorganic forms) in soil are available for plant uptake; however, each nutrient has specific aspects, such as gaseous forms of N and sulfur (S) but not of P, as well as biologically fixed N (Murrell et al., 2021; Bünemann et al., 2011; Stevenson and Cole, 1999). To address these challenges, it is important to understand the mechanisms of how cover crops optimize nutrient cycling of these most impacted macronutrients (i.e. N, S, and P).

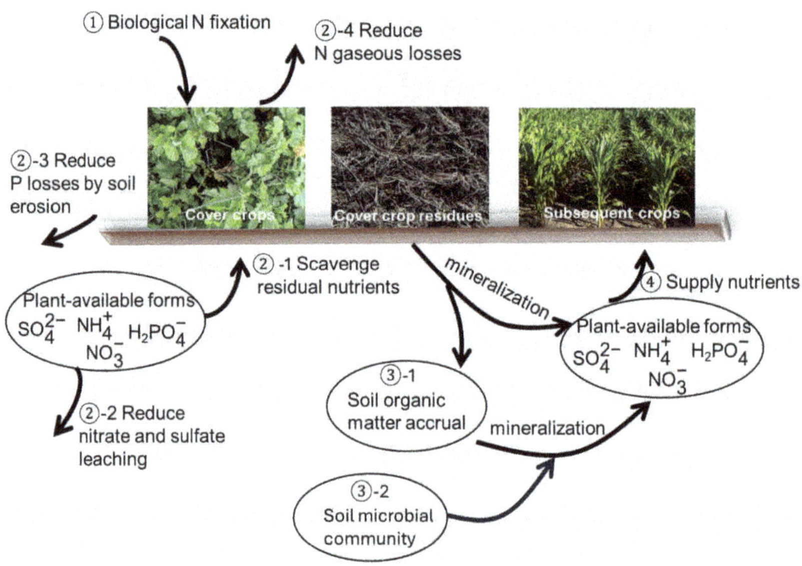

Figure 1 Graphical representation of the four major mechanisms (① to ④) of how cover crops influence nutrient cycling in agroecosystems. ① Legume cover crops can directly add nitrogen to the system via biological N fixation. Cover crops can reduce nutrient losses from the system by ②-1 scavenging soil residual nutrients, which mitigates ②-2 leaching, ②-4 nitrogen gaseous losses, and ②-3 soil erosion. After long-term use of cover crops, the positive effect on ③-1 soil organic matter content and ③-2 microbial community may improve the capacity of soil to cycle nutrients. The decomposition of cover crop residues ④ supplies nutrients to subsequent crops. Nitrogen: N, phosphorus: P, nitrate: NO_3^-, ammonium: NH_4^+, sulfate: SO_4^{2-}, dihydrogen phosphate: $H_2PO_4^-$. Note: Cover crops can cause immobilization of nutrients (e.g. N) and the increase in N gaseous emissions, under certain circumstances, which is discussed in the text but is not delineated in Figure 1 (Photo credit: Yajun Peng).

In this chapter, four primary mechanisms of how cover crops influence nutrient cycling are discussed: (1) biologically fixing atmospheric dinitrogen (N_2); (2) reducing nutrient losses by mitigating leaching, soil erosion, and gaseous forms; (3) improving soil health (soil organic matter and soil microbial community) to cycle nutrients; and (4) supplying nutrients to subsequent crops (Fig. 1). These four mechanisms are not the only means by which cover crops influence nutrient cycling, but they are dominant pathways in agroecosystems that affect crop growth. These mechanisms are neither independent nor isolated, as demonstrated via the case study of recent research. A ranking system of the performance of common cover crops towards each mechanism is provided (Table 1). Moreover, we structure this chapter by starting with N as it is the macronutrient in most demand for crop growth, followed by P and S nutrients when they apply to specific mechanisms (e.g. particulate P loss by soil erosion).

2 Adding biologically-fixed atmospheric nitrogen to systems

Nitrogen is an essential macronutrient required in large amounts for crop growth. Legume cover crops can provide large quantities of N via BNF (Herridge et al., 2008), and thereby farmers can reduce N fertilizer inputs to main crops that follow legume cover crops (Nitu et al., 2021; Soumare et al., 2020; Howard and Rees, 1996). However, the amount of N via BNF and fertilizer equivalence varies among legume cover crops (Nitu et al., 2021). Various methodologies have been used to estimate BNF (Unkovich et al., 2008). The most common approach is based on ^{15}N natural abundance, which has been widely used for estimating BNF from legume cover crops (Perrone et al., 2020; Parr et al., 2011). The proportion of N in aboveground legume cover crop biomass derived from BNF varied among species from 3% to 100% (Büchi et al., 2015). As a result, the typical levels of N fixation from legume cover crops are 30 kg N ha^{-1} yr^{-1} to 200 kg N ha^{-1} yr^{-1} (Nitu et al., 2021). Correspondingly, the N fertilizer equivalence of legume cover crops to subsequent non-legume crops also varies (e.g. 12 kg N ha^{-1} to 182 kg N ha^{-1}; Fageria et al., 2005). Given these ranges of values, there is a need for future research that provides better predictions of N fertilizer equivalence.

The wide range of BNF and N fertilizer equivalence values is mainly due to the complex interactions among legume species, planting and termination dates, soil types, and climate conditions (Perrone et al., 2020; Carranca et al., 1999). For instance, spring termination compared to fall termination of alfalfa (*Medicago sativa*) and red clover (*Trifolium pretense*) provided optimal N to the subsequent corn crop and to a greater extent than crimson clover (*Trifolium incarnatum*; Coombs et al., 2017). This could be because with spring termination the legume cover crops had (1) higher N content, (2) better synchrony between

Table 1 Ranking system of the expected performance of how select cover crops optimize nutrient cycles

Cover crops	Biologically fix nitrogen	Scavenge nutrients	Reduce soil and nutrient losses	Improve soil health
Legumes				
Clover, Crimson (*Trifolium incarnatum*)	●●●●	●	●●	●●●
Clover, Red (*Trifolium pratense*)	●●●●	●●	●●●	●●●●
Clover, White (*Trifolium repens*)	●●	●	●●●	●●
Field/Winter Pea (*Pisum sativum* subsp. *arvense*)	●●●	●	●●	●●
Hair Vetch (*Vicia villosa*)	●●●●	●	●●	●●
Grasses				
Annual Ryegrass (*Lolium multiflorum*)		●●●●	●●●●	●●●●
Barley (*Hordeum vulgare*)		●●●	●●●●	●●●
Japanese Millet (*Echinochloa* spp.)		●●●●	●●●●	●●●
Oat (*Avena sativa*)		●●●	●●●	●●●
Pearl Millet (*Pennisetum glaucum*)		●●●●	●●●●	●●●
Sudangrass (*Sorghum bicolor*)		●●●●	●●●●	●●●●
Winter Cereal Rye (*Secale cereale*)		●●●●	●●●●	●●●●
Non-legume Broadleaves				
Buckwheat (*Fagopyrum esculentum*)		●	●	●●●
Mustards (*Brassica juncea*)		●●●	●●	●●
Radish (*Raphanus sativus*)		●●●	●	●●
Turnip (*Brassica rapa*)		●●●	●●	●●

Note: The ranking system is based on the Midwest Cover Crops Field Guide (Third Edition) and Midwest Cover Crops Council (midwestcovercrops.org) and assumes expected growth to achieve the performance. More dots indicate a greater ability to optimize nutrients under the given mechanism (the maximum number of dots is four).

cover crop N mineralization and crop N demand, and (3) less N lost from the system, compared to fall termination (Coombs et al., 2017). However, growing conditions (i.e. weather and soil conditions) are a more significant driver of BNF than cover crop species and management (e.g. termination dates; Peng et al., 2023; Coombs et al., 2017; Parr et al., 2011). This is because growing conditions not only affect the establishment and growth of legumes but also their BNF and inorganic N uptake (Guinet et al., 2018). A low inorganic N condition might enhance BNF to meet the N requirements of legumes (Guinet et al., 2018). Hence, more site-specific research to understand BNF is critical to optimizing N supply to subsequent crops and increasing yields.

3 Reducing nutrient losses from the system

3.1 Scavenging post-harvest nutrients via cover crop uptake

In agricultural systems, after the harvest of main crops, residual nutrients are in the field as crop residues or the soil (Torma et al., 2018). Practically, these residual nutrients are potential resources for subsequent main crops. However, during the fallow period that follows crop harvest and before the next crop planting (i.e. no living plants in the field), residual nutrients can be lost through erosion, runoff, leaching, volatilization, and denitrification. Rather than a fallow period, cover crops can be utilized to mitigate the losses of residual nutrients (Hirsh et al., 2021; Kaye et al., 2019). Cover crops have different acquisition strategies for residual nutrients; hence, the key question is how to select cover crops to optimize the acquisition of specific nutrients.

The ability of cover crops to uptake residual nutrients largely relies on the level of plant-available residual nutrients (i.e. inorganic form), the amount of aboveground biomass, and the species of cover crops, specifically their leaf and root traits (Griffiths et al., 2022; Wendling et al., 2016; White et al., 2013). Under high nutrient conditions (fertile soil), cover crop species with the greatest production of biomass (especially shoots) have the greatest acquisition of nutrients (Wendling et al., 2016). Five nutrient acquisition strategies based on leaf and root traits of 20 cover crop species were comprehensively explored in fields with high soil fertility (Switzerland; Wendling et al., 2016). They reported that cover crop species with large root length density (large root diameter and root area) and large shoot biomass production are recommended to uptake residual nutrients. These species include sunflower (*Helianthus annuus*), white mustard (*Sinapis alba*), phacelia (*Phacelia tanacetifolia*), turnip rape (*Brassica rapa campestris*), and oilseed radish (*Raphanus sativus oleiformis*). In contrast, under low nutrient conditions (infertile soil), the acquisition of nutrients is controlled primarily by cover crop root traits (Griffiths et al., 2022; Wendling et al., 2016; White et al., 2013). Crucifer cover crops such as winter rape (*Brassica napus*), phacelia, and fodder radish (*Raphanus sativus*) with deep root penetration and fast establishment are recommended to facilitate N and S uptake from subsoil layers (>1 m soil depth; Thorup-Kristensen, 2001). These cover crops with vigorous and deep roots bring nutrients from deeper soil layers to the aboveground biomass and ultimately the soil surface, which after mineralization is an available nutrient source for subsequent crops. However, cover crops with 'topsoil foraging' root systems are recommended to take up P, because P is relatively immobile in the soil profile and is concentrated in the topsoil (Griffiths et al., 2022; Lynch et al., 2001), especially under no-tillage systems. Clearly, cover crop approaches to scavenge nutrients need to be nutrient- and species-specific.

3.2 Decreasing nitrate and sulfate leaching during the non-growing season

When cover crops scavenge residual nutrients (also water), they reduce nutrients available for losses via leaching (e.g. nitrate (NO_3^-) and sulfate; Nouri et al., 2022; Thapa et al., 2018; Eriksen and Thorup-Kristensen, 2002). Higher evapotranspiration with cover crops compared to bare soils reduces the amount of water that moves through the soil profile (Nouri et al., 2022; Meyer et al., 2019). Studies have reported that cover cropping decreased NO_3^- leaching by 40% to 70% over the non-growing season, compared to a no cover crop control, based on high-frequency weighing lysimeter systems (e.g. Lapierre et al., 2022) and global meta-analyses (Nouri et al., 2022; Thapa et al., 2018). The adoption of winter cereal rye (*Secale cereale*) in corn (*Zea mays*)-soybean (*Glycine max*) and continuous corn cropping systems in the upper midwestern United States was estimated to have the potential to decrease NO_3^- leaching in the Mississippi River Basin by 20%, improving water quality (Kladivko et al., 2014). This suggests cover cropping is a promising management practice that decreases nutrient leaching and increases water quality in rivers and lakes.

Many factors affect the effectiveness of cover crops in reducing nutrient leaching, such as cover crop species, planting date (the timing and length of cover crop growth), the time of cover crop termination, soil texture, main crop systems, tillage systems, and weather conditions (Nouri et al., 2022; Eriksen and Thorup-Kristensen, 2002; Thapa et al., 2018; Teixeira et al., 2016). The positive impact of cover crops on NO_3^- leaching reduction is greater in high-risk environments, such as sand-dominated soil (more water moving through the profile), vegetable cropping systems (fresh residues susceptible to quick mineralization), and conventional tillage systems (more homogenous distribution of soil pores within the tillage depth; Nouri et al., 2022). Moreover, non-leguminous cover crops (e.g. winter cereal rye) are generally more effective in reducing NO_3^- leaching than legume cover crops, primarily due to quicker growth, higher N demand for growth, and relatively higher C/N ratio

(Nouri et al., 2022; Thapa et al., 2018). There is significant evidence that cover crops reduce NO_3^- leaching when the cover crop is alive. However, there may be nutrient leaching losses from the field if the release of nutrients from frost- or herbicide-terminated cover crop residues coincides with water movement through the soil profile (Cober et al., 2018, 2019; Lozier et al., 2017). Research is needed on the effects of cover crop termination timing on nutrient leaching and the extent to which reducing leaching losses influences subsequent crop nutrient uptake and growth.

3.3 Decreasing surface phosphorus losses by mitigating runoff and soil erosion during the non-growing season

In addition to reducing N and S leaching, cover crops can reduce P losses. Losses of P from agricultural systems to surface waters are largely via soil erosion, surface runoff, and subsurface drainage (Hansen et al., 2002; Sharpley and Smith, 1991). This indicates that the greatest potential for P loss occurs when the soil is less covered (or uncovered) and the region has precipitation that at times exceeds the soil's infiltration capacity. Therefore, management practices that can reduce soil erosion, by protecting soil from the forces of wind and water as well as runoff by increasing water infiltration, are effective at reducing P loss from agricultural soil to water (Hansen et al., 2002). Agroecosystems with cover crops generally are less susceptible to erosion, because cover cropping extends the duration of time that soil is covered with living plants and provides more soil cover through surface plant residues (cover crops), compared to no cover crop system (Kaspar and Singer, 2011). Also, soil with cover crops has less runoff (due to increased water uptake and transpiration during growth) but greater soil structure (due to increased water infiltration), compared to soil without cover crops (Kaspar and Singer, 2011). The control of surface runoff and sediment transport by cover crops has been shown to decrease total P loss by 54% to 94% (e.g. Kaspar and Singer, 2011; Sharpley and Smith, 1991).

However, cover cropping is generally more efficient at reducing particulate P losses through soil erosion rather than dissolved P (Sharpley et al., 2000). In cold climates, frost-terminated cover crops increased dissolved reactive P load into surface runoff, compared to bare soils (Liu et al., 2019; Cober et al., 2018). However, in a region with freeze-thaw processes, most P released from cover crops was retained in surface soil rather than lost through surface or subsurface water (Cober et al., 2019). Therefore, the use of frost-tolerant cover crop species is a potential practice to mitigate P losses, especially in winter (Macrae et al., 2021). The effects of cover cropping on P losses from agricultural soil are complicated by interactions among soil types, cover crop species (frost-tolerant vs. frost-intolerant), freeze-thaw cycles, and hydrological processes (Kaspar and Singer, 2011; Cober et al., 2018, 2019). More site-specific studies are needed, such as how cover crops affect dissolved P losses via tile drainage in cold climates (Macrae et al., 2021).

3.4 Decreasing annual nitrous oxide emissions

In addition to reducing NO_3^- leaching, there is an increased interest in mitigating nitrous oxide (N_2O) emissions from agricultural systems (Liu et al., 2017; Forster et al., 2007). One pathway of N losses is nitrous oxide (and dinitrogen gas) emissions through denitrification, especially after N fertilizer/manure application, which decreases crop N use efficiency. Moreover, N_2O is a potent greenhouse gas with 298 times greater global warming potential than that of carbon dioxide and 12 times greater than that of methane.

It is a challenge to mitigate N_2O emissions in fertilized cropping systems. This is because (1) small quantities of N_2O losses occur on agricultural lands (approximately 1% of fertilizer N applied; Intergovernmental Panel on Climate Change, 2006) but scaled globally can result in substantial greenhouse gas emissions; (2) multiple processes lead to N_2O production, including nitrification, nitrifier-denitrification, chemo-denitrification, and denitrification (Venterea et al., 2012); (3) a reduction of ammonia (NH_3) and/or NO_3^- losses from the field allow for more mineral N available for direct N_2O emissions (i.e. there are trade-offs between pathways of N losses); and (4) there is high spatial and temporal variability of N_2O production in a given field, referred to as 'hot spots' (large amounts of N_2O production in specific areas) and 'hot moments' (pulses of N_2O production within a short period; Wagner-Riddle et al., 2020). Large N_2O emission pulses (hot moments) commonly occur when N is applied, and during dry-wet or frozen-thaw cycles (Eberwein et al., 2020; Wagner-Riddle et al., 2017; Teepe et al., 2001). These peak N_2O emissions may account for over 80% of annual N_2O emissions in a given environment (Foltz et al., 2021). Therefore, management practices that (1) reduce N application rates and/or (2) decrease soil mineral N levels, particularly during the occurrence of dry-wet/freeze-thaw events, are promising strategies for reducing N_2O emissions from agricultural soils (Gomes et al., 2009), for which cover crops have a role.

The effect of cover cropping on N_2O emissions is inconsistent (Zhang et al., 2022; Muhammad et al., 2019; Basche et al., 2014; Venterea et al., 2012). Cover crops can reduce N_2O emissions in the fall (after main crop harvest and before cover crop termination), primarily due to the reduction of NO_3^- available for denitrification (Foltz et al., 2021) and reducing water-saturated soil conditions. In cold climates, in late winter or early spring, the presence of cover crops increased peak N_2O fluxes, due to mineral N release from cover crops and freeze-thawing events (Gao et al., 2023; Thomas et al., 2017). However, some studies reported that cover crops reduced N_2O emissions at thaw (Foltz et al., 2021), likely because cover crop residues and trapped snowfall insulated the soil surface layer against cold temperatures and reduced soil freezing extent and depth (Lombardozzi et al., 2018; Wagner-Riddle et al., 2007). The inconsistent reports of N_2O emissions at thaw with cover crops are highly related to (1) the timing and amount of mineral N release from cover crops, which is species dependent; (2) the timing, frequency, and intensity of freeze-thaw cycles (annual weather conditions), which is influenced by insolating properties of cover crop residues on the soil surface; and (3) a function of the amount of C biomass inputs (i.e. providing C energy for denitrifiers and creating anaerobic conditions in the soil due to oxygen consumption during respiration; Farquharson and Baldock., 2008). Although cover crops can either increase or decrease N_2O emissions at thaw, when annual N_2O emissions (including the non-growing season and growing season of subsequent crops) were assessed, the effect of cover cropping on cumulative N_2O emissions was negligible (see meta-analysis by Basche et al. (2014)). Therefore, multiple years of study over the entire year (i.e. cover crop and main crop growing seasons) are essential to comprehensively assess the cover cropping effect on cumulative N_2O emissions. A recent related case study is provided below to demonstrate the role of cover crops in N_2O emissions.

3.5 *Case study: use of nitrification and urease inhibitors to mitigate N_2O emissions in diverse crop rotation with cover crops*

The effect of cover cropping alone on N_2O emissions has been discussed in Section 3.4. Other agronomic management practices, such as crop rotation diversification and the use of nitrification and urease inhibitors (NUI), also affect N_2O emissions (Ekwunife et al., 2022; Drury et al., 2021). However, there is a lack of information on the stacking effects of cover crops, diverse crop rotations, and use of NUI on annual N_2O emissions. Tariq et al. (2024) conducted a 4-year crop rotation study to determine the effect of crop rotation diversification and use of NUI on N_2O emissions, in a temperate, humid climate (Elora, Ontario, Canada, 43°39′N, 80°25′W). The N_2O flux was measured year-round using a micrometeorological method using four 4 ha fields. Two fields were under a conventional crop rotation (corn-soybean-soybean-corn). The other two fields were managed as a diverse crop rotation (corn-soybean-winter wheat (*Triticum aestivum*)-corn). In the diverse crop rotation, cover crops were either under-seeded in corn (mix of crimson clover and annual ryegrass (*Lolium perenne*)) or planted after the harvest of winter wheat (mix of cereal rye, oat (*Avena sativa*), crimson clover, and Daikon radish (*Raphanus sativus* var. *longipinnatus*)).

The effect of NUI was investigated only in the corn phase in the fourth year in both crop rotations (conventional vs. diverse). The diverse crop rotation with NUI had 13% lower annual cumulative N_2O emissions in the corn year than without NUI. However, the use of NUI did not affect annual cumulative N_2O emissions in the conventional rotation. This finding indicates the potential of using NUI to mitigate the undesired effect of increased N_2O emission from cover crops (probably due to legume cover crops without reductions of applied N to the main crop). In crop rotation with cover crops, reducing N application rates by accounting for cover crop contributions is critical to mitigate N_2O emission. This also requires improving the synchronicity between the N supply of cover crops and N uptake by the subsequent crop.

Of the 4 years, only the diverse crop rotation in the year of winter wheat followed by cover crops (the third year) reduced annual N_2O emissions by 43%, compared to the corresponding year of soybean in the conventional rotation. Three possible mechanisms were explained: (1) soils were bare for approximately 8 months for the conventional rotation (soybeans were grown from 22 May to 23 September) but for less than 1 month for the diverse rotation, reducing the amount of NO_3^- available for denitrification as it was taken up by winter wheat and cover crops; (2) soybean residue mineralization (an increase in soil mineral N due to decomposition of crop residues, roots, and nodules) during a short wet period in the fall resulted in enhanced N_2O emissions; and (3) cover crops reduced N_2O emissions in fall due to high biomass production (more cover crop biomass with more N uptake thus less mineral N susceptible to denitrification; see Section 3.1). This study's findings indicate that the inclusion of winter wheat and cover crops into the corn-soybean rotation can substantially reduce year-round N_2O emissions.

4 Improving soil organic matter and nutrient cycling

4.1 Soil organic matter accrual increases nutrient availability

Soil properties affect nutrient cycling by directly and indirectly controlling soil capacity to retain and supply nutrients to crops (Schröder et al., 2016). When plants take up nutrients (i.e. causing a decrease in nutrient concentrations in soil solution), nutrients in soil solution are replenished by surface exchange of cation/anion (e.g. soil clay surface); mineralization of crop residues, organic matter, and microbial necromass; and/or dissolution of soil minerals (e.g. P-bearing minerals; Havlin, 2020). This is referred to as soil nutrient buffer capacity, which is greater in soils with greater clay and organic matter content (Havlin, 2020; Spain et al., 1983).

In addition to being a reservoir of nutrients, soil organic matter plays an important role in improving water holding capacity, soil aggregation, and cation exchange capacity (Goulding et al., 2001). Also, soil organic matter serves as an energy source for both macro- and micro-faunal organisms (Wander, 2004), facilitating the decomposition of organic matter and the release of plant nutrients, because C concentrations are frequently in excess of organisms' assimilation needs for other nutrients (e.g. N, P, and S). Therefore, the increase of soil organic matter can potentially increase the soil's capacity to supply plant-available nutrients and the productivity of subsequent crops (e.g. N; Kuo and Jellum, 2000).

Cover crops can increase soil organic matter, largely due to the increase in plant and soil microbial biomass, and reductions in organic matter losses due to soil erosion (Hao et al., 2023; Van Eerd et al., 2023; McClelland et al., 2021; Ding et al., 2006). However, the accumulation of soil organic matter is usually detected after the long-term use of cover crops rather than its short-term use (Peng et al., 2023). The insufficient number of long-term cover cropping experiments (Hao et al., 2023) limits the study on linking soil organic matter accumulation and nutrient availability enhancement. Moreover, soil organic matter is extremely complex in terms of chemical structure and stability, which complicates cover cropping effects on nutrient cycling. For instance, cover cropping has a more pronounced effect on particulate organic matter than mineral-associated organic matter (Peng and Van Eerd, 2024; Fohrafellner et al., 2024; Wooliver and Jagadamma, 2023). Yet, how the shift in these soil organic matter characteristics affects nutrient cycling, especially under weather extremes (e.g. heat and drought), has not been addressed, let alone the cover crop influences.

4.2 Changes in soil microbial community

The role of soil microbial community in improving nutrient cycling can be categorized into three aspects. The primary role of soil microbes is the decomposition and mineralization of crop residues and organic matter, providing nutrients available for plant growth (Yadav et al., 2021; Sahu et al., 2017). Also, microbes play an important role in weathering of rocks and minerals, mobilizing nutrients, particularly P and S as well as micronutrients (Yadav et al., 2021; Hirsch and Mauchline, 2015). Moreover, microbes affect (1) the formation of organic matter by supplying components such as proteins, lipids, and polysaccharides and (2) the decomposition rates of necromass (Villarino et al., 2023; Sokol et al., 2022). Therefore, the changes in soil microbial communities under cover cropping affect nutrient cycling.

The effect of soil microbial change by cover cropping on nutrient cycling is not clear. Cover crops affect soil microbial community largely by (1) increasing substrate (e.g. root exudates) availability for soil microbes and (2) providing a better soil environment, especially soil moisture retention in the surface soil (top 15 cm depth), thereby facilitating microbial growth and biomass (Vukicevich et al., 2016). Overall, cover crops increased soil microbial abundance, activity, and diversity (Muhammad et al., 2021; Kim et al., 2020), while stimulating fungi growth more, compared to bacteria (Muhammad et al., 2021). Moreover, the improvement of soil biological processes under cover cropping positively affected soil enzyme activities and the availability of macronutrients (Hallama et al., 2022, 2019; Chavarría et al., 2016). However, the effect of cover cropping on N-cycling functional genes, including N-fixation genes (*nifH*), nitrification genes (ammonia-oxidizing archaea and bacterial *amoA*), and denitrification genes (*nirK* and *nirS*), has been inconsistent (Kim et al., 2022; Hu et al., 2021), due to different cropping systems and historical fertilization management. Therefore, more studies are needed to understand the change in soil cycling microbial community with cover cropping, especially after long-term use of cover crops across diverse crop rotations and tillage systems.

5 Supplying nutrients to subsequent crops

We have discussed ways in which cover crops can enhance nutrient storage and supply, by reducing losses of nutrients as well as increasing soil organic matter and microbial biomass. Our challenge continues to be how best to manage the system to optimize the supply of those scavenged nutrients to the subsequent crops. Nutrients contained in cover crops (i.e. organic forms) must be mineralized via microbial processes to make the nutrients (i.e. inorganic forms) available to the succeeding crops. As alluded to in the sections above, factors controlling the mineralization-immobilization of nutrients are key to understanding these cycles.

Here, we take an example of N to discuss the mineralization-immobilization of cover crop residues as N is the most demanded macronutrient of crops with the highest number of published research, compared to other nutrients. Nitrogen mineralization is defined as the transformation of organic N into inorganic N forms of ammonium (NH_4^+) and NO_3^-. Nitrogen immobilization occurs when organic material is high in C relative to N concentrations and soil microbes take up soil mineral N to meet assimilation needs. Hence, N immobilization converts inorganic N into organic forms (Jansson and Persson, 1982), reducing soil inorganic N available for crop uptake. The sum of two concurrent but opposing processes, namely gross N mineralization and gross N immobilization, is net N mineralization, which usually refers to the net increase/decrease in NH_4^+ and NO_3^- concentrations relative to blank control samples, during soil incubation assays (Myrold and Bottomley, 2008).

Different cover crops contain and release different amounts of mineral N. The decomposition of cover crop residues and N mineralization-immobilization are controlled by several factors, including the quality (e.g. C/N ratio) and chemical composition (e.g. lignin and polyphenol) of cover crop residues, soil temperature, moisture, and properties (e.g. chemical, biological, and physical), and freeze/thaw and drying-wetting events (Raimondi et al., 2023; Rosecrance et al., 2000; Thapa et al., 2021; Cabrera et al., 2005; Doran and Smith, 1991).

Apart from the selection and management of cover crop species, most of these factors are either not manageable or difficult to change. Generally, the C/N ratio of cover crops largely explained why net N mineralization was observed for most legume cover crops, such as hairy vetch (*Vicia villosa*), and crimson clover, while net N immobilization was observed for grass cover crops, such as winter cereal rye. This finding was based on various approaches, including field, mesh bag experiments, ^{15}N tracer technique, and incubation experiments (Raimondi et al., 2023; Lacey et al., 2020; Gabriel et al., 2016; Rosecrance et al., 2000; Ranells and Wagger, 1996). For example, white lupin (*Lupinus albus*) released the highest amount of mineral N (56 kg N ha^{-1}), followed by hairy vetch (40 kg N ha^{-1}), but only 10 kg N ha^{-1} for buckwheat (*Fagopyrum esculentum*; Hansen et al., 2021).

Additionally, synchronization between cover crop N mineralization and crop N uptake is critical to optimizing N supply to subsequent crops (Korsaeth et al., 2002). Specifically, crops (e.g. winter wheat, rice (*Oryza sativa*), corn, and soybean) accumulate less than 3% of their N during the seedling stage (0–30 days), but 70% to 80% during the vegetative growth stage (30–160 days; Chen et al., 2014). Thus, the mineralization of cover crop residues should match the timing of N demand for subsequent crops. The timing of cover crop termination (2–4 weeks before subsequent crop seeding) and the selection of cover crop species (e.g. legumes or legume-dominated mixture) can be adjusted to minimize the asynchrony between cover crop residue N release and crop N needs (Nyabami et al., 2023; Balkcom et al., 2015). However, the time of cover crop termination should be site-specific, year-specific, and crop rotation-specific (Nyabami et al., 2023; Nevins et al., 2020; Keene et al., 2017; Balkcom et al., 2015), leading to the need for regionally specific cover crop research on N.

Compared to N, cover crops supply less P and S to crops, while there is less opportunity for P and S deficiency (less crop demand) in the agroecosystem. The mineralization-immobilization of cover crop residues mainly depends on C/nutrient ratios and microbial assimilation needs (Nachimuthu et al., 2009; Eriksen et al., 2004; Horst et al., 2001). For instance, a C/P ratio of less than 200/1 will result in net P mineralization (Varela et al., 2017; Nachimuthu et al., 2009). Maltais-Landry et al. (2014) found that winter cereal rye is more effective in supplying available P to subsequent crops than the tested legumes, including fava bean (*Vicia faba*), vetches (*Vicia dasycarpa*, *Vicia sativa*, *Vicia benghalensis*), and pea (*Pisum sativum*), in California agricultural soils. Moreover, short-term (3 years) cover cropping had relatively little effect on soil P cycling improvement, indicating long-term studies are needed (Teles et al., 2017). Specifically, oat, winter cereal rye, and ryegrass released 4.5 kg P ha^{-1}, 4.7 kg P ha^{-1}, and 1.7 kg P ha^{-1}, respectively, which were available for soybean uptake (Varela et al., 2017). Similarly, cover crops supply a relatively low amount of S, largely due to the

Published by Burleigh Dodds Science Publishing Limited, 2026.

relatively low S accumulation. For example, oilseed radish released 3 kg S ha^{-1}, and no detectable S was mineralized from buckwheat (Hansen et al., 2021). There may be an opportunity to utilize cover crops to enhance fertilizer P and S availability to the main crop (Soltangheisi et al., 2018), but further research is needed, particularly for infertile soils.

6 Conclusion and research priorities

Four mechanisms of how cover crops can increase N, P, and S cycling in agroecosystems are: (1) biologically fixing atmospheric N_2; (2) reducing nutrient losses by leaching, soil erosion, and gaseous forms; (3) improving soil organic matter and soil microbial community to cycle nutrients; and (4) supplying nutrients to subsequent crops. Many factors affect nutrient cycling in cover crop systems, including the complex interaction among scavenging and preventing losses, delivering nutrients, and the site-specific nature of nutrient cycling, especially climate conditions, soil type, species selection, and mineralization-immobilization of cover crop residues. These interactions are further complicated by the specific cover crop grown and how it is managed. Site-specific cover crop management is clearly necessary to optimize nutrient cycling.

The grand challenge of optimizing nutrient cycling is designing research outcomes that address the specificity required to make cover crop nutrient recommendations to growers. Priority research questions for each mechanism include the following. (1) Legume cover crops can supply a substantial portion of N via BNF to crops, but what is the amount of commercial N fertilizer that can be replaced without yield penalties? How can farmers easily adjust the N fertilizer application rate based on on-farm cover crop conditions (e.g. biomass and species, soil type, and main crops)? (2) Cover crops can decrease nutrient losses through multiple processes (e.g. greenhouse gas emissions, soil erosion, runoff, and leaching). Yet, there are trade-offs among pathways of loss as minimizing runoff might lead to leaching losses. Therefore, research that integrates other management practices (i.e. tillage system, crop rotations, and NUI) rather than those that solely focus on maximizing cover crop biomass is needed to better understand multiple pathways of nutrient losses. (3) Long-term use of cover crops is important to improve nutrient cycling via soil organic matter accrual and microbial communities. There is an opportunity to explore the effect of changing climate (e.g. extreme weather season) on nutrient cycling by comparing data over multiple years under the long-term use of cover crops. (4) Optimizing nutrient use by the subsequent crop requires synchronization of nutrient mineralization of cover crops. Research evaluating both cover crop and main crop management is required. Undoubtedly, the opportunities to investigate these and other research questions are what make cover crop research exciting and valuable to researchers, farmers, and all of society.

7 Where to look for further information

- A review of cover cropping is Van Eerd, L. L., et al. (2023). Influence of cover crops at the four spheres: a review of ecosystem services, potential barriers, and future directions for North America. *Science of the Total Environment*, 858, 159990. https://doi.org/10.1016/j.scitotenv.2022.159990.
- A comprehensive introduction to nutrient cycling is Stevenson, F. J. & Cole, M. A. (1999). *Cycles of soils: carbon, nitrogen, phosphorus, sulfur, micronutrients*. John Wiley & Sons, INC.
- An advanced introduction to nitrogen cycling processes is Earl-Goulet S., et al. (2024). Advances in understanding nitrogen cycling in soil. In Jagdish, Kumar, Ladha (Eds.), *Improving nitrogen use efficiency in crop production*. Burleigh Dodds Science Publishing. https://doi/10.19103/AS.2024.0135.03.
- A soil incubation study comparing nutrient release from cover crops is Hansen, V., et al. (2021). Towards integrated cover crop management: N, P and S release from aboveground and belowground residues. *Agriculture, Ecosystems & Environment*, 313, 107392. https://doi.org/10.1016/j.agee.2021.107392.
- A study on cover cropping effect of nitrogen-cycling functional genes is Kim, N., et al. (2022). Limited impacts of cover cropping on soil N-cycling microbial communities of long-term corn monocultures. *Frontiers in Microbiology*, 13, 926592. https://doi.org/10.3389/fmicb.2022.926592.

Key organizations (North America):

- Midwest Cover Crops Council (www.midwestcovercrops.org).
- Southern Cover Crops Council (www.southerncovercrops.org).
- Northeast Cover Crops Council (www.northeastcovercrops.com).
- Sustainable Agriculture Research & Education (SARE) (www.sare.org/resources/cover-crops).
- Natural Resources Conservation Service – USDA (www.nrcs.usda.gov/resources/guides-and-instructions/cover-crop-ac-340-conservation-practice-standard).

8 Acknowledgments

This project was funded by Grain Farmers of Ontario, Fresh Vegetable Growers of Ontario, and the Ontario Agri-Food Innovation Alliance, a collaboration between the Government of Ontario and the University of Guelph.

9 References

Balkcom, K. S., et al. (2015). Timing of cover crop termination: Management considerations for the Southeast. *Crop, Forage & Turfgrass Management*, 1(1), 1–7.

Basche, A. D., et al. (2014). Do cover crops increase or decrease nitrous oxide emissions? A meta-analysis. *Journal of Soil and Water Conservation*, 69(6), 471–482.

Blanco-Canqui, H. & Ruis, S. J. (2020). Cover crop impacts on soil physical properties: a review. *Soil Science Society of America Journal*, 84(5), 1527–1576.

Blanco-Canqui, H., et al. (2015). Cover crops and ecosystem services: insights from studies in temperate soils. *Agronomy Journal*, 107(6), 2449–2474.

Büchi, L., et al. (2015). Accumulation of biologically fixed nitrogen by legumes cultivated as cover crops in Switzerland. *Plant and Soil*, 393, 163–175.

Bünemann, E. K., et al. (2011). Characterization of phosphorus forms in soil microorganisms. In E. K. Bünemann, et al. (Eds.), *Phosphorus in action: biological processes in soil phosphorus cycling* (pp. 37–57). Berlin: Springer.

Cabrera, M. L., et al. (2005). Nitrogen mineralization from organic residues: research opportunities. *Journal of Environmental Quality*, 34(1), 75–79.

Carranca, C., et al. (1999). Biological nitrogen fixation by fababean, pea and chickpea, under field conditions, estimated by the ^{15}N isotope dilution technique. *European Journal of Agronomy*, 10(1), 49–56.

Chavarría, D. N., et al. (2016). Effect of cover crops on microbial community structure and related enzyme activities and macronutrient availability. *European Journal of Soil Biology*, 76, 74–82.

Chen, B., et al. (2014). Soil nitrogen dynamics and crop residues: a review. *Agronomy for Sustainable Development*, 34, 429–442.

Cober, J. R., et al. (2018). Nutrient release from living and terminated cover crops under variable freeze–thaw cycles. *Agronomy Journal*, 110(3), 1036–1045.

Cober, J. R., et al. (2019). Winter phosphorus release from cover crops and linkages with runoff chemistry. *Journal of Environmental Quality*, 48(4), 907–914.

Coombs, C., et al. (2017). Legume cover crop management on nitrogen dynamics and yield in grain corn systems. *Field Crops Research*, 201, 75–85.

Crespo, C., et al. (2021). Improving soil organic nitrogen and sulfur pools by cover cropping and crop fertilization in soybean-based cropping systems. *Soil and Tillage Research*, 213, 105138. https://doi.org/10.1016/j.still.2021.105138.

Dabney, S. M., et al. (2010). Using cover crops and cropping systems for nitrogen management. In: Delgado, J.A. & Follett, R.F. (Eds.), Advances in nitrogen management for water quality, Soil Water Conserv. Soc., Ankeny, IA, pp. 231–282.

Ding, G., et al. (2006). Effect of cover crop management on soil organic matter. *Geoderma*, 130(3–4), 229–239.

Doran, J. W. & Smith, M. S. (1991). Role of cover crops in nitrogen cycling. In *Cover crops for clean water* (pp. 85–94). Ankeny: Soil and Water Conservation Society. https://www.swcs.org/static/media/cms/CCCW5nitrogen_80C82CC3D2751.pdf.

Drury, C. F., et al. (2021). Diverse rotations impact microbial processes, seasonality and overall nitrous oxide emissions from soils. *Soil Science Society of America Journal*, 85(5), 1448–1464.

Earl-Goulet S., et al. (2024). Advances in understanding nitrogen cycling in soil. In J. K. Ladha (Ed.), *Improving nitrogen use efficiency in crop production*. Cambridge: Burleigh Dodds Science Publishing. https://doi/10.19103/AS.2024.0135.03.

Eberwein, J. R., et al. (2020). Large nitrogen oxide emission pulses from desert soils and associated microbiomes. *Biogeochemistry*, 149, 239–250.

Ekwunife, K. C., et al. (2022). Assessing the impacts of tillage, cover crops, nitrification, and urease inhibitors on nitrous oxide emissions over winter and early spring. *Biology and Fertility of Soils*, 58, 1–12.

Eriksen, J. (2009). Soil sulfur cycling in temperate agricultural systems. *Advances in Agronomy*, 102, 55–89.

Eriksen, J. & Thorup-Kristensen, K. (2002). The effect of catch crops on sulphate leaching and availability of S in the succeeding crop on sandy loam soil in Denmark. *Agriculture, Ecosystems & Environment*, 90(3), 247–254.

Eriksen, J., et al. (2004). Plant availability of catch crop sulfur following spring incorporation. *Journal of Plant Nutrition and Soil Science*, 167(5), 609–615.

Fageria, N. K., et al. (2005). Role of cover crops in improving soil and row crop productivity. *Communications in Soil Science and Plant Analysis*, 36(19–20), 2733–2757.

Farquharson, R. & Baldock, J. (2008). Concepts in modelling N_2O emissions from land use. *Plant and Soil*, 309(1), 147–167.

Fohrafellner, J., et al. (2024). Cover crops affect pool specific soil organic carbon in cropland: meta-analysis. *European Journal of Soil Science*, 75(2), e13472. https://doi.org/10.1111/ejss.13472.

Foltz, M. E., et al. (2021). Influence of rye cover cropping on denitrification potential and year-round field N_2O emissions. *Science of the Total Environment*, 765, 144295. https://doi.org/10.1016/j.scitotenv.2020.144295.

Fontaine, S., et al. (2024). Plant–soil synchrony in nutrient cycles: learning from ecosystems to design sustainable agrosystems. *Global Change Biology*, 30(1), e17034. https://doi.org/10.1111/gcb.17034.

Forster, P., et al. (2007). Changes in atmospheric constituents and in radiative forcing. In S. Solomon, et al. (Eds.), *Climate change 2007: The physical science basis* (pp. 129–234). Cambridge: IPCC Cambridge University Press.

Gabriel, J. L., et al. (2016). Nitrogen use efficiency and fertiliser fate in a long-term experiment with winter cover crops. *European Journal of Agronomy*, 79, 14–22.

Gao, Y., et al. (2023). Nongrowing season soil nitrous oxide emissions as influenced by cover crops and fall tillage termination. *Canadian Journal of Soil Science*, 103(4), 527–537.

Gomes, J., et al. (2009). Soil nitrous oxide emissions in long-term cover crops-based rotations under subtropical climate. *Soil and Tillage Research*, 106(1), 36–44.

Goulding, K. W. T., et al. (2001). The role of soil organic matter and manures in sustainable nutrient cycling. In R. M. Rees, et al. (Eds.), *Sustainable management of soil organic matter* (pp. 221–342). Wallingford: CABI Publishing.

Griffiths, M., et al. (2022). Optimisation of root traits to provide enhanced ecosystem services in agricultural systems: a focus on cover crops. *Plant, Cell & Environment*, 45(3), 751–770.

Guinet, M., et al. (2018). Comparative effect of inorganic N on plant growth and N_2 fixation of ten legume crops: towards a better understanding of the differential response among species. *Plant and Soil*, 432, 207–227.

Hallama, M., et al. (2019). Hidden miners–the roles of cover crops and soil microorganisms in phosphorus cycling through agroecosystems. *Plant and Soil*, 434, 7–45.

Hallama, M., et al. (2022). The role of microbes in the increase of organic phosphorus availability in the rhizosheath of cover crops. *Plant and Soil*, 476(1), 353–373.

Hansen, N. C., et al. (2002). The fate and transport of phosphorus in agricultural systems. *Journal of Soil and Water Conservation*, 57(6), 408–417.

Hansen, V., et al. (2021). Towards integrated cover crop management: N, P and S release from aboveground and belowground residues. *Agriculture, Ecosystems & Environment*, 313, 107392. https://doi.org/10.1016/j.agee.2021.107392.

Hao, X., et al. (2023). Are there universal soil responses to cover cropping? A systematic review. *Science of the Total Environment*, 861, 160600. https://doi.org/10.1016/j.scitotenv.2022.160600.

Havlin, J. L. (2020). Soil: Fertility and nutrient management. In Y. Wang (Ed.), *Landscape and land capacity* (pp. 251–265). Boca Raton: CRC Press.

Herridge, D. F., et al. (2008). Global inputs of biological nitrogen fixation in agricultural systems. *Plant and Soil*, 311, 1–18.

Hirsch, P. R. & Mauchline, T. H. (2015). The importance of the microbial N cycle in soil for crop plant nutrition. *Advances in Applied Microbiology*, 93, 45–71.

Hirsh, S. M., et al. (2021). Scavenging and recycling deep soil nitrogen using cover crops on mid-Atlantic, USA farms. *Agriculture, Ecosystems & Environment*, 309, 107274. https://doi.org/10.1016/j.agee.2020.107274.

Horst, W. J., et al. (2001). Agronomic measures for increasing P availability to crops. *Plant and Soil*, 237, 211–223.

Howard, J. B. & Rees, D. C. (1996). Structural basis of biological nitrogen fixation. *Chemical Reviews*, 96(7), 2965–2982.

Hu, J., et al. (2021). Soil health management enhances microbial nitrogen cycling capacity and activity. *mSphere*, 6, e01237-20. https://doi.org/10.1128/mSphere.01237-20.

Intergovernmental Panel on Climate Change (2006). Agriculture, forestry and other land use, Volume 4. In *2006 IPCC guidelines for national greenhouse gas inventories*. Japan: Institute for Global Environmental Strategies.

Jansson, S. L. & Persson, J. (1982). Mineralization and immobilization of soil nitrogen. *Nitrogen in Agricultural Soils*, 22, 229–252.

Jian, J., et al. (2020). A meta-analysis of global cropland soil carbon changes due to cover cropping. *Soil Biology and Biochemistry*, 143, 107735. https://doi.org/10.1016/j.soilbio.2020.107735.

Kaspar, T. C. & Singer, J. W. (2011). The use of cover crops to manage soil. Soil management: Building a stable base for agriculture, 321–337.

Kaye, J., et al. (2019). Managing nitrogen through cover crop species selection in the US mid-Atlantic. *PLoS One*, 14(4), e0215448. https://doi.org/10.1371/journal.pone.0215448.

Keene, C. L., et al. (2017). Cover crop termination timing is critical in organic rotational no-till systems. *Agronomy Journal*, 109(1), 272–282.

Kim, N., et al. (2022). Limited impacts of cover cropping on soil N-cycling microbial communities of long-term corn monocultures. *Frontiers in Microbiology*, 13, 926592. https://doi.org/10.3389/fmicb.2022.926592.

Kim, N., et al. (2020). Do cover crops benefit soil microbiome? A meta-analysis of current research. *Soil Biology and Biochemistry*, 142, 107701. https://doi.org/10.1016/j.soilbio.2019.107701.

Kladivko, E. J., et al. (2014). Cover crops in the upper midwestern United States: Potential adoption and reduction of nitrate leaching in the Mississippi River Basin. *Journal of Soil and Water Conservation*, 69(4), 279–291.

Published by Burleigh Dodds Science Publishing Limited, 2026.

Korsaeth, A., et al. (2002). Temporal changes in mineralization and immobilization of N during degradation of plant material: implications for the plant N supply and nitrogen losses. *Soil Biology and Biochemistry*, 34(6), 789–799.

Kuo, S. & Jellum, E. J. (2000). Long-term winter cover cropping effects on corn (Zea mays L.) production and soil nitrogen availability. *Biology and Fertility of Soils*, 31, 470–477.

Lacey, C., et al. (2020). Carbon and nitrogen release from cover crop residues and implications for cropping systems management. *Journal of Soil and Water Conservation*, 75(4), 505–514.

Lapierre, J., et al. (2022). Cover crop mixtures: a powerful strategy to reduce post-harvest surplus of soil nitrate and leaching. *Agriculture, Ecosystems & Environment*, 325, 107750. https://doi.org/10.1016/j.agee.2021.107750.

Liu, J., et al. (2019). Impacts of cover crops and crop residues on phosphorus losses in cold climates: a review. *Journal of Environmental Quality*, 48(4), 850–868.

Liu, S., et al. (2017). A meta-analysis of fertilizer-induced soil NO and combined NO+ N_2O emissions. *Global Change Biology*, 23(6), 2520–2532.

Lombardozzi, D. L., et al. (2018). Cover crops may cause winter warming in snow-covered regions. *Geophysical Research Letters*, 45(18), 9889–9897.

Lozier, T. M., et al. (2017). Release of phosphorus from crop residue and cover crops over the non-growing season in a cool temperate region. *Agricultural Water Management*, 189, 39–51.

Lynch, J. P. & Brown, K. M. (2001). Topsoil foraging–an architectural adaptation of plants to low phosphorus availability. *Plant and Soil*, 237, 225–237.

Macrae, M., et al. (2021). One size does not fit all: Toward regional conservation practice guidance to reduce phosphorus loss risk in the Lake Erie watershed. *Journal of Environmental Quality*, 50(3), 529–546.

Maltais-Landry, G., et al. (2014). Soil phosphorus mobilization in the rhizosphere of cover crops has little effect on phosphorus cycling in California agricultural soils. *Soil Biology and Biochemistry*, 78, 255–262.

McClelland, S. C., et al. (2021). Management of cover crops in temperate climates influences soil organic carbon stocks: a meta-analysis. *Ecological Applications*, 31(3), e02278. https://doi.org/10.1002/eap.2278.

Meyer, N., et al. (2019). Cover crops reduce water drainage in temperate climates: a meta-analysis. *Agronomy for Sustainable Development*, 39, 1–11.

Muhammad, I., et al. (2019). Regulation of soil CO_2 and N_2O emissions by cover crops: a meta-analysis. *Soil and Tillage Research*, 192, 103–112.

Muhammad, I., et al. (2021). Cover cropping enhances soil microbial biomass and affects microbial community structure: a meta-analysis. *Geoderma*, 381, 114696. https://doi.org/10.1016/j.geoderma.2020.114696.

Murrell, T. S., et al. (2021). *Improving potassium recommendations for agricultural crops*. Cham: Springer Nature. https://library.oapen.org/handle/20.500.12657/46118.

Myrold, D. D. & Bottomley, P. J. (2008). Nitrogen mineralization and immobilization. *Nitrogen in Agricultural Systems*, 49, 157–172.

Nachimuthu, G., et al. (2009). Isotopic tracing of phosphorus uptake in corn from 33 P labelled legume residues and 32 P labelled fertilisers applied to a sandy loam soil. *Plant and Soil*, 314, 303–310.

Nevins, C. J., et al. (2020). The synchrony of cover crop decomposition, enzyme activity, and nitrogen availability in a corn agroecosystem in the Midwest United States. *Soil and Tillage Research*, 197, 104518. https://doi.org/10.1016/j.still.2019.104518.

Nitu, T. T., et al. (2021). Cover crops and soil nitrogen cycling. In R. Islam & B. Sherman (Eds.), *Cover crops and sustainable agriculture* (pp. 209–226). Boca Raton: CRC Press.

Nouri, A., et al. (2022). When do cover crops reduce nitrate leaching? A global meta-analysis. *Global Change Biology*, 28(15), 4736–4749.

Nyabami, P., et al. (2023). Nitrogen release dynamics of cover crop mixtures in a subtropical agroecosystem were rapid and species-specific. *Plant and Soil*, 492(1), 399–412.

Parr, M., et al. (2011). Nitrogen delivery from legume cover crops in no-till organic corn production. *Agronomy Journal*, 103(6), 1578–1590.

Peng, Y. & Van Eerd, L. L. (2024). Surface soil sampling underestimates soil carbon and nitrogen storage of long-term cover cropping. *Geoderma Regional*, 39, e00885. https://doi.org/10.1016/j.geodrs.2024.e00885.

Peng, Y., et al. (2023). Maximizing soil organic carbon stocks under cover cropping: insights from long-term agricultural experiments in North America. *Agriculture, Ecosystems & Environment*, 356, 108599. https://doi.org/10.1016/j.agee.2023.108599.

Perrone, S., et al. (2020). Nitrogen fixation and productivity of winter annual legume cover crops in Upper Midwest organic cropping systems. *Nutrient Cycling in Agroecosystems*, 117, 61–76.

Raimondi, G., et al. (2023). Maize yield and N dynamics after cover crops introduction. *European Journal of Agronomy*, 150, 126944. https://doi.org/10.1016/j.eja.2023.126944.

Ranells, N. N. & Wagger, M. G. (1996). Nitrogen release from grass and legume cover crop monocultures and bicultures. *Agronomy Journal*, 88(5), 777–882.

Rosecrance, R. C., et al. (2000). Denitrification and N mineralization from hairy vetch (*Vicia villosa Roth*) and rye (*Secale cereale* L.) cover crop monocultures and bicultures. *Plant and Soil*, 227, 283–290.

Sahu, N., et al. (2017). Strength of microbes in nutrient cycling: a key to soil health. *Agriculturally Important Microbes for Sustainable Agriculture*, I, 69–86.

Scavo, A., et al. (2022). The role of cover crops in improving soil fertility and plant nutritional status in temperate climates: a review. *Agronomy for Sustainable Development*, 42(5), 93. https://doi.org/10.1007/s13593-022-00825-0.

Schröder, J. J., et al. (2016). The elusive role of soil quality in nutrient cycling: a review. *Soil Use and Management*, 32(4), 476–486.

Sharpley, A. N. & Smith, S. J. (1991). Effects of cover crops on surface water quality. In W. L. Hargrove (Ed.), *Cover crops for clean water* (pp. 41–49). Ankeny: Soil and Water Conservation Society.

Sharpley, A., et al. (2000). Practical and innovative measures for the control of agricultural phosphorus losses to water: an overview. *Journal of Environmental Quality*, 29(1), 1–9.

Sokol, N. W., et al. (2022). Global distribution, formation and fate of mineral-associated soil organic matter under a changing climate: a trait-based perspective. *Functional Ecology*, 36(6), 1411–1429.

Soltangheisi, A., et al. (2018). Changes in soil phosphorus lability promoted by phosphate sources and cover crops. *Soil and Tillage Research*, 179, 20–28.

Soumare, A., et al. (2020). Exploiting biological nitrogen fixation: a route towards a sustainable agriculture. *Plants*, 9(8), 1011. https://doi.org/10.3390/plants9081011.

Spain, A. V., et al. (1983). Soil organic matter. In CSIRO (Eds.), *Soils: An Australian viewpoint* (pp. 551–563). Melbourne: CSIRO Publishing.

Stevenson, F. J. & Cole, M. A. (1999). *Cycles of soils: carbon, nitrogen, phosphorus, sulfur, micronutrients*. Hoboken: John Wiley & Sons, INC.

Tariq, A., et al. (2024). Increased N_2O emissions by cover crops in a diverse crop rotation can be mediated with dual nitrification and urease inhibitors. *Agriculture, Ecosystems & Environment*, 374, 109178. https://doi.org/10.1016/j.agee.2024.109178.

Teepe, R., et al. (2001). Nitrous oxide emissions from soil during freezing and thawing periods. *Soil Biology and Biochemistry*, 33(9), 1269–1275.

Teixeira, E. I., et al. (2016). Sources of variability in the effectiveness of winter cover crops for mitigating N leaching. *Agriculture, Ecosystems & Environment*, 220, 226–235.

Teles, A. P. B., et al. (2017). Do cover crops change the lability of phosphorus in a clayey subtropical soil under different phosphate fertilizers?. *Soil Use and Management*, 33(1), 34–44.

Thapa, R., et al. (2018). Cover crops reduce nitrate leaching in agroecosystems: a global meta-analysis. *Journal of Environmental Quality*, 47(6), 1400–1411.

Thapa, R., et al. (2021). Effects of moisture and temperature on C and N mineralization from surface-applied cover crop residues. *Biology and Fertility of Soils*, 57, 485–498.

Thomas, B. W., et al. (2017). Non-legume cover crops can increase non-growing season nitrous oxide emissions. *Soil Science Society of America Journal*, 81(1), 189–199.

Thorup-Kristensen, K. (2001). Are differences in root growth of nitrogen catch crops important for their ability to reduce soil nitrate-N content, and how can this be measured?. *Plant and Soil*, 230, 185–195.

Tilman, D. (1999). Global environmental impacts of agricultural expansion: the need for sustainable and efficient practices. *Proceedings of the National Academy of Sciences*, 96(11), 5995–6000.

Torma, S., et al. (2018). Residual plant nutrients in crop residues–an important resource. *Acta Agriculturae Scandinavica, Section B-Soil & Plant Science*, 68(4), 358–366.

Unkovich, M., et al. (2008). Measuring plant-associated nitrogen fixation in agricultural systems. ACIAR Monograph No. 136. Australian Centre for International Agricultural Research, Canberra, Australia.

Van Eerd, L. L., et al. (2023). Influence of cover crops at the four spheres: a review of ecosystem services, potential barriers, and future directions for North America. *Science of the Total Environment*, 858, 159990. https://doi.org/10.1016/j.scitotenv.2022.159990.

Varela, M. F., et al. (2017). Biomass decomposition and phosphorus release from residues of cover crops under no-tillage. *Agronomy Journal*, 109(1), 317–326.

Venterea, R. T., et al. (2012). Challenges and opportunities for mitigating nitrous oxide emissions from fertilized cropping systems. *Frontiers in Ecology and the Environment*, 10(10), 562–570.

Villarino, S. H., et al. (2023). A large nitrogen supply from the stable mineral-associated soil organic matter fraction. *Biology and Fertility of Soils*, 59(7), 833–841.

Vukicevich, E., et al. (2016). Cover crops to increase soil microbial diversity and mitigate decline in perennial agriculture. A review. *Agronomy for Sustainable Development*, 36, 1–14.

Wagner-Riddle, C., et al. (2007). Intensive measurement of nitrous oxide emissions from a corn–soybean–wheat rotation under two contrasting management systems over 5 years. *Global Change Biology*, 13(8), 1722–1736.

Wagner-Riddle, C., et al. (2017). Globally important nitrous oxide emissions from croplands induced by freeze–thaw cycles. *Nature Geoscience*, 10(4), 279–283.

Wagner-Riddle, C., et al. (2020). Mitigation of nitrous oxide emissions in the context of nitrogen loss reduction from agroecosystems: managing hot spots and hot moments. *Current Opinion in Environmental Sustainability*, 47, 46–53.

Wander, M. (2004). Soil organic matter fractions and their relevance to soil function. In F. Magdoff & R. R. Weil (Eds.), *Soil organic matter in sustainable agriculture* (pp. 67–102). Boca Raton: CRC Press.

Wendling, M., et al. (2016). Influence of root and leaf traits on the uptake of nutrients in cover crops. *Plant and Soil*, 409, 419–434.

White, P. J., et al. (2013). Root traits for infertile soils. *Frontiers in Plant Science*, 4, 193. https://doi.org/10.3389/fpls.2013.00193.

Wooliver, R. & Jagadamma, S. (2023). Response of soil organic carbon fractions to cover cropping: a meta-analysis of agroecosystems. *Agriculture, Ecosystems & Environment*, 351, 108497. https://doi.org/10.1016/j.agee.2023.108497.

Yadav, A. N., et al. (2021). Biodiversity, and biotechnological contribution of beneficial soil microbiomes for nutrient cycling, plant growth improvement and nutrient uptake. *Biocatalysis and Agricultural Biotechnology*, 33, 102009. https://doi.org/10.1016/j.bcab.2021.102009.

Zhang, Z., et al. (2022). Cover crops and N fertilization affect soil ammonia volatilization and N_2O emission by regulating the soil labile carbon and nitrogen fractions. *Agriculture, Ecosystems & Environment*, 340, 108188. https://doi.org/10.1016/j.agee.2022.108188.

Published by Burleigh Dodds Science Publishing Limited, 2026.

Chapter 2

The role of cover crops in improving soil biological health

Michelle Wander, University of Illinois, USA

1 Introduction

Cover crops are grown primarily to build and maintain soil health. They can increase the productivity of soil and its resilience to stress by increasing soil organic matter, improving soil structure (via improving aeration, water holding capacity, and drainage as well as preventing soil loss), retaining and recycling nutrients, suppressing diseases and pests (including weeds), and thus helping to promote subsequent cash crop growth. The many benefits of their use explain why cover cropping is a core strategy for soil health management initiatives in the United States, promoted by the USDA Natural Resources Conservation Service (NRCS) (http://www.nrcs.usda.gov/wps/portal/nrcs/main/soils/health/mgnt/).

When integrating cover crops into a rotation, farmers do not commonly cite soil biology or soil life as the main reason for their use (Das et al., 2022).

http://dx.doi.org/10.19103/AS.2025.0162.04

Instead, they typically focus on overall promotion of soil health even though the connection between soil life and soil health is widely recognized. The Food and Agriculture Organization's (FAO) definition of soil health (https://www.fao.org/agriculture/crops/thematic-sitemap/theme/spi/soil-biodiversity/the-nature-of-soil/what-is-a-healthy-soil/en/) attests that: 'the capacity of soil to function as a living system, within ecosystem and land use boundaries, to sustain plant and animal productivity, maintain or enhance water and air quality, and promote plant and animal health. Healthy soils maintain a diverse community of soil organisms that help to control plant disease, insect and weed pests; form beneficial symbiotic associations with plant roots; recycle essential plant nutrients; improve soil structure with positive repercussions for soil water and nutrient holding capacity; and ultimately improve crop production'.

Untangling some of the complexity associated with managing biology in cover cropped systems may help farmers overcome constraints that limit cover crop use (Golden et al., 2023). The scientific literature links soil productivity to biology indirectly by tying cover crop management to crop performance. Figure 1 summarizes the results from a search for literature reviews considering 'cover crops' and 'soil biology' that identified 117 peer-reviewed articles. Bibliometric analysis of these works identifies links between cover crops and soil biology via organic matter and its management. Management of organic matter likely emerges because cumulative benefits of cover crop use are more persistent and better understood (Krupek et al., 2022) than more nuanced and dynamic influences on soil organisms that are typically expressed with or immediately after cover crop growth (Junod et al., 2024).

The complexity and diversity of cover crop systems make it difficult to make broad statements about their direct influence on soil biology. Cover crops can be grown in annual cropped systems where they are sown in rotation or as companion crops, or in perennial systems where they are planted as understory, intercrops, or strips between cropped areas (Castellano-Hinojosa and Strauss, 2020; Drost et al., 2020; Quintarelli et al., 2022). Their effects are therefore highly context dependent and vary in both time and space.

The selection of species, period of cultivation, management (including any use of fertilizers and pesticides), and method of termination all influence outcomes in any given year, and their influence on plant and soil environments varies with the establishment, growth, and termination phases of the cover crops as well as with subsequent or adjacent cash cropping phases (Van Bruggen et al., 2021; Quintarelli et al., 2022; Cazzangia et al., 2023; Junod et al., 2024). Attempts to match cover crops with the desired benefits or services need to fit within the constraints set by the farming system and its biophysical and economic environments (Daryanto et al., 2018).

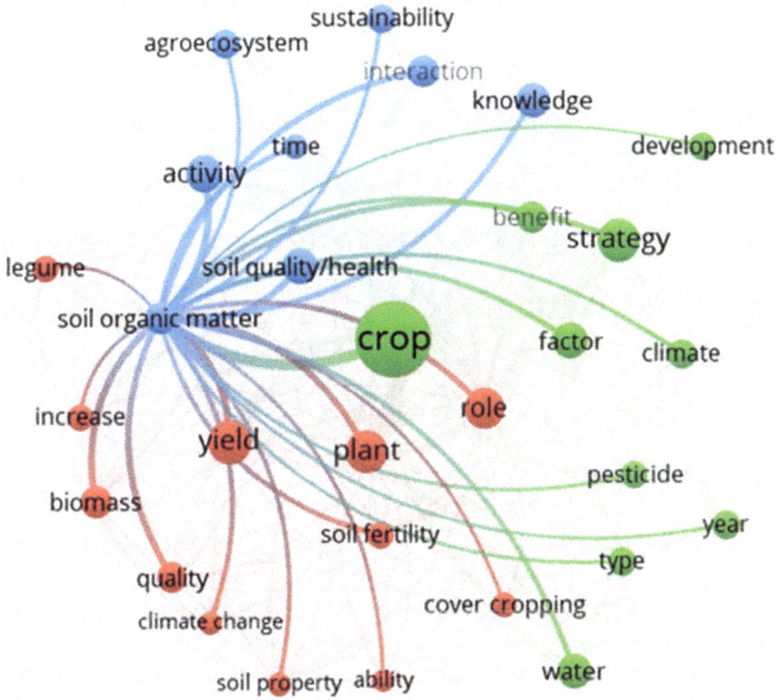

Figure 1 Visual summary of terms and their linkages produced using VosViewer. Terms were cited more than 10 times in the titles, abstracts, and keyword lists of 117 review articles containing both 'cover crops' and 'soil biology' that were recovered using Web of Science.

To navigate this complexity, this review will consider the methods used to assess soil organisms and their function before summarizing their transient and cumulative effects on belowground biology impacting crop yield.

2 Soil biota and soil function

The vast range of soil biological life, which is thought to account for almost 60% of all life on earth (Antony et. al., 2023.), includes earthworms, nematodes, protozoa, fungi, and bacteria (Coleman et al., 2017) that are important contributors to, and indicators of, soil quality (Lavelle, 2006; Geisen et al., 2018). Soil flora, in particular Bacteria, Actinomycetes, and Fungi, receive the most attention in soils because they account for the largest number of individuals, biomass, and heterotrophic activity in soils (Table 1).

Efforts to group and classify soil-dwelling organisms have long sought to resolve meaningful categories. In his seminal text, Animal Ecology, 1927, Charles Elton noted that '...Progress in ecology depends on accurate identification

Table 1 Estimates of abundance and mass of soil organisms commonly found in surface soils

	#/gm soil	kg/ha
Bacteria	10^8–10^9	400–5000
Actinomycetes	10^7–10^8	400–5000
Fungi	10^5–10^6	1000–15 000
Algae	10^4–10^5	10–500
Protozoa	10^4–10^5	20–200
Nematodes	10–10^2	10–150
Mites	1–10	5–150
Springtails	1–10	5–150
Earthworms	0.001	100–1500

Source: Adapted from James Nardi's The World Beneath Our Feet: A Guide to Life in the Soil (2003).

and upon the existence of sound systematic groundwork for all groups of animals' ... *and; this,* 'is the essential basis of the whole thing; without it the ecologist is helpless, the whole of their work can be rendered useless.' Efforts to classify organisms in useful ways extend well beyond taxonomic groupings (Table 2). Evaluating the importance of biodiversity to biogeochemical cycling will ultimately depend on how we define biodiversity, and on our ability to characterize it functionally and identify the critical biotic and abiotic factors that regulate it. Approaches vary in their strengths, weaknesses, and the insights they provide (Norton et al., 2026).

Historically, soil organisms have been grouped based on morphological traits (e.g. size or taxa) or using other easily observable characteristics, including size, which is simple to understand and related to methods needed to identify individuals but lumps organisms that rely on very different life strategies. Unfortunately, the range and complexity of the belowground community make it difficult for even the most well-trained soil ecologist to fully identify even a fraction of the population. Feeding relationships that can be viewed or assessed by dissecting stomach contents, examining mouth structures, etc., and activity measures are attractive metrics because, as Charles Elton (2001) emphasized, 'Food is the burning question in animal society, and the whole structure and activities of the community are dependent upon questions of food-supply'. Food web studies are tremendously valuable but, due to the difficulty associated with isolation and identification of individuals and their feeding relationships, these types of descriptions are often incomplete and oversimplified.

Efforts to understand the contributions of biotic diversity to soil health ideally consider not only the abundance of species but also their ability to exploit different habitats and thus their functional class, or guilds (Ferris and Tuomisto, 2015). Soil habitats, which include the detritusphere, drilosphere, porosphere,

Table 2 An incomplete summary of some of the ways soil biota can be assessed.

Approach	Measurement	Comments
Size		
	Macrofauna: over 2 millimeters (mm)	Includes earthworms (Oligochaete) or beetles (Coleoptera) or arachnids such as spiders
	Mesofauna: 0.1 to 2 mm	Includes mites (Acari) and springtails (Collembola)
	Microfauna: 0.1 mm to 5 μm (micrometers – one thousandth of a millimeter)	Examples include: nematodes (Nematoda), Protozoa, water bears (Tardigrades), and Rotifera
	Microorganisms: under 5 μm um	Includes Bacteria, Fungi and Protozoa
Feeding relationships		
	Producer	Plants, Autotrophs
	Primary consumer	Saprophytes, Detritovores, Shredders, Parasites, Symbiotic associates
	Secondary consumer	Corprophages, Grazers, Predators
	High level consumers	Predators, Scavengers
Activity		
	Respiration rates (CO_2 evolution)	Inexpensive, integrative measure of heterotrophic activity, sensitive to temperature, moisture, can be done in field or in lab
	C and N mineralization rates	Can be done in lab or field to understand rates of decay and N supply
	Mass Loss or disappearance	Typically done in field with litter bags or materials added (cotton strips) to assess decay rates; simple, semi quantitative, low cost
	Enzyme activity (β-glucosidase, phosphatase, urease)	Done in lab to assess the potential activity
	Soil adenosine triphosphate (ATP)	Snapshot of microbial biomass and activity, sensitive to temperature, water content, and organic matter
Abundance	Microbial biomass C chloroform fumigation or substrate induced respiration	Integrative lab method, estimate of C or N released by microbial community, sensitive to organic matter and soil texture

(*Continued*)

Table 2 *Continued*

Approach	Measurement	Comments
	Quantitative Polymerase Chain Reaction (qPCR)	Lab method that amplifies DNA and RNA sequences from target organisms or genes; is sensitive and group-specific but depends on primers and gene copy assumptions
	Phospholipid Fatty Acid analysis (PLFA)	Estimate of living biomass, can provide coarse taxonomic description
	Colony Forming Units (CFUs) and Gram negative and Gram positive bacteria	Estimates culturable microbes (<1% of population). Cell wall staining reveals gram-positive copiotrophic bacteria that grow rapidly on labile organic matter and gram-negative bacteria that tolerate stress. Their ratio is an indicator of soil stress and carbon availability.
Metabolism		
	Metabolic quotient (qCO2)	Measures metabolic efficiency using the ratio of basal respiration to microbial biomass carbon. Indicates how much energy microbes use for maintenance versus growth.
	Allochthonous/Autochthonous growth	Autochthonous organisms are autotrophs and produce organic matter via photosynthesis, chemosynthesis, allochthonous organisms grow on organic matter (e.g., dead leaves, wood, or other organic inputs from adjacent areas). Commonly measured using isotopes or biomarkers
	Trophic state (Oligotrophic/Eutrophic)	Oligotrophs grow in nutrient poor and Eutrophs grow in nutrient-rich environments. Estimated in numerous ways including biomass and activity. Nutrient availability (C, N, P, etc.) influences the composition of the microbial community and its activity.

Table 2 *Continued*

Approach	Measurement	Comments
	r and K selected organisms (r/K) and selection theory	Ratio of opportunistic fast growing 'r-selected' to slower growing 'K-selected' species reveals how microbial metabolism responds to the environment and resource availability.
Community characteristics		
	Composition and diversity assessed with numerous methods (DNA metabarcoding (16S, ITS, 18S) • Metagenomics/ metatranscriptomics • PLFA community profiling • Morphological identification (nematodes, microarthropods) • Community-level physiological profiling (CLPP/Biolog))	DNA based methods quantify fragments from unculturable, living and dead organisms, metagenomics can reveal function as well as composition, faunal morphology (nematodes) is laborious, can reveal food-web structures, CLPP reveals potential metabolic diversity
	Functional guilds can be measured in numerous ways including functional genes, or CLPP, stable isotope probing.	Metabolic functional guilds are groups that have the same biogeochemical or ecological functions (e.g., nitrogen fixation or chitin degradation)

aggregatusphere, and the rhizosphere, possess distinct characteristics that regulate the organisms and processes occurring within (Beare et al., 1995).

Diversity within taxonomic groups is commonly associated with functional redundancy and metabolic competence. Molecular techniques are increasing, affording us the ability to assess the presence of multiple species, but their ties to function can be difficult to make. For example, using eDNA, Koninger et al. (2025) found soil biodiversity was greater in croplands than in grasslands or woodlands whereas morphologically based methods found the opposite.

Integrative assays that require less expertise than morphology-based techniques commonly measure microbial abundance or heterotrophic activity (e.g. respiration or mineralization rates). Biochemical assays of microbial biomass or abundance are more complete than culture-based techniques that recover only a small fraction of the community. Measures including enzyme assays that tie to nutrient cycling are indicative of soil fertility but are typically not tied to specific organisms.

Functional classification schemes, of which there are many, vary in complexity and their relationships to soil health. Lavelle et al. (2006), for example, divided soil organisms into different functional groups: symbionts, decomposers, elemental transformers, ecosystem engineers, controllers of pests and diseases. Newer 'omic' assays are providing detailed insights into the structure and composition of the soils' biotic community. Despite these many tools, we continue to lack a full understanding of the soil microbiome, its complex interactions, and how these relate to functions in soil (Mendes et al., 2013). For example, it remains difficult to relate genetic diversity and community structure of the microbiome to decomposition (Nannipieri et al., 2017).

While the growing number of assessment tools provides new insights, variable and sometimes contradictory results can complicate efforts to understand how to select and manage covers to promote organisms and activities that increase soil health. Both timeliness and tradeoffs should be considered when considering when and how cover cropping influences biota known to impact crop productivity (Table 3).

Table 3 Microorganisms widely known to influence crop productivity.

Microbial taxa	Reputed function in soil health
Rhizobium spp.	Nitrogen fixation in legume root nodules; improves plant growth and fertility
Azotobacter spp.	Free-living nitrogen fixation; enhances soil fertility and organic matter cycling
Bacillus spp.	Produces antibiotics and enzymes; suppresses pathogens; promotes plant growth via phytohormones
Pseudomonas spp.	Biocontrol of soil-borne pathogens; siderophore production; nutrient solubilization
Actinobacteria (e.g., Streptomyces)	Decomposition of complex organic matter; production of antibiotics; disease suppression
Mycorrhizal fungi (Glomeromycota)	Symbiotic nutrient exchange (especially phosphorus); improves soil aggregation and plant resilience
Trichoderma spp.	Antagonistic to plant pathogens; promotes root growth; enhances nutrient uptake
Cyanobacteria	Nitrogen fixation in soils; contributes to soil crust formation and carbon cycling
Firmicutes (general group)	Organic matter decomposition; stress tolerance; contributes to soil resilience under disturbance
Core soil taxa (e.g., Burkholderia, Flavobacterium)	Regulate microbial competition; influence plant health under environmental stress

3 Cover crops, microbial activity and community composition

Cover crops commonly increase microbial biomass and stimulate soil biotic activity and metabolism, which are assumed to be positively related to soil health (King and Hofmockel, 2017; Sharma et al., 2018; Crookston et al., 2023; Bagnall et al., 2023). Regardless of cover crop type, their use generally increases soil microbial biomass (Blanco-Canqui et al., 2022; Van Eerd, et al., 2023; King and Hofmockel, 2017), enzymatic activities of soil microbes (Sürücü et al., 2014), and their relative abundance in soils (Li et al., 2020a; Schmidt et al., 2018). For example, in a meta-analysis of data from 81 studies to compare soil microbial community abundance and composition in cover cropped and non-cover cropped soils, cover cropping increased microbial biomass C and N (MBC and MBN) by 40% and 51%, respectively, as well as PLFA abundance, which is a measure of living biomass, by 24% (Muhammed et al., 2021). Total soil bacteria and fungi increased by 7–31% relative to non-cover cropped treatments. In another meta-analysis of 60 studies comparing cover cropped and fallow soils, cover crops significantly increased microbial abundance (27%), activity (22%), and community diversity (2.5%) (Kim et al., 2020).

Other studies also commonly report an increase in microbial diversity associated with cover cropping (Chavarria et al., 2016; Sanchez et al., 2019). The influence of cover cropping on microbial diversity interacts with cover crop management practices. For example, Schmidt et al. (2018) found cover crop production shifted microbial communities toward organisms with wider metabolic capacities and slower metabolic rates, particularly in no-till systems. This is consistent with Delitte et al. (2021), who reported reduced tillage systems favor greater microbial diversity, including more oligotrophic lifestyles. Tillage of covers produces bursts of nutrient availability and niche homogenization that can increase C supply (Schmidt et al., 2018). Disruption of soil structure can reduce the cohesion and diversity of soil microbial network and favor opportunistic commensal and copiotrophic microbes (Delitte et al., 2021).

Other practices commonly used with cover crops like herbicides can also influence microbial activity and community composition. For example, cover crop termination with rolling or herbicide applications (e.g. eliminating tillage) had a positive influence on total bacterial community abundance compared to tillage-based termination (Romdhane et al., 2019). Glyphosate-based herbicides, which are widely used to terminate covers, impact soil flora (plants, bacteria, fungi, and algae) by blocking the shikimate pathway that converts carbohydrates into chorismate, which is the precursor for essential aromatic amino acids. Addition of glyphosate-based herbicides is widely observed to stimulate soil respiration and often shifts microbial community composition (Aslam et al.,2024, and numerous references therein). An in-depth review of the impact of glyphosate on biota by Van Bruggen et al. (2021) found that most studies have considered short-term treatment effects on microbial biomass or general community composition at high taxonomic levels and found little effect. The review also found studies using deep sequencing methods suggest plant growth-promoting rhizobacteria are commonly negatively affected and that pathogenic bacteria and fungi are often enhanced. Brillon et al. (2025) investigated the effects of cover crops and glyphosate-based herbicide rates using metagenomic analysis of 16S and 18S rRNA and found that use of covers helped to mitigate the negative impacts of glyphosate on microbial community composition. Using community-level physiological profiles, Ortega et al. (2021) found microbial communities were highly sensitive to glyphosate applications. They found diversity indices varied with cover crop type, but abundance and functional diversity were greatest before glyphosate application and declined significantly after glyphosate application and at soybean harvest.

4 Cover crops, N_2 fixation and soil nitrogen

Cover crops are commonly used to increase soil nitrogen supply potential and retention (Abdalla et al., 2019). Use of nitrogen-fixing cover crops to supply nitrogen to the following crop is a key tactic used to increase sustainability by reducing purchased inputs and improving soil health. The symbiotic relationship between legumes and rhizobia relies on cold-hardy legume cover crops that include crimson clover, hairy vetch, winter pea, red clover, white clover, and sweet clover, and warm-season clover cover crops such as cowpea, and sunn hemp. Symbiotic associations with Rhizobia, wherein plants exchange carbohydrates (energy) for ammonium after bacteria convert atmospheric nitrogen (N_2) to ammonia (Poole et al., 2018; Burghardt, 2020; Castellano-Hinojosa et al., 2022), commonly supply N in ranges from 21 kg ha^{-1} to 389 kg ha^{-1} (Vasconcelos et al., 2020). The amount of N fixed and supplied by legumes varies depending on the species and soil conditions. Perrone et al. (2020) found that vetch provided 211 kg N ha^{-1} from aboveground biomass and that N fixation provided between 57% and 75% of that N. Both the amount of biomass and N fixed by legumes are commonly suppressed when mineral N is abundant (Blesh, 2018). High levels of available N suppress communication between rhizobia and legumes, chemotaxis, root exudation patterns, and nodulation and can inhibit nitrogenase activity (Abd-Alla et al., 2023).

Efforts to increase N_2 fixation potential that have long sought to improve the compatibility and efficiency of the symbiotic partners using inoculation find crop-bacterial specificity and inoculation methods are important, but results vary widely with field management history. Cover crops may also have important legacy effects on nitrogen cycling through the microbial populations they enrich. Legume cover crop legacy depends on host partner specificity as some legumes nodulate only with a particular bacterial species or subspecies, and others are promiscuous. Failure of inoculation with rhizobial strains that have high N-fixing capacity to form symbioses has long been attributed to the competitive abilities of resident microbes. Increasingly, work suggests that there is strong partner selectivity by the legume host (Clouse et al., 2025). They also show legume cover crops can enhance the effectiveness of rhizobial strains that partner with leguminous cash crops grown after cover termination.

Grass–legume mixtures are commonly recommended as a way to optimize soil N management by adding N via N fixation, enhancing N availability by leveraging nitrogen from symbiotic and non-symbiotic sources, and reducing N leaching (Drinkwater et al., 2000; de Notaris et al., 2021; Reiss and Drinkwater, 2020). By exploiting several mechanisms, cover crop mixes often supply more nitrogen than legume monocultures (Nyfeler et al., 2011). The ability of grasses to scavenge and retain N allows mixtures to adapt to N supply (White et al., 2017).

When soil N levels are low, grasses can also partner with a range of microbes to acquire N through biological N_2 fixation (Chalk, 1991; Boddey and Dobereiner, 1995; Perrone et al., 2020). Associative N_2-fixing rhizosphere organisms (*Azotobacter paspali*, *Beijerinkia* spp.), and facultative (*Azospirillum*, except *Azospirillum halopraeferans*) and obligate endophytes (*Acetobacter diazotrophicus*, *Herbaspirillum* spp., and *Azoarcus* spp.) (Baldani et al., 1997) can be agronomically important for unfertilized grasslands and pastures and bioenergy crops (Carlsson et al., 2003). Rhizobia, *Frankia*, and *Azospirillum*, which can be important microsymbionts of grasses like wheat, are often considered to be plant growth-promoting organisms because they do more than just increase N_2 fixation (Li et al., 2020b).

5 Cover crops and arbuscular mycorrhizal fungi (AMF)

Cover cropping can also promote symbioses between crops and arbuscular mycorrhizal fungi (AMF) that enhance crop productivity and soil health (Fall et al., 2022). By acting as hosts for AMF, cover crops can increase AMF abundance of hyphae, arbuscules, vesicles, moniliform hyphae, and microsclerotia and colonization of subsequent crop plants (Bowles et al., 2017; Bromley and Rintoul-Hynes, 2025). AMF, which are the most widespread fungal mutualists, include both the phylum Glomeromycota and the clade of Endogonomycete, fine root endophytes within the phylum Mucoromycota. These obligate biotrophic fungi supply inorganic nutrients and water to plants in return for photosynthates in a symbiotic relationship (Lutz et al., 2024). In addition to nutritional symbionts, AMF are also considered pathogen antagonists (Azcón-Aguilar and Barea, 1997; Cameron et al., 2013). Competition for space on root surfaces is one way that AM fungi can limit root disease (Cameron et al., 2013).

Positive crop growth responses to AMF inoculation are associated with the presence of pathogenic fungi (Lutz et al., 2023). AMF fungi may also inhibit pathogenic fungi through induced systemic resistance (ISR), which is an enhanced immune response in the host plant (Pozo et al., 2008). Finally, AM fungal hyphae in the soil foster microbial communities (Scheublin et al., 2010) that include fungi and bacteria that play a role in protection from soilborne pathogens (Filion et al., 1999).

AMF respond rapidly to site conditions and cropping practices. The ability of cover crops to stimulate AMF growth can occur after growth of a single cover crop (Galvez et al., 1995) and is greatest in unfertilized soils (Junod et al., 2024). High nitrogen and/or phosphorus availability reduces mycorrhizal colonization (Han et al., 2020; Wang et al., 2011). Cropping sequence and practices like tillage or fertilization both influence AMF by influencing inoculum presence and viability. Somenahally et al. (2018) found the legacy effects of wheat cover crops exerted more influence than tillage on AMF colonization of subsequent crops. A meta-analysis shows winter cover cropping combined with reduced tillage similarly increased AMF colonization of summer annual cash crop roots by ~30%, and the use of legume cover crops had the greatest benefit to AMF colonization (Bowles et al., 2017). Most covers can be affiliated with AMF. For example, Soti et al. (2023) showed that the growth of summer cover crops sunn hemp (*Crotalaria juncea*), sorghum (*Sorghum drummondii*), and cowpea (*Vigna unguiculata*) all increased AMF spore abundance 2.5–3-fold. AMF are commonly abundant beneath oat and cereal rye cover crops (Finney et al., 2017). However, production of non-mycorrhizal cover crops such as canola/rapeseed or radish in the Brassicacea and Amaranthacea families can suppress mycorrhizal colonization of the following crop (Koide and Peoples, 2012; Hill, 2006). Growing these cover crops in mixtures with mycorrhizal cover crops can allay the negative effects of non-mycorrhizal cover crops on AMF (Cloutier et al., 2020).

The legacy effects of covers on AMF colonization of crops vary. For example, mycorrhizal infection of maize, which is commonly correlated with maize growth and yield (Boswell et al., 1998), was increased more by a barley cover crop compared to fallow, but there was no similar benefit after a vetch cover (Hontoria et al., 2019). Variability may be due to complex interactions with the microbial community and AMF diversity, which is often mixed (Junod et al., 2024). Initial understanding of AMF as a promiscuous associate with low species diversity and limited host specificity suggests functional diversity present within individual species that enables responses to different soil conditions (Munkvold et al., 2004; Edlinger et al., 2022). This explains why cropping practices and management influence AMF in nuanced and context-specific ways. For example, Lahijanian et al. (2025) found a complex interplay between AMF diversity and microbial community composition and seed inoculation, field site characteristics, and agricultural management. They found greater AMF and bacterial diversity in an organic grain system than in adjacent conventional treatments and that seed inoculation significantly increased maize yield in the high-fertility conventional field, but cover cropping improved yield in the organic and transitional fields. Frontiers for managing cover crops to leverage AMF affiliates must consider cultivar or species selection, network interactions, and their interaction with management.

6 Cover crops and plant growth-promoting bacteria (PGPB)

Growth of cover crops such as legumes, C4 and C3 grasses, and non-leguminous forbs and brassicas can increase the abundance of beneficial microbes such as PGPB by increasing plant functional group richness (Vukicevich et al., 2016). The main types of plant growth-promoting bacteria (PGPB) include free-living microbes, such as *Pseudomonas*, *Bacillus*, *Azotobacter*, *Azospirillum*, *Actinomycetes* (for improved organic matter decomposition and nutrient release), and symbiotic nitrogen-fixers, such as Rhizobia and endophytes, including *Trichoderma* fungi (effective for biocontrol), living in plant tissue.

PGPB can be major drivers of soil health (Trivedi et al., 2016), enhancing plant growth by production of phytohormones and siderophores, increasing plant nutrient uptake, pest and disease suppression, and increasing resistance to abiotic stress (Timofeeva et al., 2023). As previously noted, rhizobia can also form associative relationships with non-leguminous cover crops (Mehboob et al., 2009) and contribute to nitrogen fixation (Dent et al., 2017). They can also stimulate plant growth through phytohormone production which stimulates root and shoot growth, promoting solubilization of nutrients such as phosphorus, stimulating plant systemic resistance to infection as well as competitive exclusion of pathogens (Fahde et al., 2023).

Different cover crop species can exert distinct selection effects on PGPR that influence the crop grown after cover crop termination by increasing the microbial network complexity relative to a fallow control (Cazziangia et al., 2023). That work found oilseed radish (Brassicaceae), in particular, can promote Pseudomonadaceae and lentil (Fabaceae) stimulated fungal taxa, including Trichocomaceae and members of the Glomerales order. Studies of PGPB promotion include apple orchards intercropped with crown vetch (*Coronilla varia*) that increase the number of PGPB species (Castellano-Hinojosa and Straus, 2020). Qi et al. (2020) found wheat has the greatest ability to increase the growth of plant-beneficial bacteria, followed by rapeseed. Cover crops can also influence the efficacy of PGPB. For example, oat mulch promoted the greatest total antagonistic effect of *Bacillus* spp. and *Pseudomonas* spp. toward *Altenaria alternata*, *F. culmorum*, *F. oxysporum*, *H. haematococca*, *P. irregulare*, and *T. cucumeris*, and the smallest total antagonistic effect occurred with tansy phacelia mulch (Patkowska and Konopiński, 2014).

As with true for N management, planting multispecies cover crop mixtures is commonly recommended because it can increase the diversity and abundance of beneficial soil microbes (*Azospirillum*, *Azotobacter*, *Bacillus*, and *Pseudomonas*), mycorrhizal fungi (*Acaulospora morrowiae*, *Archaeospora trappei*, *Gigaspora gigantea*, and *Scutellospora calospora*), as well as Acidobacteria, Burholderiales, Sphingobacteria, and Thermomicrobia abundance (Dube et al., 2019; Xu et al., 2020). Advanced tools suggest that consortia rather than single species of PGPBs are important contributors to productivity. Increased understanding of how PBGBs interact in the environment and benefit plants will allow us to harness their potential.

7 Cover crops and nematodes

Cover crops vary in their ability to alter nematode communities, which include both plant pathogenic nematode (PPN) and free-living beneficial species (Wang et al., 2007). Factors such as context, cover crop type, and host-specificity of nematode species within the crop rotation must be considered to optimize cover crop selection and management. The effects of covers can be complex and include trade-offs.

Cover crops can be grown to break pest cycles, but the effectiveness varies with cover and pest type. A meta-analysis by Daryanto et al. (2018) found a 29% decrease in the abundance of plant pathogenic nematodes from use of cover crops in the Brassicaceae, Fabaceae, and Poaceae families. A study of *Brassica* and rye cover crops found biofumigant and allelopathic effects in suppressing PPN, although these effects could also extend to beneficial species (Stocking Gruver et al., 2010). Biofumigation can also be an important mechanism for suppression of PPN using marigold (*Tagetes patula* – but not all marigold species), which produces compounds such as α-terthienyl that are allelopathic to many species of plant-parasitic nematodes. The antagonistic effect of nematode suppressive cover crops carried over to reduce infestation of plant-parasitic nematodes in the following crop (Hooks et al, 2010).

Winter cover crops of cereal rye (*Secale cereale*), barley (*Hordeum vulgare*), and oat (*Avena sativa*) can also promote nematode community abundance and diversity with fewer PPN than found in fallow (Akanwari et al, 2024). The influence on PPNs varied with some cover crop mixtures increasing abundance, for example, of the plant-parasitic stunt nematode *Tylenchorhynchus*.

Many legume cover crops (cowpea, crotalaria, jointvetch, sunn hemp) are very poor hosts or non-hosts of parasitic species such as root knot nematodes (McSorley, 1999). For example, *Crotalaria* (sunn hemp) is a poor host to many plant-parasitic nematodes including *Meloidogyne* spp., *Rotylenchulus reniformis*, *Radopholus similis*, *Belonolaimus longicaudatus*, and *Heterodera glycines* (Wang et al, 2002). Crimson clover (*Trifolium incarnatum* L.) has potential to manage the soybean cyst nematode (*Heterodera glycines*) by acting as a trap crop that stimulates hatching and/or attracting hatched SCN juveniles that may penetrate roots but are unable to feed (Harbach et al., 2021; Van Eerd et al., 2023). Annual ryegrass and daikon radish reduced SCN egg numbers to a greater extent than other cover crops in naturally infested soils by 65% and 67 %, respectively (Acharya et al., 2021).

Nematode communities and foodwebs, which are sensitive to fertility status and disturbance, respond to cover crop management and are excellent assays of soil health (Martin and Sprunger, 2022). The relative abundance of bacterivorous, fungivorous, and herbivorous nematodes reflects carbon flow through resource channels (Ferris and Bongers, 2006). Many studies show that cover cropping increases food web complexity and, thus, species richness and abundance. For example, both grass and legume cover crops increased soil food web structure and complexity, but the influence of specific types of cover crops varied within the growing season and between study years (Leslie et al., 2017). Additionally, Italian/annual ryegrass (*L. multiflorum*) fostered greater nematode abundance and more complex and dynamic food webs with higher abundance and diversity (Zhang et al., 2022). A grass-legume mixture was found to be particularly effective in promoting nematode numbers, diversity, and dynamics, with beneficial effects on N and C cycling. However, this increased numbers of plant-parasitic nematodes (*Pratylenchus neglectus* and *Merlinius brevidens*), indicates the trade-offs involved in using cover crops to promote nematode species.

The quality of cover crop residues influences nematode community structure. Dupont et al. (2009) found increased nematode numbers in tomato systems in cover crop treatments with low to mid C:N ratios, with higher nitrogen content leading to corresponding nematode enrichment. Garba et al. (2024) also found the abundance of free-living nematodes was sensitive to cover crop residue quality (C:N ratio and cellulose content) but less so to cover crop quantity (biomass). Gruver et al. (2010) reported that organic matter added by cover crops such as forage radish, rapeseed, and rye promoted nematode species. However, working in carrot production systems, Grabau et al. (2017) found delayed effects of rape and radish-oat covers on nematode communities, which needed a year to start to have an impact, suggesting that changes in the soil community following cover cropping may be gradual (Jones et al., 1969).

Effective use of cover crops to suppress PPN and increase the complexity of soil food webs rests upon our ability to manage the interactions between crop and cover crop species grown in rotation or combination with perennial crops.

8 Cover crops and plant pathogenic bacteria and fungi

Most practices that maintain or improve soil health, including use of cover crops, also have a positive effect on reducing soilborne plant diseases by improving biotic diversity, which then creates a more competitive environment for pathogens (Abawi and Widmer, 2000). As noted earlier, suppressive effects include biofumigation resulting from the biocidal activity of specific compounds, notably glucosinolates (GSLs) (glucose- and sulfur-containing organic anions) and isothiocyanates (ITCs). Most GSL-containing genera are clustered within the Brassicaceae, Capparaceae, and Caricaceae families (Rodman, 1981).

Incorporation of *Brassica* cover crops has been shown to reduce multiple soilborne diseases, particularly in potato cropping systems (Baysal-Gurel and Liyanapathiranage, 2019; Larkin, 2021; Hemkemeyer et al., 2024). Cereal cover crops such as rye, barley, and wheat also suppress disease in potato-based production (Rittl et al., 2023). Manici et al. (2018) found similar effects in use of vetch in suppressing plant-pathogenic bacteria (PPB), which then boosted yields of tomato and zucchini. Both canola and rapeseed combined with winter rye reduced black scurf (*R. solani*) disease severity by 35–41% and common scab (*Streptomyces scabies*) by 20–33% compared with no cover crop (West, 2018). Incorporation of red clover and wheat cover crops reduced disease severity and pathogen recovery in *Rhizoctonia solani* and *Phytophthora nicotianae* infested soils (Dawadi et al., 2019). Both cover crops increased the presence of Pseudomonads, again a PGPB, which are thought to be antagonistic to soilborne pathogens. Grass cover crops are commonly found to suppress fungal diseases. A sudangrass cover crop suppressed *Fusarium* spp. in both barley and soybean. However, oilseed radish had the opposite effect, increasing disease incidence in both barley and soybean (Aiyer et al., 2024). Grass species such as rye can suppress root rot caused by *Rhizoctonia solani* and *Fusarium virguliforme* in soybean (Wen et al., 2017).

Disease suppression can also occur when saprotrophic fungi opportunistically colonize roots as endophytes and aid in controlling pathogenic fungi. For example, *Trichoderma* can parasitize plant-pathogenic bacteria in potato using cell wall-degrading enzymes and antibiotic compounds (Harman, 2024). Other protective fungi include *Penicillium* (Hossain et al., 2014), *Clonostachys* (Luongo et al., 2005), and non-pathogenic strains of *Fusarium oxysporum* (Benhamou et al., 2002). These fungi can also promote increased root and plant growth (Harman, 2024).

However, the efficacy of cover crops can vary, sometimes having no effect (Araldi-Da-Silva et al., 2022). A Triticale cover crop, for example, was ineffective in suppressing plant pathogenic *Fusarium verticillioides* (Ray et al., 2022). Mustard cover cropping did not effectively suppress the black root rot complex in strawberry cultivation (Rahman et al., 2023). Winter cereal rye increased pathogen populations of *Fusarium graminearum* and *Pythium* spp. in corn seedling roots, leading to corn yield decreases (Bakker et al., 2016; Acharya et al., 2018; Van Eard et al., 2023). Cover crops may also act as green bridges for some pathogens where, nonhost crop or fallow break instead may have effectively broken a disease cycle. Kobayashi-Leonel et al. (2017), for example, found legume cover crops can act as *Fusarium* hosts. Delaying the planting of the crop can ameliorate this problem. Our ability to more consistently leverage disease suppression via cover cropping systems will increase with our understanding of the genetic and molecular mechanisms regulating plant-soil feedbacks (Wang et al., 2025).

9 Cover crops and soil organic matter

The addition of cover crops to farming systems leads to persistent benefits that result from cumulative effects derived mostly from gains in organic matter quantity and quality (Martínez-García et al., 2018) and/or improved soil structure (Haruna et al., 2020). Cover crops increased organic matter by an average of 13% in temperate climates, according to a meta-analysis (McClleland et al., 2021). In a global analysis of cover crop use (93 studies), increases in soil organic matter ranged from 7% to 33% with a related 33% average increase in soil microbial biomass (Hu et al., 2023). This is confirmed by other studies (Zhou et al., 2023; Kucerik et al., 2024). For example, after 10 years of use, cover cropping increased the abundance of all microbial groups and total catabolic activity, associated with a higher concentration of dissolved organic matter (Martínez-García et al., 2018).

Soil organic matter influences physical, chemical, and biological aspects of soils that interact in numerous ways. For example, soil physical structure, which results from affiliation of particle size separates (sand, silt, and clay) into aggregates, or peds, that contain pore spaces within and produce drainage in interspaces that collectively affect soil biology by providing a habitat for various organisms and adequate aeration and moisture to support their growth (Shi, 2025). Cover crops improve structure by preventing soil erosion, soil crusting, and compaction, by promoting aggregate and pore formation (particularly macropores) and aggregate stability that improve aeration and water infiltration (Haruna et al., 2020; Saleem et al., 2020). A meta-analysis by Yan and Arthur (2025) found on average that cover crops reduced soil bulk density by 3.2%, penetration resistance by 11.8% and, increased water stable aggregates by 15.9%, total porosity by 6.1%, and water infiltration by 37.2%.

The benefits from cover crop use are often linked to soil organic matter's contribution to improved nutrient availability and plant nutritional status (Shen et al., 2021; Scavo et al., 2022).

Additionally, organic matter added by cover crops acts as a source of food and energy for soil microbes (Finney et al., 2017; Kong and Six, 2012). Heterotrophic activity, including the rate of organic matter decay, is determined by microbial dynamics (Mooshammer et al., 2014), which is often associated with increased functional diversity and activity of microbes in cover cropped soils (Nivelle et al., 2016). One way of understanding the interactions between organic matter and soil microorganisms is by looking at carbon to nitrogen (C:N) ratios, which provide a simple measure of the biochemical composition of organic matter from cover crops (Cao et al., 2021). Even though we still do not fully understand exactly how individual microorganisms alter their metabolic strategies in response to substrate stoichiometry (Graham and Hofmockel, 2022), different C:N ratios appear to promote different microbial groups in the soil food web (Dupont et al., 2009; Zhang et al., 2022).

In general, lower residue-based C:N ratios (>15–25) with a higher N content promote nitrogen mineralization by soil microorganisms, which then makes the N more immediately available, for example, to a following cash crop. Higher C:N ratios (with a lower N content) mean more C (energy) available than N (protein) which slows down decomposition by soil microorganisms, promotes N scavenging (e.g. from deeper soil layers) and N immobilization by soil microorganisms, releasing N more slowly and potentially making it available to a following cash crop over a longer period (Li et al., 2020a; Xu et al., 2024). Leguminous cover crops tend to have lower C:N ratios (<25) than grasses (>25–40) and so decay more rapidly in the short term. Legume cover crops are associated with the accumulation of mineral-associated N-rich residues (Strickland et al., 2019; Hu et al., 2023; Gan and Drinkwater, 2025). Legume cover crops had an N replacement value of 62 kg N ha^{-1} in corn-based rotations and produced similar or higher yields than the use of N fertilizer (Marcillo et al., 2024). Legume cover crops in South African citrus production systems added 67 to 209 kg N ha^{-1} through N fixation and increased fruit yield by 7.6% to 64% (Silwana, 2025).

With a lower C:N ratio and more rapidly decaying biomass, leguminous cover crops are thought to promote 'r strategist' (copiotrophic) bacteria that thrive on easily available nutrients (Rousk et al., 2015; Gao et al., 2025). Dominant r-strategists specialize in the rapid decay of easily accessible, labile organic compounds like simple sugars, proteins, and cellulose. They include Proteobacteria (including Gammaproteobacteria), Bacteroidetes, and certain fungi such as Mortierellomycota.

In contrast, grasses tend to have finer roots and produce more root exudates and have a larger biomass, which decays more slowly (Wong et al., 2024; Opoku, 2022). Non-leguminous cover crops (such as grasses and brassicas) favor 'K-strategist' (oligotrophic) microorganisms adapted to high-carbon, low-nutrient environments that are able to degrade more recalcitrant compounds.

They include fungi, specifically Basidiomycota and certain saprotrophic fungi, and bacterial phyla like Actinobacteria, Acidobacteria, and Chloroflexi.

Managing cover crops for optimal decay is tricky, particularly in rain-fed production systems. Nutrient release by decay does not always coincide with plant demand as nutrient cycling is coordinated with nutrient immobilization/sequestration mediated by microbes (Fontaine et al., 2024). Asynchrony between plant demand and nutrient supply can occur when, for example, N immobilization by rye residues reduces soil N during peak maize demand (Nevins et al., 2020). Even though challenges posed by variable weather are on the rise, improved use of practices like N crediting and split N applications can increase the nutrient use efficiency of cover cropped systems (Adeyemi et al., 2020).

10 Conclusion

By improving the physical and chemical properties of soil, as well as by developing symbiotic relationships with some species, cover crops can have significant beneficial effects on soil biology. This produces important benefits for overall soil health, which then feeds through to more resilient, sustainable, and productive farming systems. However, the effectiveness of cover crops is highly context dependent and subject to a complex range of variables. Similarly, soil microbiomes are complex, variable, and highly sensitive to their environments. Methods to measure soil biology are still evolving, and researchers recognize we still have much to learn about how soil food webs operate.

Improvement in methods to understand how cover crops and their interactions with management influence soil properties and associated biotic communities should enable us to better design systems and select cover crops that increase or maintain soil productivity. Research has shown that, while the influence of specific cover crop systems on biotic communities can be significant, effects vary over time. To increase the consistency of desired outcomes, we need to appreciate variables such as the characteristics of a particular site, including the legacy effects of prior management. Cover crop mixtures are recommended because they include more functional diversity and thus can adapt to a wider range of conditions. High input production systems suppress plant-microbe associations that facilitate productivity and resource use efficiency.

Ideally, we can decrease the need for chemical inputs by developing cover crop cultivars that modulate biotic communities for soil and crop health. Promising leads to follow include tactics to build soil organic matter using legumes for nutrient capture and supply, selecting specific grasses, brassicas (and to a lesser extent legumes) for disease suppression, and maximizing biomass returned to the soil to fully synchronize with plant nutrient demand.

11 Acknowledgments and declaration of interests

Efforts related to this work have been supported by USDA Hatch Grant U.S. Department of Agriculture (USDA)-Grant: 1026327. I declare that I have no known competing financial interests or personal relationships that could have appeared to influence the work reported.

12 Where to look for further information

- An overview of the ways soil biology contributes to soil health and ecosystem services is provided by Vasquez et al. (2026).
- The differing roles of soil microbiomes and nitrogen cycling are explored in Clough (2025).
- An overview of carbon cycling is provided by Plante and Pett-Ridge (2025).
- For a summary of pathogen and pest control, see Kandaswamy et al. (2025).

13 References

Abawi, G. S. and Widmer, T. L. (2000). Impact of soil health management practices on soil borne pathogens, nematodes and root diseases of vegetable crops. *Applied Soil Ecology*, 15, 37–47. https://doi.org/10.1016/S0929-1393(00)00070-6.

Abdalla, M., et al. (2019). A critical review of the impacts of cover crops on nitrogen leaching, net greenhouse gas balance and crop productivity. *Global Change Biology*, *25*(8), 2530–2543.

Abd-Alla, M. H., et al. (2023). Enhancing rhizobium–legume symbiosis and reducing nitrogen fertilizer use are potential options for mitigating climate change. *Agriculture*, 13(11), 2092.

Acharya, K., et al. (2021). Effects of cover crops on population reduction of soybean cyst nematode (Heterodera glycines). *Plant Disease*, 105(4), 764–769.

Acharya, J., et al. (2018). Effects of fungicide seed treatments and a winter cereal rye cover crop in no till on the seedling disease complex in corn. *Canadian Journal of Plant Pathology*, 40(4), 481–497.

Adeyemi, O., et al. (2020). Effect of wheat cover crop and split nitrogen application on corn yield and nitrogen use efficiency. *Agronomy*, 10(8), 1081.

Aiyer, H. S., et al. (2024). Select cover crop residue and soil microbiomes contribute to suppression of fusarium root and crown rot in Barley and Soybean. *Microorganisms*, 12(2), 404. https://doi.org/10.3390/microorganisms12020404.

Akanwari, J., et al. (2024). The impact of winter cover crops on soil nematode communities and food web stability in corn and Soybean Cultivation. *Microorganisms*, 12(10), 2088.

Araldi-Da-Silva, G., et al. (2022). Field studies on the effect of rye cover crop on soybean root disease and productivity. *PhytoFrontiers™*, 2(3), 192–201.

Aslam, S., et al. (2024). Microbial activity, community composition and degraders in the glyphosate-spiked soil are driven by glycine formation. *Science of the Total Environment*, 907, 168206.

Azcón-Aguilar, C. and Barea, J. M. (1997). Arbuscular mycorrhizas and biological control of soil-borne plant pathogens–an overview of the mechanisms involved. *Mycorrhiza*, 6(6), 457–464.

Bagnall, D. K., et al. (2023). A minimum suite of soil health indicators for North American agriculture. *Soil Security*, 10, 100084.

Baysal-Gurel, F. and Liyanapathiranage, P. (2019). Pathogenicity of rhizoctonia solani and phytophthora nicotianae to brassicaceae cover crops. *Archives of Phytopathology and Plant Protection*, 52(3–4), 288–302.

Bakker, M. G., et al. (2016). The terminated cover crop as a "green bridge" to facilitate Pythium damping-off in corn. *Phytopathology*, 106, 1276–1285.

Bakker, M. G., et al. (2016). The potential for cereal rye cover crops to host corn seedling pathogens. *Phytopathology®*, 106(6), 591–601.

Baldani, J., et al. (1997). Recent advances in BNF with non-legume plants. *Soil Biology and Biochemistry*, 29(5–6), 911–922.

Beare, M. H., et al. (1995). A hierarchical approach to evaluating the significance of soil biodiversity to biogeochemical cycling. *Plant and Soil*, 170(1), 5–22.

Benhamou, N., et al. (2002). Ability of nonpathogenic Fusarium oxysporum strain Fo47 to induce resistance against Pythium ultimum infection in cucumber. *Applied and Environmental Microbiology*, 68(8), 4044–4060.

Bernier Brillon, J., et al. (2025). Cross-effect between cover crops and glyphosate-based herbicide application on microbiote communities in field crops soils. *Agriculture*, 15(4), 432.

Blesh, J. (2018). Functional traits in cover crop mixtures: biological nitrogen fixation and multifunctionality. *Journal of Applied Ecology*, 55(1), 38–48.

Boddey, R. M. and Dobereiner, J. (1995). Nitrogen fixation associated with grasses and cereals: recent progress and perspectives for the future. *Fertilizer Research*, 42(1), 241–250.

Boswell, E. P., et al. (1998). Winter wheat cover cropping, VA mycorrhizal fungi and maize growth and yield. *Agriculture, Ecosystems and Environment*, 67(1), 55–65.

Bowles, T. M., et al. (2017). Ecological intensification and arbuscular mycorrhizas: a meta-analysis of tillage and cover crop effects. *Journal of Applied Ecology*, 54(6), 1785–1793.

Bromley, P. L. and Rintoul-Hynes, N. L. J. (2025). Cover cropping increases the abundance of mycorrhizal and endophytic fungi structures associated with ecosystem functioning. *Agronomy Journal*, 117, e70075. https://doi.org/10.1002/agj2.70075.

Burghardt, L. T. (2020). Evolving together, evolving part: measuring the fitness of rhizobial bacteria in and out of symbiosis with leguminous plants. *New Phytologist*, 228(1), 28–34.

Cameron, D. D., et al. (2013). Mycorrhiza-induced resistance: more than the sum of its parts?. *Trends in plant science*, *18*(10), 539–545.

Cao, Y., et al. (2021). Organic-C quality as a key driver of microbial nitrogen immobilization in soil: A meta-analysis. *Geoderma*, *383*, 114784.

Carlsson, G., and Huss-Danell, K. (2003). Nitrogen fixation in perennial forage legumes in the field. *Plant and soil*, *253*(2), 353–372.

Castellano-Hinojosa, A. and Strauss, S. L. (2020). Impact of cover crops on the soil microbiome of tree crops. *Microorganisms*, 8(3), 328. https://doi.org/10.3390/microorganisms8030328.

Castellano-Hinojosa, A., et al. (2022). Native rhizobia improve plant growth, fix N_2, and reduce greenhouse emissions of sunn hemp more than commercial rhizobia inoculants in Florida citrus orchards. *Plants (Basel)*, 11(22), Article 3011.

Cazzaniga, S. G., et al. (2023). On the legacy of cover crop-specific microbial footprints. *Soil Biology and Biochemistry*, 184, 109080.

Chavarría, D. N., et al. (2016). Effect of cover crops on microbial community structure and related enzyme activities and macronutrient availability. *European Journal of Soil Biology*, 76, 74–82.

Chalk, P. M. (1991). The contribution of associative and symbiotic nitrogen fixation to the nitrogen nutrition of non-legumes. *Plant and Soil*, 132(1), 29–39.

Clough, T. (2025). Advances in understanding the role of the soil microbiomes's effects on nitrogen cycling. In Dunfield, K. E. (ed.). *Understanding and utilising soil microbiomes for a more sustainable agriculture*. Burleigh Dodds Science Publishing, Cambridge, UK.

Clouse, K. M., et al. (2025). Cover crop microbiomes affect legume cash crop growth but not consistently through enriching nitrogen-fixing rhizobia. *bioRxiv*, 2025-10.

Cloutier, M. L., et al. (2020). Fungal community shifts in soils with varied cover crop treatments and edaphic properties. *Scientific Reports*, 10(1), 6198.

Coleman, D. C., et al. (2017). *Fundamentals of soil ecology*. Academic Press, London.

Crookston, B. S., et al. (2023). Microbial respiration gives early indication of soil health improvement following cover crops. *Journal of Soil and Water Conservation*, 78(3), 272–281. https://doi.org/10.2489/jswc.2023.00015.

Daryanto, S., et al. (2018). Quantitative synthesis on the ecosystem services of cover crops. *Earth-Science Reviews*, 185, 357–373.

Das, K. B., et al. (2022). Soil health,cover crop, and fertility management: Nebraska producers' perspectives on challenges andadoption, *Journal of Soil and Water Conservation*, 77(2), 126–134. https://doi.org/10.2489/jswc.2022.00058

Dawadi, S., et al. (2019). Impact of cover crop usage on soilborne diseases in field nursery production. *Agronomy*, 9(11), 753.

Delitte, M., et al. (2021). Plant microbiota beyond farming practices: a review. *Frontiers in Sustainable Food Systems*, 5, 624203.

De Notaris, C., et al. (2021). Cover crop mixtures including legumes can self-regulate to optimize N2 fixation while reducing nitrate leaching. *Agriculture, Ecosystems and Environment*, 309, 107287.

Dent, D., et al. (2017). Establishing symbiotic nitrogen fixing in cereals and other non-legume crops. *Agriculture and Food Security*, 6, Article 7.

Drinkwater, L. E., et al. (2000). Effects of tillage intensity on nitrogen dynamics and productivity in legume-based grain systems. *Plant and Soil*, 227(1), 99–113.

Drost, S. M., et al. (2020). Decomposition of mixtures of cover crop residues increases microbial functional diversity. *Geoderma*, 361, 114060.

Dube, J. P., et al. (2019). Differences in bacterial diversity, composition and function due to long-term agriculture in soils in the eastern free State of South Africa. *Diversity*, 11(4), 61.

DuPont, S. T., et al. (2009). Effects of cover crop quality and quantity on nematode-based soil food webs and nutrient cycling. *Applied Soil Ecology*, 41(2), 157–167.

Edlinger, A., et al. (2022). Agricultural management and pesticide use reduce the functioning of beneficial plant symbionts. *Nature Ecology and Evolution*, 6(8), 1145–1154. https://doi.org/10.1038/s41559-022-01799-8.

Elton, C. S. (2001). *Animal ecology*. University of Chicago Press, Chicago.

Fahde, S., et al. (2023). Rhizobia: a promising source of plant growth-promoting molecules and their non-legume interactions: examining applications and mechanisms. *Agriculture*, 13(7), Article 1279.

Fall, A. F., et al. (2022). Roles of arbuscular mycorrhizal fungi on soil fertility: contribution in the improvement of physical, chemical, and biological properties of the soil. *Frontiers in Fungal Biology*, 3, 723892.

Ferris, H. and Bongers, T. (2006). Nematode indicators of organic enrichment. *Journal of Nematology*, 38(1), 3.

Ferris, H. and Tuomisto, H. (2015). Unearthing the role of biological diversity in soil health. *Soil Biology and Biochemistry*, 85, 101–109.

Filion, M., et al. (1999). Direct interaction between the arbuscular mycorrhizal fungus Glomus intraradices and different rhizosphere microorganisms. *The New Phytologist*, *141*(3), 525–533.

Finney, D. M., et al. (2017). Living cover crops have immediate impacts on soil microbial community structure and function. *Journal of Soil and Water Conservation*, *72*(4), 361–373.

Fontaine, S., et al. (2024). Plant–soil synchrony in nutrient cycles: learning from ecosystems to design sustainable agrosystems. *Global Change Biology*, 30(1), e17034.

Gan, H. and Drinkwater, L. E. (2025). Leguminous green manure promotes N accrual in labile and persistent soil organic matter pools. *Plant Soil*, 507, 763–781 (2025). https://doi.org/10.1007/s11104-024-06764-x.

Galvez, L., et al. (1995). An overwintering cover crop increases inoculum of VAM fungi in agricultural soil. *American Journal of Alternative Agriculture*, *10*(4), 152–156.

Gao, Y., et al. (2025). The dominance of K-strategy microbes enhances the potential of soil carbon decomposition under long-term warming. *Applied Soil Ecology*, 206, 105854.

Garba, I. I., et al. (2024). Cover crop functional types alter soil nematode community composition and structure in dryland crop-fallow rotations. *Applied Soil Ecology*, 194, 105196.

Geisen, S., et al. (2018). Soil protists: a fertile frontier in soil biology research. *FEMS Microbiology Reviews*, 42(3), 293–323.

Golden, L. A., et al (2023). Benefits, barriers, and use of cover crops in the western United States: regional survey results. *Journal of Soil and Water Conservation*, 78(3), 260–271. https://doi.org/10.2489/jswc.2023.00170.

Grabau, Z. J., et al. (2017). Effects of cover crops on Pratylenchus penetrans and the nematode community in carrot production. *Journal of Nematology*, 49(1), 114.

Graham, E. B., and Hofmockel, K. S. (2022). Ecological stoichiometry as a foundation for omics-enabled biogeochemical models of soil organic matter decomposition. *Biogeochemistry*, *157*(1), 31–50.

Gruver, L. S., et al. (2010). Brassicaceous and rye cover crops altered free-living soil nematode community composition. *Applied Soil Ecology*, 45(1), 1–12.

Han, Y., et al. (2020). Responses of arbuscular mycorrhizal fungi to nitrogen addition: a meta-analysis. *Global Change Biology*, 26(12): 7229–7241.

Harbach, C. J., et al. (2021). A mechanistic approach to assessing the potential for cover crops to serve as trap crops for the soybean cyst nematode. *Plant Disease*, 105(4), 1136–1142.

Harman, G. E. (2024). Integrated benefits to agriculture with Trichoderma and other endophytic or root-associated microbes. *Microorganisms*, 12(7), 1409.

Haruna, S. I., et al. (2020). Improving soil physical properties through the use of cover crops: a review. *Agrosystems, Geosciences and Environment*, 3(1), e20105.

Hemkemeyer, M., et al. (2024). Potato yield and quality are linked to cover crop and soil microbiome, respectively. *Biology and Fertility of Soils*, 60(4), 525–545.

Hill, J. (2006). Inhibition of vesicular–arbuscular mycorrhizae on soybean roots following brassica cover crop. *Journal of Natural Resources and Life Sciences Education*, 35(1), 158–160.

Hooks, C. R., et al. (2010). Using marigold (Tagetes spp.) as a cover crop to protect crops from plant-parasitic nematodes. *Applied Soil Ecology*, 46(3), 307–320.

Hontoria, C., et al. (2019). The cover crop determines the AMF community composition in soil and in roots of maize after a ten-year continuous crop rotation. *Science of the Total Environment*, 660, 913–922.

Hu, Q., et al. (2023). Soil organic carbon fractions in response to soil, environmental and agronomic factors under cover cropping systems: a global meta-analysis. *Agriculture, Ecosystems and Environment*, 355, 108591. https://doi.org/10.1016/j.agee.2023.108591.

Hu, Q., et al. (2024). Legume cover crops sequester more soil organic carbon than non-legume cover crops by stimulating microbial transformations. *Geoderma*, 450, 117024.

Jones, F. G. W., et al. (1969). The influence of soil structure and moisture on nematodes, especially xiphinema, longidorus, trichodorus and heterodera spp. *Soil Biology and Biochemistry*, 1(2), 153–165.

Junod, M. F., et al. (2024). Cover crops in cereal rotations: a quantitative review. *Soil and Tillage Research*, 238, 105997.

Kandaswamy, R., et al. (2025). Advances in understanding and exploiting the role of soil microbiomes in protecting crops against pathogens and pests. In Dunfield, K. E. (ed.). *Understanding and utilising soil microbiomes for a more sustainable agriculture*. Burleigh Dodds Science Publishing, Cambridge, UK.

Kim, N., et al. (2020). Do cover crops benefit soil microbiome? A meta-analysis of current research. *Soil Biology and Biochemistry*, 142, 107701.

King, A. E. and Hofmockel, K. S. (2017). Diversified cropping systems support greater microbial cycling and retention of carbon and nitrogen. *Agriculture, Ecosystems and Environment*, 240, 66–76.

Kobayashi-Leonel, R., et al. (2017). Susceptibility of cover crop plants to Fusarium virguliforme, causal agent of soybean sudden death syndrome, and Heterodera glycines, the soybean cyst nematode. *Journal of Soil and Water Conservation*, 72(6), 575–583.

Koide, R. T. and Peoples, M. S. (2012). On the nature of temporary yield loss in maize following canola. *Plant and Soil*, 360(1–2), 259–269.

Kong, A. Y. and Six, J. (2012). Microbial community assimilation of cover crop rhizodeposition within soil microenvironments in alternative and conventional cropping systems. *Plant and Soil*, 356(1), 315–330.

Koninger, J., et al. (2025). Molecular and morphological identification methods show contrasting trends in soil fauna diversity along land-use intensity gradients. *Applied Soil Ecology*, 216, 106499.

Krupek, F. S., et al. (2022). Assessing how cover crops close the soil health gap in on-farm experiments. *Agricultural and Environmental Letters*, 7, e20088. https://doi.org/10.1002/ael2.20088.

Kucerik, J., et al. (2001). Utilization of diversified cover crops as green manure-enhanced soil organic carbon, nutrient transformation, microbial activity, and maize growth. *Agronomy*, 14, 2001. https://doi.org/10.3390/agronomy14092001.

Lahijanian, S., et al. (2025). Effects of cover crops and microbial inoculants in different farming systems on soil microbial communities and yield of maize. *Biology and Fertility of Soils*, 1–18. https://doi.org/10.1007/s00374-025-01929-x.

Larkin, R. P. (2021). Use of crop rotations, cover crops and green manures for disease suppression in potato cropping systems. *Global Journal of Agricultural Innovation, Research and Development*, 8, 153–68. https://avanti-journals.com/index.php/gjaird/article/view/1135.

Lavelle, P., et al. (2006). Soil invertebrates and ecosystem services. *European Journal of Soil Biology*, 42, S3–S15.

Leslie, A. W., et al. (2017). Influence of cover crops on arthropods, free-living nematodes, and yield in a succeeding no-till soybean crop. *Applied Soil Ecology*, 117, 21–31.

Li, F., et al. (2020a). Carbon and nitrogen mineralization differ between incorporated shoots and roots of legume versus non-legume based cover crops. *Plant and Soil*, 446(1), 243–257.

Li, Y., et al. (2020b). Phosphate solubilizing bacteria stimulate wheat rhizosphere and endosphere biological nitrogen fixation by improving phosphorus content. *PeerJ*, 8, e9062.

Luongo, L., et al. (2005). Potential of fungal antagonists for biocontrol of Fusarium spp. in wheat and maize through competition in crop debris. *Biocontrol Science and Technology*, 15(3), 229–242.

Lutz, S., et al. (2024). Global richness of arbuscular mycorrhizal fungi. *Fungal Ecology*, 74, 101407. https://doi.org/10.1016/j.funeco.2024.101407.

Lutz, S., et al. (2023). Soil microbiome indicators can predict crop growth response to large-scale inoculation with arbuscular mycorrhizal fungi. *Nature Microbiology*, 8(12), 2277–2289.

Manici, L. M., et al. (2018). The impact of legume and cereal cover crops on rhizosphere microbial communities of subsequent vegetable crops for contrasting crop decline. *Biological Control*, 120, 17–25. https://doi.org/10.1016/j.biocontrol.2016.11.003.

Marcillo, G. S., et al. (2024). The nitrogen value of cover crops: How much N can cover crops replace? *Agricultural and Environmental Letters*, 9, e70006. https://doi.org/10.1002/ael2.70006.

Martin, T. and Sprunger, C. D. (2022). Soil food web structure and function in annual row-crop systems: How can nematode communities infer soil health?. *Applied Soil Ecology*, *178*, 104553.

Martínez-García, L. B., et al. (2018). Organic management and cover crop species steer soil microbial community structure and functionality along with soil organic matter properties. *Agriculture, Ecosystems and Environment*, 263, 7–17.

McClelland, S. C., et al. (2021). Management of cover crops in temperate climates influences soil organic carbon stocks: A meta-analysis. *Ecological Applications*, 31(3), e02278.

McSorley, R. (1999). Host suitability of potential cover crops for root-knot nematodes. *Journal of Nematology*, 31(4S), 619.

Mehboob, I., et al. (2009). Rhizobial association with non-legumes: mechanisms and applications. *Critical Reviews in Plant Sciences*, 28(6), 432–456.

Mendes, R., et al. (2013). The rhizosphere microbiome: significance of plant beneficial, plant pathogenic, and human pathogenic microorganisms. *FEMS Microbiology Reviews*, 37, 634–663. https://doi.org/10.1111/1574-6976.12028.

Mooshammer, M., et al. (2014). Adjustment of microbial nitrogen use efficiency to carbon: nitrogen imbalances regulates soil nitrogen cycling. *Nature Communications*, 5(1), 3694.

Muhammad, I., et al. (2021). Cover cropping enhances soil microbial biomass and affects microbial community structure: a meta-analysis. *Geoderma*, 381,114696. https://doi.org/10.1016/j.geoderma.2020.114696.

Munkvold, L., et al. (2004). High functional diversity within species of arbuscular mycorrhizal fungi. *New Phytologist*, 164, 357–364. https://doi.org/10.1111/j.1469-8137.2004.01169.x.

Nannipieri, P., et al. (2017). Microbial diversity and soil functions. *European Journal of Soil Science*, 68, 12–26. https://doi.org/10.1111/ejss.4_12398.

Nardi, J. B. (2003). *The world beneath our feet: a guide to life in the soil*. Oxford University Press, Oxford.

Nevins, C. J., et al. (2020). The synchrony of cover crop decomposition, enzyme activity, and nitrogen availability in a corn agroecosystem in the Midwest United States. *Soil and Tillage Research*, 197, 104518.

Nivelle, E., et al. (2016). Functional response of soil microbial communities to tillage, cover crops and nitrogen fertilization. *Applied Soil Ecology*, 108, 147–155.

Norton, J., et al. (eds). (2026). *Measuring and assessing the biological health of soils*. Burleigh Dodds Science Publishing, Cambridge, UK.

Nyfeler, D., et al. (2011). Grass–legume mixtures can yield more nitrogen than legume pure stands due to mutual stimulation of nitrogen uptake from symbiotic and non-symbiotic sources. *Agriculture, Ecosystems and Environment*, 140(1–2), 155–163.

Opoku, A. (2022). *Cover crop systems impact on biomass production, carbon-to-nitrogen ratio, forage quality, and soil health indices in a semiarid environment* (Master's thesis, University of Nevada, Reno).

Ortega, J. S., et al. (2021). Impact of management of cover crop–soybean agroecosystems on rhizosphere microbial communities. *European Journal of Soil Science*, 72(3), 1154–1176.

Patkowska, E. and Konopiński, M. (2014). Antagonistic bacteria in the soil after cover crops cultivation. *Plant, Soil and Environment*, 60(2), 69–73.

Perrone, S., et al. (2020). Nitrogen fixation and productivity of winter annual legume cover crops in Upper Midwest organic cropping systems. *Nutrient Cycling in Agroecosystems*, 117(1), 61–76.

Plante, A. F. and Pett-Ridge, J. (2025). Advances in understanding the role of soil microbiomes in carbon cycling in soil. In Dunfield, K. E. (ed.). *Understanding and utilising soil microbiomes for a more sustainable agriculture*. Burleigh Dodds Science Publishing, Cambridge, UK.

Poole, P., et al. (2018). Rhizobia: from saprophytes to endosymbionts. *Nature Reviews Microbiology*, 16(5), 291–303.

Pozo, M. J., et al. (2008). Priming plant defence against pathogens by arbuscular mycorrhizal fungi. In *Mycorrhizas-functional processes and ecological impact* (pp. 123–135). Springer Berlin Heidelberg.

Qi, G., et al. (2020). Cover crops restore declining soil properties and suppress bacterial wilt by regulating rhizosphere bacterial communities and improving soil nutrient contents. *Microbiological Research*, 238, 126505.

Quintarelli, V., et al. (2022). Cover crops for sustainable cropping systems: a review. *Agriculture*, 12, 2076.

Rahman, M., et al. (2023). Probiotic bacteria, anaerobic soil disinfestation, and mustard cover crop biofumigation suppress soilborne disease and increase yield of strawberry in a perennial organic production system. *Plant Disease*, 107(8), 2490–2499.

Ray, S., et al. (2022). Cover crop selection affects maize susceptibility to the fungal pathogen Fusarium verticillioides. *Pedobiologia*, 91, 150806.

Reiss, E. R. and Drinkwater, L. E. (2020). Ecosystem service delivery by cover crop mixtures and monocultures is context dependent. *Agronomy Journal*, 112(5), 4249–4263.

Rittl, T. F., et al. (2023). Effects of organic amendments and cover crops on soil characteristics and potato yields. *Acta Agriculturae Scandinavica B*, 73, 13–26.

Rodman, J. E. (1981). Divergence, convergence and parallelism in phytochemical characters: the glucosinolate–myrosinase system. In: Young, D. A. and Seigler, D. S. (eds). *Phytochemistry and angiosperm phylogeny*. Praeger, New York, pp. 43–79.

Romdhane, S., et al. (2019). Cover crop management practices rather than composition of cover crop mixtures affect bacterial communities in no-till agroecosystems. *Frontiers in Microbiology*, 10, 1618.

Rousk, J., et al. (2015). Priming of the decomposition of ageing soil organic matter: concentration dependence and microbial control. *Functional Ecology*, 29(2), 285–296.

Saleem, M., et al. (2020). Cover crop diversity improves multiple soil properties via altering root architectural traits. *Rhizosphere*, 16, Article 100248.

Sanchez, I. I., et al. (2019). Soil biological response to integration of cover crops and nitrogen rates in a conservation tillage corn production system. *Soil Science Society of America Journal*, 83(5), 1356–1367.

Scavo, A., et al. (2022). The role of cover crops in improving soil fertility and plant nutritional status in temperate climates: a review. *Agronomy for Sustainable Development*, 42(5), 93.

Scheublin, T. R., et al. (2010). Characterisation of microbial communities colonising the hyphal surfaces of arbuscular mycorrhizal fungi. *The ISME journal*, 4(6), 752–763.

Schmidt, R., et al. (2018). Long-term use of cover crops and no-till shift soil microbial community life strategies in agricultural soil. *PloS one*, 13(2), e0192953.

Sharma, P., et al. (2018) The role of cover crops towards sustainable soil health and agriculture–a review paper. *American Journal of Plant Sciences*, 9, 1935–1951. https://doi.org/10.4236/ajps.2018.99140.

Shen, W., et al. (2021). Microbial deterioration and restoration in greenhouse-based intensive vegetable production systems. *Plant and Soil*, 463(1), 1–18.

Shi, W. (2025). Understanding and optimizing the physical properties of soil to promote soil microbiomes. In Dunfield, K. E. (ed.). *Understanding and utilising soil microbiomes for a more sustainable agriculture*. Burleigh Dodds Science Publishing, Cambridge, UK.

Silwana, S. (2025). *The use of cover crops for sustainable citrus orchard performance* (Doctoral dissertation, University of the Western Cape).

Somenahally, A., et al. (2018). Microbial communities in soil profile are more responsive to legacy effects of wheat-cover crop rotations than tillage systems. *Soil Biology and Biochemistry*, 123, 126–135.

Soti, P., et al. (2023). Effective farm management promotes native AMF and benefit organic farming systems. *Agriculture, Ecosystems and Environment*, 342, 108240.

Strickland, M. S., et al. (2019). Short-term effects of cover crops on soil microbial characteristics and biogeochemical processes across actively managed farms. *Agrosystems, Geosciences & Environment*, *2*(1), 1–9.

Sürücü, A., et al. (2014). Effects of green manuring on soil enzyme activity. *Fresenius Environmental Bulletin*, 23(9), 2126–2132.

Timofeeva, A. M., et al. (2023). Plant growth-promoting bacteria of soil: designing of consortia beneficial for crop production. *Microorganisms*, 11(12), 2864.

Trivedi P., et al. (2016). Response of soil properties and microbial communities to agriculture: implications for primary productivity and soil health indicators. *Frontiers in Plant Science*, 7, 990. https://www.frontiersin.org/journals/plant-science/articles/10.3389/fpls.2016.00990.

Van Bruggen, A. H., et al. (2021). Indirect effects of the herbicide glyphosate on plant, animal and human health through its effects on microbial communities. *Frontiers in Environmental Science*, 9, 763917.

Van Eerd, L. L., et al. (2023). Influence of cover crops at the four spheres: a review of ecosystem services, potential barriers, and future directions for North America. *The Science of the Total Environment*, 858, 159990.

Vasconcelos, M. W., et al. (2020). The biology of legumes and their agronomic, economic, and social impact. In Hasanuzzaman, M., et al. (eds). *The plant family fabaceae*. Singapore, Springer, pp. 3–25. https://doi.org/10.1007/978-981-15-4752-2_1.

Vasquez, C., et al. (2026). Soil biology, soil health and ecosystem services: an overview. In Norton, J. et al. (eds). *Measuring and assessing the biological health of soils*. Burleigh Dodds Science Publishing, Cambridge, UK.

Vukicevich, E., et al. (2016). Cover crops to increase soil microbial diversity and mitigate decline in perennial agriculture. A review. *Agronomy for Sustainable Development*, 36(3), 48.

Wang, Q., et al. (2007). Influence of cover crops on populations of soil nematodes. *Nematropica*, 37, 79–92.

Wang, K. H., et al. (2002). Crotalaria as a cover crop for nematode management: a review. *Nematropica*, 32, 35–58.

Wang, G., et al. (2025). Steering plant-soil feedback for sustainable agriculture. *Science*, 389(6758), eads2506.

Wang, X., et al. (2011). Effects of co-inoculation with arbuscular mycorrhizal fungi and rhizobia on soybean growth as related to root architecture and availability of N and P. *Mycorrhiza*, 21(3), 173–181.

Wen, L., et al. (2017). Suppression of soilborne diseases of soybean with cover crops. *Plant Disease*, 101(11), 1918–1928.

West, B. L. (2018). *Impacts of cover crop mixtures on productivity of cropping systems*. Graduate Theses/Dissertations. 3334. https://bearworks.missouristate.edu/theses/3334.

White, C. M., et al. (2017). Managing the trade off between nitrogen supply and retention with cover crop mixtures. *Agriculture, Ecosystems and Environment*, 237, 121–133.

Wong, K. M., et al. (2024). Cover crop cultivars and species differ in root traits potentially impacting their selection for ecosystem services. *Plant Soil*, 500, 279–296. https://doi.org/10.1007/s11104-023-06431-7.

Xu, J., et al. (2020). Organic amendmentschange soil organic C structure and microbial community but not total organicmatter on sub-decadal scales. *Soil Biology and Biochemistry*. 150, 107986 https://doi.org/10.1016/J.SOILBIO.2020.107986.

Xu, Q., et al. (2024). Priming and balance of soil organic carbon differ with additive C: N ratios and long-term green manuring. *Applied Soil Ecology*, 201, 105495.

Yan, F. and Arthur, E. (2025). Cover crops alter soil physicochemical properties: a global meta-analysis. *Geoderma*, 460, Article 117436.

Zhang, C., et al. (2022). Afterlife effect of cover crops on soil nematode food web: implications from the plant ecological strategy. *Biology and Fertility of Soils*, 58(8), 937–947.

Zhou, R., et al. (2023). Microbial necromass in cropland soils: a global meta-analysis of management effects *Global Change Biology*, 29, 1998–2014. https://doi.org/10.1111/gcb.16613.

Zheng, W., et al. (2018). Changes in the soil bacterial community structure and enzyme activities after intercrop mulch with cover crop for eight years in an orchard. *European Journal of Soil Biology*, 86, 34–41.

Zheng, W., et al. (2018). Effects of cover crop in an apple orchard on microbial community composition, networks, and potential genes involved with degradation of crop residues in soil. *Biology and Fertility of Soils*, 54, 743–759.

Chapter 3

The role of cover crops in improving biodiversity

Deirdre A. Prischmann-Voldseth and Keiran Hyte, North Dakota State University, USA; Rachel Mallinger, University of Florida, USA; and Jarrad Prasifka, USDA-ARS-ETSARC, USA

1 Introduction: defining and measuring biodiversity

Biological diversity (i.e. biotic diversity or biodiversity) is defined as the 'biological variety in an environment as indicated by numbers of different species of organisms' (Merriam-Webster 2024). However, it can refer to variation at other scales (Duelli and Obrist 2003, Gaston and Spicer 2013, Diaz and Malhi 2022, Dufresnes et al. 2023) and a more comprehensive definition is as follows: 'The variability among living organisms from all sources including terrestrial, marine and other aquatic ecosystems and the ecological complexes of which they are a part. This includes variation in genetic, phenotypic, phylogenetic, and functional attributes, as well as changes in abundance and distribution over time and space within and among species, biological communities and ecosystems' (IPBES 2024).

Farnham (2007) and Diaz and Malhi (2022) discuss the rise of interest in biological diversity in the 1980s, its emergence as a key concept in conservation, and its modern-day prominence. Initial focus remained on the value of wilderness, endangered species, environmental crises, and taxonomic documentation; however, this later shifted toward the value and loss of biodiversity, as well as understanding factors driving species' declines and

http://dx.doi.org/10.19103/AS.2025.0162.05

mitigation efforts (Josephson 1982; Soulé 1985; Wilson 1985,1989; Farnham 2007). Tilman et al. (2014) and Ali (2023) review the history of biodiversity–ecosystem-functioning research and key points of debate, but in general, higher biodiversity is perceived as beneficial and desirable (Christie et al. 2006, Duffy 2009, Tilman et al. 2014, Burch-Brown and Archer 2017, Köninger et al. 2022). A growing emphasis area is understanding biodiversity beyond the species level – whether characterized as functional diversity, phylogenetic diversity, intraspecific diversity, population richness and/or genetic diversity within and among populations (Taberlet 1998, Raffard et al. 2019, Lawrence and Fraser 2020, Smith et al. 2020). Biodiversity is used by researchers and policy-makers to understand and evaluate various aspects of natural and managed ecosystems (e.g. composition, structure, function, productivity, stability, resilience) and ecosystem multifunctionality, with connections to conservation, agriculture, and human health, to name a few (Hector et al. 2001, Catovsky et al. 2002, Srivastava and Vellend 2005, Costanza et al. 2007, Tilman et al. 2014, Oliver et al. 2015, Romanelli et al. 2015, Dudley and Alexander 2017, Marselle et al. 2021). Interest in the effects of cover crops on insect biodiversity within agricultural systems began building in the 1980s (Altieri and Letourneau 1982, Altieri et al. 1984) and now encompasses effects on soil microbes (Kim et al. 2020), insect natural enemies such as predators and parasitoids (Abad et al. 2021, Beaumelle et al. 2021), pollinators (discussed in more detail later; Mallinger et al. 2019, Triquet et al. 2024), and the main focus of this chapter, soil arthropods.

Various indices (or metrics, variables, indicators) are used to quantify the current and changing status of diversity within ecosystems, to understand the effects of harmful practices and mitigation efforts, and to justify the use of resources and policies to protect and restore biodiversity (Duelli and Obrist 2003, Moonen and Bàrberi 2008, Pereira et al. 2013, Brummitt et al. 2017). However, due to the complex nature of biodiversity, it is difficult to select one metric to fully evaluate a system (Purvis and Hector 2000, Hill et al. 2016, Santini et al. 2017, Gatti et al. 2020), especially when considering differences in biodiversity among spatial scales, sometimes referred to as alpha (local biodiversity), beta (differences among communities), and gamma diversity (differences among communities across a larger geographical scale, like a landscape) (Whittaker 1960, Andermann et al. 2022). Species richness, or the total number of species present, is one of the simplest metrics, but using only one metric, especially species richness, may be inappropriate or misleading, as it can remain the same while other metrics are changing (Jennings et al. 2008, Hill et al. 2016, Santini et al. 2017, Gatti et al. 2020).

Simpson and Shannon developed some of the best-known diversity indices, which use species richness (the number of different species in a particular area) and species evenness (the number of individuals in a particular species relative to the total number of individuals in all species) to assess diversity (Simpson 1949, Shannon and Weaver 1949, Duelli and Obrist 2003, Gatti et al. 2020). Key differences between them are that the Simpson index is more sensitive to abundant species that dominate a community, while the Shannon index gives more weight to rare species and considers the evenness of species distribution across the community, making it more responsive to changes in the relative abundance of all species present (Nagendra 2002). The Sørensen similarity index is common beta-level biodiversity metric that considers the number of shared and unique species among communities (Sørensen 1948), although a limitation is that it only compares two communities and does not take into consideration species abundance, only using presence and absence data, which gives undue weight to less common species over more abundant ones (Chao et al. 2006). Research efforts are underway to evaluate, prioritize, and coordinate biodiversity assessments and metrics (Pereira et al. 2013, Brummitt et al. 2017, Santini et al. 2017, Lumbierres and Kissling 2023) and integrate them with remote sensing (Skidmore et al. 2021, Reddy et al. 2024).

Interest in using functional diversity rather than species-based metrics like richness has increased (McGill et al. 2006, Hillebrand and Matthiessen 2009, Cadotte et al. 2011, Mammola et al. 2021), as the former impacts many aspects of ecosystem functioning (Tilman et al. 2014, Finney and Kaye 2017, Huang et al. 2020, Schmera et al. 2023). Functional diversity has been defined as the range of traits, niches, or 'functions' that organisms in a community or ecosystem perform, although the definition and value of the term have been debated, as well as what it encompasses, and how to quantify it (Petchey and Gaston 2006, Schleuter et al. 2010, Cadotte et al. 2011, Mammola et al. 2021, Streit and Bellwood 2023, Schmera et al. 2023). Petchey and Gaston (2006) and Streit and Bellwood (2023) delve into myriad different traits, how traits can relate to functional classification, and what traits may (or may not be) ecologically relevant. There are also different aspects of functional diversity (e.g. functional richness, functional evenness, functional regularity, functional divergence) and ways of measuring and interpreting these metrics (Petchey et al. 2004, 2009; Mouchet et al. 2010; Schleuter et al. 2010; Pla et al. 2011; Schmera et al. 2023).

Researchers distinguish between a species diversity gradients with declining biodiversity at increasing altitudes (Rahbek 1995) and a latitudinal diversity gradient with increasing biodiversity when moving toward the equator, although some species have higher biodiversity in temperate areas (Willig et al. 2003, Brown 2014). However, within these general patterns, landscape identity, size, and heterogeneity can impact biodiversity (Duelli 1997, Keil and Chase 2019). When evaluating biodiversity, consider spatial scale, or extent (sampling universe being considered), coverage (amount of the universe sampled), and grain or focus (sampling units) (Wiens 1989, Bunnell and Huggard 1999, Rahbek 2005, Du Toit 2010, Dornelas et al. 2013, Spake et al. 2021). Many studies focus on specific portions of a single ecosystem, such as a particular habitat (e.g. soil, tree, forb canopy), taxa (e.g. microbes, plants, arthropods, or vertebrates) or trophic group (e.g. predators, herbivores, pollinators). Environments change over the course of months and years, and there is growing interest in temporal biodiversity information (Willis and Birks 2006, Costa and Magnusson 2010, Dornelas et al. 2013), including the use of long-term datasets (Magurran et al. 2010) and the examination of challenges in comparing historical and modern data (Dornelas et al. 2013, Douda et al. 2023).

Because the direction, degree, and processes driving changes in biodiversity can depend on the context, more attention is being paid to aspects of sampling, including scope and scale, both in space and time (Bunnell and Huggard 1999, Chase et al. 2018, Keil and Chase 2019, Gonzalez et al. 2020, Levin 2000). Environments change over the course of months and years through a myriad of processes, and there is growing interest in temporal biodiversity information (Willis and Birks 2006, Costa and Magnusson 2010, Dornelas et al. 2013), including long-term datasets (Magurran et al. 2010), the Biodiversity Exploratories (Fischer et al. 2010), and assessments of data quality, including the examination of challenges in comparing historical and modern data (Dornelas et al. 2013, Douda et al. 2023).

2 Cover crops and plant diversity

We have discussed some of the complexities associated with measuring, contextualizing, and interpreting biodiversity research. Studies of cover crop effects on biodiversity must take these into account. The use of cover crops in agriculture has a long history (Groff 2015), and there has been a surge of interest in this practice within the U.S. and globally (Wallander et al. 2021, Deines et al. 2023). As the name implies, cover crops cover the soil, either in the absence of the cash crop (e.g. during the spring or winter or in fallow years) or in the space not occupied by cash crop plants (e.g. interseeding, overseeding, living mulch, green carpet, or planting green) (Magdoff and Van Es 2021, Wallander et al. 2021). Cover crops are typically killed or managed (e.g. mowed, grazed, tilled, or sprayed) so they do not compete with the cash crop (USDA-NRCS 2014, Magdoff and Van Es 2021). They were initially used to improve the soil and prevent erosion (Groff 2015), although now they are used for multiple purposes, like soil nutrient management, the increase of soil organic matter, the alleviation of soil compaction, environmental modification, pest management, and the provision of habitat for beneficial arthropods (Lu et al. 2000, Magdoff and Van Es 2021, Quintarelli et al. 2022, Scavo et al. 2022). Cover crops fall into several major plant categories: grasses (e.g. cereal rye, ryegrass, barley), legumes (e.g. vetch, clovers), and non-legumes (e.g. radishes, turnips, buckwheat) (Magdoff and Van Es 2021). Different cover crop species are used for different objectives, and mixtures are often employed to meet management objectives (Ramírez-García et al. 2015, Chapagain et al. 2020, Haruna et al. 2020, McKenzie-Gopsill et al. 2022).

Focusing on cover crop functional diversity, rather than simply species richness, can lead to improved ecosystem services (Finney and Kaye 2017). Organisms can be organized somewhat subjectively by their functional group. For example, plants could be grouped by their dispersal strategy, root structure, ability to host beneficial insects, or any other number of variables (Moonen and Bàrberi 2008). Cover crops can affect multiple abiotic and biotic components of the environment while they are living and through their residues (Magdoff and Van Es 2021). They increase plant biodiversity within a cropping system, spatially and/or temporally.

There are several different mechanisms by which increasing plant diversity could affect ecosystem function. Eisenhauer et al. (2024) describe a list of several different situations and effects that can occur with a change in diversity. Of note, the broadest effects are:

- Complementarity: when co-occurring species divide the use of resources in either space, time, or both, in order to maximize their collective growth, defense, and reproduction. For example, adding plants to fill an area during a non-cropping period where the field would otherwise be left fallow (Loreau and de Mazancourt 2008, Barry et al. 2019).
- Facilitation: when the presence of different species mitigates stresses on the ecosystem, or facilitates positive effects through mutualisms. For example, legumes using rhizobia to fix nitrogen for other plants, or greater space between plants of the same species in a high diversity area reducing the impact of pathogens on that species (Wright et al. 2017).
- Insurance: when overall ecosystem services vary less in areas with high diversity due to species with similar functions compensating for one another when an individual population is negatively impacted (Loreau et al. 2021).

While most of the effects described are observed primarily in high-diversity ecosystems with dozens of different species coexisting and compensating for one another, this does not mean that changes do not occur with the addition of a single new species. One of the most famous examples is the 'Three Sisters' intercropping system of corn, beans, and squash practiced by Native Americans; despite being a relatively low-diversity system consisting of only three different plant species, the complementarity of different functions provides significant benefits to soil nutrient uptake and efficiency, as well as changes to the overall root architecture of the community (Zhang et al. 2014, Kapayou et al. 2023). This demonstrates that even in systems with relatively low diversity, the strategic combination of species can drive ecological changes, given the selection of species with complementary functions.

Recent research highlights positive outcomes associated with increasing plant biodiversity, whether directly or indirectly (Cappelli et al. 2022), e.g. increasing the biodiversity of other organisms and promoting plant productivity at a landscape scale (Oehri et al. 2017, Estrada-Carmona et al. 2022), and benefiting ecosystem functioning under global environmental stressors (Hong et al. 2022), enhancing soil fertility (Furey and Tilman 2021), and decreasing negative impacts of phytophagous arthropods on plants (Barnes et al. 2020). This chapter primarily focuses on how cover crops in agricultural systems impact the biodiversity of arthropods. However, a key issue in assessing cover crop research, whether looking at impacts on arthropods or other biotic or abiotic factors, is the high level of variation, including differences in: environmental context and setting, main cropping system, crop management practices (e.g. tillage, fertilizers), crop rotations, cover crop species, timing of planting and termination, and factors related to sampling (Junod et al. 2024).

3 Soil arthropods and cover crops

3.1 Key soil arthropods

While studies on foliar or aerial arthropods like pollinators, insect pests, and natural enemies are often the focus of biodiversity research (Abad et al. 2021, Beaumelle et al. 2021, Triquet et al. 2024), arthropods that dwell in and around the soil are arguably just as important. Soil arthropods are sensitive to environmental changes and have major impacts on the soil food web, decomposition, and chemical and physical properties of the soil (Swift et al. 1979, Eisenbeis and Wichard 2012, Crowther et al. 2019, Menta and Remelli 2020). Their activities also affect plants and above-ground processes (Wardle 2002, Bezemer and van Dam 2005) and their biodiversity impacts above-ground biodiversity and ecosystem functioning, which can benefit natural and agricultural systems (Bardgett and van der Putten 2014). Documenting, understanding, and protecting soil biodiversity has been a major research endeavor for several decades (Usher et al. 2006, Guerra et al. 2021, van der Putten et al. 2023).

Research on soil arthropod diversity usually distinguishes between two distinct but equally important environments, subterranean (underground, soil-dwelling) and epigeal (surface-active) areas. Additionally, soil arthropods can be categorized in various other ways, including: taxonomically, according to their size (e.g. meso- and macrofauna; Swift et al. 1979), by trophic group or functional role (Briones 2014, Hedde et al. 2022, Potapov et al. 2022) and by life stages (Villani et al. 1999). Hemiedaphic arthropods (e.g. ants) only spend part of their life cycle in the soil, while all life stages of edaphic arthropods are associated with the soil, with epedaphic organisms associated with the soil surface or litter and euedaphic organisms living within soil pores or voids (Villani et al. 1999, Eisenbeis and Wichard 2012).

While there are many taxa of subterranean arthropods that are discussed in depth elsewhere, here the focus is on the mites (Chelicerata: Araneae: Acari) and the springtails (Hexapoda: Entognatha: Collembola), which are generally the most diverse and abundant non-eusocial groups (Dindal 1991, Eisenbeis and Wichard 2012, Culliney 2013, Menta and Remelli 2020). Soil-dwelling mites break down plant matter or consume other organisms, with mites in the taxonomic groups Oribatida and Astigmata serving as primary consumers of decomposing organic material, and mites in the Prostigmata and Mesostigmata making up the bulk of the predatory mesofauna (Luxton 1982, Behan-Pelletier 1999, Ruf and Beck 2005, Culliney 2013). Springtails are primarily fungivores, but there are also detritivorous and carnivorous species that consume nematodes and other microbiota (Rusek 1998, Culliney 2013, Coleman and Wall 2015). Some species are more restricted to subterranean environments, while others occur more broadly and are often found on the soil surface (Querner and Bruckner 2010). Researchers have used edaphic microarthropod taxa as bioindicators of soil quality or health (Ruf et al. 2003, Menta and Remelli 2020) and have developed various indices, including the QBS (or QBS-ar) index, which sums eco-morphological index/scores (EMI) based on the arthropod's adaptation to the soil environment, with higher QBS scores indicating a greater number of well-adapted soil arthropods are present, and thus improved soil health (Parisi et al. 2005, Menta et al. 2018, Elmquist et al. 2023).

Ground beetles (Insecta: Coleoptera: Carabidae) are a key group of epigeal arthropods frequently studied due to their importance as generalist predators (Ekschmitt et al. 1997, Kromp 1999, Holland and Luff 2000), although certain species also eat weed seeds (Shearin et al. 2008, Ward et al. 2011). While adults are typically found on the soil surface and mostly move on foot, adults can fly and immature stages of most species live underground (Lövei and Sunderland 1996, Kromp 1999). Ground beetle communities are used as bioindicators to provide insight into the overall health and biodiversity of ecosystems, since, as generalists, they are attracted to areas of greater prey and plant diversity, exhibit high species richness, and are sensitive to environmental change (Holland and Luff 2000, Kotze et al. 2011, Cameron and Leather 2012, Burgio et al. 2016, Pizzolotto et al. 2018, Menta and Remelli 2020, Maksimovich et al. 2023).

3.2 How cover crops impact soil arthropods

Living cover crops, their residues, and their management (e.g. tillage) affect soil conditions (Magdoff and Van Es 2021), which in turn impact soil arthropods.

Tillage, especially more intensive types of tillage, tends to have detrimental impacts on soil-dwelling arthropods and arthropods that periodically inhabit the soil, e.g. immature life stages of ground-nesting pollinators (Rowen et al. 2020, Betancur-Corredor et al. 2022, Müller et al. 2022, Tschanz et al. 2024). The biology and movement of subterranean, or euedaphic, arthropods are primarily affected by soil properties, including texture, pores, moisture, and temperature (Villani et al. 1999, Eisenbeis and Wichard 2012, Vanolli et al. 2024). Subterranean arthropods are highly sensitive to moisture levels. Soil pores act as channels that allow them to avoid overly dry or wet conditions and can also serve as refuges from predators (Villani et al. 1999, Eisenbeis and Wilchard 2012). Soil properties, and therefore subterranean arthropods, are influenced by living or dead/dying plants (i.e. litter, residue), whether cash crops or cover crops (Kaspar and Singer 2011, Scheunemann et al. 2015, Koudahe et al. 2022, Lucas et al. 2022). Cover crops can directly reduce soil compaction, increase soil porosity, and increase water infiltration (Blanco-Canqui and Ruis 2020, Haruna et al. 2020). Plant taxonomic and functional diversity affects the variety of root structures (e.g. taproots, fibrous roots, root hairs), and subsequently soil structure, chemical processes, and biology (Griffiths et al. 2022, Lucas et al. 2022). For example, cover crops can support higher earthworm populations (Roarty et al. 2017); earthworm burrowing breaks up the soil and alters soil pores used by subterranean arthropods (Blanchart et al. 1999, Lamandé et al. 2003, Fischer et al. 2014, Lucas et al. 2022). Fiorini et al. (2022) suggest that leguminous cover crops such as vetch and clover provide the most benefit to earthworms and other detritivores, perhaps due to the high nitrogen content and increased food quality of legume residues compared to those of other cover crops.

Cover crops can indirectly affect soil arthropods via soil moisture and temperature (e.g. through shading, absorption or reflectance of radiation, water use, evapotranspiration), although effects can depend on other factors, like soil type, plant species, termination timing, and season (Kahimba et al. 2008, Alonso-Ayuso et al. 2014, Kaye and Quemada 2017, Blanco-Canqui and Ruis 2020, Wang et al. 2021). Other studies have shown a link between the presence of cover crops and the biodiversity of soil microorganisms (e.g. bacteria, fungi, and nematodes) which are prey for some subterranean arthropods (Eisenbeis and Wichard 2012, Vukicevich et al. 2016, Finney et al. 2017, Karimi et al. 2020). Litter identity and quality can impact subterranean soil arthropods. Legumes are particularly beneficial for springtail abundance and diversity, possibly due to a higher abundance of microbes and fungi on which springtails feed (Salamon et al. 2004, Birkhofer et al. 2011). Root exudates and plant defensive compounds can affect interactions among soil organisms, particularly the biofumigation potential of brassica cover crops (e.g. mustard and radish), as they can suppress fungal, bacterial, and herbivorous insect populations in the soil (Sarwar et al. 1998, Szczygłowska et al. 2011).

In contrast to subterranean arthropods, epigeal arthropods such as ground beetles are fast-moving and highly mobile; they readily move to favorable environments. Changes in the aboveground canopy structure, especially within more diverse habitats at field edges or when cover crops are present, can impact the biology, behavior (e.g. movement, foraging, nesting), biodiversity, and functional diversity of epigeal arthropods (Woodcock et al. 2010, Gayer et al. 2019), although these effects can be landscape dependent (Beaumelle et al. 2021). Cover crops can alter temperature and humidity at the soil surface, which affects movement and capture rates in pitfall traps (Holland and Luff 2000, Saska et al. 2013). Prolonged drought conditions have been shown to dramatically impact ground beetle abundance, with decreases of 51% in European forests (Weiss et al. 2024). Certain cover crop residues, such as hairy vetch and rye, stabilize temperatures at the soil surface and prevent the decline of soil water during drought periods, making them good candidates for the preservation of epigeal arthropods (Teasdale and Mohler 1993).

Ground beetles are generalist predators that arrive early in the season to nest and hunt, attracted by the increased abundance and diversity of prey found in the cover crop canopy (Carmona and Landis 1999, Gallé et al. 2019, Triquet et al. 2023). Planting a mix of cover crops may be more beneficial than a single species (Kahimba et al. 2008, Hojati et al. 2024), because the heterogeneity increases spatial diversity, as well as resources for multiple types of prey. In addition, cover crops or their residues that persist over the winter can provide shelter for both predators and their prey (Dupre et al. 2021) or affect ground beetle movement (Davis et al. 2009).

3.3 Effects of cover crops on biodiversity of subterranean soil arthropods

Research on how cover crops affect soil arthropod biodiversity is often embedded within studies that deal with broader aspects of crop management, such as different agricultural land use types, cropping systems, organic versus conventional management, tillage regimes, or weed or nutrient management; however, these studies do not always manipulate cover crop treatments independently from other factors. For example, cover crops are manipulated simultaneously with tillage or chemical inputs (Holland and Luff 2000, Lalonde et al. 2012, Sapkota et al. 2012, Patterson et al. 2019, Menta et al. 2020, Boeraeve et al. 2022). Such research often focuses on finding taxa that could serve as bioindicators, or on how management practices affect arthropod abundance, or it uses multivariate techniques to assess community composition, rather than actually assessing biodiversity or functional biodiversity. Research more specifically focused on biodiversity often examines epigeal or foliar arthropods within perennial systems, including vineyards, orchards, olive groves, or forest ecosystems, while some examine ground cover rather than cover crops (Bugg and Waddington 1994, Minor et al. 2004, Carpio et al. 2019, De Pedro et al. 2020, Beaumelle et al. 2021, Chapela-Oliva et al. 2022).

Several studies have found positive impacts of various cover crops or increased organic matter on soil arthropod abundance when focusing on fauna extracted from soil (e.g. using Berlese or Tullgren funnels); however, others have not found any impact (Fiorini et al. 2022). In a continuous corn system, a rye cover crop (*Secale cereale*) positively affected the abundance of certain taxonomic groups of mites (i.e. prostigmatid, mesostigmatid, and oribatid mites) but not of other types of mites (i.e. astigmatid mites), while positive effects on springtails (mainly Isotomidae and Onychiuridae) were sporadic (Reeleder et al. 2006). Miyazawa et al. (2002) saw an increase in mite populations, particularly predatory mesostigmatid mites, when organic matter from green manure was added to field plots – including oats (*Avena sativa*), phacelia (*Phacelia tanacetifolia*), red clover (*Trifolium pratense*), hairy vetch (*Vicia villosa*) or to weedy plots – but only when paired with reduced tillage. Springtail populations were higher in plots with more added organic matter from vegetation. Strip-tilled cover cropping with Sunn hemp (*Crotalaria juncea*) and French marigold (*Tagetes patula*) in bitter melon (*Momordica charantia*) field plots increased numbers of predatory soil mesoarthropods (i.e. Astigmata, Prostigmata, Mesostigmata, and spiders), and fungivorous and detritivorous organisms (i.e. oribatid mites and springtails), although positive impacts varied through time (Marahatta et al. 2010). Trying to link effects of management practices on soil arthropods to consistent changes in specific soil properties has proved challenging (Miyazawa et al. 2002, Reeleder et al. 2006, Elmquist et al. 2023).

Other studies looked at both abundance and biodiversity of subterranean soil arthropods. In general, effects of cover crops varied, and depended on cover crop and microarthropod species. In plots with a winter cover crop with black oat (*Avena strigosa*) and hairy vetch established in a long-term corn-soybean agroecosystem, there was an increase in abundance and biodiversity of subterranean arthropods in areas with black oat, especially in the subsurface soil layers under no-till conditions (Ramos et al. 2023). Elmquist et al. (2023) investigated how diversifying a traditional wheat production system by replacing fallow or a spring-planted legume cash crop with a winter pea or forage cover crop affected subterranean arthropods, with most identified to the level of order or family. The diversified systems enhanced abundance and diversity of soil arthropods relative to the crops they replaced, with novel species of detritivores and predators associated with specific crops strongly impacting community composition.

Crotty et al. (2015) investigated cover crops that differed in their rooting structure: perennial ryegrass (*Lolium perenne*; shallow, extensive root system), chicory (*Cichorium intybus*; deep tap roots), red clover (*Trifolium pratense*; tap root), and white clover (*T. repens*; short tap root and stolons). Mites and springtails dominated the microarthropod community, and effects of cover crops varied by taxonomic group, with the abundance of detritivorous Poduromorpha higher in clover and chicory, herbivorous Symphypleona higher in ryegrass, and no impact on Entomobryomorpha and Neelipleona. There was no effect of cover crops on prostigmatid or oribatid mites, although predatory mesostigmatid mites tended to be more abundant in red clover plots. Cover crop species affected microarthropod biodiversity, with Simpson's index of diversity lowest in minimally fertilized rye, intermediate in chicory, and highest in plots with clovers.

Some studies did not find significant or consistent effects of cover crops on subterranean arthropods. In a willow planting, winter rye did not have significant effects on mite (Gamasida and Oribatida) communities. In fact, the highest abundance and diversity was in weedy plots where the willows didn't establish (Minor et al. 2004). Fiorini et al. (2022) investigated effects of four cover crop treatments, including 1) cereal rye, 2) phacelia plus white mustard (*Sinapis alba*) 3) Italian ryegrass, crimson clover (*Lolium multiflorum*) and Persian clover (*T. resupinatum*), and 4) hairy vetch plus crimson clover in a 3-yr maize-soybean rotation. Subterranean arthropods were identified to order (or class for Myriapoda) and the community was composed primarily of mites and springtails. Arthropod abundance and biodiversity – measured using Pielou's evenness – were similar between control and cover crop plots, although biodiversity tended to be lower in phacelia and white mustard plots, which was significantly lower than hairy vetch and crimson clover plots.

3.4 Effects of cover crops on biodiversity of epigeal (aboveground) soil arthropods

Cover crop effects on the abundance and biodiversity of epigeal soil arthropods are measured using methods such as pitfall traps or sweep nets in perennial systems (e.g. orchards, vineyards), annual field crops, and systems transitioning from conventional to more sustainable management practices. Research in perennial systems indicates that cover crops have positive effects on epigeal arthropod diversity and abundance compared to bare soil or areas lacking cover crops, although in some cases disking or tillage was used, and this disturbance can impact soil arthropods independently of cover crop destruction (Altieri and Schmidt 1985, Carpio et al. 2019, Sáenz-Romo et al. 2019, De Pedro et al. 2020, Eckert et al. 2020, Geldenhuys et al. 2021). In two studies, springtails were more abundant in non-cover cropped plots, perhaps because predatory beetles were less abundant in those areas (Carpio et al. 2019, De Pedro et al. 2020).

Cover crop species can affect biodiversity metrics, with a natural mix of grasses and legumes disked in the winter and plots seeded with common vetch (*Vicia sativa*) planted every winter having lower arthropod diversity (epigeal plus above-ground) than permanent cover of fescue (*Festuca arundinacea*) plus alfalfa (*Medicago sativa*) or strawberry clover (*T. fragiferum*), but only when using the Shannon–Weiner index; biodiversity among groups was similar when using Simpson's index (Fernández et al. 2008). Another study showed that increasing cover crop species diversity from low (perennial ryegrass and red fescue, *Festuca rubra*) to high (mix of 20 plant species) did not translate into higher arthropod diversity, although it did boost arthropod abundance (Beaumelle et al. 2021).

In a field crop system transitioning to organic management, Jabbour et al. (2016) investigated how planting a field in cover crops, cereal rye followed by hairy vetch, or oat followed by timothy (*Phleum pretense*) plus red clover for one year prior to planting soybean and corn affected abundance (i.e. activity-density) and diversity of soil arthropods captured in pitfall traps. They identified most arthropods to order or family, and ground and tiger beetles (Coleoptera: Carabidae) to species. There were no effects of cover crops on the activity-density of the entire soil arthropod community, although adult carabid beetles had higher abundance in rye versus oat. Cover crops did affect community composition, and species evenness was higher in oat plots, although effects depended on year, tillage, and presence of weeds. Rivers et al. (2018) investigated effects of two cover crops, hairy vetch plus triticale (*Triticosecale*) and cereal rye terminated at three different times in a corn-soybean-wheat system transitioning to organic. Cover crop species, termination date, and condition (live or dead) affected activity-density and community composition of epigeal arthropods, with ground beetles being more abundant (activity-density) in the hairy vetch and triticale plots.

Soares et al. (2023) conducted an experiment in a wheat system with a winter cover crop (black oat, turnip rape (*Brassica rapa* subsp. *oleifera*), black oat + turnip rape) compared to fallow, with various N fertilizer treatments. Soil arthropods were sampled with pitfall traps and identified to class or order, with springtails (primarily Entomobryidae) dominating the epigeal arthropod community. In unfertilized plots, biodiversity and species evenness were highest in the black oat + turnip rape plots, whereas in the high-fertilizer plots (87.5 kg/ha), biodiversity was higher in fallow plots. However, Lundgren and Fergen (2010) investigated a winter slender wheatgrass (*Elymus trachycaulus*) cover crop within a corn system. They focused primarily on epigeal soil arthropods, and identified most key groups to species, although they did not consider mites. Predator densities increased in cover cropped maize plots compared to bare soil, but there was no impact on community biodiversity (total species captured, Shannon Index, and evenness).

In perennial systems like vineyards and orchards, crop plants and cover crops are grown simultaneously, although grape or tree canopies are typically physically distant from cover crops (i.e. they are higher and thus separated spatially), while in annual systems, living cover crops are often present in the absence of the cash crop (e.g. during the winter or fallow periods), although some cover crops are grown as living mulches and cover crop residues are often left in the field. Similar to perennial systems, cover crops can increase the diversity and abundance of epigeal arthropods in annual agricultural systems (Prasifka et al. 2006, Pullaro et al. 2006, Kelly et al. 2021), although factors related to cover crop species and management (e.g. disturbance and fertilization) affect their activity and density (see review by Inveninato Carmona et al. 2021). Effects can also depend on time of year, with winter cover crops having a greater impact than spring cover crops on ground beetles (Holland and Luff 2000).

The landscape can serve as a sink or a source for arthropods (Pizzolotto et al. 2018, Lichtenberg et al. 2023). Non-crop habitat surrounding cropping systems can benefit arthropods by providing an undisturbed environment, escape from harsh environmental conditions or chemicals, shelter, overwintering habitats, dispersal sites, oviposition sites, and prey (Holland and Luff 2000, Landis et al. 2000). Effects of cover crops on arthropods within cropping systems are context-dependent. Weeds are another source of plant diversity that can impact soil arthropods, for example (Minor et al. 2004, Davis et al. 2009, Sanguankeo et al. 2011, Jabbour et al. 2016), while crops in adjacent fields can also affect epigeal arthropod diversity (Molina et al. 2014). Therefore, cover crops can have a greater impact on arthropods in homogeneous landscapes (Beaumelle et al. 2021). Even when considering one family of arthropods such as the ground beetles (Carabidae), certain species have distinct habitat preference; some prefer crop habitat, others prefer habitat outside the managed crop, and some move back and forth (reviewed in Holland and Luff 2000, Shearin et al. 2008). For example, Davis et al. (2009) noted that cover crops increased abundance of two specific ground beetle genera, *Harpalus* and *Poecilus*, but other genera were not affected. Thus, the taxonomic rank at which biodiversity is assessed and the available pool of species in the area should be considered when studying the effects of cover crops on the biodiversity of highly mobile species.

4 Effects of cover crops on biodiversity of insect pollinators

Pollinators are the second most widely studied functional group of insects, second only to pest insects, due to their significant impact on the pollination (and thus yield) of most of the world's fruit, seed, and nut crops (Klein et al. 2006). It is because of this importance that their recent decline has alarmed researchers. A large part of their decline has been attributed to intensification of agricultural practices, with pesticide usage and heavy tillage having the greatest impact (Winfree 2010, Koh et al. 2016, Potts et al. 2016). Pollinators can be split into two major groups of study based on taxonomy: fly pollinators and bee pollinators, with the latter of the two being far more widely studied. Although other groups, such as beetles and moths, can serve as pollinators, their impacts are largely niche cases compared to flies and bees (Bryan et al. 2021). Bees in particular can be further divided into categories based on their sociality. Eusocial bees consist of honey bees (*Apis* spp.), bumble bees (*Bombus* spp.), and stingless bees (tribe Meliponini); these species form large colonies with a queen and worker castes (Michener 1974). Many species within the other bee taxa (e.g. sweat bees, leafcutter bees, and carpenter bees) are non-social or solitary, with each female making their own nest. Bumble bees and many species of solitary bees create their nests either underground or near the soil surface in vegetation; the intensification of agriculture has led to a loss of habitat and forage for these bees (Kells and Goulson 2003, Danforth et al. 2019).

Pollinator diversity is largely influenced by two factors: the abundance of floral resources and structural diversity. Increasing available pollen and nectar provides more food for pollinators to forage, whereas increasing structural diversity increases the available nesting sites for solitary and ground-nesting bees (Goulson 2003, Bretagnolle and Gaba 2015, Kwaiser and Hendrix 2008). Floral resources (nectar and pollen) provide the vast majority of carbohydrates and proteins that make up the diet of pollinating insects (Brodschneider and Crailsheim 2010, Vaudo et al. 2015). Many studies have shown that planting flowering cover crops along the edges of agricultural fields can significantly increase the diversity of pollinators in the area (Decourtye et al. 2010, Hevia et al. 2016, Orford et al. 2016), whereas a monoculture landscape reduces the abundance and species richness of bee pollinators (Connelly et al. 2015). A diverse array of flowering cover crops has more positive effects on pollinating insects than single species cover crops (Kratschmer et al. 2018, 2019).

The increased structural diversity in cover crop systems also benefits pollinators. Although grasses are mostly wind-pollinated and provide few floral resources for pollinators, planting grasses intermixed with flowering species can provide the necessary structural diversity for solitary and ground-nesting bees to overwinter (Sardiñas et al. 2018). Several studies have shown that increasing landscape diversity has a positive effect on bee diversity, even when said landscapes had lower overall floral resources (Carré et al. 2009, Kratschmer et al. 2019). While only one of many management decisions, these findings highlight the value of diverse cover cropping strategies in increasing pollinator diversity. Another management strategy to increase pollinator diversity is to reduce tillage intensity and pesticide use, allowing ground and solitary pollinator nests to persist throughout multiple seasons (Nicholls and Altieri 2013).

A wide diversity of cover crop species is ideal for attracting pollinators, with flowering plants to provide floral resources and grasses to provide additional structural diversity. Some species that perform exceedingly well due to their abundant floral resources include clovers, sunflower (*Helianthus*), and buckwheat (*Fagopyrum esculentum*) (Mallinger et al. 2019, Bryan et al. 2021, Lami et al. 2024). Oats have been used in combination with flowering plants to provide more structural diversity to attract bees (Mallinger et al. 2019, Bryan et al. 2021). When planting cover crops to improve the pollination of adjacent cash crops, select cover crops that flower at different times than the cash crop, thereby increasing the availability of floral resources; many fruit trees and crops flower for a relatively short period of time (Martins et al. 2018, O'Brien and Arathi 2021). By selecting diverse cover crop species and aligning their flowering times with the needs of pollinators, farmers can create more resilient, pollinator-friendly landscapes, and in doing so, contribute to the long-term sustainability of our agricultural ecosystems.

5 Case study

Cover crops are not usually selected for pollinator enhancement; therefore, cover crop plantings may consist of a single plant species or a mixture of plant species that do not necessarily provide floral resources for pollinators. Relatively simple cover crop mixtures can provide sustained floral cover that leads to high rates of bee visitation. However, the most effective plant species will depend on the specific pollinator enhancement goals, particularly whether the goal is to support native pollinators or non-native honey bees. Furthermore, even diverse cover crop mixtures are unlikely to support the full diversity of pollinators observed in the surrounding non-crop landscape.

Research by Mallinger et al. (2019) demonstrates both the potential and the limitations of cover crops to support wild and managed pollinators. Field trials conducted in 2016–2017 were designed to explore the potential for different cover crop treatments to provide floral cover and attract bee pollinators, while providing other agroecosystem services typically associated with cover crops. The Northern Great Plains serves as a primary summer habitat for honey bees (Gallant et al. 2014) and is home to significant plantings of pollinator-dependent crops (Koh et al. 2016); therefore, identifying cover crops that can support pollinators while providing additional ecosystem services is essential.

The general objectives of the research were to assess several cover crop treatments that varied in plant diversity for their ability to provide season-long

(June–August) floral resources and attract a diversity of bee pollinators. The attractiveness of cover crop plants to bee species of conservation concern was also studied, and the diversity of bees found in the cover crops was compared to that found in adjacent grassland habitat to document what proportion of the broader bee community was visiting the cover crop treatments.

Cover crop treatments included monocultures (single-species) as well as two-, four-, and six-species mixtures. The mixtures were not designed exclusively for pollinators, but rather included plants traditionally used for soil benefits in the Northern Great Plains. These mixtures also included warm- and cool-season plants to ensure establishment of season-long ground cover. Small (28 m^2) cover crop plantings of each treatment were established in a randomized complete block design with four replicates. Weekly observations of floral density and bee visitation rates were made in all plots over the summer months and over two years. Bee-plant interactions in surrounding grassland habitat were also documented.

Two key results emerged from the data. First, high floral cover generally led to high total bee visitation rates; however, bee preferences for certain plant species remained even after accounting for floral cover. For example, simple mixes of phacelia alone (Fig. 1) or a two-species mix of phacelia and buckwheat attracted the most bees in general, and also produced a dense, season-long floral display (Fig. 1). The annual wildflower mixture was expected to provide the greatest diversity of resources for bees; however, poor establishment of several species led to minimal floral cover and reduced total bee visitation.

The second key result was that the cover crop treatment that provided the greatest benefit for managed honey bees differed from the treatment providing the greatest benefit for native solitary bees (Fig. 2). Monoculture and two-species mixtures containing phacelia were most attractive to honey bees as well as social bumble bees, whereas more diverse mixtures containing sunflower

Figure 1 A monoculture phacelia treatment plot.

Figure 2 A native solitary bee, *Melissodes* sp., on a cultivated sunflower.

or native wildflowers were most attractive to native (primarily solitary) bees. Furthermore, while both phacelia and buckwheat established and flowered in abundance, they primarily attracted bees that are not of conservation concern (honey bees, bumble bees); in contrast, sunflower attracted species considered declining or threatened (based on analysis by Kopec and Burd 2017). Consequently, different flowering species are appropriate for a cover crop intended to benefit apiculture (including honey production), versus a crop cover intended to support wild pollinator conservation. Finally, none of the cover crop treatments (individually or combined) could recreate the pollinator diversity in the surrounding environment, with 30 species found in the experimental plots and 48 species in the broader surrounding landscape. The limited ability of the cover crop treatments to fully support the broader bee community is likely due to the small selection of annual cover crop species; furthermore, pollinators that specialize on native perennial plants may not be supported by an annual cover crop species or mixture.

The primary implication of Mallinger et al. (2019) and other research on pollinator preferences is that efforts to support pollinators through cover crops should have specific goals. Not only do the goals help inform the choice of plant species or location of plantings within the broader landscape, but they also assist in documenting the success or failure of efforts, allowing for potential improvements in the future. In particular, this case study demonstrated that there is often a trade-off between high plant and flower diversity, which could better support a diverse bee community, and a dense, uniform floral display, which results in greater honey bee recruitment. This trade-off further argues for the use of specific goals to inform plant species selection.

6 Conclusion and future trends

Increased arthropod biodiversity within a system is perceived to be beneficial; however, it is important to understand the limitations of how biodiversity is measured. Functional diversity is thought to be more important than biodiversity per se. Arthropod biology and behavior can vary greatly within families or even genera; furthermore, many arthropod taxa, especially soil arthropods, are difficult to identify to lower taxonomic ranks. Arthropods are often highly mobile and exhibit preferences for specific habitat types. Arthropod biodiversity within cropping systems is affected by the surrounding landscape and the time scale at which it is being assessed. There is a wide variety of cover-crop species selections and management practices, each of which impacts arthropods. Research-based knowledge about the effects on soil arthropods, insect pollinators, and other organisms in agricultural systems can be used when setting goals and selecting and managing cover crops. Further research designed to test specific hypotheses regarding the effects of cover crops on biodiversity is required.

7 Where to look for further information

The following articles provide good overview of their respective subjects:

- Ali (2023) provides a comprehensive review on biodiversity–ecosystem-functioning research.
- Hedde et al. (2022) review how soil fauna have been classified and provide information on how to group organisms using functional traits.
- Inveninato Carmona et al. (2021) review how cover crops affect arthropods in soybean and corn agroecosystems.

Key research in this area is conducted by the following organizations:

- Sustainable Agriculture Research and Education (SARE, sare.org), including the article 'Cover Crop Impacts on Pollinators,' https://www.sare.org/publications/cover-crops-ecosystem-services/cover-crop-impacts-on-pollinators/.

- Xerces Society (xerces.org), including the article on 'Cover Cropping for Pollinators and Beneficial Insects,' https://xerces.org/publications/guidelines/cover-cropping-for-pollinators-and-beneficial-insects.

8 References

Abad, J., et al. (2021), "Cover crops in viticulture. A systematic review (1): implications on soil characteristics and biodiversity in vineyard." *Oeno One*, Vol. 55 No. 1, pp. 295-312, https://doi.org/10.20870/oeno-one.2021.55.1.3599.

Ali, A. (2023), "Biodiversity–ecosystem functioning research: brief history, major trends and perspectives", *Biological Conservation*, Vol. 285, p. 110210, https://doi.org/10.1016/j.biocon.2023.110210.

Alonso-Ayuso, M., et al. (2014), "The kill date as a management tool for cover cropping success", *PLOS ONE*, Public Library of Science, Vol. 9 No. 10, p. e109587, https://doi.org/10.1371/journal.pone.0109587.

Altieri, M.A. and Letourneau, D.K. (1982), "Vegetation management and biological control in agroecosystems", *Crop Protection*, Vol. 1 No. 4, pp. 405–430, https://doi.org/10.1016/0261-2194(82)90023-0.

Altieri, M.A. and Schmidt, L.L. (1985), "Cover crop manipulation in northern california orchards and vineyards: effects on arthropod communities", *Biological Agriculture & Horticulture*, Taylor & Francis, Vol. 3 No. 1, pp. 1–24, https://doi.org/10.1080/01448765.1985.9754453.

Altieri, M.A., et al. (1984). "Vegetation diversity and insect pest outbreaks", *Critical Reviews in Plant Sciences*, Vol. 2 No. 2, pp. 131–169, https://doi.org/10.1080/07352688409382193.

Andermann, T., et al. (2022), "Estimating alpha, beta, and gamma diversity through deep learning", *Frontiers in Plant Science*, Frontiers, Vol. 13, p. 839407, https://doi.org/10.3389/fpls.2022.839407.

Bardgett, R.D. and van der Putten, W.H. (2014), "Belowground biodiversity and ecosystem functioning", *Nature*, Vol. 515 No. 7528, pp. 505–511, https://doi.org/10.1038/nature13855.

Barnes, A.D., et al. (2020), "Biodiversity enhances the multitrophic control of arthropod herbivory", *Science Advances*, Vol. 6 No. 45, p. eabb6603, https://doi.org/10.1126/sciadv.abb6603.

Barry, K.E., et al. (2019), "The future of complementarity: disentangling causes from consequences", *Trends in Ecology & Evolution*, Vol. 34 No. 2, pp. 167–180, https://doi.org/10.1016/j.tree.2018.10.013.

Beaumelle, L., et al. (2021), "Benefits of increased cover crop diversity for predators and biological pest control depend on the landscape context", *Ecological Solutions and Evidence*, Vol. 2 No. 3, p. e12086, https://doi.org/10.1002/2688-8319.12086.

Behan-Pelletier, V.M. (1999), "Oribatid mite biodiversity in agroecosystems: role for bioindication", *Agriculture, Ecosystems & Environment*, Vol. 74 No. 1, pp. 411–423, https://doi.org/10.1016/S0167-8809(99)00046-8.

Betancur-Corredor, B., et al. (2022), "Reducing tillage intensity benefits the soil micro-and mesofauna in a global meta-analysis", *European Journal of Soil Science*, Vol. 73 No. 6, p. e13321, https://doi.org/10.1111/ejss.13321.

Bezemer, T.M. and van Dam, N.M. (2005), "Linking aboveground and belowground interactions via induced plant defenses", *Trends in Ecology & Evolution*, Vol. 20 No. 11, pp. 617–624, https://doi.org/10.1016/j.tree.2005.08.006.

Birkhofer, K., et al. (2011), "Soil fauna feeding activity in temperate grassland soils increases with legume and grass species richness", *Soil Biology and Biochemistry*, Vol. 43 No. 10, pp. 2200–2207, https://doi.org/10.1016/j.soilbio.2011.07.008.

Blanchart, E., et al. (1999), "Effects of earthworms on soil structure and physical properties", *Earthworm Management in Tropical Agroecosystems*, Vol. 5, pp. 149–171.

Blanco-Canqui, H. and Ruis, S.J. (2020), "Cover crop impacts on soil physical properties: a review", *Soil Science Society of America Journal*, Vol. 84 No. 5, pp. 1527–1576, https://doi.org/10.1002/saj2.20129.

Boeraeve, F., et al. (2022), "Combining organic and conservation agriculture to restore biodiversity? Insights from innovative farms in Belgium and their impacts on carabids and spiders", *Frontiers in Sustainable Food Systems*, Vol. 6, https://doi.org/10.3389/fsufs.2022.1003637.

Bretagnolle, V. and Gaba, S. (2015), "Weeds for bees? A review", *Agronomy for Sustainable Development*, Vol. 35 No. 3, pp. 891–909, https://doi.org/10.1007/s13593-015-0302-5.

Briones, M.J.I. (2014), "Soil fauna and soil functions: a jigsaw puzzle", *Frontiers in Environmental Science*, Frontiers, Vol. 2, https://doi.org/10.3389/fenvs.2014.00007.

Brodschneider, R. and Crailsheim, K. (2010), "Nutrition and health in honey bees", *Apidologie*, Vol. 41 No. 3, pp. 278–294, https://doi.org/10.1051/apido/2010012.

Brown, J.H. (2014), "Why are there so many species in the tropics?", *Journal of Biogeography*, Vol. 41 No. 1, pp. 8–22, https://doi.org/10.1111/jbi.12228.

Brummitt, N., et al. (2017), "Taking stock of nature: Essential biodiversity variables explained", *Biological Conservation*, Vol. 213, pp. 252–255, https://doi.org/10.1016/j.biocon.2016.09.006.

Bryan, C.J., et al. (2021), "Efficacy of cover crops for pollinator habitat provision and weed suppression", *Environmental Entomology*, Vol. 50 No. 1, pp. 208–221, https://doi.org/10.1093/ee/nvaa159.

Bugg, R.L. and Waddington, C. (1994), "Using cover crops to manage arthropod pests of orchards: a review", *Agriculture, Ecosystems & Environment*, Vol. 50 No. 1, pp. 11–28, https://doi.org/10.1016/0167-8809(94)90121-X.

Bunnell, F.L. and Huggard, D.J. (1999), "Biodiversity across spatial and temporal scales: problems and opportunities", *Forest Ecology and Management*, Vol. 115 No. 2, pp. 113–126, https://doi.org/10.1016/S0378-1127(98)00392-2.

Burch-Brown, J. and Archer, A. (2017), "In defence of biodiversity", *Biology & Philosophy*, Vol. 32 No. 6, pp. 969–997, https://doi.org/10.1007/s10539-017-9587-x.

Burgio, G., et al. (2016), "Habitat management of organic vineyard in Northern Italy: the role of cover plants management on arthropod functional biodiversity", *Bulletin of Entomological Research*, Vol. 106 No. 6, pp. 759–768, https://doi.org/10.1017/S0007485316000493.

Cadotte, M.W., et al. (2011), "Beyond species: functional diversity and the maintenance of ecological processes and services", *Journal of Applied Ecology*, Vol. 48 No. 5, pp. 1079–1087, https://doi.org/10.1111/j.1365-2664.2011.02048.x.

Cameron, K.H. and Leather, S.R. (2012), "How good are carabid beetles (Coleoptera, Carabidae) as indicators of invertebrate abundance and order richness?", *Biodiversity and Conservation*, Vol. 21 No. 3, pp. 763–779, https://doi.org/10.1007/s10531-011-0215-9.

Cappelli, S.L., et al. (2022), "Plant biodiversity promotes sustainable agriculture directly and via belowground effects", *Trends in Plant Science*, Vol. 27 No. 7, pp. 674–687, https://doi.org/10.1016/j.tplants.2022.02.003.

Carmona, D.M. and Landis, D.A. (1999), "Influence of refuge habitats and cover crops on seasonal activity-density of ground beetles (Coleoptera: Carabidae) in field crops", *Environmental Entomology*, Vol. 28 No. 6, pp. 1145–1153, https://doi.org/10.1093/ee/28.6.1145.

Carpio, A.J., et al. (2019), "Arthropod biodiversity in olive groves under two soil management systems: presence versus absence of herbaceous cover crop", *Agricultural and Forest Entomology*, Vol. 21 No. 1, pp. 58–68, https://doi.org/10.1111/afe.12303.

Carré, G., et al. (2009), "Landscape context and habitat type as drivers of bee diversity in European annual crops", *Agriculture, Ecosystems & Environment*, Vol. 133 No. 1, pp. 40–47, https://doi.org/10.1016/j.agee.2009.05.001.

Catovsky, S., et al. (2002), "Biodiversity and ecosystem productivity: implications for carbon storage", *Oikos*, Vol. 97 No. 3, pp. 443–448, https://doi.org/10.1034/j.1600-0706.2002.970315.x.

Chao, A., et al. (2006), "Abundance-based similarity indices and their estimation when there are unseen species in samples", *Biometrics*, Vol. 62 No. 2, pp. 361–371, https://doi.org/10.1111/j.1541-0420.2005.00489.x.

Chapagain, T., et al. (2020), "The potential of multi-species mixtures to diversify cover crop benefits", *Sustainability*, Multidisciplinary Digital Publishing Institute, Vol. 12 No. 5, p. 2058, https://doi.org/10.3390/su12052058.

Chapela-Oliva, C., et al. (2022), "Edaphoclimatic drivers of the effect of extensive vegetation management on ecosystem services and biodiversity in vineyards", *Agriculture, Ecosystems & Environment*, Vol. 339, p. 108115, https://doi.org/10.1016/j.agee.2022.108115.

Chase, J.M., et al. (2018), "Embracing scale-dependence to achieve a deeper understanding of biodiversity and its change across communities", *Ecology Letters*, Vol. 21 No. 11, pp. 1737–1751, https://doi.org/10.1111/ele.13151.

Christie, M., et al. (2006), "Valuing the diversity of biodiversity", *Ecological Economics*, Vol. 58 No. 2, pp. 304–317, https://doi.org/10.1016/j.ecolecon.2005.07.034.

Coleman, D.C. and Wall, D.H. (2015), "Chapter 5 - Soil fauna: occurrence, biodiversity, and roles in ecosystem function", in Paul, E.A. (Ed.), *Soil Microbiology, Ecology and Biochemistry (Fourth Edition)*, Academic Press, Boston, pp. 111–149, https://doi.org/10.1016/B978-0-12-415955-6.00005-0.

Connelly, H., et al. (2015), "Landscape simplification decreases wild bee pollination services to strawberry", *Agriculture, Ecosystems & Environment*, Vol. 211, pp. 51–56, https://doi.org/10.1016/j.agee.2015.05.004.

Costa, F.R.C. and Magnusson, W.E. (2010), "The need for large-scale, integrated studies of biodiversity - the experience of the program for biodiversity research in Brazilian Amazonia", Natureza a Conservacao, https://doi.org/10.4322/natcon.00801001.

Costanza, R., et al. (2007), "Biodiversity and ecosystem services: a multi-scale empirical study of the relationship between species richness and net primary production", *Ecological Economics*, Vol. 61 No. 2, pp. 478–491, https://doi.org/10.1016/j.ecolecon.2006.03.021.

Crotty, F.V., et al. (2015), "Assessing the impact of agricultural forage crops on soil biodiversity and abundance", *Soil Biology and Biochemistry*, Vol. 91, pp. 119–126, https://doi.org/10.1016/j.soilbio.2015.08.036.

Crowther, T.W., et al. (2019), "The global soil community and its influence on biogeochemistry", *Science*, Vol. 365 No. 6455, p. eaav0550, https://doi.org/10.1126/science.aav0550.

Culliney, T.W. (2013), "Role of arthropods in maintaining soil fertility", *Agriculture*, Multidisciplinary Digital Publishing Institute, Vol. 3 No. 4, pp. 629–659, https://doi.org/10.3390/agriculture3040629.

Danforth, B.N., et al. (2019), *The Solitary Bees: Biology, Evolution, Conservation*, Princeton University Press, Princeton.

Davis, H.N., et al. (2009), "Impact of land management practices on carabids (Coleoptera: Carabidae) and other arthropods on the western high plains of North America", *Southwestern Entomologist*, Vol. 34 No. 1, pp. 43–59, https://doi.org/10.3958/059.034.0104.

de Pedro, L., et al. (2020), "The effect of cover crops on the biodiversity and abundance of ground-dwelling arthropods in a Mediterranean pear orchard", *Agronomy*, Vol. 10 No. 4, p. 580, https://doi.org/10.3390/agronomy10040580.

Decourtye, A., et al. (2010), "Landscape enhancement of floral resources for honey bees in agro-ecosystems", *Apidologie*, EDP Sciences, Vol. 41 No. 3, pp. 264–277, https://doi.org/10.1051/apido/2010024.

Deines, J.M., et al. (2023), "Recent cover crop adoption is associated with small maize and soybean yield losses in the United States", *Global Change Biology*, Vol. 29 No. 3, pp. 794–807, https://doi.org/10.1111/gcb.16489.

Díaz, S. and Malhi, Y. (2022), "Biodiversity: concepts, patterns, trends, and perspectives", *Annual Review of Environment and Resources*, Vol. 47 No. Volume 47, 2022, pp. 31–63, https://doi.org/10.1146/annurev-environ-120120-054300.

Dindal, D.L. (1991), *Soil Biology Guide*, Wiley, New York.

Dornelas, M., et al. (2013), "Quantifying temporal change in biodiversity: challenges and opportunities", *Proceedings of the Royal Society B: Biological Sciences*, Royal Society, Vol. 280 No. 1750, p. 20121931, https://doi.org/10.1098/rspb.2012.1931.

Douda, J., et al. (2023), "Historical sampling error: a neglected factor in long-term biodiversity change research", *Biological Conservation*, Vol. 286, p. 110317, https://doi.org/10.1016/j.biocon.2023.110317.

Du Toit, J.T. (2010), "Considerations of scale in biodiversity conservation", *Animal Conservation*, Vol. 13 No. 3, pp. 229–236, https://doi.org/10.1111/j.1469-1795.2010.00355.x.

Dudley, N. and Alexander, S. (2017), "Agriculture and biodiversity: a review", *Biodiversity*, Vol. 18 No. 2–3, pp. 45–49, https://doi.org/10.1080/14888386.2017.1351892.

Duelli, P. (1997), "Biodiversity evaluation in agricultural landscapes: an approach at two different scales", *Agriculture, Ecosystems & Environment*, Vol. 62 No. 2, pp. 81–91, https://doi.org/10.1016/S0167-8809(96)01143-7.

Duelli, P. and Obrist, M.K. (2003), "Biodiversity indicators: the choice of values and measures", *Agriculture, Ecosystems & Environment*, Vol. 98 No. 1, pp. 87–98, https://doi.org/10.1016/S0167-8809(03)00072-0.

Duffy, J.E. (2009), "Why biodiversity is important to the functioning of real-world ecosystems", *Frontiers in Ecology and the Environment*, Vol. 7 No. 8, pp. 437–444, https://doi.org/10.1890/070195.

Dufresnes, C., et al. (2023), "Acknowledging more biodiversity without more species", *Proceedings of the National Academy of Sciences*, Vol. 120 No. 40, p. e2302424120, https://doi.org/10.1073/pnas.2302424120.

DuPre, M.E., et al. (2021), "Impacts of dryland cropping systems on ground beetle communities (Coleoptera: Carabidae) in the northern Great Plains", *Journal of Insect Science*, Vol. 21 No. 1, p. 19, https://doi.org/10.1093/jisesa/ieab008.

Eckert, M., et al. (2020), "Native cover crops enhance arthropod diversity in vineyards of the Cape Floristic Region", *Journal of Insect Conservation*, Vol. 24 No. 1, pp. 133–149, https://doi.org/10.1007/s10841-019-00196-0.

Eisenbeis, G. and Wichard, W. (2012), *Atlas on the Biology of Soil Arthropods*, Springer Science & Business Media, Berlin.

Eisenhauer, N., et al. (2024), "The multiple-mechanisms hypothesis of biodiversity–stability relationships", Basic and Applied Ecology, Vol. 79, pp. 153–166, https://doi.org/10.1016/j.baae.2024.07.004.

Ekschmitt, K., et al. (1997), "Spiders, carabids, and staphylinids: the ecological potential of predatory macroarthropods", in Benckiser, G. (Ed.), *Fauna in Soil Ecosystems*, CRC Press, New York.

Elmquist, D.C., et al. (2023), "Linking agricultural diversification practices, soil arthropod communities and soil health", *Journal of Applied Ecology*, Vol. 60 No. 9, pp. 1952–1963, https://doi.org/10.1111/1365-2664.14453.

Estrada-Carmona, N., et al. (2022), "Complex agricultural landscapes host more biodiversity than simple ones: a global meta-analysis", *Proceedings of the National Academy of Sciences*, Vol. 119 No. 38, p. e2203385119, https://doi.org/10.1073/pnas.2203385119.

Farnham, T.J. (2007), *Saving Nature's Legacy: Origins of the Idea of Biological Diversity*, Yale University Press, New Haven.

Fernández, D.E., et al. (2008), "Effect of different cover crops on the presence of arthropods in an organic apple (*Malus domestica* Borkh) Orchard", *Journal of Sustainable Agriculture*, Vol. 32 No. 2, pp. 197–211, https://doi.org/10.1080/10440040802170624.

Finney, D.M. and Kaye, J.P. (2017), "Functional diversity in cover crop polycultures increases multifunctionality of an agricultural system", *Journal of Applied Ecology*, Vol. 54 No. 2, pp. 509–517, https://doi.org/10.1111/1365-2664.12765.

Finney, D.M., et al. (2017), "Living cover crops have immediate impacts on soil microbial community structure and function", *Journal of Soil and Water Conservation*, Vol. 72 No. 4, pp. 361–373, https://doi.org/10.2489/jswc.72.4.361.

Fiorini, A., et al. (2022), "Driving crop yield, soil organic C pools, and soil biodiversity with selected winter cover crops under no-till", *Soil and Tillage Research*, Vol. 217, p. 105283, https://doi.org/10.1016/j.still.2021.105283.

Fischer, C., et al. (2014), "How do earthworms, soil texture and plant composition affect infiltration along an experimental plant diversity gradient in grassland?", Edited by Hector, A. *PLoS One*, Vol. 9 No. 6, p. e98987, https://doi.org/10.1371/journal.pone.0098987.

Fischer, M., et al. (2010), "Implementing large-scale and long-term functional biodiversity research: the Biodiversity Exploratories", *Basic and Applied Ecology*, Vol. 11 No. 6, pp. 473–485, https://doi.org/10.1016/j.baae.2010.07.009.

Furey, G.N. and Tilman, D. (2021), "Plant biodiversity and the regeneration of soil fertility", *Proceedings of the National Academy of Sciences*, Vol. 118 No. 49, p. e2111321118, https://doi.org/10.1073/pnas.2111321118.

Gallant, A.L., et al. (2014), "Mapping large-area landscape suitability for honey bees to assess the influence of land-use change on sustainability of national pollination services", *PLoS One*, Vol. 9 No. 6, p. e99268, https://doi.org/10.1371/journal.pone.0099268.

Gallé, R., et al. (2019), "Landscape configuration, organic management, and within-field position drive functional diversity of spiders and carabids", *Journal of Applied Ecology*, Vol. 56 No. 1, pp. 63–72, https://doi.org/10.1111/1365-2664.13257.

Gaston, K.J. and Spicer, J.I. (2013), *Biodiversity: An Introduction*, John Wiley & Sons, Chichester.

Gatti, R.C., et al. (2020). "Estimating and comparing biodiversity with a single universal metric", *Ecological Modelling*, 424, p. 109020, https://doi.org/10.1016/j.ecolmodel.2020.109020.

Gayer, C., et al. (2019), "Carabid functional diversity is enhanced by conventional flowering fields, organic winter cereals and edge habitats", *Agriculture, Ecosystems & Environment*, Vol. 284, p. 106579, https://doi.org/10.1016/j.agee.2019.106579.

Geldenhuys, M., et al. (2021), "Diverse herbaceous cover crops promote vineyard arthropod diversity across different management regimes", *Agriculture, Ecosystems & Environment*, Vol. 307, p. 107222, https://doi.org/10.1016/j.agee.2020.107222.

Gonzalez, A., et al. (2020), "Scaling-up biodiversity-ecosystem functioning research", *Ecology Letters*, Vol. 23 No. 4, pp. 757–776, https://doi.org/10.1111/ele.13456.

Goulson, D. (2003), "Conserving wild bees for crop pollination", *Journal of Food Agriculture and Environment*, Vol. 1, pp. 142–144.

Griffiths, M., et al. (2022), "Optimisation of root traits to provide enhanced ecosystem services in agricultural systems: a focus on cover crops", *Plant, Cell & Environment*, Vol. 45 No. 3, pp. 751–770, https://doi.org/10.1111/pce.14247.

Groff, S. (2015), "The past, present, and future of the cover crop industry", *Journal of Soil and Water Conservation*, Vol. 70 No. 6, pp. 130A-133A, https://doi.org/10.2489/jswc.70.6.130A.

Guerra, C.A., et al. (2021), "Tracking, targeting, and conserving soil biodiversity", *Science*, Vol. 371 No. 6526, pp. 239–241, https://doi.org/10.1126/science.abd7926.

Haruna, S.I., et al. (2020), "Improving soil physical properties through the use of cover crops: a review", *Agrosystems, Geosciences & Environment*, Vol. 3 No. 1, p. e20105, https://doi.org/10.1002/agg2.20105.

Hector, A., et al. (2001), "Conservation implications of the link between biodiversity and ecosystem functioning", *Oecologia*, Vol. 129 No. 4, pp. 624–628, https://doi.org/10.1007/s004420100759.

Hedde, M., et al. (2022), "A common framework for developing robust soil fauna classifications", *Geoderma*, Vol. 426, p. 116073, https://doi.org/10.1016/j.geoderma.2022.116073.

Hevia, V., et al. (2016), "Bee diversity and abundance in a livestock drove road and its impact on pollination and seed set in adjacent sunflower fields", *Agriculture, Ecosystems & Environment*, Vol. 232, pp. 336–344, https://doi.org/10.1016/j.agee.2016.08.021.

Hill, S.L.L., et al. (2016), "Reconciling biodiversity indicators to guide understanding and action", *Conservation Letters*, Vol. 9 No. 6, pp. 405–412, https://doi.org/10.1111/conl.12291.

Hillebrand, H. and Matthiessen, B. (2009), "Biodiversity in a complex world: consolidation and progress in functional biodiversity research", *Ecology Letters*, Vol. 12 No. 12, pp. 1405–1419, https://doi.org/10.1111/j.1461-0248.2009.01388.x.

Hojati, S., et al. (2024), "Quantifying the spatial heterogeneity of topsoil microstructure in a horticultural cover cropping system from Southwestern Ontario, Canada", *Geoderma Regional*, Vol. 37, p. e00789, https://doi.org/10.1016/j.geodrs.2024.e00789.

Holland, J. and Luff, M. (2000), "The effects of agricultural practices on Carabidae in temperate agroecosystems", *Integrated Pest Management Reviews*, Vol. 5, pp. 109–129, https://doi.org/10.1023/A:1009619309424.

Hong, P., et al. (2022), "Biodiversity promotes ecosystem functioning despite environmental change", *Ecology Letters*, Vol. 25 No. 2, pp. 555–569, https://doi.org/10.1111/ele.13936.

Huang, M., et al. (2020), "Functional and phylogenetic diversity explain different components of diversity effects on biomass production", *Oikos*, Vol. 129 No. 8, pp. 1185–1195, https://doi.org/10.1111/oik.07032.

Inveninato Carmona, G., et al. (2021). "Does cover crop management affect arthropods in the subsequent corn and soybean crops in the United States? A systematic review", *Annals of the Entomological Society of America*, 114(2), pp. 151–162, https://doi.org/10.1093/aesa/saaa049.

IPBES (Intergovernmental Science-Policy Platform on Biodiversity and Ecosystem Services). 2024. https://ipbes.net/glossary/biodiversity. Accessed 7-17-24.

Jabbour, R., et al. (2016), "Cover crop and tillage intensities alter ground-dwelling arthropod communities during the transition to organic production", *Renewable Agriculture and Food Systems*, Vol. 31 No. 4, pp. 361–374, https://doi.org/10.1017/S1742170515000290.

Jennings, M.D., et al. (2008), "A comparative measure of biodiversity based on species composition", *Biodiversity and Conservation*, Vol. 17 No. 4, pp. 833–840, https://doi.org/10.1007/s10531-008-9322-7.

Josephson, J. (1982), "Why maintain biological diversity?", *Environmental Science & Technology*, Vol. 16 No. 2, pp. 94A–97A, https://doi.org/10.1021/es00096a717.

Junod, M., et al. (2024), "Cover crops in cereal rotations: a quantitative review", *Soil and Tillage Research*, Vol. 238, p. 105997, https://doi.org/10.1016/j.still.2023.105997.

Kahimba, F., et al. (2008), "Cover crop effects on infiltration, soil temperature, and soil moisture distribution in the Canadian prairies", *Applied Engineering in Agriculture*,Vol. 24 No. 3, pp. 321–333, https://doi.org/10.13031/2013.24502.

Kapayou, D.G., et al. (2023), "Reuniting the Three Sisters: collaborative science with Native growers to improve soil and community health", *Agriculture and Human Values*, Vol. 40 No. 1, pp. 65–82, https://doi.org/10.1007/s10460-022-10336-z.

Karimi, B., et al. (2020), "A meta-analysis of the ecotoxicological impact of viticultural practices on soil biodiversity", *Environmental Chemistry Letters*, Vol. 18 No. 6, pp. 1947–1966, https://doi.org/10.1007/s10311-020-01050-5.

Kaspar, T.C. and Singer, J.W. (2011), "The use of cover crops to manage soil", in *Soil Management: Building a Stable Base for Agriculture*, John Wiley & Sons, Ltd, Hoboken, pp. 321–337, https://doi.org/10.2136/2011.soilmanagement.c21.

Kaye, J.P. and Quemada, M. (2017), "Using cover crops to mitigate and adapt to climate change: a review", *Agronomy for Sustainable Development*, Vol. 37 No. 1, p. 4, https://doi.org/10.1007/s13593-016-0410-x.

Keil, P. and Chase, J.M. (2019), "Global patterns and drivers of tree diversity integrated across a continuum of spatial grains", *Nature Ecology & Evolution*, Vol. 3 No. 3, pp. 390–399, https://doi.org/10.1038/s41559-019-0799-0.

Kells, A.R. and Goulson, D. (2003), "Preferred nesting sites of bumblebee queens (Hymenoptera: Apidae) in agroecosystems in the UK", *Biological Conservation*, Vol. 109 No. 2, pp. 165–174, https://doi.org/10.1016/S0006-3207(02)00131-3.

Kelly, C., et al. (2021), "Winter cover crops and no-till promote soil macrofauna communities in irrigated, Mediterranean cropland in California, USA", *Applied Soil Ecology*, Vol. 166, p. 104068, https://doi.org/10.1016/j.apsoil.2021.104068.

Kim, N., et al. (2020), "Do cover crops benefit soil microbiome? A meta-analysis of current research", *Soil Biology and Biochemistry*, Vol. 142, p.107701, https://doi.org/10.1016/j.soilbio.2019.107701.

Klein, A.-M., et al. (2006), "Importance of pollinators in changing landscapes for world crops", *Proceedings of the Royal Society B: Biological Sciences*, Vol. 274 No. 1608, pp. 303–313, https://doi.org/10.1098/rspb.2006.3721.

Koh, I., et al. (2016), "Modeling the status, trends, and impacts of wild bee abundance in the United States", *Proceedings of the National Academy of Sciences*, Proceedings of the National Academy of Sciences, Vol. 113 No. 1, pp. 140–145, https://doi.org/10.1073/pnas.1517685113.

Köninger, J., et al. (2022), "In defence of soil biodiversity: towards an inclusive protection in the European Union", *Biological Conservation*, Vol. 268, p. 109475, https://doi.org/10.1016/j.biocon.2022.109475.

Kopec, K. and Burd, L.A. (2017), *Pollinators in Peril*, Center for Biological Diversity, Tucson. https://www.biologicaldiversity.org/campaigns/native_pollinators/pdfs/Pollinators_in_Peril.pdf.

Kotze, D.J., et al. (2011), "Forty years of carabid beetle research in Europe – from taxonomy, biology, ecology and population studies to bioindication, habitat assessment and conservation", *ZooKeys*, No. 100, pp. 55–148, https://doi.org/10.3897/zookeys.100.1523.

Koudahe, K., et al. (2022), "Critical review of the impact of cover crops on soil properties", *International Soil and Water Conservation Research*, Vol. 10 No. 3, pp. 343–354, https://doi.org/10.1016/j.iswcr.2022.03.003.

Kratschmer, S., et al. (2018), "Tillage intensity or landscape features: what matters most for wild bee diversity in vineyards?", *Agriculture, Ecosystems & Environment*, Vol. 266, pp. 142–152, https://doi.org/10.1016/j.agee.2018.07.018.

Kratschmer, S., et al. (2019), "Response of wild bee diversity, abundance, and functional traits to vineyard inter-row management intensity and landscape diversity across Europe", *Ecology and Evolution*, Vol. 9 No. 7, pp. 4103–4115, https://doi.org/10.1002/ece3.5039.

Kromp, B. (1999), "Carabid beetles in sustainable agriculture: a review on pest control efficacy, cultivation impacts and enhancement", *Agriculture, Ecosystems & Environment*, Vol. 74 No. 1, pp. 187–228, https://doi.org/10.1016/S0167-8809(99)00037-7.

Kwaiser, K.S. and Hendrix, S.D. (2008), "Diversity and abundance of bees (Hymenoptera: Apiformes) in native and ruderal grasslands of agriculturally dominated landscapes", *Agriculture, Ecosystems & Environment*, Vol. 124 No. 3, pp. 200–204, https://doi.org/10.1016/j.agee.2007.09.012.

Lalonde, O., et al. (2012), "Carabid beetle communities after 18 years of conservation tillage and crop rotation in a cool humid climate", *The Canadian Entomologist*, Vol. 144 No. 5, pp. 645–657, https://doi.org/10.4039/tce.2012.55.

Lamandé, M., et al. (2003), "Changes of pore morphology, infiltration and earthworm community in a loamy soil under different agricultural managements", *CATENA*, Vol. 54 No. 3, pp. 637–649, https://doi.org/10.1016/S0341-8162(03)00114-0.

Lami, F., et al. (2024), "Sustaining multiple ecosystem functions in agricultural landscapes: effect of summer cover crops on weed control, soil quality and support to pollinators", *Crop Protection*, Vol. 184, p. 106832, https://doi.org/10.1016/j.cropro.2024.106832.

Landis, D.A., et al. (2000), "Habitat management to conserve natural enemies of arthropod pests in agriculture", *Annual Review of Entomology,* Vol. 45 No. 1, pp. 175–201.

Lawrence, E.R. and Fraser, D.J. (2020), "Latitudinal biodiversity gradients at three levels: -linking species richness, population richness and genetic diversity", *Global Ecology and Biogeography*, Vol. 29 No. 5, pp. 770–788, https://doi.org/10.1111/geb.13075.

Levin, S.A. (2000), "Multiple scales and the maintenance of biodiversity", *Ecosystems*, Vol. 3 No. 6, pp. 498–506, https://doi.org/10.1007/s100210000044.

Lichtenberg, E.M., et al. (2023), "Differential effects of soil conservation practices on arthropods and crop yields", *Journal of Applied Entomology*, Vol. 147 No. 10, pp. 931–940, https://doi.org/10.1111/jen.13188.

Loreau, M. and de Mazancourt, C. (2008), "Species synchrony and its drivers: neutral and nonneutral community dynamics in fluctuating environments", *The American Naturalist*, Vol. 172 No. 2, pp. E48–E66, https://doi.org/10.1086/589746.

Loreau, M., et al. (2021), "Biodiversity as insurance: from concept to measurement and application", *Biological Reviews*, Vol. 96 No. 5, pp. 2333–2354, https://doi.org/10.1111/brv.12756.

Lövei, G.L. and Sunderland, K.D. (1996), "Ecology and behavior of ground beetles (Coleoptera: Carabidae)", *Annual Review of Entomology*, Vol. 41 No. 41, 1996, pp. 231–256, https://doi.org/10.1146/annurev.en.41.010196.001311.

Lu, Y.-C., et al. (2000), "Cover crops in sustainable food production", *Food Reviews International*, Vol. 16 No. 2, pp. 121–157, https://doi.org/10.1081/FRI-100100285.

Lucas, M., et al. (2022), "Cover crop influence on pore size distribution and biopore dynamics: enumerating root and soil faunal effects", *Frontiers in Plant Science*, Vol. 13, pp. 928569, https://doi.org/10.3389/fpls.2022.928569.

Lumbierres, M. and Kissling, W.D. (2023), "Important first steps towards designing the freshwater, marine and terrestrial Essential Biodiversity Variable (EBV) workflows for the European Biodiversity Observation Network", *Research Ideas and Outcomes*, Vol. 9, p. e109120.

Lundgren, J.G. and Fergen, J.K. (2010), "The effects of a winter cover crop on *Diabrotica virgifera* (Coleoptera: Chrysomelidae) populations and beneficial arthropod communities in no-till maize", *Environmental Entomology*, Vol. 39 No. 6, pp. 1816–1828, https://doi.org/10.1603/EN10041.

Luxton, M. (1982), "The biology of mites from beech woodland soil", *Pedobiologia*, Vol. 23 No. 1, pp. 1–8, https://doi.org/10.1016/S0031-4056(23)03608-9.

Magdoff, F. and van Es, H. (2021), *Building Soils for Better Crops: Ecological Management for Healthy Soils*, Sustainable Agriculture Research and Education Program, Beltsville.

Magurran, A.E., et al. (2010), "Long-term datasets in biodiversity research and monitoring: assessing change in ecological communities through time", *Trends in Ecology & Evolution*, Vol. 25 No. 10, pp. 574–582, https://doi.org/10.1016/j.tree.2010.06.016.

Maksimovich, K.Y., et al. (2023), "Species composition and ecological structure of ground beetles (Coleoptera, Carabidae) communities as biological indicators of the agro-environmental sustainability", *Environmental Research*, Vol. 234, p. 116030, https://doi.org/10.1016/j.envres.2023.116030.

Mallinger, R.E., et al. (2019), "Annual cover crops for managed and wild bees: optimal plant mixtures depend on pollinator enhancement goals", *Agriculture, Ecosystems & Environment*, Vol. 273, pp. 107–116, https://doi.org/10.1016/j.agee.2018.12.006.

Mammola, S., et al. (2021), "Concepts and applications in functional diversity", *Functional Ecology*, Vol. 35 No. 9, pp. 1869–1885, https://doi.org/10.1111/1365-2435.13882.

Marahatta, S.P., et al. (2010), "Strip-tilled cover cropping for managing nematodes, soil mesoarthropods, and weeds in a bitter melon agroecosystem", *Journal of Nematology*, Vol. 42 No. 2, pp. 111–119.

Marselle, M.R., et al. (2021), "Pathways linking biodiversity to human health: a conceptual framework", *Environment International*, Vol. 150, p. 106420, https://doi.org/10.1016/j.envint.2021.106420.

Martins, K.T., et al. (2018), "Complementary crops and landscape features sustain wild bee communities", *Ecological Applications*, Vol. 28 No. 4, pp. 1093–1105, https://doi.org/10.1002/eap.1713.

McGill, B.J., et al. (2006), "Rebuilding community ecology from functional traits", *Trends in Ecology & Evolution*, Vol. 21 No. 4, pp. 178–185, https://doi.org/10.1016/j.tree.2006.02.002.

McKenzie-Gopsill, A., et al. (2022), "The importance of species selection in cover crop mixture design", *Weed Science*, Vol. 70 No. 4, pp. 436–447, https://doi.org/10.1017/wsc.2022.28.

Menta, C. and Remelli, S. (2020), "Soil health and arthropods: from complex system to worthwhile investigation", *Insects*, Vol. 11 No. 1, p. 54, https://doi.org/10.3390/insects11010054.

Menta, C., et al. (2020), "Soil arthropod responses in agroecosystem: implications of different management and cropping systems", *Agronomy*, Vol. 10 No. 7, p. 982, https://doi.org/10.3390/agronomy10070982.

Menta, C., et al. (2018), "Soil biological wuality index (QBS-ar): 15 years of application at global scale", *Ecological Indicators*, Vol. 85, pp. 773–780, https://doi.org/10.1016/j.ecolind.2017.11.030.

Merriam-Webster. (2024). Biodiversity. In Merriam-Webster.com dictionary. Retrieved July 2, 2024, from https://www.merriam-webster.com/dictionary/biodiversity.

Michener, C.D. (1974), *The Social Behavior of the Bees: A Comparative Study*, Harvard University Press, Cambridge.

Minor, M.A., et al. (2004), "Effects of site preparation techniques on communities of soil mites (Acari: Oribatida, Acari: Gamasida) under short-rotation forestry plantings in New York, USA", *Applied Soil Ecology*, Vol. 25 No. 3, pp. 181–192, https://doi.org/10.1016/j.apsoil.2003.10.002.

Miyazawa, K., et al. (2002), "The effects of cropping systems and fallow managements on microarthropod populations", *Plant Production Science*, Vol. 5 No. 3, pp. 257–265, https://doi.org/10.1626/pps.5.257.

Molina, G.A.R., et al. (2014), "Epigeal arthropod communities in intensively farmed landscapes: effects of land use mosaics, neighbourhood heterogeneity, and field position", *Agriculture, Ecosystems & Environment*, Vol. 192, pp. 135–143, https://doi.org/10.1016/j.agee.2014.04.013.

Moonen, A.-C. and Bàrberi, P. (2008), "Functional biodiversity: an agroecosystem approach", *Agriculture, Ecosystems & Environment*, Vol. 127 No. 1, pp. 7–21, https://doi.org/10.1016/j.agee.2008.02.013.

Mouchet, M.A., et al. (2010), "Functional diversity measures: an overview of their redundancy and their ability to discriminate community assembly rules", *Functional Ecology*, Vol. 24 No. 4, pp. 867–876, https://doi.org/10.1111/j.1365-2435.2010.01695.x.

Müller, P., et al. (2022), "Tillage effects on ground beetles in temperate climates: a review", *Agronomy for Sustainable Development*, Vol. 42, p. 65, https://doi.org/10.1007/s13593-022-00803-6.

Nagendra, H. (2002), "Opposite trends in response for the Shannon and Simpson indices of landscape diversity", *Applied Geography*, Vol. 22 No. 2, pp. 175–186, https://doi.org/10.1016/S0143-6228(02)00002-4.

Nicholls, C.I. and Altieri, M.A. (2013), "Plant biodiversity enhances bees and other insect pollinators in agroecosystems. A review", *Agronomy for Sustainable Development*, Vol. 33 No. 2, pp. 257–274, https://doi.org/10.1007/s13593-012-0092-y.

O'Brien, C. and Arathi, H.S. (2021), "If you build it, they will come–agroecosystem-based management practices support pollinators", *Annals of the Entomological Society of America*, Vol. 114 No. 3, pp. 322–328, https://doi.org/10.1093/aesa/saaa037.

Oehri, J., et al. (2017), "Biodiversity promotes primary productivity and growing season lengthening at the landscape scale", *Proceedings of the National Academy of Sciences*, Vol. 114 No. 38, pp. 10160–10165, https://doi.org/10.1073/pnas.1703928114.

Oliver, T.H., et al. (2015), "Biodiversity and resilience of ecosystem functions", *Trends in Ecology & Evolution*, Elsevier, Vol. 30 No. 11, pp. 673–684, https://doi.org/10.1016/j.tree.2015.08.009.

Orford, K.A., et al. (2016), "Modest enhancements to conventional grassland diversity improve the provision of pollination services", *Journal of Applied Ecology*, Vol. 53 No. 3, pp. 906–915, https://doi.org/10.1111/1365-2664.12608.

Parisi, V., et al. (2005), "Microarthropod communities as a tool to assess soil quality and biodiversity: a new approach in Italy", *Agriculture, Ecosystems & Environment*, Vol. 105 No. 1, pp. 323–333, https://doi.org/10.1016/j.agee.2004.02.002.

Patterson, E.S.P., et al. (2019), "Soil tillage reduces arthropod biodiversity and has lag effects within organic and conventional crop rotations", *Journal of Applied Entomology*, Vol. 143 No. 4, pp. 430–440, https://doi.org/10.1111/jen.12603.

Pereira, H.M., et al. (2013), "Essential biodiversity variables", *Science*, Vol. 339 No. 6117, pp. 277–278, https://doi.org/10.1126/science.1229931.

Petchey, O.L. and Gaston, K.J. (2006), "Functional diversity: back to basics and looking forward", *Ecology Letters*, Vol. 9 No. 6, pp. 741–758, https://doi.org/10.1111/j.1461-0248.2006.00924.x.

Petchey, O., et al. (2009), "A functional guide to functional diversity measures", *Biodivers. Ecosyst. Funct. Hum.Wellbeing Ecol. Econ. Perspect.*, Vol. 85, https://doi.org/10.1093/acprof:oso/9780199547951.003.0004.

Petchey, O.L., Hector, A. and Gaston, K.J. (2004), "How do different measures of functional diversity perform?", *Ecology*, Vol. 85 No. 3, pp. 847–857, https://doi.org/10.1890/03-0226.

Pizzolotto, R., et al. (2018), "Ground beetles in Mediterranean olive agroecosystems: their significance and functional role as bioindicators (Coleoptera, Carabidae)", *PLoS One*, Vol. 13 No. 3, p. e0194551, https://doi.org/10.1371/journal.pone.0194551.

Pla, L., et al. (2011), *Quantifying Functional Biodiversity*, Springer Science & Business Media, Berlin.

Potapov, A.M., et al. (2022), "Feeding habits and multifunctional classification of soil-associated consumers from protists to vertebrates", *Biological Reviews*, Vol. 97 No. 3, pp. 1057–1117, https://doi.org/10.1111/brv.12832.

Potts, S.G., et al. (2016), "Safeguarding pollinators and their values to human well-being", *Nature*, Vol. 540 No. 7632, pp. 220–229, https://doi.org/10.1038/nature20588.

Prasifka, J.R., et al. (2006), "Effects of living mulches on predator abundance and sentinel prey in a corn–soybean–forage rotation", *Environmental Entomology*, Vol. 35 No. 5, pp. 1423–1431, https://doi.org/10.1093/ee/35.5.1423.

Pullaro, T.C., et al. (2006), "Effects of killed cover crop mulch on weeds, weed seeds, and herbivores", *Agriculture, Ecosystems & Environment*, Vol. 115 No. 1, pp. 97–104, https://doi.org/10.1016/j.agee.2005.12.021.

Purvis, A. and Hector, A. (2000), "Getting the measure of biodiversity", *Nature*, Vol. 405 No. 6783, pp. 212–219, https://doi.org/10.1038/35012221.

Querner, P. and Bruckner, A. (2010), "Combining pitfall traps and soil samples to collect Collembola for site scale biodiversity assessments", *Applied Soil Ecology*, Vol. 45 No. 3, pp. 293–297, https://doi.org/10.1016/j.apsoil.2010.05.005.

Quintarelli, V., et al. (2022), "Cover crops for sustainable cropping systems: a review", *Agriculture*, Vol. 12 No. 12, p. 2076, https://doi.org/10.3390/agriculture12122076.

Raffard, A., et al. (2019), "The community and ecosystem consequences of intraspecific diversity: a meta-analysis", *Biological Reviews*, Vol. 94 No. 2, pp. 648–661, https://doi.org/10.1111/brv.12472.

Rahbek, C. (1995), "The elevational gradient of species richness: a uniform pattern?", *Ecography*, Vol. 18 No. 2, pp. 200–205, https://doi.org/10.1111/j.1600-0587.1995.tb00341.x.

Rahbek, C. (2005), "The role of spatial scale and the perception of large-scale species-richness patterns", *Ecology Letters*, Vol. 8 No. 2, pp. 224–239, https://doi.org/10.1111/j.1461-0248.2004.00701.x.

Ramírez-García, J., et al. (2015), "Multicriteria decision analysis applied to cover crop species and cultivars selection", *Field Crops Research*, Vol. 175, pp. 106–115, https://doi.org/10.1016/j.fcr.2015.02.008.

Ramos, R.F., et al. (2023), "Edaphic fauna in soil profile after three decades of different soil management and cover crops in a subtropical region", *Anais Da Academia Brasileira de Ciências*, Vol. 95, p. e20201667, https://doi.org/10.1590/0001-3765202320201667.

Reddy, C.S., et al. (2024), "Harnessing essential biodiversity variables and remote sensing of earth observations - synthesizing biodiversity insights", *Spatial Information Research*, Vol. 32 No. 3, pp. 265–276, https://doi.org/10.1007/s41324-023-00558-6.

Reeleder, R.D., et al. (2006), "Impacts of tillage, cover crop, and nitrogen on populations of earthworms, microarthropods, and soil fungi in a cultivated fragile soil", *Applied Soil Ecology*, Vol. 33 No. 3, pp. 243–257, https://doi.org/10.1016/j.apsoil.2005.10.006.

Rivers, A.N., et al. (2018), "Cover crop species and management influence predatory arthropods and predation in an organically managed, reduced-tillage cropping system", *Environmental Entomology*, Vol. 47 No. 2, pp. 340–355, https://doi.org/10.1093/ee/nvx149.

Roarty, S., et al. (2017). Earthworm populations in twelve cover crop and weed management combinations. *Applied Soil Ecology*, 114, 142–151. https://doi.org/10.1016/j.apsoil.2017.02.001.

Romanelli, C., et al. (2015), "Connecting global priorities: biodiversity and human health: a state of knowledge review", World Health Organistion/Secretariat of the UN Convention on Biological Diversity. https://www.who.int/publications/i/item/connecting-global-priorities-biodiversity-and-human-health.

Rowen, E.K., et al. (2020), "Is tillage beneficial or detrimental for insect and slug management? A meta-analysis", *Agriculture, Ecosystems & Environment*, Vol. 294, p.106849, https://doi.org/10.1016/j.agee.2020.106849.

Ruf, A. and Beck, L. (2005), "The use of predatory soil mites in ecological soil classification and assessment concepts, with perspectives for oribatid mites", *Ecotoxicology and Environmental Safety*, Vol. 62 No. 2, pp. 290–299, https://doi.org/10.1016/j.ecoenv.2005.03.029.

Ruf, A., et al. (2003), "A biological classification concept for the assessment of soil quality: 'biological soil classification scheme' (BBSK)", *Agriculture, Ecosystems & Environment*, Vol. 98 No. 1, pp. 263–271, https://doi.org/10.1016/S0167-8809(03)00086-0.

Rusek, J. (1998), "Biodiversity of Collembola and their functional role in the ecosystem", *Biodiversity & Conservation*, Vol. 7 No. 9, pp. 1207–1219, https://doi.org/10.1023/A:1008887817883.

Sáenz-Romo, M.G., et al. (2019), "Ground cover management in a Mediterranean vineyard: Impact on insect abundance and diversity", *Agriculture, Ecosystems & Environment*, Vol. 283, p. 106571, https://doi.org/10.1016/j.agee.2019.106571.

Salamon, J.-A., et al. (2004), "Effects of plant diversity on Collembola in an experimental grassland ecosystem", *Oikos*, Vol. 106 No. 1, pp. 51–60, https://doi.org/10.1111/j.0030-1299.2004.12905.x.

Sanguankeo, P.P. and León, R.G. (2011), "Weed management practices determine plant and arthropod diversity and seed predation in vineyards", *Weed Research*, Vol. 51 No. 4, pp. 404–412, https://doi.org/10.1111/j.1365-3180.2011.00853.x.

Santini, L., et al. (2017), "Assessing the suitability of diversity metrics to detect biodiversity change", *Biological Conservation*, Vol. 213, pp. 341–350, https://doi.org/10.1016/j.biocon.2016.08.024.

Sapkota, T.B., et al. (2012), "Fifteen years of no till increase soil organic matter, microbial biomass and arthropod diversity in cover crop-based arable cropping systems", *Agronomy for Sustainable Development*, Vol. 32 No. 4, pp. 853–863, https://doi.org/10.1007/s13593-011-0079-0.

Sardiñas, H., et al. (2018), *Maintaining Diverse Stands of Wildflowers Planted for Pollinators: Ongoing Management of Pollinator Habitat*. The Xerces Society for Invertebrate Conservation, Portland. 52 pp. https://xerces.org/sites/default/files/publications/16-028_02_MaintainingDiverseWildflowers_web_final.pdf.

Sarwar, M., et al. (1998), "Biofumigation potential of brassicas", *Plant and Soil*, Vol. 201 No. 1, pp. 103–112, https://doi.org/10.1023/A:1004381129991.

Saska, P., et al. (2013), "Temperature effects on pitfall catches of epigeal arthropods: a model and method for bias correction", *Journal of Applied Ecology*, Vol. 50 No. 1, pp. 181–189, https://doi.org/10.1111/1365-2664.12023.

Scavo, A., et al. (2022), "The role of cover crops in improving soil fertility and plant nutritional status in temperate climates. A review", *Agronomy for Sustainable Development*, Vol. 42 No. 5, p. 93, https://doi.org/10.1007/s13593-022-00825-0.

Scheunemann, N., et al. (2015), "The role of shoot residues *vs.* crop species for soil arthropod diversity and abundance of arable systems", *Soil Biology and Biochemistry*, Vol. 81, pp. 81–88, https://doi.org/10.1016/j.soilbio.2014.11.006.

Schleuter, D., et al. (2010), "A user's guide to functional diversity indices", *Ecological Monographs*, Vol. 80 No. 3, pp. 469–484, https://doi.org/10.1890/08-2225.1.

Schmera, D., et al. (2023), "Components of functional diversity revisited: a new classification and its theoretical and practical implications", *Ecology and Evolution*, Vol. 13 No. 10, p. e10614, https://doi.org/10.1002/ece3.10614.

Shannon, C.E. and Weaver, W. (1949), *The Mathematical Theory of Communication*, University of illinois Press Champaign, IL.

Shearin, A.F., et al. (2008), "Cover crop effects on the activity-density of the weed seed predator harpalus rufipes (Coleoptera: Carabidae)", *Weed Science*, Vol. 56 No. 3, pp. 442–450, https://doi.org/10.1614/WS-07-137.1.

Simpson, E.H. (1949), "Measurement of diversity", *Nature*, Vol. 163 No. 4148, pp. 688–688, https://doi.org/10.1038/163688a0.

Skidmore, A.K., et al. (2021), "Priority list of biodiversity metrics to observe from space", *Nature Ecology & Evolution*, Vol. 5 No. 7, pp. 896–906, https://doi.org/10.1038/s41559-021-01451-x.

Smith, S.D., et al. (2020), "Phylogenetics is the new genetics (for most of biodiversity)", *Trends in Ecology & Evolution*, Vol. 35 No. 5, pp. 415–425, https://doi.org/10.1016/j.tree.2020.01.005.

Soares, E.F., et al. (2023), "Poultry litter and different cover crops and their effects on the edaphic fauna community of an Oxisol", *DELOS: Desarrollo Local Sostenible, Curitiba*, Vol. 16 No. 43, pp. 683–698, https://doi.org/10.55905/rdelosv16.n43-013.

Sorenson, T., 1948. A method of establishing groups of equal amplitude in plant sociology based on similarity of species content and its application to analyses of the vegetation on Danish commons. *Biologiske Skrifter*, *5*, 1–34.

Soulé, M.E. (1985), "What is conservation biology?", *BioScience*, Vol. 35 No. 11, pp. 727–734, https://doi.org/10.2307/1310054.

Spake, R., et al. (2021), "Implications of scale dependence for cross-study syntheses of biodiversity differences", *Ecology Letters*, Vol. 24 No. 2, pp. 374–390, https://doi.org/10.1111/ele.13641.

Srivastava, D.S. and Vellend, M. (2005), "Biodiversity-ecosystem function research: is it relevant to conservation?", *Annual Review of Ecology, Evolution, and Systematics*, Vol. 36 No. 36, 2005, pp. 267–294, https://doi.org/10.1146/annurev.ecolsys.36.102003.152636.

Streit, R.P. and Bellwood, D.R. (2023), "To harness traits for ecology, let's abandon 'functionality'", *Trends in Ecology & Evolution*, Vol. 38 No. 5, pp. 402–411, https://doi.org/10.1016/j.tree.2022.11.009.

Swift, M.J., et al. (1979), *Decomposition in Terrestrial Ecosystems*, University of California Press, Berkeley.

Szczygłowska, M., et al. (2011), "Use of brassica plants in the phytoremediation and biofumigation processes", *International Journal of Molecular Sciences*, Vol. 12 No. 11, pp. 7760–7771, https://doi.org/10.3390/ijms12117760.

Taberlet, P. (1998), "Biodiversity at the intraspecific level: the comparative phylogeographic approach", *Journal of Biotechnology*, Vol. 64 No. 1, pp. 91–100, https://doi.org/10.1016/S0168-1656(98)00106-0.

Teasdale, J.R. and Mohler, C.L. (1993), "Light transmittance, soil temperature, and soil moisture under residue of hairy vetch and rye", *Agronomy Journal*, Vol. 85 No. 3, pp. 673–680, https://doi.org/10.2134/agronj1993.00021962008500030029x.

Tilman, D., et al. (2014), "Biodiversity and ecosystem functioning", *Annual Review of Ecology, Evolution, and Systematics*, Vol. 45 No. 45, 2014, pp. 471–493, https://doi.org/10.1146/annurev-ecolsys-120213-091917.

Triquet, C., et al. (2023), "In-field cover crop strips support carabid communities and shape the ecological trait repartition in maize fields", *Agricultural and Forest Entomology*, Vol. 25 No. 1, pp. 152–163, https://doi.org/10.1111/afe.12540.

Triquet, C., et al. (2024), "Undestroyed winter cover crop strips support wild bee abundance and diversity in intensive cropping systems", *Biodiversity Conservation*, Vol. 33, pp. 179–204, https://doi.org/10.1007/s10531-023-02741-5.

Tschanz, P., et al. (2024), "A review of soil tillage impacts on ground-nesting wild bees–mechanisms, implications, and future research perspectives", *Agriculture, Ecosystems & Environment*, Vol. 375, p.109224, https://doi.org/10.1016/j.agee.2024.109224.

USDA-NRCS (U.S. Department of Agriculture, Natural Resources Conservation Service). 2014. Revised Cover Crop Termination Guidelines. NRCS National Bulletin NB 450-15-1 TCH.

Usher, M.B., et al. (2006), "Understanding biological diversity in soil: the UK's Soil Biodiversity Research Programme", *Applied Soil Ecology*, Vol. 33 No. 2, pp. 101–113, https://doi.org/10.1016/j.apsoil.2006.03.006.

van der Putten, W.H., et al. (2023), "Soil biodiversity needs policy without borders", *Science*, Vol. 379 No. 6627, pp. 32–34, https://doi.org/10.1126/science.abn7248.

Vanolli, B.S., et al. (2024), "Edaphic mesofauna responses to land use change for sugarcane cultivation: insights from contrasting soil textures", *Frontiers in Ecology and Evolution*, Vol. 11, p. 1305115, https://doi.org/10.3389/fevo.2023.1305115.

Vaudo, A.D., et al. (2015), "Bee nutrition and floral resource restoration", *Current Opinion in Insect Science*, Vol. 10, pp. 133–141, https://doi.org/10.1016/j.cois.2015.05.008.

Villani, M.G., et al. (1999), "Adaptive strategies of edaphic arthropods", *Annual Review of Entomology*, Vol. 44 No. Volume 44, 1999, pp. 233–256, https://doi.org/10.1146/annurev.ento.44.1.233.

Vukicevich, E., et al. (2016), "Cover crops to increase soil microbial diversity and mitigate decline in perennial agriculture: a review", *Agronomy for Sustainable Development*, Vol. 36 No. 3, p. 48, https://doi.org/10.1007/s13593-016-0385-7.

Wallander, S., et al. (2021), "Cover crop trends, programs, and practices in the United States". efaidnbmnnnibpcajpcglclefindmkaj/https://www.ers.usda.gov/sites/default/files/_laserfiche/publications/100551/EIB-222.pdf?v=23008.

Wang, J., et al. (2021), "A meta-analysis on cover crop impact on soil water storage, succeeding crop yield, and water-use efficiency", *Agricultural Water Management*, Vol. 256, p. 107085, https://doi.org/10.1016/j.agwat.2021.107085.

Ward, M.J., et al. (2011), "Cover crops and disturbance influence activity-density of weed seed predators *Amara aenea* and *Harpalus pensylvanicus* (Coleoptera: Carabidae)", *Weed Science*, Vol. 59 No. 1, pp. 76–81, https://doi.org/10.1614/WS-D-10-00065.1.

Wardle, D.A. (2002), *Communities and Ecosystems: Linking the Aboveground and Belowground Components*, Princeton University Press, Princeton.

Weiss, F., et al. (2024), "Evidence for regional-scale declines in carabid beetles in old lowland beech forests following a period of severe drought", *Landscape Ecology*, Vol. 39 No. 7, p. 123, https://doi.org/10.1007/s10980-024-01920-1.

Whittaker, R.H. (1960), "Vegetation of the Siskiyou Mountains, Oregon and California", *Ecological Monographs*, Vol. 30 No. 3, pp. 279–338, https://doi.org/10.2307/1943563.

Wiens, J.A. (1989), "Spatial Scaling in Ecology", *Functional Ecology*, Vol. 3 No. 4, pp. 385–397, https://doi.org/10.2307/2389612.

Willig, M.R., et al. (2003), "Latitudinal gradients of biodiversity: pattern, process, scale, and synthesis", *Annual Review of Ecology, Evolution, and Systematics*, Annual Reviews, Vol. 34 No. 34, pp. 273–309, https://doi.org/10.1146/annurev.ecolsys.34.012103.144032.

Willis, K.J. and Birks, H.J.B. (2006), "What is natural? The need for a long-term perspective in biodiversity conservation", *Science*, Vol. 314 No. 5803, pp. 1261–1265, https://doi.org/10.1126/science.1122667.

Wilson, E.O. (1985), "The biological diversity crisis", *BioScience*, Vol. 35 No. 11, pp. 700–706, https://doi.org/10.2307/1310051.

Wilson, E.O. (1989), "Threats to biodiversity", *Scientific American*, Vol. 261, pp. 108–116, https://doi.org/10.1038/scientificamerican0989-108.

Winfree, R. (2010), "The conservation and restoration of wild bees", *Annals of the New York Academy of Sciences*, Vol. 1195 No. 1, pp. 169–197, https://doi.org/10.1111/j.1749-6632.2010.05449.x.

Woodcock, B.A., et al. (2010), "Impact of habitat type and landscape structure on biomass, species richness and functional diversity of ground beetles", *Agriculture, Ecosystems & Environment*, Vol. 139 No. 1, pp. 181–186, https://doi.org/10.1016/j.agee.2010.07.018.

Wright, A.J., et al. (2017), "The overlooked role of facilitation in biodiversity experiments", *Trends in Ecology & Evolution*, Vol. 32 No. 5, pp. 383–390, https://doi.org/10.1016/j.tree.2017.02.011.

Zhang, C., et al. (2014), "Root foraging elicits niche complementarity-dependent yield advantage in the ancient 'three sisters' (maize/bean/squash) polyculture", *Annals of Botany*, Vol. 114 No. 8, pp. 1719–1733, https://doi.org/10.1093/aob/mcu191.

Chapter 4

Cover crops for arthropod pest management

Nicola A. Irvin, University of California-Riverside, USA; Ivan Milosavljević, University of California-Riverside and Citrus Research Board, USA; and Mark S. Hoddle, University of California-Riverside, USA

1 Introduction

Over the last six decades, agriculture has intensified, in part through the widespread use of agrochemicals, which has produced higher yields, but has also resulted in significant long-term negative consequences such as environmental pollution; development of pesticide-resistant insects, pathogens, and weeds; and loss of biodiversity due to monocultures, which collectively is expressed as weakened ecosystem resilience (Rockstrom et al., 2017). The reliance on pesticides has caused contamination of soil and water, disruption of ecosystem services provided by non-target organisms, and potential health risks to humans through direct exposure during applications and indirectly via residues in food (Claeys et al., 2011).

As research increasingly focuses on sustainable agricultural methods, cover crops have emerged as a promising tool for pest management,

http://dx.doi.org/10.19103/AS.2025.0162.06

motivated by the need to reduce reliance on insecticides, sequester carbon to combat climate change, and prevent soil health decline (Kaye and Quemada, 2017; Quintarelli et al., 2022). Cover cropping can be defined as the deliberate planting of annual or perennial non-crop plants, before, during, or after the primary crop, which cover the soil for a single season or throughout the entire year (González-Chang et al., 2019). While traditionally used to improve soil health, cover crops are gaining increased recognition for their multifaceted role in pest management, offering a natural and environmentally acceptable alternative to synthetic pesticides.

This chapter explores the evolving role of cover crops as a tool for managing arthropod pests (i.e. insects and mites), examines the underlying ecological mechanisms that drive the effectiveness of this approach, and explores the potential benefits and challenges associated with implementing cover crops for pest control. By enriching habitat complexity and increasing biodiversity, cover crops can serve as a cornerstone of integrated pest management (IPM) programs because of the enhancement of natural enemies that attack crop pests. This chapter will delve into the diverse strategies through which cover crops contribute to suppression of pest populations, including habitat provision for beneficial insects, trap cropping, allelopathic effects, and the disruption of pest behavior and reproduction. In addition, we will discuss practical considerations like cover crop species selection, timing of seed sowing, and plant management (e.g. mowing, irrigation, termination) to illustrate how cover crops can be integrated effectively and optimized within various cropping systems for pest control.

To illustrate these concepts with real-world applications, this chapter presents a case study from commercial citrus orchards in southern California, showcasing specific mechanisms, challenges, and outcomes of cover crop use in an arid perennial cropping system. This chapter provides insights into the strategic use of cover crops for arthropod pest management and offers recommendations for future research directions.

2 How cover crops affect pest populations: mechanisms

Agricultural crops are often grown in monocultures, creating ecosystems with significantly simpler and depauperate structures compared with natural biodiverse ecosystems (Crews et al., 2018). These simplified agroecosystems tend to support conditions favorable to agricultural pests which can be due to limited species diversity. Ecosystem resilience is directly linked to the diversity of beneficial insects, which help maintain ecological balance, including moderation of pest population densities (Rusch et al., 2016). Increasing habitat diversity has been proposed as a strategy to favor beneficial insects, such as pollinators, and natural enemies that regulate pest populations, which in turn assist with restoration and diversification of ecosystem services (Wilkinson and Landis, 2005; Tittonell, 2014). In terms of agroecosystem diversification, cover crops provide a management tool for growers that can be used to enhance habitat diversity and ecological complexity within the cropping system, which can promote suppression of pest populations.

Cover crops can influence plant feeding pests via two key mechanisms. The 'resource concentration hypothesis' (Root, 1973), also known as 'bottom-up' effects, can promote pest population growth because of abundant concentrated food sources (Hunter and Price, 1992). In contrast, the 'enemies hypothesis,' also known as 'top-down' control (Hunter and Price, 1992), suppresses pest population growth and can be achieved through the enhancement of natural enemies that consume pests (Russell, 1989). When cover crops are integrated into agricultural systems, bottom-up or top-down effects can occur independently or simultaneously, depending on interrelated factors such as plant, pest, and natural enemy species, environmental conditions, and management practices (Gratton and Denno, 2003; Wang et al., 2020).

2.1 The enemies hypothesis

Natural enemies of arthropods pests include predators, parasitoids, and pathogens that can exert sufficient 'top-down' pressure and suppression of pest population growth (Fig. 1). Many predator species are generalists, feeding on a variety of pest species. Examples are coccinellid beetles (ladybugs, ladybird beetles) that feed on aphids (Aphididae), whiteflies (Aleyrodidae), scales (e.g. Diaspididae), mites (e.g. Tetranychidae), and mealybugs (Pseudococcidae). Predatory hoverfly larvae (Syrphidae) feed on a variety of soft-bodied insects, and some, like *Allograpta obliqua* (Say), feed on psyllids (Liviidae and Triozidae). Predatory mites (Phytoseiidae), another important predator guild, feed on pestiferous mites, thrips (Thripidae), and fungus gnat larvae (Mycetophilidae).

Wasps (Hymenoptera) can be predatory or parasitic. For example, predatory crabronid wasps paralyze a variety of insect prey and transport them back to nests where they provide food for wasp larvae. Parasitoids lay eggs in or on a host, and larvae develop by feeding on the host, eventually killing it. Common examples of parasitoids include various *Trichogramma* spp. that attack eggs of caterpillars, *Encarsia* and *Eretmocerus* spp., which are parasitoids of whiteflies, and *Tamarixia* spp. that attack psyllids (Fig. 1).

The enemies hypothesis predicts that the inclusion of non-crop resources in agricultural systems, like cover crops, can increase the abundance, diversity, and efficacy of natural enemies, and as a consequence of enhanced predator and parasitoid populations, pest populations are reduced through predation and parasitism, which also increases natural enemy populations (Landis et al.,

Figure 1 Natural enemies on sown cover crops: (a) predatory crabronid wasp on *Plantago lanceolata* (Photo credit: W. B. Westbury); (b) hoverfly (*Toxomerus marginatus*) on an alyssum leaf (Mike Lewis); (c) whitefly parasitoid, *Eretmocerus eremicus* (Mike Rose); (d) Asian citrus psyllid parasitoid, *Tamarixia radiata*, feeding on buckwheat nectar (Mike Lewis); (e) seven-spotted ladybug (*Coccinella septempunctata*) on buckwheat flowers (Mike Lewis); (f) hoverfly (*Allograpta obliqua*) feeding on an alyssum flower (Mike Lewis).

2000; Balmer et al., 2013; Gurr et al., 2017). Resources needed to enhance natural enemy populations in cropping systems include **s**helter (refugia for overwintering); a sugar source, like **n**ectar, that provides energy; sources of protein, such as **a**lternative food (prey or hosts); and **p**ollen that promotes egg development. These requirements have been summarized as the acronym SNAP (Landis et al., 2000; Barnes et al., 2009; Wäckers and van Rijn, 2012).

Cover crops that promote SNAP can play an important role in attracting and sustaining natural enemies before the arrival of key insect pests in agricultural fields. Certain generalist predators, such as spiders (Araneae), ground beetles (Carabidae), and rove beetles (Staphylinidae), which are not very mobile and produce only one or two generations per year, benefit from undisturbed vegetation and alternative prey near the crop which allows them to build up populations prior to moving into crops once pests arrive (Reddy, 2017).

Once established, natural enemy resources, like floral and extrafloral nectar and pollen, help increase the longevity, fecundity, searching activity, and parasitism/predator rates of most natural enemies, and natural enemy sex ratios, especially for parasitoids that can become more female biased which increases rates of natural enemy population growth (Berndt and Wratten, 2005; Kost and Heil, 2005; Irvin et al., 2006; Hogg et al., 2011a). Additionally, non-prey foods, like nectar and pollen, support survival, diapause, dispersal, and reproduction of some predators such as ladybugs (Lundgren, 2009). Honeydew, a sugary waste product excreted by sap-sucking hemipterans like aphids, which sometimes inhabit cover crops, can also be consumed by natural enemies, thereby enhancing their fitness (Tylianakis et al., 2004; Irvin and Hoddle, 2007).

Furthermore, structural diversity provided by cover crops can moderate crop microclimates that are unfavorable to natural enemies by providing shade between crop rows, lowering ground temperature, and raising humidity under the foliage canopy (Diehl et al., 2012). For example, grasses forming dense tussocks, such as orchard grass (*Dactylis glomerata* L.) and tall fescue (*Schedonorus (Festuca) arundinaceus* L.), can be grown between rows of citrus trees to provide microclimate, shelter, and alternative prey and hosts for natural enemies (Mockford et al., 2023, 2024) (Fig. 2).

Additionally, an insulating habitat provided by cover crops enables natural enemies to overwinter more successfully, and the greater number of survivors

Figure 2 Grasses that form dense tussocks, such as orchard grass (*Dactylis glomerata* L.) and tall fescue (*Schedonorus (Festuca) arundinaceus* L.), provide favorable microclimate, shelter, and alternative prey and hosts for natural enemies in an orange orchard in Spain (Mockford 2023, 2024). Photo credit: W. B. Westbury.

can quickly colonize crop plants at the beginning of each growing season (Lys et al., 1994; Boinot et al., 2019). For example, grass cover crop strips between fields of annual crops like corn (*Zea mays* L.) can provide critical overwintering sites for natural enemy species like ground beetles, ensuring their presence in the following growing season (Reddy, 2017). By creating favorable habitats and providing essential SNAP resources, cover crops not only attract and retain natural enemies in agricultural systems but also improve natural enemy fitness, efficacy, and overwintering survivorship rates (see '2.3.1 Flowering insectary plants as cover crops').

2.2 The resource concentration hypothesis

The resource concentration hypothesis states that insect pests are more likely to thrive and cause damage in areas where their host plants are concentrated. This hypothesis posits that when these resources (i.e. the crop) are found in dense, concentrated patches, insect pests are more likely to locate and exploit them. Therefore, increasing concentrations of preferred resources attracts pests, leading to higher rates of infestation and increased pest damage (Root, 1973). Additionally, crop monocultures emit stronger and more uniform chemical signals or volatiles (kairomones) that attract herbivorous pests, facilitating pest location, exploitation, and population growth (Bruce, 2005). With a large, uninterrupted food source, pests experience ideal conditions for feeding and reproduction, and they can more easily move from plant to plant, leading to rapid pest population growth.

Resource concentration can be reduced by increasing diversity within cropping systems. Increased plant diversity, for example, makes it more difficult for herbivores to locate their preferred hosts, which reduces damage to the primary crop (Root, 1973). This occurs because the concentration of the preferred 'resource' – the crop – is effectively diluted, creating less favorable conditions for pest population growth. The movement and host-finding ability of herbivores can be disrupted by a mixture of plant structures, colors, textures, leaf shapes, and volatile odors, which can be important cues that pests need to find their preferred host plant, the crop (Bruce, 2005; Reddy, 2017). Cover crops can visually or physically conceal the primary crop, making it more difficult for pests to locate their host plants (Reddy, 2017).

Fragmenting concentrated resources can have additional deleterious effects on pests. For example, herbivorous insects require specific conditions before egg laying occurs, such as regular contact with a preferred host plant in a specific growth phase. The addition of non-host vegetation which is frequently encountered by gravid female pests can detrimentally alter their oviposition behavior (Hooks and Johnson, 2003). For example, in the presence of non-host cover crops, female pests can change the arrangement, location, or number of eggs oviposited on the primary crop (Hern et al., 1996). In diverse systems, pests are more likely to land on unsuitable plants and possibly move on from the field if preferred hosts are not encountered frequently enough (Finch and Collier, 2000). The additional time spent by pests searching and probing non-crop plants can reduce damage to the primary crop and increase the likelihood that pests will be detected and consumed by predators before finding suitable host plants to exploit (Straub et al., 2014; Reddy, 2017).

Cover crops can also alter the microclimate or the quality of the primary crop, making it less favorable for the pest. For example, a native perennial grass cover crop planted in northern California vineyards contributed to a reduction in leafhopper pests by reducing vine petiole nitrate levels through competition for soil nitrogen, with no apparent effects on berry weight or sugar content (Daane et al., 2018).

Diverse non-crop plantings can also influence insect behavior through the release of plant volatiles which can attract or repel specific herbivores and natural enemies (Finch, 1976; Hooks and Johnson, 2003). Volatiles released by plants such as onion (*Allium cepa*), garlic (*Allium sativum*), marigolds (*Tagetes* spp.), billygoat weed (*Ageratum conyzoides* L.), *Ageratum* spp., summer savory (*Satureja hortensis* L.), and catnip (*Nepeta cataria* L.) can interfere with host plant location by masking host plant odors (Reddy, 2017; Huang et al., 2020; Song and Han, 2020).

Masking of host plants by volatiles released by non-crop plants was demonstrated for carrot rust fly, *Sila rosae* (F.). Semiochemicals (chemicals produced by organisms to communicate with each other) in the surface wax of carrot leaves stimulate carrot rust fly oviposition (egg-laying) (Degen et al., 1999a, 1999b). This oviposition stimulant is masked by non-crop aromatic plants, such as coriander (*Coriandrum sativum* L.), which reduces the abundance of this pest (Jankowska and Wojciechowicz-Żytko, 2016). Similarly, planting coriander with carrot decreased numbers of pestiferous aphids and *Lygus* spp. on leaves by up to 68% and reduced carrot root damage by up to 97% (Jankowska and Wojciechowicz-Żytko, 2016).

Cover crops can 'trap' pests. Trap crops are plants grown alongside the primary crop to attract, intercept, or retain insect pests (see '2.3.2 Trap cropping'). Trap crops can also indirectly reduce pest abundance by disrupting the continuity of host plants that pests rely on for feeding and reproduction (Clark, 2007).

2.3 Mechanisms associated with flowering insectary plants and trap cropping

Flowering insectary plants and trap cropping can reduce arthropod pest populations through various mechanisms, as described below.

2.3.1 Flowering insectary plants as cover crops

Insectary plants are nectar-providing plant species that attract, retain, and enhance natural enemies by potentially providing SNAP. Insectary plantings have been used in perennial orchards and vineyards, and in annual crops like vegetables, legumes, and cereals (Berndt et al., 2002; Lee and Heimpel, 2005; Irvin et al., 2006; Gurr et al., 2016; Zhu et al., 2018; Park et al., 2024). Flowering cover crops can also provide additional ecosystem services such as attracting pollinators to the crop, weed control, aesthetics, and an overall sense of rural well-being (Wratten et al., 2012). Increased aesthetics and an enhanced sense of well-being can lead to added crop value as a result of increased consumer perceptions of improved environmental quality associated with farming practices that support biodiversity, sustainability, and conservation (Warner, 2007; Delmas and Grant, 2014).

An increase in the abundance of parasitoids occurred when flowering insectary plants, such as buckwheat (*Fagopyrum esculentum* Moench) (Fig. 1e), alyssum (*Lobularia maritima* L. Desv.) (Fig. 1f), and coriander, were incorporated in apple (*Malus domestica* [Suckow] Borkh.) orchards and vineyards (Irvin, 1999; Berndt et al., 2002), leading to increased levels of parasitism of pest lepidopteran eggs (Begum et al., 2006). When key natural enemies had access to nectar-producing insectary plants in laboratory studies, the increased parasitism was attributed to enhanced longevity and/or fecundity of the natural enemies (Johanowicz and Mitchell, 2000; Vattala, et al., 2006; Begum et al., 2006).

In legume production, buckwheat grown in strips in soybean (*Glycine max* [L.] Merr.) fields in the Republic of Korea (i.e. South Korea) led to enhanced levels of minute pirate bug predators (*Orius* spp.) and crab spiders (Thomisidae), lower levels of pestiferous adult alydids [*Riptortus pedestris* (F.)], but higher levels of two other pests, agromyzid leafminers and cicadellid leafhoppers (Park et al., 2024). Since buckwheat is known to increase the oviposition rate and survival of minute pirate bugs (Hinds and Barbercheck, 2020), buckwheat strips may have enhanced the reproduction of minute pirate bugs or attracted predators to the cover crop treatment plots.

Interestingly, only numbers of alydid adults differed between the buckwheat treatment and monoculture controls, while there was no difference in alydid nymph density. This selective life stage–specific reduction may imply a potential 'bottom-up' effect due to enhanced plant diversity or to semiochemicals emitted from buckwheat that reduced the pest's ability to locate its host plant (Park et al., 2024). Since alydids typically locate soybeans by detecting host plant volatiles (Song et al., 2022), buckwheat strips may have created 'confusion' in the host plant recognition behavior of this insect, reducing the number of adults finding host plants in soybean fields (Park et al., 2024).

In cereal production, rice (*Oryza* spp.) farmers in Vietnam, Thailand, and China sow non-crop plants on rice bunds, the levee built around rice fields, to enhance biological control of planthoppers and lepidopteran pests (Gurr et al., 2016; Zhu et al., 2018). Zhu et al. (2018) found that sesame (*Sesamum indicum* L.) increased parasitoid populations, *Anagrus* spp., and lowered populations of the brown planthopper (*Nilaparvata lugens* Stål.) compared to control plots without sesame. Egg parasitoids of planthoppers have been observed to increase parasitism rates when they feed on sesame flowers (Zhu et al., 2013). Furthermore, planting sesame with vetiver grass (*Vetiveria zizanioides* L.) was associated with a fourfold increase in *Anagrus* spp. parasitoids and reduced populations of rice stem borers (*Chilo suppressalis*) (Lu et al., 2015).

Stem borer adults are attracted to the vetiver grass where females then lay eggs. However, this plant does not support complete development, making it an effective trap crop (see '2.3.2 Trap cropping'). This mixed system combines nectar-producing plants ('top-down') and trap cropping ('bottom-up') which can reduce grower reliance on insecticides with concomitant increases in crop yields. In a cost-benefit analysis in Asian rice production areas that spanned 16 sites over 4 years, this approach reduced insecticide applications by 70%, increased grain yields by 5%, and delivered a 7.5% financial advantage when compared to typical rice farm practices (Gurr et al., 2016).

2.3.2 Trap cropping

Trap cropping is growing non-crop plants next to the primary crop to attract, intercept, or retain insect pests that threaten the primary crop. Trap cropping was practiced over 200 years ago in Europe where trap trees or wooden logs were used to attract and trap the spruce bark beetle, *Ips typographus* DeGeer (Bakke and Riege, 1982). More recently, trap cropping has been commonly used to target beetles (Coleoptera), flies (Diptera), caterpillars (Lepidoptera), thrips (Thysanoptera), grasshoppers (Orthoptera), and sap-sucking bugs (Hemiptera) (Hokkanen, 1991; Rea et al., 2002; Reddy, 2017; Sarkar et al., 2018; Tiwari et al., 2018; Badenes-Pérez, 2019).

Trap cropping for diamondback moth (*Plutella xylostella* L.) has been widely studied, and insecticide inputs have been significantly reduced through growing trap crops around the perimeter of primary crops susceptible to this pest (Wan et al., 2016; Badenes-Pérez, 2019). Mitchell et al. (2000) found that diamondback moth infestations in cabbage (*Brassica oleracea* L.) grown in Florida were effectively controlled by using a non-crop collard (*B. oleracea* var. *acephala* L.) as a trap crop planted around field margins. Female diamondback moths significantly preferred to oviposit in trap crop collards rather than the primary crop, cabbage, thereby preventing movement into the cabbage crop (Mitchell et al., 2000). Additionally, parasitism by *Diadegma insulare* (Cresson) was higher in diamondback moth larvae collected in the trap collards than for larvae in cabbage (Mitchell et al., 1997). In cabbage fields protected by trap crops, the number of insecticide applications was reduced by up to 63%, and this translated to an insecticide treatment savings of up to $156 per hectare compared to conventional fields without trap crops (Mitchell et al., 2000).

Control of *Lygus* spp. in lettuce, *Lactuca sativa* (L.), is difficult because this pest is protected between the leaves of lettuce making them difficult to reach with insecticides (Rämert et al., 2001). In Sweden, trap crops of either red clover (*Trifolium pratense* L.), lucerne (*Medicago sativa* L.), or common vetch (*Vicia sativa* L.) contained up to 40-fold higher *Lygus* spp. populations than in iceberg lettuce, making these plants ideal trap crops (Rämert et al., 2001). Modeling studies indicated that a fivefold difference in preference between the trap crop and primary crop can yield a 460% reduction in pest densities (Banks and Ekbom, 1999).

Trap crops can be established within fields, at the farm scale, or across landscapes (Singhal et al., 2020), and pest populations are reduced via different mechanisms that are dependent on the scale of the trap cropping approach. At the within-field scale, trap rows or patches can intercept pests as they move through the crop, concentrating them in localized areas where they can be destroyed by applying targeted insecticide applications to the trap crop, thus eliminating application on the primary crop, or by physical destruction of the trap crop and associated pests through tillage, burning (Reddy, 2017), or vacuuming (Wells et al., 2020). At the farm scale, larger blocks of trap crops can act as 'sinks,' drawing pests away from the primary crop and reducing the need for broad-spectrum sprays. At the landscape scale, coordinated use of trap crops among neighboring farms can slow regional pest spread and provide area-wide suppression, especially for pests with limited dispersal ability. Trap crops are most suitable for herbivores which are reasonably sedentary, compared to highly mobile pests or those that are wind dispersed when disturbed (Reddy, 2017).

For example, minimal (< 1% of total crop area) plantings of squash trap crop plants treated with systemic insecticides can be used to attract and kill squash bug, *Anasa tristis* (De Geer), and cucumber beetle, *Acalymma vittatwn* (F.), as they emerge from overwintering sites, effectively protecting the primary watermelon (*Citrullus lanatus* (Thunb.) Matsum. and Nakai), squash (*Cucurbita* spp.), and cantaloupe (*Cucumis melo* L.) crops (Pair, 1997).

Trap crops intended to divert pests from primary crops can sometimes result in unintended consequences if pests reach high densities on trap crops and then spill over onto the primary crop (Luther et al., 1996). This spill over can be mitigated through the selection of 'dead end' trap crops that are highly attractive to pests but do not support offspring development (Shelton and Badenes-Pérez, 2006). Dead-end trap crops are usually located at crop edges and are highly preferred for egg laying by lepidopteran pests (Thompson and Pellmyr, 1991).

For example, the 'push-pull' system developed for maize (*Zea mays* L.) production in East Africa uses two grass species belonging to the family Poacae as trap crops to control maize stem borer *Chilo partellus* Swinhoe. Napier grass, *Pennisetum purpureum* Schumach, or Sudan grass, *Sorghum vulgare sudanese* Pers., is preferred for oviposition, therefore diverting maize stem borer females from ovipositing in maize crops (pull factor) (Khan et al., 2001, 2010, 2014). Offspring emerging on Napier grass did not survive because the plant is of low nutritive value for the larvae and produces a gummy substance that immobilizes larvae and prevents feeding (Khan et al., 2014). Molasses grass, *Melinis* spp., was selected as a repellent (push) crop since it neither attracted oviposition by stemborer moths nor supported survival of the young larvae (Khan et al., 2014). The enemies hypothesis also contributed to pest control since this push-pull system resulted in increased abundance and diversity of natural enemies, partially due to the reduced use of insecticides needed to control maize stem borer (Khan et al., 2014).

In some cases, a single plant species can serve as both a trap crop for pests and a food source for natural enemies. For example, yellow rocket (*Barbarea vulgaris* W.T. Aiton) functions as a floral resource for the predatory hoverfly *Sphaerophoria scripta* L. while also acting as a trap crop for its prey, the diamondback moth (Badenes-Pérez 2019). Detailed and updated lists of target insect species, their potential trap crops, and application strategies have been published by Hokkanen (1991), Shelton and Badenes-Pérez (2006), Piñero and Manandhar (2015), Reddy (2017), Sarkar et al. (2018), and Badenes-Pérez (2019).

3 Selecting a cover crop for arthropod pest management

When selecting and planning a cover crop for pest management in perennial and annual cropping systems, consider the following:

1 Identification of key pests affecting the primary crop and determination of the phenology of these pests and their associated natural enemy species. Select a cover crop that will provide ecosystem services during critical phenological times when pest control is needed. Consider whether a cool season or warm season cover crop is required. Information on growth time required to flowering and duration of nectar production, for example, may be important for synchronizing plant produced resources with the phenology of natural enemies attacking key pests infesting the primary crop (Irvin et al., 2014, 2021).

2 Determine whether the goal is to repel pests, provide a physical or chemical barrier to pests, or attract and enhance natural enemies (see '2 How cover crops affect pest populations: mechanisms').
3 Evaluate site conditions such as water availability and soil type and quality, since different cover crops thrive in specific soil types and pH ranges (McCauley et al., 2017). Cover-crop mixes have been assessed for their adaptability to different soil types and topographies (Breil, 1999).
4 Consider the local climate and how the cover crop will grow throughout the year. For example, cover crop species recommendations change depending on geographic location (Cover Crop Resources, 2024a). Choose a cover crop species that matches the climate zone of the site (Sarrantonio, 2007; USDA, 2023a) and ensure the cover crop doesn't have weed status in the area of interest. For example, Indian mustard (*Brassica juncea* (L.) Czern.) and *Desmodium intortium* have been used to control lepidopteran pests in East Africa and Sweden, respectively, but are considered weeds in Australia (Oram et al., 2005; Rizvi et al., 2022).
5 Assess compatibility with the primary crop including the height of the cover crop and its competitiveness to ensure it fits well into the primary crop maintenance schedule. Plant height information, for example, can be useful when selecting cover crops with the aim of preventing moisture-dependent diseases developing in the primary crop. Tall cover crops can impede air movement, leading to increased disease or frost risk (McGourty, 2004). The age of a perennial primary crop can also affect cover crop choice and compatibility. Vineyards, for example, may require the cover crop to increase or decrease vegetative growth depending on the age of the vineyard, vegetative state of the vines, and time of year (McGourty, 2004).
6 Determine the previous use of herbicides and whether pre-emergent applications can interfere with sowing a cover crop. Some cover crop species, such as buckwheat, are sensitive to herbicide residues in soil (Irvin et al., 2022, see '7 Case study: the use of cover crops in citrus orchards in Southern California'). Research in corn and soybean shows that cereal rye (*Secale cereale* L.), wheat (*Triticum aestivum* L.), oats (*Avena sativa* L.), and camelina (*Camelina sativa* [L.] Crantz) tend to tolerate many commonly used herbicides, while establishment of red clover can be poor due to sensitivity to herbicide residues (Stahl, 2022). For herbicide-sensitive cover crop species, the time between herbicide applications and cover crop seeding should be lengthened to minimize injury to the cover crop (Stahl, 2022).
7 Determine the cover crop sowing date, flowering date, and termination goals and ensure these complement the life cycle of the primary crop

and disadvantage pest population growth. Species of cover crops that flower early or quickly such as 'Abruzzi' rye, crimson clover (*Trifolium incarnatum* L.), and buckwheat can be useful for situations requiring quick-growing cover crops that provide resources to natural enemies (Adamson et al., 2016). Leaving non-living cover crops in the field at the end of their life cycle can provide habitat for ground-dwelling predators such as spiders, ground and rove beetles, big-eyed bugs (*Geocoris* spp.), and minute pirate bugs (Adamson et al., 2016).

8 Determine if the cover crop will be left to set seed, or whether it will be tilled or mowed prior to seed set to ensure seeds do not set, which could result in the cover crop becoming weedy (Adamson et al., 2016) (see '4.2 Cover crop termination').

9 Consider the factors that affect placement of the cover crop such as competitiveness of the cover crop with the primary crop and whether the goal is to repel pests, provide a physical or volatile barrier to pests, or attract and enhance natural enemies. The cover crop can be placed in between crop rows, in the tree line of a perennial orchard, around field margins, or covering an entire field between rotations of an annual crop (see '4.1 Cover crop placement and timing'). Growers should sow cover crops in a place that limits competition with the primary crop and monitor soil moisture or primary crop stress to minimize competition for water between the cover crop and primary crop (Almond Board of California, 2020).

10 Consider the labor and resource costs for sowing and management of the cover crop. Select a cover crop that can be easily sown and maintained relatively cheaply, given equipment availability and primary crop management (Cover Crop Resources, 2024a, 2024b).

3.1 Perennial vs. annual agricultural cropping systems

Cover crop species recommendations often vary depending on whether the primary cropping system is perennial (e.g. orchards or vineyards) or annual/rotation-based (e.g. row crops). Each system presents unique challenges and opportunities for integrating cover crops effectively. Perennial cover crops offer several advantages over their annual counterparts when used in perennial agricultural systems. Unlike annuals, perennial cover crops do not require yearly sowing and can provide continuous habitat and SNAP resources to natural enemies throughout the year. They also develop more resource-rich environments over time and typically exhibit longer flowering periods compared to annual cover crops (Fiedler and Landis, 2007a). In contrast, the effectiveness of annual cover crops in enhancing natural enemy populations may rely on the proximity of permanent undisturbed habitats. Annual cover crops require spillover of natural enemies from semi-natural habitats that offer critical overwintering sites and other resources not provided by the seasonal nature of annual cover crops (Landis et al., 2000; Haenke et al., 2014).

This spillover effect was demonstrated with *Anagrus* spp., key egg parasitoids of the western grape leafhopper (*Erythroneura elegantula* Osborn) in northern California vineyards. The western grape leafhopper overwinters as an adult in reproductive diapause. Therefore, to successfully overwinter, *Anagrus* spp. must find an alternate leafhopper host that overwinters in an egg stage in nearby non-crop landscapes (Wilson et al., 2016). Therefore, while flowering cover crops in vineyards attracted natural enemies, the proximity of the vineyard to *Anagrus* spp. overwintering habitat was most important for enhancing early-season abundance of *Anagrus* spp. and leafhopper biocontrol (Wilson et al., 2017).

In annual cropping systems, cover crops can be used to disrupt pest and disease life cycles, and sowing must be timed to ensure growth fits within the growing season so planting or harvesting schedules are not disrupted. Quick-growing cover crops, such as buckwheat, annual ryegrass (*Lolium multiflorum* Lam.), and mustard (white mustard, *Sinapis alba* L., and Indian mustard), are particularly well-suited for these systems due to their rapid establishment, which provides SNAP benefits in a short timeframe (Clark, 2007; Irvin et al., 2014). The timing of cover crop planting is crucial since planting early relative to the primary crop can enhance pest control by providing a sufficiently tall canopy to camouflage the crop from foraging herbivorous pests (Weiss et al., 1994; Hooks and Johnson, 2003).

Other attributes to consider when choosing a cover crop in an annual or perennial system are the ease, reliability, and cost of attaining seed; the costs of establishing, maintaining, and potentially irrigating the cover crop; whether the cover crop species are easy to grow; their potential to become a weed and attract pests or host pathogens; whether the cover crop is of value and could be harvested and sold; and how easily the cover crop can fit into existing farming schemes (Norris and Kogan, 2000; Buchanan et al., 2018; Irvin et al., 2021).

3.2 Considerations for selecting insectary plants as cover crops

3.2.1 Screening cover crops for their attractiveness and resource quality to natural enemies

When selecting insectary plants that benefit natural enemies, it is important to screen plants for their attractiveness and the resources they provide to predators and parasitoids (Fig. 3). Some floral resources can inadvertently harm natural enemies (Goleva and Zebitz, 2013) or fail to satisfy the nutritional needs of predators and parasitoids (Davies et al., 2004, Karp et al., 2018). Key floral

Figure 3 Field (a) and laboratory (b) experiments at the University of California, Riverside, evaluating cover crop species to assess their suitability as nectar-providing plants for enhancing natural enemies of citrus pests. Photo credit: N. A. Irvin.

traits such as flower color, size, structure, nectar quality, floral area, and timing of nectar availability play a crucial role in determining a plant's attractiveness and the benefits it may provide to natural enemies (Fiedler and Landis, 2007b; Lundin et al., 2019). For example, the length of the corolla or floral tube can restrict access to nectar, preventing certain natural enemies from accessing nectar, and some nectars may be nutritionally inadequate and fail to benefit natural enemies (Patt et al., 1997; Wäckers, 2001; Irvin et al., 2021).

Research demonstrates these nuances: Buckwheat (open flower) can significantly enhance the longevity and fecundity of *Cosmocomoidea* (formerly *Gonatocerus*) spp., mymarid egg parasitoids of glassy-winged sharpshooter, *Homalodisca vitripennis* (Germar), while *Phacelia tanacetifolia* Benth. (tubular flower) provides minimal support, performing no better than water alone because parasitoids are unable to reach nectar (Irvin and Hoddle, 2007; Irvin et al., 2007). Similarly, a review of 70 studies concluded that flowers with open or exposed nectaries are more likely to prolong longevity of natural enemies (He et al., 2021). For further exploration of these key principles, refer to '7 Case study: the use of cover crops in citrus orchards in Southern California.'

Ideally, evaluation of floral resources provided by potential cover crop species, such as floral attractiveness and resource accessibility (e.g. nectar), should be conducted by assessing effects on key natural enemies in the specific crop and region of interest, grown under prevailing field conditions, so that abiotic influences on natural nectar secretion rhythms, for example, can be determined (Fig. 3). Most flowers do not secrete nectar continuously, but only during fixed periods of the day or the night. For example, buckwheat secretes floral nectar in the morning, which aligns with peak insect activity and ceases secretion in the afternoon (Lee and Heimpel, 2002).

3.2.2 The role of extrafloral nectar in enhancing natural enemy fitness

Cover crops such as faba bean (*Vicia faba* L.) and vetch (*Vicia* spp.) contain extrafloral nectaries which are nectar-secreting glands located outside of flowers (Jones et al., 2017) (Fig. 4). Unlike floral nectar, extrafloral nectar is often secreted in larger volumes, over longer periods, and is more accessible to natural enemies, making it a highly reliable resource (Koptur, 2005; Irvin et al., 2014). Importantly, extrafloral nectar is often available well before flowering occurs, potentially giving plants with extrafloral nectaries an advantage in selection of cover crop species over plant species that only produce nectar in flowers (Koptur, 1994; Géneau et al., 2013; Lu et al., 2014). Early nectar provision by extrafloral nectaries can help bolster natural enemy populations at a critical time when pest populations are beginning to establish, giving predators and

Figure 4 Cahaba white vetch: (a) extrafloral nectaries at the base of each stipule along the stem provide nectar to a wide range of natural enemies; (b) supports populations of alternative prey, which in turn enhance predator populations, such as this ladybug larva. Photo credit: N. A. Irvin.

parasitoids a head start in suppressing pests and minimizing the likelihood of potential pest outbreaks during the early stages of the growing season.

3.2.3 Risks of benefiting pests while enhancing natural enemies

Even after careful consideration to select insectary plants for enhancing natural enemy abundance and fitness in the field, the provision of non-crop plants can benefit pest species by increasing their longevity and fecundity (Wäckers et al., 2007). Zhao et al. (1992) reported that larvae of the imported cabbageworm (*Pieris rapae* [L.]) and diamondback moth were much higher in broccoli (*B. oleracea* var. *italica*) interplanted with nectar-producing anise hyssop (*Agastache foeniculum*) than in broccoli planted alone. Some species of adult lepidopterans feed extensively on floral nectar (Kevan and Baker, 1984), and Zhao et al. (1992) observed cabbageworm butterflies visiting anise hyssop flowers then moving into the broccoli crop. Feeding on nectar may have led to an increase in fitness of this pest and increased crop damage (Baggen and Gurr, 1998). In such cases, plants with extrafloral nectar may be a preferable cover crop, as their concentrated nectar can limit use by pests like Lepidoptera because their narrow tube-like proboscis cannot efficiently imbibe the thicker liquid when compared to lower viscosity flower produced nectar (Géneau et al., 2012; Géneau et al., 2013).

In order to maximize biological control, plant screening trials can identify plant species that selectively improve parasitoid and predator fitness without benefiting pest species. For example, buckwheat, cornflower (*Centaurea cyanus* L.), and common vetch enhanced the longevity and parasitism rates of a parasitoid of cabbage moth (*Mamestra brassicae* [L.]) by up to 1300% compared to controls without floral resources, and no increased fitness of the cabbage moth was observed (Géneau et al., 2012). Similarly, alyssum selectively attracted and enhanced the performance of *Cotesia vestalis* (Halliday), a parasitoid of diamondback moth without benefiting the target pest (Chen et al., 2020).

3.2.4 Avoiding intraguild predation and benefits to the fourth trophic level

Plant screening trials can also determine which cover crop species are of limited or no benefit to the fourth trophic level. The first trophic level consists of plants that herbivore pests feed on, the second level includes the herbivore pests themselves, the third level comprises the pest's natural enemies (i.e. parasitoids and predators), and the fourth level consists of hyperparasitoids that attack natural enemies. For example, in a 4-year study conducted in cabbage fields in Minnesota, USA, buckwheat borders significantly enhanced parasitism rates of two parasitoids (*Voria ruralis* [Fallen] and *Cotesia rubecula* [Marshall]) of lepidopteran pests without increasing hyperparasitism rates of these parasitoid species (Lee and Heimpel, 2005).

Within the third trophic level, intraguild predation occurs when a predator kills (and sometimes consumes) a competing natural enemy of a different species (Lucas and Maisonhaute, 2019). Intraguild predation can be direct, involving physical attacks on a vulnerable species, or alternatively, the competition is indirect and lack physical contact between species, and instead superior competition by one species reduces resource availability for another species. The main factor influencing intraguild predation is the relative size and mobility of the two competing species, which can vary throughout different life stages of competing species. Molecular gut content analyses in aphidophagous ladybugs have revealed that over 50% of larvae collected in soybean fields have DNA of other ladybug species, with 13% containing DNA from more than one other competitor species (Gagnon et al., 2011).

Enhancing agricultural landscapes with insectary plants and/or other cover crops increases habitat complexity, which has been shown to reduce intraguild predation by lowering encounter rates between predators. This reduction fosters coexistence and improves biological control efficiency (Finke and Denno, 2002). Balmer et al. (2013) showed that planting cornflowers alongside cabbage significantly increased predation and parasitism rates of cabbage moths without interference with generalist predators like ground beetles, rove beetles, and spiders.

3.3 Promising cover crop options

The most common types of cover crops are winter annuals that are grown during the off-season winter months and summer annuals or perennials that are grown during part or all of the cropping season (e.g. cover crop strips between rows of orchard trees) (Lu et al., 2000). Examples of common cover crop species are listed in Tables 1 and 2. Table 1 contains common warm and cool season cover crops used in pest management and their general benefits and potential disadvantages. Table 2 contains details of natural enemy enhancement and/or pest control for specific research studies that investigated the effects of cover crops on pest populations in various agricultural crops around the world. Snapp et al. (2005) also list promising cover species for four cover cropping niches in the USA: northern winter, northern summer, southern winter, and southern summer cover crops.

3.3.1 Grasses

Rye is among the most reliable cereal cover crop due to its excellent winter hardiness and ability to continue to grow in the spring (Lu et al., 2000).

Table 1 Common warm season and cool season cover crops and their potential benefits and challenges

Warm season cover crop name	Potential cover crop benefits and challenges	References
Buckwheat	• Provides floral nectar and pollen to natural enemies (NEs). Flowers have shallow floral tubes which increases the availability of nectar to NEs. • Flowers within 3–4 weeks of sowing and continue for up to 10 weeks. Plant will regrow if cut before it reaches 25% bloom. • Very attractive to predatory hoverflies. Attracts more hymenoptera and ladybugs than alyssum. Very attractive to honeybees. • Has been successfully used to control pests in a variety of cropping systems including cabbage, cotton, cucumber, soybean, apple, and grape. • Can suppress weeds if planted thickly. • Scavenges phosphorous and minor nutrients, residue decomposes quickly releasing nutrients to the next crop. • Can harbor insect pests such as *Lygus* spp. and tarnish plant bugs. • Can become a weed in some crops. Sets seed within 70–90 days.	Lee and Heimpel, 2005; Lu et al., 2020; Platt et al., 1999; Park et al., 2024; Berndt et al., 2002; Irvin et al., 2006; Phatak and Diaz-Perez, 2007; Bugg and Ellis, 1990; Bugg, 1991; Irvin et al., 2014, 2021; Clark, 2007; SARE Outreach, 2015
Alyssum[3] *Lobularia maritima* L.	• Provides floral nectar and pollen to NEs. Flowers have shallow floral tubes which increase the availability of nectar to NEs. • Enhances populations of parasitoids, hoverflies, ladybugs, and minute pirate bugs, leading to reductions in pest populations. • Enhances the longevity and fecundity of many parasitoids and NEs. • Attracts more predatory hoverflies than buckwheat. • Shows potential as a trap crop for the management of the wheat bug, *Nysius huttoni* White.	Irvin et al., 2021; Buchanan et al., 2018; Ribeiro and Gontijo, 2017; Mena and Gospodarek, 2024; Tiwari et al., 2018; Badenes-Pérez, 2019
Cowpea[1] *Vigna unguiculata*	• Provides additional extra-floral nectar to NEs. Attracts large numbers of beneficial hymenoptera, ladybugs, and other NEs. • Suppresses weeds and prevents erosion. Consistently enhances yield in subsequent crops. • Can have high populations of cowpea aphid (*Aphis craccivora* Koch) in the summer. • Stink bugs may reproduce heavily on cowpea in the summer which may damage nut crops.	Adamson et al., 2016; Bugg, 1991; Clark, 2007; SARE Outreach, 2015; Snapp et al., 2005

(*Continued*)

Table 1 *Continued*

Warm season cover crop name	Potential cover crop benefits and challenges	References
Sunflower	• Attracts large numbers of pollinators and NEs. • Provides additional extra-floral nectar to NEs. • Scavenges nitrogen and suppresses weeds.	SARE Outreach, 2015; Adamson et al., 2016
Sorghum-sudangrass hybrid *Sorghum bicolor* x *S. bicolor* var. *sudanese*	• Tall, fast-growing annual grass. • Attractive to grass-specific aphids which may provide alternative prey and hosts for natural enemies. Harbor NEs such as ladybugs and lacewings. Sorghum pollen enhances reproduction of ladybugs. • Alternative host of pestiferous corn aphids. • Contains sorgoleone which reduces weed emergence. Suppresses weeds and nematodes. • Builds soil organic matter and improves soil compaction. Contributes to soil carbon. • Border plantings of sorghum around capsicum field harbored generalist predators (lacewing, ladybug) and impeded the movement of pests (e.g. aphids) into the primary crop.	Michaud and Grant, 2005; Clark, 2007; SARE Outreach, 2015; Hewlett et al., 2019; Snapp et al., 2005; Duke et al., 2000
Sunn hemp[1] *Crotalaria juncea* L.	• Tropical annual legume. • Can fill a narrow niche between harvest of a summer crop and planting of a fall crop and is especially fitting to vegetable production. • Attracts wild carpenter and leafcutter bees in tropical farm systems and supports parasitoids of corn earworm in the Pacific Islands. • Increased yield in tomatoes. • Prevents soil erosion, increases soil organic matter, and suppresses nematodes and soil mites. • Effective at scavenging herbicide concentrations in ground water. • Seed is expensive.	Clark, 2007; SARE Outreach, 2015; Zannopoulos et al., 2024
Coriander *Coriandrum sativum* L.	• Aromatic annual can mask the scent of primary crop, leading to reduced pest abundance. • Provides floral nectar and pollen to NEs. • Enhances populations of parasitoids, lacewings, ladybugs, rove beetles, and spiders, leading to reductions in pest populations. • Enhances the longevity of parasitoids and hoverfly oviposition.	Jankowska et al., 2016; Irvin et al., 2007; Baggen and Gurr, 1998; Mena and Gospodarek, 2024

Table 1 *Continued*

Warm season cover crop name	Potential cover crop benefits and challenges	References
Sesbania[1] *Sesbania* spp.	• Fast-growing annual legume. • Provides alternative prey and habitat to NEs. • Stink bugs may reproduce heavily on sesbania in the summer which can damage nut crops. • Can support high populations of cowpea aphid.	Bugg, 1991; SAREP, 2021; Bugg and Waddington, 1994
Medics,[1] *Medicago* spp.	• There are winter annual, summer annual, or perennial medics. Annual medics include 35 known species that vary in plant habitat, maturity date, and cold tolerance. • Medics grow well in mixtures with grasses and clovers, but don't perform well with red clover. • Provides alternative prey such as *Lygus* spp. to NEs. • Alfalfa (*Medicago sativa* L.) sown as a cover crop in soybean and corn enhanced ground-dwelling predators, such as ground beetles, harvestmen (Opiliones), and spiders, and increased consumption of lepidopteran pupae. • Suppresses weeds and prevents erosion. • Enhances soil water infiltration. • Tolerate mowing or grazing. Can be used as a pasture or forage crop. • Perennials easily become weeds. • Can harbor two-spotted spider mite which may move into primary crop.	Clark, 2007; Snapp et al., 2005; Prasifka et al., 2006; MSU Extension, 2024
Cool season cover crop name	**Potential cover crop benefits and challenges**	**References**
Clover[1] crimson *Trifolium incarnatum*, red *T. pratense*, white *T. repens*, subterranean *Trifolium subterraneum*	• Clovers provide alternative prey and habitat to NEs. • Crimson clover is attractive to predatory hoverflies. • Subterranean clover and white clover suppressed pests in cabbage. • White clover is more flood tolerant than red and annual clovers. White clover produces more nitrogen, while red clover attracts more NEs. Crimson clover is self-reseeding and suppresses weeds by forming a thick mulch. Subterranean clover does well in acid soils. • Clovers demonstrated to increase yield in vineyards and vegetable crops. • Reduce or prevent erosion and build organic matter. • Pestiferous big-eyed bugs may build up in subterranean clover which may move into vegetable crops when clovers die in early summer.	Carroll, 2017; Phatak and Diaz-Perez, 2007; McGourty, 2004; Abad et al., 2021; Bugg et al., 1990; Bugg et al., 1991; Clark, 2007; Theunissen et al., 1995; Dutcher et al., 2010; Snapp et al., 2005

(*Continued*)

Table 1 *Continued*

Warm season cover crop name	Potential cover crop benefits and challenges	References
Vetch[1] hairy, common *Vicia villosa*, lana, purple, cahaba white *Vicia sativa* x *V. cordata*	• Provides aphids, thrips, and mites as alternative prey or hosts for predators and parasitoids. • Provides extra-floral nectar to attract and enhance NEs. • Cahaba and hairy vetch are attractive to predatory hoverflies. • Hairy vetch makes an excellent cover crop because of its broad adaptability, fast growth, high efficiency in fixing nitrogen, and production of high-quality biomass for mulch. • Cahaba vetch suppresses root-knot nematodes. • *Lygus* spp. and potato mirid (*Calocoris norvegicus* Leston) can be extremely abundant on vetches, especially common vetch. Spider mites on vetch can move into crop if cover crop senesces.	Carroll, 2017; Adamson et al., 2016; Teasdale, 2007; Bugg, 1991; Bugg et al., 1990; Clark, 2007; Dutcher et al., 2010
Sweetclover[1] *Melilotus* spp.	• Biennial spring or fall sown legume. • Flowers attract tachinid flies and large predatory wasps. • Reduces or prevents erosion, suppresses weeds, adds organic matter, and aerates subsoil. Sweetclover residue is allelopathic against some weeds.	Carroll, 2017; Clark, 2007
Radish,[3] *Raphanus sativus* L.	• Daikon or forage radish is tall growing and forms a large taproot that helps to open up clay soils and alleviates soil compaction. • Flowers profusely and provides habitat for NEs and their prey. • Prevents erosion and scavenges nutrients. • Bio-toxicity activity can stunt growth of primary crop if crop is direct drilled into just-killed residue of brassicas. • Should not be planted in rotation with other brassica crops because they are susceptible to similar diseases.	McGourty, 2004; Clark, 2007; Bugg, 1991

Table 1 *Continued*

Warm season cover crop name	Potential cover crop benefits and challenges	References
Mustard[3] *Brassica* spp. *Sinapis* spp.	• Provides aphids as alternative prey or hosts for predators and parasitoids. • White mustard (*Sinapis alba* L.) enhances the abundance of hoverflies, parasitoids, ladybugs, and ground beetles and increases the fecundity of hoverflies and the longevity and fecundity of some parasitoids. • Indian mustard (*Brassica juncea* [L.] Czern.) and white mustard are used as trap crops for diamondback moth. Indian mustard can also be used as a trap crop for the cabbage webworm (*Hellula undalis* Fabicius), the cabbage head caterpillar (*Crocidolomia pavonana* Fabricius), and the harlequin bug (*Murgantia histrionica* Hahn). Field mustard (*Brassica rapa* L.) has been used as a trap crop for cabbage seedpod weevil (*Ceutorhynchus obstrictus* Marsham) and cabbage fly (*Delia radicum* L.). • Prevents erosion and scavenges nutrients. • Bio-toxicity activity can stunt growth of primary crop if crop is direct drilled into just-killed residue of brassicas. • Should not be planted in rotation with other brassica crops because they are susceptible to similar diseases. • Can be weedy and invasive in California. • *Lygus* spp. and potato mirid (*Calocoris noregicus*) can be extremely abundant on mustard. May be an alternative host for pestiferous flea beetles.	Bugg, 1991; Clark, 2007; Mena and Gospodarek, 2024; Badenes-Pérez, 2019
Ryegrass[2] *Lolium* spp.	• Provides habitat for spiders. • May have weed potential if left to set seed. • Prevents erosion, improves soil structure and drainage, scavenges excess nutrients, and adds organic matter. • Heavy user of moisture and nitrogen and performs poorly during drought. • Ryegrass can be annual or perennial depending on the cultivar. Annual ryegrass establishes faster than perennial ryegrass but is less cold-hardy. Annual ryegrass is about half as expensive as perennial ryegrass.	Clark, 2007; Bugg, 1991; SARE Outreach, 2015

(*Continued*)

Table 1 *Continued*

Warm season cover crop name	Potential cover crop benefits and challenges	References
Cereal rye[2] *Secale cereale*	• Flowers provide pollen for predators such as ladybugs, minute pirate bug, and soldier beetles. Can provide grain aphids as alternative prey or hosts for predators and parasitoids. Provides habitat for spiders and other NEs. • Can reduce pest problems in crop rotations. • Scavenges excess nitrogen, prevents erosion, and adds organic matter. • Controls weeds through allelopathy. • Ability to reduce soil-borne diseases and nematodes. • Rye is taller and quicker growing than wheat – can serve as a windbreak and trap snow or hold rainfall over winter. Rye is more cold and drought tolerant than wheat. • Very winter hardy and can continue to grow into the spring.	Adamson et al., 2016; Phatak and Diaz-Perez, 2007; Clark, 2007; Bugg, 1991; Wingard, 1996; Bowers et al., 2020; Snapp et al., 2005
Oats[2] *Avena sativa*	• Flowers provide pollen for predators such as ladybug, minute pirate bug, and soldier beetles. Provides habitat for spiders. • Increases yield in vineyards. • Suppresses weeds, prevents erosion, scavenges excess nutrients, adds biomass, and can be used as a nurse crop. • Allelopathic compounds in oat roots and residue can hinder weed growth for a few weeks. Bio-toxicity activity can stunt germination or growth of lettuce, cress, rice, wheat, and peas if crop is direct drilled into just-killed residue of oats. • Less prone to insect problems than oats or barley. • Oats winterkill (dies in cold temperatures when winter comes).	Adamson et al., 2016; Abad et al., 2021; Clark, 2007; Lu et al., 2000
Barley[2]	• Provides habitat for spiders and pollen for predators such as ladybug, minute pirate bug, and soldier beetles. Provides non-pest aphids as alternative prey and enhances lacewing larvae in lettuce. • Can suppress the incidence of leafhoppers, aphids, armyworms, and root-knot nematodes. • Prevents erosion, suppresses weeds, scavenges excess nutrients, and adds organic matter. • Tolerates alkaline soils better than any other cereal.	Adamson et al., 2016; Clark, 2007; Hogg et al., 2023

Table 1 *Continued*

Warm season cover crop name	Potential cover crop benefits and challenges	References
Wheat[2] *Triticum aestivum*	• Flowers provide pollen for predators such as ladybug, minute pirate bug, and soldier beetles. Provides habitat for spiders. • The presence of wheat stubble from no-till plots of crucifers reduced flea beetle (*Phyllotreta cruciferae* Goeze) due to greater structural diversity of the no-till plots interfering with host plant location. • Prevents erosion, scavenges excess nutrients, and adds organic matter. • Ability to reduce soil-borne diseases and nematodes. • Control weeds, but not as effective as rye due to lower biomass and less allelopathic effects. • Less likely than cereal rye and barley to become a weed problem, but more susceptible to insects and disease.	Adamson et al., 2016; Phatak and Diaz-Perez, 2007; Clark, 2007; Snapp et al., 2005;
Medics[1] *Medicago* spp.	• There are winter annual, summer annual, or perennial medics. Annual medics include 35 known species that vary in plant habitat, maturity date, and cold tolerance. • (see warm season Medics characteristics)	Clark, 2007
Faba or fava bean[1] *Vicia faba* L.	• Provides floral nectar and extra-floral nectar to NEs. • Provides alternative prey. • Attractive to Ichneumonidae parasitoids and other parasitic Hymenoptera. • Prevents soil erosion.	Bugg et al., 1989; Herz et al., 2019

[1] legumes: Legumes have nitrogen fixing bacteria – when legume decomposes, nitrogen is released into the soil.
[2] cereal grains;
[3] Brassica: Brassicas have been shown to release allelopathic compounds that inhibit bacteria, fungi, insects, nematodes, and weeds.

Table 2 Effects of common cover crops on arthropod pests and their natural enemies: results and mechanisms for studies in various agricultural crops around the world

Cover crop species	Country where study was conducted	Crop	Experimental results	Mechanism(s) behind effect on pest populations	References
Alyssum, *Lobularia maritima* L. Desv	Brazil	Collard greens, *Brassica oleracea* (L.)	Densities of spiders, ladybugs, hoverflies, and minute pirate bug increased. Densities of green peach aphid (*Myzus persicae* Sulzer) and diamondback moth decreased up to 83% and 75%, respectively.	Enhanced natural enemy abundance	Ribeiro and Gontijo, 2017
Buckwheat, *Fagopyrum esculentum* Moench	Minnesota, USA	Cabbage, *Brassica oleracea* L.	Buckwheat floral borders consistently increased some parasitism rates over the 4-year study. Overall, parasitism of larvae of cabbage looper, *Trichoplusia ni* (Hübner), and imported cabbageworm was increased in buckwheat plots. Overall, no significant decrease in pest densities (cabbage looper, imported cabbageworm, or diamondback moth). In year 2, densities of imported cabbageworm were significantly lower in buckwheat treatments.	Enhanced natural enemy abundance	Lee and Heimpel, 2005
Buckwheat, *Fagopyrum esculentum* Moench	Republic of Korea	Soybean, *Glycine max* [L.] Merr	Buckwheat strips increased levels of minute pirate bug and crab spiders. Lowered levels of adult alydid *Riptortus pedestris* (F.), but increased levels of two other pests, agromyzid leafminers and cicadellid leafhoppers.	Enhanced natural enemy abundance and reproduction. Semiochemicals emitted from buckwheat may have also reduced host plant location.	Park et al., 2024
Barley, *Hordeum vulgare* L.	California, USA	Lettuce, *Lactuca sativa* L.	Barley contained the highest numbers of non-pest aphids as alternative prey and lacewing larvae out of nine candidate plants tested. There was no significant effect of barley on populations of the pestiferous currant-lettuce aphid, *Nasonovia ribisnigri* Mosley.	Alternative prey fed and attracted predators	Hogg et al., 2023

Lucerne, *Medicago littoralis* Rohne ex Loisel	Sweden	Carrot, *Dacus carota* L.	Lucerne significantly reduce damage by carrot rust fly, *Psila rosae* (F.) compared to monocultures in all experiments and in both years. The experiment showed this reduction was independent of predator abundance. Significantly more carrot rust fly eggs were laid in monocultures where host plant location is easier.	Reduces resource concentration	Rämert and Ekbom, 1996
Coriander, *Coriandrum sativum* L.	Poland	Carrot, *Dacus carota* L.	Significantly lower numbers of pestiferous carrot psyllid (*Trioza viridula* Zett) larvae and adults, aphids, leafhoppers, and planthoppers (Auchenorrhyncha) in coriander treatments compared with monoculture in both years. Significantly lower numbers of pestiferous *Lygus* sp. in coriander treatments in year 2. In both years, percentage of roots infested by carrot rust fly and pestiferous aphids was significantly higher in monoculture compared with coriander treatment.	Reduces resource concentration	Jankowska and Wojciechowicz-Żytko, 2016
			Numbers of predatory hoverflies and ladybugs were significantly higher in coriander treatment in both years.	Enhanced natural enemy abundance	
Cornflower, *Centaurea cyanus* L.	Switzerland	White cabbage, *Brassica oleracea*	Predation of eggs of the cabbage moth was up to 160% higher in cabbage with cornflower than without. Parasitism of cabbage moth larvae by *Trichogramma* spp. and *Microplitis mediator* (Haliday) was significantly higher on cabbage in cornflower plots. Parasitoid DNA rarely found in predator guts indicating ground beetles, rove beetles, and spiders do not interfere with biocontrol.	Enhanced natural enemy abundance	Balmer et al., 2013

(*Continued*)

Table 2 *Continued*

Cover crop species	Country where study was conducted	Crop	Experimental results	Mechanism(s) behind effect on pest populations	References
Desmodium uncinatum Jacq. and Napier grass, *Pennisetum purpureum* Schumach	East Africa	Maize, *Zea mays* L.	Highly significant reductions in infestation levels of fall armyworm *Spodoptera frugiperda* (Smith) (average of 82.7%) and proportions of maize plants damaged by the larvae (average of 86.7%) which increased yield by up to 2.7 times.	Push-pull trap crop system	Tian et al., 2012
Coriander, dill (*Anethum graveolens* L.), and buckwheat	New Jersey, USA	Bell pepper, *Capsicum annum* L.	Predation of European corn borer (*Ostrinia nubilalis* Hubner) eggs by insidious flower bugs (*Orius insidiosus* Say) was significantly improved in the flower-mix treatment in four out of five field seasons. This capacity may depend on the abundance of green peach aphid as alternative prey which negatively affected predation of the second generation of European corn borer eggs.	Enhanced natural enemy abundance Alternative prey negatively affected predation	Bickerton and Hamilton, 2012
Sesame, *Sesamum indicum* L.	Vietnam, Thailand, and China	Rice, *Oryza* spp.	Increased populations of the parasitoid, *Anagrus* spp., and significantly lower populations of the brown planthopper compared to monoculture controls.	Enhanced natural enemy abundance	Zhu et al., 2018
Anise hyssop, *Agastache foeniculum*	Michigan, USA	Broccoli, *B. oleracea* var. *italica*	Pupae of the braconid parasitoid *Cotesia rubecula* were significantly higher in broccoli interplanted with anise hyssop than in broccoli monoculture. Eggs and larvae of the imported cabbageworm and larvae of the diamondback moth were much higher in broccoli interplanted with anise hyssop than in broccoli planted alone (negative effect – see mechanism column). Density of cabbage looper larvae was similar among treatments.	Possible enhancement of fitness of imported cabbageworm since butterflies were observed visiting anise hyssop flowers	Zhao et al., 1992

Red clover, *Trifolium pratense* L.	Sweden	Iceberg lettuce, *Lactuca sativa* (L.)	Abundance of *Lygus* spp. were 7–30 times higher in red clover than in lettuce making it a good trap crop to reduce pest population in lettuce fields.	Trap crop	Rämert et al., 2001
Rye, *Secale cereal* L.	Georgia, USA	Cotton, *Gossypium hirsutum* L.	A winter rye cover crop harbored up to 1000% higher densities of predators at seedling stage of cotton compared to no-cover conventional plots. Rye had significantly lower numbers of thrips on cotton seedlings at 21 days after cotton planting, with no differences among treatments 14 days after planting. During stink bug pressure, cotton boll injury was 200% lower in rye plots compared to no cover controls. There were no negative effects of rye on cotton yield or quality of fiber.	Enhanced natural enemy abundance	Bowers et al., 2000
White mustard, *Sinapis alba* L.	Poland	Faba bean, *Vicia faba* L.	Sowing white mustard between rows of faba bean reduced the abundance of pestiferous aphids (*Aphis fabae* Scop.) on faba bean plants up to sevenfold compared with the faba bean monoculture. Hoverfly eggs appeared earlier in intercropped treatment compared with faba bean monoculture and a positive influence of mustard vicinity on predator-prey ratio was recorded.	Enhanced natural enemy abundance	Gospodarek, 2021
Black mustard, *Brassica nigra* (L.)	New Zealand	Corn, *Zea mays* (L.)	A border of black mustard reduced the mean number of adult green vegetable bug (*Nezara viridula* [L.]) in sweet corn crops by 1567% and reduced the percentage of damaged sweet corn cobbs by 16%, which was below the processing threshold.	Trap crop	Rea et al., 2002

(*Continued*)

Table 2 *Continued*

Cover crop species	Country where study was conducted	Crop	Experimental results	Mechanism(s) behind effect on pest populations	References
A mixture of 11 flowering annuals	Switzerland	Potato, *Solanum tuberosum* L.	Abundance of adult hoverflies, ladybugs, and lacewings was increased up to 339% in flower strips compared with potato controls. The number of eggs deposited by hoverflies and lacewings was 127% and 48% higher, respectively, in flower strips, and this translated to a 75% reduction in the number of aphids in adjacent potato crops.	Enhanced natural enemy abundance	Tschumi et al., 2016

It provides habitat for generalist predators, alternative prey or hosts for predators and parasitoids, and pollen for predators, including ladybugs, minute pirate bugs, and soldier beetles (Cantharidae) (Table 1). Since rye is taller and quicker growing than wheat, it can serve as a windbreak which helps to protect natural enemies from harsh environmental conditions (Clark, 2007). Research in Georgia (USA) cotton fields demonstrated that a winter rye cover crop enhanced predator populations, reduced the abundance of thrips, and decreased the proportion of cotton boll injury from stink bugs (Pentatomidae) (Bowers et al., 2020) (Table 2) (Fig. 5).

Various annual and perennial grasses provide wind-blown pollen in late winter and early spring which feed predatory mites and promote predator population growth (Bugg, 1991; Kennett et al., 1979). Predatory mites can be important biological control agents of phytophagous mites (e.g. spider mites, *Tetranychus urticae* Koch), thrips, whiteflies, and scale crawlers. Populations of naturally occurring predatory phytoseiid mites that benefit from windborne pollens have provided some suppression of avocado brown mite (*Oligonychus punicae* [Hirst]), citrus thrips (*Scirtothrips citri* [Moulton]), and citrus red mite (*Panonychus citri* [McGregor]) (Bugg, 1991; McMurtry et al., 2013) in commercial orchards.

Sorghum (*Sorghum bicolor* [L.] Moench) is a tall, fast-growing annual grass (Table 1). When combined with sunflower (*Helianthus annuus* L.) as a border around capsicum (*Capsicum annuum* L.) fields, sorghum acts as a trap crop, reducing pestiferous stink bugs and significantly lowering damage to the capsicum crop (Mathews et al., 2017). Similarly, border plantings of sorghum around capsicum harbored generalist predators (lacewings [Chrysopidae] and ladybugs) and impeded the movement of aphids into the primary crop (Hewlett et al., 2019).

Grass cover crops like tussock or bunch grasses (cocksfoot, *Dactylis glomerata* L., or tall oat grass, *Arrhenatherum elatius* [L.] Presl.) provide overwintering habitat for generalist predators such as ground and rove beetles (Andersen, 1997; Collins et al., 2003; Kajak and Lukasiewicz, 1994). These predators can move from grass/clover strips into nearby crops where they can provide biological control of pests (Kajak and Lukasiewicz, 1994).

3.3.2 Brassicas

Brassica cover crops, such as mustard (*Brassica* spp. and *Sinapis* spp.) and radish (*Raphanus sativus* L.) (Fig. 5), are valued for their rapid fall growth, nutrient scavenging capabilities, and release of bio-toxic compounds that inhibit bacteria, fungi, insects, nematodes, and weeds (Table 1). These allelopathic compounds, primarily glucosinolates, are released when plant cells are ruptured, making practices like mowing and incorporating brassica

Figure 5 Examples of cover crops: (a) cereal rye (*Secale cereale*) in rotation on a farm in Maryland, USA (Photo credit: Edwin Remsberg and USDA-SARE); (b) radish (*Raphanus sativus*) producing allelopathic compounds that inhibit bacteria, fungi, insects, nematodes, and weeds (Natalie Lounsbury, Jack Gurley, and USDA-SARE); (c) barley grown between raspberry rows in Oxnard, California, USA (Ben A. Faber); (d) *Phacelia tanacetifolia* and alyssum sown between vine rows in northern California (Houston Wilson).

residues into the soil a strategy for pest suppression (Clark, 2007). Timing the incorporation of brassica residues to coincide with the most vulnerable life stages of target pests further enhances their efficacy (Clark, 2007).

White mustard significantly enhances populations of natural enemies, including hoverflies, parasitoids, ladybugs, and ground beetles, while also increasing hoverfly fecundity and parasitoid longevity (Mena and Gospodarek, 2024). One study demonstrated that sowing white mustard between rows of faba bean significantly reduced the abundance of black bean aphid (*Aphis fabae* Scop.) on faba bean plants, achieving pest suppression similar to chemical treatments (Gospodarek, 2021) (Table 2). This reduction in aphid populations was possibly attributable to two mechanisms: a masking effect, where the dense vegetation made it more difficult for aphids to locate host plants in the intercropped treatment, and enhanced biological control from hoverflies and ladybugs (Table 2). Thinning white mustard plants at the flowering bud stage of faba beans is critical to avoid impairing flowering and pod formation, ensuring both pest suppression and optimal crop performance are achieved (Gospodarek, 2021).

3.3.3 Legumes

Legumes, including cowpea (*Vigna unguiculata* [L.] Walp.), sunn hemp (*Crotalaria juncea* L.), sesbania (*Sesbania* spp.), medics (*Medicago* spp.), clover (*Trifolium* spp.), and vetch (*Vicia* spp.), exhibit diverse seasonality, growth habits, and life cycles (Table 1). These plants are beneficial for supporting populations of alternative prey or hosts such as aphids, which, in turn, sustain and enhance populations of parasitoids and predators (Bugg, 1991; Clark, 2007) (Fig. 5). By providing a consistent food source, alternative prey species enable natural enemies to establish and maintain their populations even when pest aphids are absent from the crop. This early establishment ensures that natural enemies are available to respond quickly to pest outbreaks, helping to suppress emerging pest populations more effectively (Gardarin, 2023).

While these alternative prey may not necessarily cause problems in primary crops (Dutcher et al., 2010), caution is necessary when the alternative prey itself is a pest in vulnerable primary crops (Bugg, 1991). In some cases, this dynamic has been leveraged intentionally, such as with the use of legumes, like red clover, that acts as a trap crop for *Lygus* spp. in lettuce, thereby reducing *Lygus* spp. populations in lettuce fields (Table 2).

3.3.4 Flowering herbaceous plants

Establishing flowering strips is one of the most common ways to promote biological control and other ecosystem services, like pollination (Landis et al., 2000; Gurr et al., 2017) (Fig. 5). Common flowering plants used as cover crops include alyssum, buckwheat, cornflower, coriander, and dill (*Anethum graveolens* L.) (Rizvi et al., 2022) (Tables 1 and 2). These plants attract and retain natural enemies locally while enhancing their fitness by potentially providing SNAP (see '2.3.1 Flowering insectary plants as cover crops'). The increased abundance and fitness of natural enemies increases their efficacy, resulting in reduced pest population densities in the primary crop.

Buckwheat, coriander, and alyssum are particularly effective for supporting natural enemies due to their shallow floral tube (corolla), which makes nectar easily accessible to parasitoids and predators (van Rijn and Wäckers, 2016). Furthermore, several traits contribute to their success as insectary plants, including rapid germination to flowering times, high-quality nectar and pollen, extended flowering periods, fast growth rates, and short vegetative growth stages (Bowie et al., 1995; Chaney, 1998; Hogg et al., 2011b; Irvin et al., 2014, 2021).

Alyssum, for example, enhances populations of hoverflies, ladybugs, and minute pirate bugs, as well as the longevity of parasitoids and predators (Mena and Gospodarek, 2024). Similarly, coriander supports populations of lacewings, ladybugs, rove beetles, and spiders, while enhancing parasitoid longevity and hoverfly oviposition (Mena and Gospodarek, 2024).

Buckwheat has been extensively utilized to increase biodiversity and enhance populations of natural enemies which results in pest suppression across a range of crops, including cabbage (Lee and Heimpel, 2005), cotton (Li et al., 2019), cucumber (*Cucumis sativus* L.) (Platt et al., 1999), soybean (Park et al., 2024), apple (Irvin et al., 2006), and grape (*Vitis vinifera* L.) (Berndt et al., 2002). Despite these successes, Philips et al. (2014) found no significant pest reduction by buckwheat, emphasizing the importance of evaluating the effectiveness of cover crops within specific cropping systems and the regions within which they are grown.

Although spiders can benefit from the nectar and pollen provided by flowering herbaceous plants, alternative prey and increased habitat complexity offered by these plants can be more critical in supporting spider populations (Herz et al., 2019). In addition to their pest management roles, buckwheat, coriander, and alyssum have easily decomposable residues that support soil biota and quickly release nutrients to subsequent crops (Thies and Abawi, 2009) (Table 1). They also contribute to wider ecosystem services, such as resource support for bees and soil health improvement (Table 1).

González-Chang et al. (2019) summarized 44 studies using 24 insectary plant cover crops and list the target pests and effects on natural enemies, while Herz et al. (2019) offer a table of 25 families of herbaceous flowering plants used as cover crops in apple orchards.

3.3.5 Mixtures of cover crop species

Mixtures of two or more cover crop species often outperform single species plantings by combining complementary benefits provided by different species, such as varying flowering phenologies and longevities, and growth habits. For example, including a drought-tolerant plant into a perennial mix can enhance persistence during dry periods, while combining the weed suppression of grasses with the nitrogen production of legumes can achieve multiple goals simultaneously (Clark, 2007).

A mix of cover crop species can have sequential flowering cycles, thereby providing SNAP to natural enemies year-round (Phatak and Diaz-Perez, 2007). Several selection criteria were used by Pålsson et al. (2022) and Mockford et al. (2024) to choose flower and grass species as a mixed cover crop for enhancing biological control services in Spanish citrus orchards and Swedish apple orchards, respectively. Plants were chosen based on sequential flowering and overlapping nectar availability, accessibility of nectar and pollen, and support of alternative prey and hosts for natural enemies. Other considerations included plant tolerances to varying microclimate, longevity, vegetation

Figure 6 A novel seed mix sown between rows of orange trees in Spain, consisting of 12 forb species and two tussock-forming grass species, selected to enhance floral structure diversity, provide sequential flowering with overlapping nectar availability, support alternative prey and hosts, and create varied vegetation structure, height, and microclimates. Photo credit: W. B. Westbury.

structure and height, low maintenance requirements, and diversity of flower structures. Selected cover crop species resulted in overlapping and sequential flowering periods that provided nectar to natural enemies for 9 months, a time period when natural enemy services were most required (Mockford et al., 2024) (Fig. 6). Grass species in the cover crop mixtures offered stability against weed invasion and durability against tractor traffic moving between trees (Pålsson et al., 2022).

In Swiss potato (*Solanum tuberosum* L.) fields, Tschumi et al. (2016) selected flower species based on previous studies that demonstrated positive effects on the abundance and performance of key natural enemies of aphids, such as hoverflies, ladybugs, lacewings, and parasitoids. In this study, a mix of 11 annuals sown in a 3-meter strip positioned at the edge of each potato field increased the number of eggs deposited by hoverflies and lacewings by 127% and 48%, respectively, and reduced aphid numbers by 75% in adjacent potato crops.

Mixed cover crop species can also be useful for trap crops (Muthiah, 2003). For example, marigolds, sunflower, and Chinese cabbage (*Brassica campestris* L. subsp. *pekinensis*) have been used as a mixed plant species trap crop for pollen beetle (*Meligethes aeneus* F.) in cauliflower (*B. oleracea* var. *botrytis*) fields in Finland (Hokkanen, 1989).

SARE Outreach (2015) lists three examples of mixtures of cover crop species tailored to various seasons and regions and includes resources for rotation sequences. Additional guidance on using mixed cover crops for pest management can be found in Section 9 'Where to look for further information' of this chapter.

4 Placement, timing, termination and mowing of cover crops

4.1 Cover crop placement and timing

When cover crops are used to improve soil quality and health, seeds are typically sown directly into the soil containing the primary crop or in rotation with the primary crop (Myers, 2017). However, for pest management, alternative placement strategies can be employed. Mobile natural enemies can feed on SNAP resources provided by cover crops and subsequently move from this resource into the nearby primary crop to control pests. This has been demonstrated in wheat fields, almond (*Prunus dulcis* [Mill.] D.A. Webb) orchards, and vegetable crops where rubidium chloride was used to mark parasitoids, lacewings, and ladybugs. These natural enemies labeled with rubidium chloride moved up to 76.2 m from insectary plantings into nearby crops (Freeman-Long et al., 1998).

Cover crops can be strategically grown between crop rows, along the entire length of the row, in small patches dispersed throughout fields, in orchard tree lines, around field margins, on rice bunds, in a strip on one side of a field, or across an entire field between rotations of an annual crop (Freeman-Long et al., 1998; Baggen and Gurr, 1998; Irvin et al., 2006; Bugg et al., 2008; Tschumi et al., 2016; Zhu et al., 2018) (Fig. 7). In cotton, legume cover crops such as clover

Figure 7 Two cover crop placements in vineyards for different purposes: (a) 'Pollinator Pathway Mix' (Buzz Cover Crop Seeds), a diverse blend of drought-tolerant, short-statured, self-reseeding annual and perennial flowers and grasses, sown between rows to enhance natural enemies in Salem, Oregon, USA; (b) spring-sown buckwheat planted on a vineyard headland for natural enemy habitat and erosion control in Willamette Valley, Oregon. Photo credit: R. M. Sweet.

and vetch can be planted along field borders, around water reservoirs, near irrigation lines, and in other non-cropped areas close to cotton fields (Phatak and Diaz-Perez, 2007). In potato crops, field margin planting of buckwheat and coriander increased the fitness of parasitoids, promoting the biological control of the potato tuber moth, *Phthorimaea operculella* Zeller (Baggen and Gurr, 1998).

The placement of cover crops often depends on the mechanism of pest suppression and mobility of natural enemies. In Michigan (USA) asparagus (*Asparagus officinalis* L.) fields, alyssum border plantings were associated with reduced thrips populations only at the field edge and not at distances of 5–15 meters from the flowering strips (Buchanan et al., 2018). This study indicated the importance of placement – specifically the arrangement, size, and position of cover crops, relative to crop field and surrounding habitat. In such cases, planting patches of insectary plants dispersed throughout crop fields can better facilitate the movement and dispersal of natural enemies across cropping systems.

In arid regions like southern California, where spring and summer rainfall is insufficient to sustain cover crops without supplementary irrigation, placing flowering insectary plants in dispersed patches rather than down entire rows can be a more viable approach when incorporating warm-season cover crop species in vineyards. Irvin et al. (2016) demonstrated that flowering buckwheat grown between vine rows with irrigation attracted natural enemies but also increased pestiferous leafhopper densities on grape leaves by up to 240% compared to non-irrigated control plots without buckwheat. This surge in leafhopper populations was likely due to their preference for the more vigorous, irrigated vines, which exhibited a 222% higher mean cane weight (a standard measure of vine vigor based on the biomass of pruned canes) than non-irrigated controls. Supplemental irrigation also had adverse effects on berry quality, including increased berry scarring by thrips and a reduction in sugar content of berries by 3.2° Brix compared to non-irrigated controls. Placing patches of flowering plants, rather than along entire rows, may help minimize the potential negative impacts of introducing insectary plants into sensitive agricultural systems.

The placement of cover crops can also influence the selection of suitable cover crop species. For example, planting within orchard tree lines may challenge sun-loving species due to shading from mature trees, reducing their growth and effectiveness (Hubbard, 2023). Furthermore, when applying agrochemicals such as insecticides, fungicides, or herbicides in fields with cover crops or associated crops, it is essential to carefully time applications to avoid periods of peak insect activity, thereby minimizing harm to beneficial insects (Adamson et al., 2016). For example, applying chemicals to corn or other grasses during flowering should be avoided, as beneficial insect species, like natural enemies, feed on cereal flower pollen (Adamson et al., 2016).

4.2 Cover crop termination

Cover crops may need to be terminated before they compete with the primary crop or to make room for the primary crop as part of the rotation sequence. Termination of cover crops can be achieved mechanically or with herbicides (Sundermeier, 2016). Mechanical tilling of the cover crop may be advantageous for both termination and incorporating nitrogen into the soil (McGourty, 2004). However, tillage can disrupt and kill the larval stages of beetle predators, which have limited resistance to disturbance (Blubaugh and Kaplan, 2015), and may degrade soil structure and reduce organic matter, if disturbance is excessive (McGourty, 2004).

After termination, leaving cover crop residues on the soil surface as a mulch can extend weed suppression benefits (Lu et al., 2000) and provide ongoing habitat and protection for natural enemies such as rove and ground beetles and spiders (Altieri and Schmidt, 1985). In annual rotations, planting the primary crop directly into the cover crop residue allows natural enemies to remain established and ready to attack pests as the primary crop develops (Fig. 8).

Research in Georgia, Alabama, and Mississippi in the USA has demonstrated that summer vegetables planted into residues of cool-season cover crops benefited from the presence of established natural enemies, which suppressed crop pests as the vegetables grew (Phatak and Diaz-Perez, 2007). Cabbage planted into herbicide-killed rye had significantly lower populations of pestiferous lepidopteran larvae and aphids than rye subjected to conventional tillage (Bottenberg et al., 1997). In the mid-western USA, growers may utilize

Figure 8 Cover crop residue creates habitat for natural enemies, enhancing their ability to control pests more quickly in the primary crop: (a) a no-till corn planter sowing seeds directly into cover crop residue (Photo credit: Edwin Remsberg and USDA-SARE); (b) pumpkins growing after being planted into the residue of a rye cover crop in Western Maryland, USA (Edwin Remsberg and USDA-SARE).

winter-hardy cover crops, like canola (*Bassica napus* L.), which provide extended soil protection, enhanced soil health, and overwintering habitat for natural enemies (Myers, 2018) (Fig. 9).

In perennial systems, timing of cover crop termination can be important. For example, legumes planted in pecan (*Carya illinoinensis* [Wangenh.] K. Koch) orchards can be beneficial by attracting predators that feed on aphids. As aphid densities on legumes decline or the legumes senesce, these predators move into the tree canopy in search of another food source, helping to control pestiferous aphids and other pests (Carroll, 2017). However, poor timing of cover crop termination in pecan systems can have unintended consequences. Legumes can attract stink bugs, which damage pecans by feeding on the nuts. Delaying the termination or mowing of cover crops until pecan ripening (i.e. shuck separation) reduces unwanted movement of stink bugs from the cover crop to trees where they can cause feeding damage to pecans (Carroll, 2017).

4.3 Cover crop mowing

Some cover crops may require mowing to encourage stem development from the crown of the parent plant (tillering) (McGourty, 2004). Wildflower strips can benefit from a single cutting to enhance vegetation cover and extend the flowering period (Mockford et al., 2023). During establishment, mowing legumes at the flowering stage, and grasses before flowering, tends to enhance rates of soil coverage (Colugnati et al., 2004). Additionally, mowing can enhance

Figure 9 Canola flowering in Harford County, Maryland. Photo credit: Edwin Remsberg and USDA-SARE.

pest control in certain cases, such as with brassica cover crops, which release bio-toxic compounds from cells damaged by cutting, and these compounds inhibit bacteria, fungi, insects, nematodes, and weeds (Clark, 2007).

In some situations, mowing may benefit the primary crop directly. Tall cover crops can impede air circulation, which, for example, increases the risk of disease or frost. Mowing can mitigate these issues (McGourty, 2004). Mowing can also encourage natural enemies to move from cover crops to the adjacent primary crop, thereby enhancing pest control within the cropping system (Goller et al., 1997). However, cutting cover crops can sometimes have unwanted adverse effects, such as killing natural enemies (Bugg and Waddington, 1994), or reducing the ability of the cover crop to support beneficial insects. This may lead to excessive or premature dispersal of predatory insects where they either spread out too much or abandon the area instead of concentrating where pest suppression is needed (Bugg, 1991; Mockford et al., 2024). For example, in walnut (*Juglans regia* L.) orchards, if ground cover harboring alternative prey is mown too early, before walnut aphid is present in trees, ladybugs may disperse from the area resulting in diminished biological control (Sluss, 1967). Similarly, in pear (*Pyrus* spp.) orchards, reducing mowing regimes from 2–3 times to once per month increased predator and parasitoid numbers due to an increase in alternative prey within the ground cover (Horton et al., 2003).

Leaving unmown strips of a cover crop when most of the cover crop is mown or incorporated into the soil can provide a continuing refuge and food source for natural enemies which might otherwise die or leave the area if the cover crop was completely destroyed (Phatak and Diaz-Perez, 2007; Fountain, 2022). Employing sickle-bar mowers or setting a flail or rotary mower to an increased cutting height can permit better survival of natural enemies (Bugg, 1991). Furthermore, adjusting the timing of mowing to allow natural enemies to mature or disperse can enhance the ecological benefits of well-planned cover crop management (Bugg, 1991).

5 Cover cropping as part of an integrated pest management program

Cover cropping can be enhanced by integrating other strategies to optimize pest management. Combining cover cropping with additional pest management approaches can help minimize shortcomings that may be associated with enhancing biodiversity in agricultural systems (Hooks and Johnson, 2003). Complementary tools in an integrated pest management (IPM) framework include selective pesticides, genetically modified crops expressing toxins from Bt crops, host plant resistance, augmentative releases of biological control agents, and semiochemicals, such as sex-pheromone mating disruption and herbivore-induced plant volatiles (Hooks and Johnson, 2003; Tian et al.,

2015; Torres and de Bueno, 2018). Augmentative biological control is a pest management strategy that involves the supplemental release of mass-reared natural enemies to enhance resident existing populations at times when naturally occurring populations are low, with the expectation that enhanced levels of biological control result.

Synthetic herbivore-induced plant volatiles, like methyl salicylate, can be deliberately deployed to attract natural enemies to crops from surrounding habitats (Orre et al., 2010). This strategy complements cover crops well because non-crop vegetation provides food and shelter to retain and support natural enemies attracted by these volatiles. This novel conservation biological control approach is called 'attract and reward.' It combines synthetic herbivore-induced plant volatiles to attract beneficial arthropods into crops, and floral resources (the reward), to maintain populations of immigrating natural enemies (Simpson et al., 2011).

Cover cropping can also complement augmentative releases of natural enemies by providing SNAP and improving the retention of released parasitoids and predators, thereby maximizing the benefits of augmentative biological control. For example, cornflower has been effectively used in Swiss cabbage fields to support augmentative releases of the egg parasitoid *Trichogramma brassicae* Bezdenko (Balmer et al., 2013).

Ant control is another strategy in IPM programs with cover cropping. Several ant species can protect citrus by preying upon and repelling pests. However, the presence of ants, especially the invasive Argentine ant (*Linepithema humile* [Mayr]) in citrus, has been associated with population outbreaks of honeydew-producing pests (Anjos et al., 2021). Ants feed on honeydew excreted by these hemipteran pests and protect them from their natural enemies, thereby exacerbating pest populations and hindering effective pest management (Daane et al., 2007; Buczkowski and Bennett, 2008; McCalla et al., 2023).

For an example of how Argentine ants in citrus orchards can be managed by low toxicity liquid baiting programs or biodegradable hydrogel ant baits to enhance biological control of hemipteran pests, see '7 Case study: the use of cover crops in citrus orchards in Southern California.'

6 Challenges associated with cover cropping

Farmers remain hesitant to adopt cover cropping practices due to several challenges, including the complexity of planning and managing cover cropping systems, the relative ease and reliability of prophylactic pesticide use, the logistical difficulties of implementing cover cropping practices, the additional demands of land, labor and resources, inconsistent pest control outcomes, as well as limited knowledge of effective management practices and access to suitable supporting extension services (Snapp et al., 2005; Godfray and Garnett,

2014; Tracey, 2014; SARE Outreach, 2015; González-Chang et al., 2019; Singhal et al., 2020). Concerns about potential yield reductions in the primary crop, increased disease risks, insufficient time for planting, or an inadequate growing season further contribute to the reluctance to adopt cover cropping (Snapp et al., 2005). These perceptions can hinder adoption despite evidence of the potential benefits cover crops can provide in sustainable farming systems.

6.1 Variable pest outcomes

A significant limitation of cover cropping for pest management is that increasing biodiversity does not necessarily guarantee a reduction in pest populations or associated crop damage (Andow and Risch, 1985; Tscharntke et al., 2016; Fountain, 2022). The effectiveness of cover crops can vary significantly based on local factors, such as soil type, pest levels and life stages, and the available pool of beneficial arthropod species, as well as the abundance of natural enemies in surrounding landscapes (Lundin et al., 2019; Gurr et al., 2017). For example, Herrera et al. (2022) demonstrated that the proximity of arable farms in Sweden to surrounding stands of trees positively impacted the abundance of predatory communities.

Efforts to increase biodiversity through cover cropping can have unintended consequences, affecting not only pests and their natural enemies but also other species in the agroecosystem under management. These dynamics can lead to unpredictable pest management outcomes, with some pests benefiting from the cover crop habitat (Irvin et al., 2016; Schaffers et al., 2008, van Schalkwyk et al., 2021) (see '3.2.3 Risks of benefiting pests while enhancing natural enemies'). For example, pests may use cover crops for shelter or feeding, potentially exacerbating pest populations and causing increased crop damage (Begum et al., 2006; Winkler at al., 2010).

In an organic apple orchard in Sweden, perennial wildflowers and grasses grown as a cover crop between tree rows enhanced the abundance of predatory mites, ladybugs, lacewings, and hoverflies, but more importantly, benefited overwintering populations of pestiferous leafrollers (tortricids) (Pålsson et al., 2022). However, if natural enemy fitness can be enhanced more than the pest, especially the search efficiency and prey conversion rates of natural enemies, cover crops can still be useful for enhancing ecosystem services that support pest management (Kean et al., 2003).

To help address the unpredictability of the impacts of cover crops on natural enemy and pest abundance, systematic reviews and meta-analyses have been conducted to determine overall impacts of cover crops. Letourneau et al. (2011) conducted a meta-analysis using hundreds of case studies involving the effect of plant diversification schemes on natural enemies and herbivores and found that herbivores were suppressed on average by 23%, and natural enemies increased by 44%, in cropping systems with higher diversity plantings compared with lower diversity plantings (diversity measurements were based on calculated Hedges' *d* values). Abad et al. (2021) showed that 95% of studies in vineyards reported no increase in pest abundance in the presence of cover crops and 50% showed a decrease in pest populations, most notably for mites and leafhoppers, the latter due to the presence of *Anagrus* spp. egg parasitoids.

A meta-analysis of cover crops used in apple orchards showed that ground covers and flowering cover crops generally resulted in a moderate positive effect on the abundance of natural enemies, with parasitoids being especially favored by flowers and predators by ground cover (Judt et al., 2023). Rowen and Blubaugh (unpublished data) reviewed 59 papers published between 1986 and 2023 that compared cover cropped and non-cover cropped treatments and quantified arthropod pests and predators. Overall, cover cropping reduced herbivore densities across cropping systems and increased natural enemy populations, particularly in perennial systems.

6.2 Reduction of crop yields

Another significant drawback of cover cropping is competition with primary crops, which can reduce yields (Letourneau et al., 2011). For example, using lucerne as a cover crop in soybean enhanced populations of beneficial predators and improved aphid control, but resulted in a 26% yield reduction due to competition for nutrients (Schmidt et al., 2007). Reduced yields are particularly problematic for short-duration, high-value crops like vegetables, where even minor losses can significantly impact premium pricing and profitability (Singhal et al., 2020).

In vineyards, cover crops often compete with vines for water and nutrients, leading to yield reductions in 56% of the 68 studies reviewed by Abad et al. (2021). However, this reduction in vine vegetative growth was also commonly associated with the benefit of reduced incidence of key fungal diseases (Abad et al., 2021).

To mitigate competition between cover crops and primary crops, several strategies have been proposed, such as ensuring adequate fertilization and irrigation of the primary crop, using less competitive or low growing cover crops, growing more vigorous cultivars of the primary crop, increasing the distance between the primary crop and cover crop, adjusting planting of the primary crop so that it is planted into a senescing cover crop, planting the cover crop in narrower strips or as dispersed patches, or using management practices such as mowing and tillage to suppress cover crop vigor (Hooks and Johnson, 2003).

Reductions in yield from cover cropping are not always inevitable (Abad et al., 2021). In some cases, cover cropping can improve yields by enhancing pest control and soil health. For example, cotton planted simultaneously with crimson clover produced yields over 150% higher than cotton monocultures (Yancey, 1994). This increase was partly attributed to enhanced soil fertility from nitrogen fixation, and high levels of pest control in cover crop fields, whereas monoculture cotton suffered infestations of beet armyworms (Noctuidae) and whiteflies, requiring 8–12 insecticide applications for management (Yancey, 1994). Some yield reductions attributed to cover crops may be explained by reduced crop density, as cover crops can reduce the area planted in the primary crop (Letourneau et al., 2011).

6.3 Other challenges and trade-offs

Other challenges with implementing cover crops for pest management include increased frost risk due to changes in microclimate caused by the cover crop canopy (Shields et al., 2019), lower soil temperatures which may slow down seedling emergence of the primary crop (Fanadzo et al., 2018), increased disease pressure on the primary crop (Rieux et al., 1999; Clark, 2007; Bone et al., 2009), potential for cover crop to compete with the primary crop for pollinators (Free, 1993), the risk of the cover crop becoming a weed (Gillespie and Wratten, 2017), and increased water use to support cover crops and possible reductions in crop quality due to this supplemental water use (Irvin et al., 2016). Allelopathic effects from certain cover crops can inhibit seed germination or stunt the growth of the primary crop if not managed appropriately (Clark, 2007; Zhang et al., 2007).

In some cases, cover crops can negatively affect primary crop quality and size (Abad et al., 2021). Cover crops and associated management practices in vineyards and orchards can affect soluble sugars, titratable acidity, soluble solids, and other fruit characteristics, which can cause delays in harvest (Irvin et al., 2016; Giacalone et al., 2021). A delay in harvest may not necessarily be a disadvantage in some cropping systems. One study in fresh market tomatoes (*Solanum lycopersicum* L) showed that a hairy vetch (*Vicia villosa* Roth.) cover crop resulted in later production, but higher yields coincided with increased late season prices (Lu et al., 2000).

Cover crops can also inadvertently attract non-target pests, such as spider mites, which can migrate into the primary crop when the cover crop senesces or becomes drought-stressed (Bugg, 1991) (Fig. 10). Cover crops can affect non-arthropod pest populations, causing increased rodent populations and crop damage from gophers (Guerra and Steenwerth, 2012; Herz et al., 2019), or the cover crop may attract rabbits and deer in vineyards and orchards (Irvin et al., 2016).

Competition between the cover crop and the primary crop can lead to reduced vegetative growth of the primary crop which was a negative outcome

Figure 10 Cahaba vetch can harbor pestiferous spider mites, which may migrate into the primary crop as the cover crop senesces or experiences drought-stress. Photo credit: N. A. Irvin.

in apple orchards intersown with white clover (*Trifolium repens* L.) and a grass cover crop (Bugg, 1991). However, this can be an advantage in vineyards where cover crops reduced high-vigor growth which eliminated the need for some vineyard operations such as leaf removal and fruit thinning (Guerra and Steenwerth, 2012).

Poor establishment of the cover crop or lack of synchronization with the primary crop's growth cycle can undermine effectiveness (Irvin et al., 2016; Tschumi et al., 2016). Furthermore, some cover crops may support higher trophic levels, such as hyperparasitoids, and intraguild predation can increase, both of which can disrupt pest control (Stephens et al., 1998) (see '3.2.4 Avoiding intraguild predation and benefits to the fourth trophic level').

While these challenges may discourage growers from adopting cover cropping, they highlight the importance of careful planning and monitoring when using cover crops for pest management, as well as the need for ongoing research to better understand the effects of cover crops on pest complexes and primary crop yields. Tailored management practices and monitoring can help mitigate potential negative effects and maximize the ecological and economic benefits of cover cropping in pest management systems.

For a detailed evaluation of the trade-offs associated with cover cropping, see review by Gillespie and Wratten (2017). Uyttenbroeck et al. (2016) also reviewed 31 studies that discuss farmer perspectives on herbaceous flower strips and cover cropping challenges in different agroecosystems.

7 Case study: the use of cover crops in citrus orchards in Southern California

Citrus is a significant economic crop worldwide with an annual production of 143.8 million tons (FAO, 2021). In recent years, consumer demand for pesticide-free, low chemical residue, and organic produce has grown substantially (Shafie and Rennie, 2012). The citrus industry in California plays a critical role in the state's economy, producing about 79% of fresh citrus in the USA, with an estimated value of $2.15 billion (USDA, 2023b). To meet consumer preferences, IPM programs for citrus in California emphasize the use of selective insecticides or the application of broad-spectrum insecticides at minimal rates and/or frequencies (Grafton-Cardwell et al., 2015; Monzó and Stansly, 2020). This approach aims to reduce negative impacts of insecticides on natural enemies, thereby optimizing biological control of citrus pests.

California citrus is affected by several economically important hemipteran pests, including Asian citrus psyllid (*Diaphorina citri* Kuwayama), brown soft scale (*Coccus hesperidum* L.), California red scale (*Aonidiella aurantii* [Maskell]; this hard scale does not produce honeydew), and citrus mealybug (*Planococcus citri* [Risso]) (Sorribas and Garcia-Mari, 2010; Grafton-Cardwell, 2015; UCIPM, 2020). The honeydew produced by some of these hemipteran pests promotes the growth of sooty mold, which degrades fruit during the packing process (UCIPM, 2020). Also, Asian citrus psyllid vectors the bacterium *Candidatus* Liberibacter asiaticus (*C*Las), the putative causative agent of a lethal citrus disease, huanglongbing (HLB), or citrus greening disease. HLB manifests as leaf yellowing, small misshapen fruit with bitter juice, and tree death within 5–8 years of infection (CDFA, 2022).

Argentine ant is a significant pest in citrus orchards because they disrupt biological control of honeydew-producing hemipteran pests by forming mutualistic food-for-protection relationships with these pests (McCalla et al., 2023; Milosavljević et al., 2024). Honeydew, a sugary waste product produced by sap-sucking hemipteran pests, results from a phloem feeding diet, is coveted by Argentine ants, and in return for this food, ants protect hemipteran pests from parasitoids and predators (Hoddle et al., 2022a; McCalla et al., 2023; Milosavljević et al., 2024). Argentine ants also disperse honeydew-producing pests to more suitable sites throughout citrus groves which can result in benefits such as increased survival, fecundity, and maturation for both species and collectively result in higher rates of population growth and economic damage (Styrsky and Eubanks, 2007; Anastasio et al., 2021; McCalla et al., 2023).

By protecting hemipterans from their natural enemies, and by stimulating feeding rates, fecundity, survival, and dispersal, ant-tending intensifies the negative effects of honeydew-producing pests on their host plants and results in increased crop damage, stunted plant growth, reduced leaf area, and the introduction of plant pathogens (e.g. *C*Las) (Anastasio et al., 2021; Milosavljević et al., 2021). Even armored scales, like California red scale, that do not produce honeydew, benefit from the presence of ants in citrus orchards because they coexist with honeydew producers and indirectly benefit from ants searching for honeydew which results in the accidental disruption of natural enemies of hard scales (Anjos et al., 2021; McCalla et al., 2023).

Research in commercial citrus orchards in southern California is exploring ways to enhance IPM programs by incorporating ant management strategies and promoting biodiversity and ecosystem services through incorporating flowering cover crops into citrus pest management programs. Here we outline advances in the selection of cover crop species, placement of cover crops, and recent research investigating the effect of flowering cover crops and ant management on natural enemy abundance, ant abundance, and populations of key hemipteran pests in southern California citrus.

7.1 Selection of cover crop species

Eight flowering plant species were tested in the laboratory to evaluate their impact on the longevity, survivorship, and reproduction of *Tamarixia radiata* (Waterston) (Fig. 1d), a parasitoid of Asian citrus psyllid nymphs. Among the plant species tested, buckwheat significantly enhanced the survival and fecundity of *T. radiata* compared with the water-only treatment (Table 3). No other floral resources enhanced egg load compared with water only. These results suggest that buckwheat is a strong candidate for use as a flowering cover crop that exhibits potential to improve the fitness of *T. radiata* in citrus orchards (Irvin and Hoddle, 2021).

Conversely, desertbells (*Phacelia campanularia* A. Gray), a drought-tolerant plant native to California, performed poorly in this laboratory study. *Tamarixia radiata* only lived for 1 day when provided access to desertbells, likely due to the plant's long floral tube, which restricts nectar access by parasitoids (Jervis et al., 1993; Baggen et al., 1999). Additionally, outward-pointing hairs on the style and ovary, along with scale-like appendages within the floral tubes, may deter feeding or obstruct access for *T. radiata* (Baggen et al., 1999; Jacob and Evans, 2000). Such structures can induce extensive grooming in this parasitoid species, further reducing its ability to feed effectively (Patt and Rohrig, 2017). Interestingly, female *T. radiata* demonstrated equivalent fitness when females were provided orange blossoms or water alone highlighting that even abundant floral resources on citrus trees can fail to enhance survival or reproduction of natural enemies (Irvin and Hoddle, 2021).

In southern California, the larvae of the hoverfly species (*Allograpta* spp.) are dominant predators of Asian citrus psyllid, accounting for approximately 60% of recorded nymph predation (Kistner et al., 2017). *Allograpta obliqua* (Fig. 1f), in particular, is a voracious predator, with individual larvae consuming on average 420 Asian citrus psyllid nymphs over the course of their development (Irvin et al., 2021). Field studies within close proximity of citrus orchards in southern California demonstrated that predatory hoverflies prefer feeding from buckwheat and alyssum flowers compared with phacelia and native California poppy (*Eschscholzia californica* Cham.) (Table 3) (Fig. 3a) (Irvin et al., 2021). These preferences appear to be attributable to differences in flower morphology. Buckwheat and alyssum, with their shallow floral tubes, offer easier access to nectar, while the deep corolla of phacelia limits nectar use by certain natural enemies (van Rijn and Wäckers, 2016). Furthermore, blue and purple flowers, such as those of phacelia, are less effective in stimulating feeding by hoverflies (van Rijn and Wäckers, 2016), while California poppy flowers provide pollen, but little or no nectar (Percival, 1955).

In contrast, studies outside of California have shown that hoverflies are attracted to phacelia and feed on pollen produced by this flower (Sengonça and Frings, 1988; Hickman and Wratten, 1996; Colley and Luna, 2000). These differences may reflect variations in hoverfly species' floral preferences across regions. In the arid climate of southern California, nectar may serve as the only source of water in the environment (Petanidou, 2007), thereby influencing hoverfly flower preferences. Alternatively, the inability of phacelia to attract feeding hoverflies, ladybugs, parasitoids, or predatory wasps may have been attributable to the limited flowering period, averaging only 15 days in southern California (Irvin et al., 2021). In contrast, in regions like New Zealand, phacelia blooms for 3 months (Bowie et al., 1995). The difference in the length of flowering by phacelia between these two studies demonstrates the value of conducting plant screenings in the specific region of interest as flowering periods need to overlap at times when beneficial insects are present and responsive to these resources.

In California field studies, alyssum was more effective than buckwheat in attracting feeding adult hoverflies and parasitoids, while buckwheat attracted greater numbers of ladybugs and predatory wasps (Table 3). These findings highlight the potential benefits of using mixed species plantings in cover crop systems, since different plant species may support distinct natural enemy communities within the same primary crop system (Chaney, 1998; Hogg et al., 2011b).

California poppy flowered for up to 90 days longer than buckwheat and alyssum which illustrates its potential to provide pollen to natural enemies for long periods (Table 3). However, California poppy was only marginally attractive to feeding adult hoverflies and ineffective at attracting adult ladybugs, parasitoids, or predatory wasps (Table 3) (Irvin et al., 2021). In southern California, buckwheat and alyssum have the advantage of rapid flowering, requiring approximately 30 days from sowing to flowering when sown in March

Table 3 Comparative results from several studies investigating the use of flowering cover crops for enhancing natural enemies of several economically important hemipteran pests, including Asian citrus psyllid (*Diaphorina citri* Kuwayama) (ACP), in southern California citrus orchards

	Alyssum, *Lobularia maritima*	Buckwheat, *Fagopyrum esculentum*	Poppy, *Eschscholzia californica*	Phacelia, *Phacelia tanacetifolia*
Days to flowering (sown in Mar 2018)[1]	Flowered in ~30 days	Flowered in ~30 days	Flowered ~25 days after buckwheat and alyssum	Flowered ~30 days after buckwheat and alyssum
Length of flowering (sown in Mar 2018)[1]	Flowered for ~75 days	Flowered for ~65 days	Flowered ~80–90 days longer than alyssum and buckwheat	Flowered ~50–60 days shorter than alyssum and buckwheat
Attraction of feeding hoverflies during Apr–Aug 2018[1]	Attracted 80% more than buckwheat and 1857% more than poppy	Attracted 985% more than poppy		Attracted none
Attraction of feeding ladybugs during Apr–Aug 2018[1]		Attracted 1080% more than alyssum	Attracted none	Attracted none
Attraction of feeding parasitoids during Apr–Aug 2018[1]	Attracted 700% more than buckwheat		Attracted none	Attracted none
Attraction of feeding predatory wasps during Apr–Aug 2018[1]		Attracted 313% more than alyssum	Attracted none	Attracted none
Effect on the longevity and fitness of *Tamarixia radiata*, a parasitoid of ACP, in the laboratory in 2017[2]	Not included	Enhanced *T. radiata* survival by 29 days and increased egg load by ~5 mature eggs compared with the water-only treatment.	Not included	Not included

(*Continued*)

Table 3 *Continued*

	Alyssum, *Lobularia maritima*	Buckwheat, *Fagopyrum esculentum*	Poppy, *Eschscholzia californica*	Phacelia, *Phacelia tanacetifolia*
Effect on hoverfly oviposition and predator abundance on potted citrus trees infested with ACP in the field Jun–Jul 2018[1]	Enhanced hoverfly oviposition by 247% and predator abundance by 155% compared to controls lacking flowering alyssum	Not included	Not included	Not included
Effect on the survival of ACP nymphs on potted citrus trees in the field Jun–Jul 2018[1]	0% of ACP nymphs survived on potted plants in the alyssum treatments, which was significantly lower than 10% survival of nymphs in the controls	Not included	Not included	Not included
Effect on the abundance of hoverflies counted during 2-min visual counts every month from May to Oct 2022[3]	When combined with ant bait hydrogels, a buckwheat and alyssum cover crop enhanced hoverfly abundance by up to 1852% compared with control plots without buckwheat and alyssum.		Not included	Not included
Effect on the abundance of hemipteran pests on foliage samples removed in Oct 2022[3]	When combined with ant bait hydrogels, a buckwheat and alyssum cover crop reduced the abundance of hemipteran pests by 57–60% compared with control plots without buckwheat and alyssum.		Not included	Not included

[1]Irvin et al., 2021
[2]Irvin and Hoddle, 2021
[3]Milosavljević, Irvin, and Hoddle, unpublished data

(Table 3). Alyssum, when sown in February, began flowering at the same time as buckwheat, but continued blooming for 42 days longer (Irvin et al., 2021).

Together, these studies illustrate the importance of conducting plant screenings to assess: (1) floral attractiveness and resource accessibility by key natural enemies and (2) the growing and flowering phenology of plant candidates in the specific region of interest to ensure nectar availability coincides with natural enemy presence at the times pest control is needed (Fig. 3). Combining quick-flowering species with those offering extended nectar production in mixed-species sowings can provide extended periods of nutrition availability to support natural enemies and potentially enhance their effectiveness as biological control agents of key pests in primary crops.

7.2 The placement of cover crops in citrus orchards

Globally, the alleyways between rows of citrus trees are typically maintained as bare soil using herbicides or cultivation to control naturally occurring vegetation (Monzó et al., 2020). In southern California, hot and dry weather during spring and summer further suppresses natural vegetation. To address the resulting low plant diversity and to provide floral resources for natural enemies, the placement of a mixed cover crop of alyssum and buckwheat was evaluated in citrus orchards in this region for enhancing biological control of hemipteran pests.

In 2021, buckwheat and alyssum seed were sown at the recommended rate (buckwheat: 4.46 g/m or 3 lb per 1,000 feet; alyssum: 0.93 g/m or 1 ounce per 1,000 feet) in 10 plots throughout a citrus orchard in Mentone, San Bernardino County, California, on March 29 in the open center between tree rows. Supplemental irrigation was provided using a separate water supply from the existing tree irrigation system. However, the sown cover crop failed to establish due to pre-emergent herbicide residues in the orchard soil, which negatively affected seed germination (Figs. 11 and 12a), and herbivory by deer and rabbits, which consumed germinated buckwheat seedlings (Fig. 12b) (Irvin et al., 2022). Attempts to address these challenges by transplanting alyssum seedlings directly into the soil also failed in some plots since drought conditions in mid-July 2021 restricted water use and significantly reduced the amount of available water to irrigate cover crops (Fig. 12a).

In 2022, several adjustments were made to improve cover crop establishment. Seeds were sown earlier (February 28) to take advantage of cooler spring temperatures, at twice the recommended sowing rate, and within tree rows where there is partial shade and, presumably, less herbicide residue due to leaching from tree irrigation. These modifications eliminated the need for supplemental watering, improved germination rates, and resulted in successful establishment of cover crops (Fig. 12c).

Figure 11 A greenhouse trial comparing cover crop seed germination rates in pots with normal potting mix (a) and orchard soil (b) suggested that low germination rates in the orchard soil may have resulted from pre-emergent herbicide residues, which inhibited seed germination. Photo credit: N. A. Irvin.

Figure 12 Causes of cover crop failure: (a) pre-emergent herbicide preventing buckwheat and alyssum seed emergence, followed by drought conditions and lack of irrigation killing alyssum transplants; (b) rabbit and deer damage to buckwheat seedlings. Moving the buckwheat and alyssum cover crop from the middle of the row to the tree line of citrus orchards (c), where partial shade and higher herbicide residue leaching occur, eliminated the need for supplemental watering and enabled successful germination and establishment. Photo credit: N. A. Irvin.

7.3 The timing of cover crops in citrus orchards

Selecting the most suitable insectary plants for a specific pest and crop scenario requires careful consideration of the flower attractiveness, accessibility, and phenology, as well as the phenology of the natural enemies and pest complex (Patt et al., 1997). For instance, flowers should bloom during periods of low prey availability, during pest outbreaks, and peak foraging activity of natural enemies on nectar and insect prey (Bickerton and Hamilton, 2012). Optimizing this timing can involve adjusting planting dates of cover crops composed of single or mixed plant species to align with the predicted activity periods of pests and their natural enemies (Bickerton and Hamilton, 2012).

In southern California citrus orchards, establishment of insectary plants should be timed so that plants bloom during the active periods of key natural enemies. To ensure that cover crop flowering is synchronized with periods of natural enemy activity, adult hoverflies and ladybugs were monitored from May through October in 2022 by conducting monthly 2-min timed visual counts of hoverflies and ladybugs observed along six 15 m x 1 m transects distributed throughout a citrus orchard located in Mentone, San Bernardino County, California. A total of 173 hoverflies and 7 ladybugs were observed during the visual counts. Adult hoverflies, a key predator of Asian citrus psyllid, were present every month from May through October (Fig. 13), coinciding with the two peak activity periods of Asian citrus psyllid in spring and fall (Milosavljević et al., 2021).

In a separate study conducted in southern California from June through November, Irvin et al. (2021) demonstrated that hoverflies oviposited between 1 and 20 eggs on potted citrus trees artificially infested with Asian citrus psyllid,

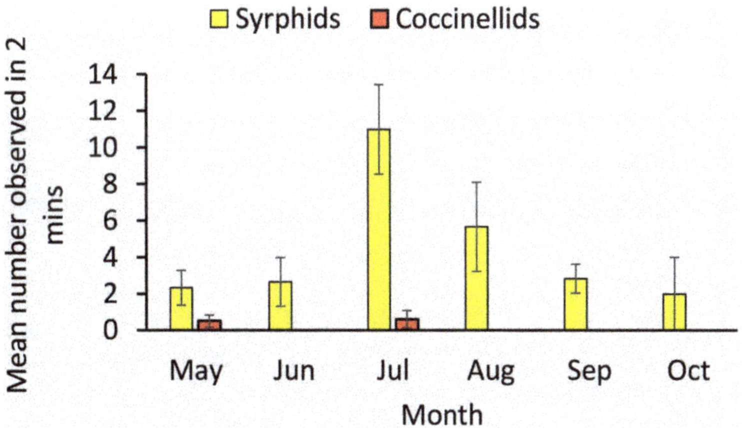

Figure 13 The mean (±SEM) monthly number of hoverflies (syrphids) and ladybugs (coccinellids) counted during 2-min timed visual observations of six 15 m x 1 m transects located in six alyssum and buckwheat cover crop plots positioned throughout a citrus orchard in Mentone, California, USA.

every month for the duration of the experiment. Collectively, these two studies illustrate that hoverfly abundance and oviposition are consistent throughout the growing season and coincide with peak activity of a key citrus pest, Asian citrus psyllid. Therefore, accessible floral resources are required by these key predators from May through November.

Flowering weeds represent a potential alternative to establishing and maintaining insectary plantings that could provide SNAP resources for natural enemies in commercial citrus groves in southern California. The flowers of some weedy herbaceous plant species have been shown to act as food resources for natural enemies, while flowering weeds can influence the diversity and abundance of arthropod communities in orchards (Liang and Huang, 1994; Norris and Kogan, 2000; Serée et al., 2023). To better understand weed biodiversity and density under normal orchard practices of herbicide use and mowing in southern California citrus orchards, monthly transect walks between May and November 2021 were conducted at an orchard in Mentone, San Bernardino County, California, to determine the identity and number of flowering weeds.

Peak flowering of weeds occurred during May and sharply declined in summer and again in fall (Fig. 14) because of hot dry conditions. This suggests that it may be more important to sow nectar-providing cover crops in June through November when naturally occurring nectar resources are lowest. Although on average 26 flowers/m^2 were present throughout the citrus orchard in May, not all flowering resources function as food sources for natural enemies due to flower morphology that may prevent natural enemies from accessing nectar, or because nectar produced by weed species may be of poor quality (Karp et al., 2018) (see '3.2 Considerations for selecting insectary plants as cover crops').

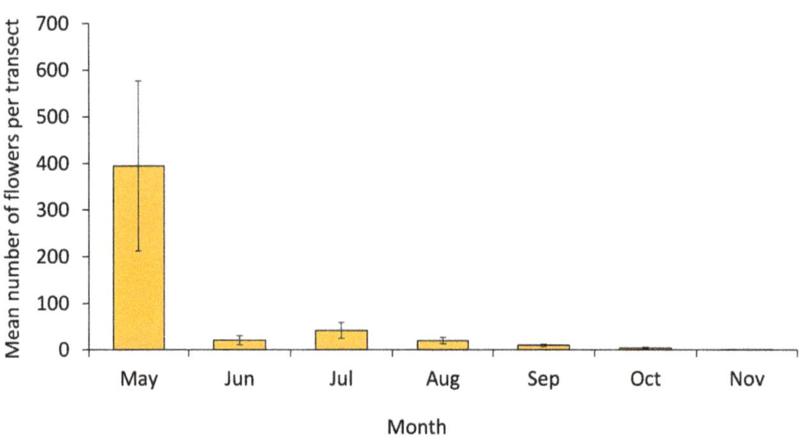

Figure 14 The mean (±SEM) number of flowers on weeds observed along 40 15 m x 1 m long transects each month in a citrus orchard in Mentone, California. Figure from Irvin et al. (2022); reproduced with permission from California Association of Pest Control Advisors.

Hoverflies were frequently observed feeding from mustard (*Brassica nigra* [L.] Koch) and sow thistle (*Sonchus* sp.) flowers. Therefore, it is likely these weeds served as nectar sources in the orchard. However, it is unknown whether weed density was high enough to enhance survival and fitness of resident natural enemies throughout the year. Mustard and sow thistle only accounted for 20% and 8% of flowering weeds in May, with scarlet pimpernel (*Lysimachia arvensis* [L.] U. Manns and Anderb.), a poor resource, being dominant (Irvin et al., 2022).

Based on the results and challenges of these research projects on the selection, timing, and placement of cover crops in southern California citrus orchards, it is recommended to sow a buckwheat and an alyssum cover crop in the tree line, using the existing tree irrigation schedule, and to stagger sowings to provide sequential flowering of cover crops from April through November. In addition, further research is needed on incorporating Argentine ant management strategies into an IPM program for southern California citrus.

7.4 Reducing Argentine ant densities in citrus orchards

Cover crop efficacy can be enhanced within the context of IPM programs. With respect to natural enemies, the increased benefits from cover crops may be unsuccessful if parasitoids and predators are unable to inflict sufficient control of pest populations due to the presence of antagonists, such as Argentine ant. Insecticide-based ant management programs need to be conducted in a targeted manner that delivers toxicants to ants while minimizing insecticide exposure to non-target organisms, especially pollinators and natural enemies. This can be achieved with low toxicity liquid baiting programs.

To manage ants with minimal insecticide use, a biodegradable alginate hydrogel bead laced with a minute amount of insecticide, such as thiamethoxam (0.001%) or spinosad (0.01%), in a 25% sucrose solution (Fig. 15), is a readily deployable ant control option that reduces ant activity by up to 90% after three applications spaced 3 weeks apart in commercial citrus orchards in southern

Figure 15 Argentine ant feeding from biodegradable alginate hydrogel beads laced with a minute amount of insecticide in a 25% sucrose solution. Photo credit: Mike Lewis.

California (McCalla et al., 2020; Milosavljević et al., 2024). When Argentine ants are suppressed with low insecticide baiting techniques, natural enemies provide exceptional levels of control of hemipteran pests in citrus orchards (McCalla et al., 2023).

A large-scale multi-site field trial was conducted at two southern California commercial grapefruit (*Citrus paradisi* Macfad) and orange (*Citrus sinensis* (L.) Osbeck) orchards to investigate the effect of establishing a flowering cover crop from April through November and/or applying hydrogel beads on populations of natural enemies, ants, and hemipteran pests. Six replicates of four treatments were set up in 2022 across the two field sites: (1) buckwheat and alyssum cover crop; (2) alginate hydrogel bead baits; (3) cover crop plus hydrogel bead baits; (4) control, or normal orchard practices without cover crops or hydrogel bead baits.

Cover crop plots consisting of buckwheat and alyssum were sown on February 28, 2022, in a 6 m x 1 m strip, located in the tree line in the center of the plot, and irrigated via sprinklers installed within the existing tree irrigation. Since buckwheat has a short flowering period, additional staggered sowings of buckwheat seed were made every month by removing buckwheat plants that had gone to seed and sowing cleared areas with new seed.

For the two treatments containing hydrogel bead baits, alginate hydrogel beads containing 0.01% spinosad (insecticide) and 25% sugar solution were applied at the base of each of 64 trees in each plot at a rate of 250 g/tree. Seven deployments of hydrogels were conducted between April and September; the first two applications were spaced 2 weeks apart, the next three applications were monthly, while the last two applications were separated by 2 weeks. Control plots were maintained under normal orchard practices without cover crops or hydrogel bead baits.

Adult hoverflies and ladybugs were monitored monthly between May and October by conducting 2-min timed visual counts of natural enemies visiting weeds and vegetation along a 15 m x 1 m transect, randomly selected within each treatment plot. Ants were monitored by infrared sensors mounted onto irrigation pipes (Hoddle et al., 2022b). Each plot had three sensors randomly allocated to irrigation lines for a total of 36 sensors at each site. Sensors self-activated every hour on the hour and recorded the number of ants breaking the IR beam spanning the irrigation pipe for 1 min. Ants were counted in plots over 10 days for pre-counts (prior to the deployment of hydrogels) and then over a period of 10 days following each hydrogel application.

To determine the abundance of key citrus pests in experimental plots, eight branch (~30 cm) and eight flush (branch tips containing new growth) foliage samples were removed from four study trees positioned at a diagonal crossing the center of each treatment plot (total of 64 samples plot). Samples were removed from trees in April (before hydrogel applications), May, and

October (2 weeks after the last hydrogel application). In the laboratory, foliage samples were examined under a dissecting microscope to estimate numbers of brown soft scale, California red scale, Asian citrus psyllid, and citrus mealybug.

Preliminary results from the Mentone orchard, pooled across all dates, demonstrated that the number of hoverflies observed during visual counts was over 3.5 times higher in the cover crop and 'cover crop + hydrogel' treatments compared to the hydrogel-only treatment and control plots (Fig. 16) (Table 3). This indicates that cover crops were attracting these key natural enemies and enhancing their abundance in cover crop treatment plots. These findings align with prior research demonstrating that flowering cover crops are a valuable tool for attracting and enhancing populations of predatory hoverflies in southern California citrus orchards (Irvin et al., 2021, 2022, 2024).

Mean ant counts and pest hemipteran counts in October were more than halved in the two treatments containing hydrogels when compared with control plots and cover crop plots without hydrogel deployments (Figs. 17 and 18) (Table 3) (Milosavljević, Irvin, and Hoddle, unpublished data). These results suggest that hydrogels effectively suppressed ant populations, which

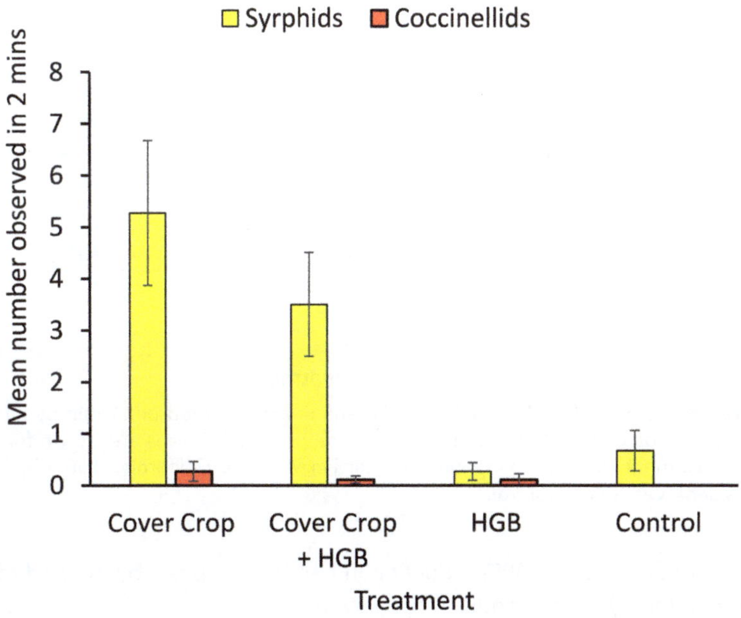

Figure 16 The mean number of hoverflies (syrphids) and ladybugs (coccinellids) counted during 2-min timed visual observations along a 15 m x 1 m transect in each of four treatment plots set up in a citrus orchard in Mentone, California; means are calculated over all observations carried out monthly from May through October 2022; HGB = hydrogel beads deployed for ant control.

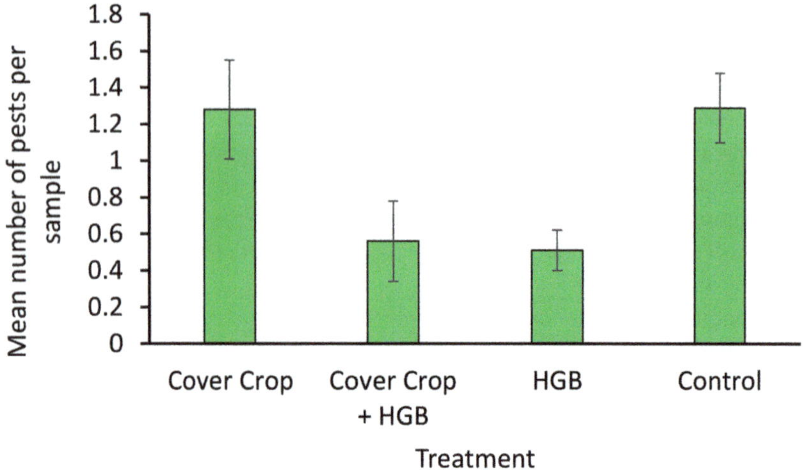

Figure 17 The mean number of hemipteran pests counted on twig and flush samples removed in October 2022 from trees in four treatment plots set up in a citrus orchard in Mentone, California; HGB = hydrogel beads deployed for ant control.

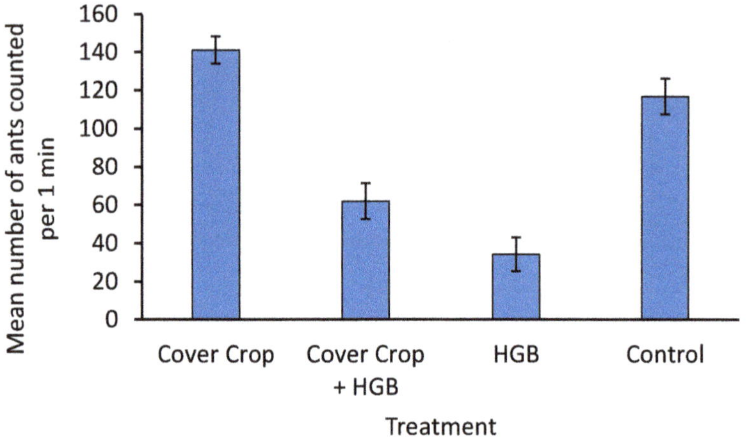

Figure 18 The mean (±SEM) number of Argentine ants counted per 1 min by infrared sensors recording every hour for 10 days in October 2022. Means are shown for each of four treatment plots set up in a citrus orchard in Mentone, California; HGB = hydrogel beads deployed for ant control.

subsequently led to a ~60% reduction in hemipteran pests by the end of the hydrogel deployment periods.

Interestingly, Argentine ants were observed feeding on nectar from buckwheat and alyssum flowers in cover crop plots, and there were 20% more ants counted in cover crop plots compared to control plots lacking flowering cover crops (Fig. 18). This indicates that ants utilized cover crop flowers as a sugar resource, which may have inadvertently subsidized ant populations in those plots. Many ant species, including Argentine ant, utilize floral and extrafloral nectar produced by a variety of flowering plant species (Koptur and Truong, 2005; Lach, 2005).

Collectively, the preliminary results from the Mentone citrus orchard indicate that while an alyssum and buckwheat cover crop successfully enhanced populations of hoverflies and ladybugs, this did not translate to a notable reduction in hemipteran pests unless hydrogels were also deployed for ant control. This suggests that the natural enemy-pest complex in southern California citrus orchards is strongly influenced by the presence of Argentine ants, which protect hemipteran pests from natural enemies (McCalla et al., 2023). This is similar to previous research that demonstrated that ant management in orchards and vineyards in California significantly reduces infestation by hemipteran pests (Cooper et al., 2008; Schall et al., 2018;McCalla et al., 2023; Milosavljević et al., 2024) and associated fruit damage (Daane et al., 2008). Similarly, reducing ants in Spanish citrus orchards enhanced the efficacy of generalist predators, reduced populations of pestiferous mealybugs, and decreased the amount of damaged fruit at harvest (Plata et al., 2023).

For example, a multi-season low toxicity liquid baiting program targeting Argentine ants in southern California resulted in a 96% reduction in ant activity in treated plots, accompanied by an 84–99% decrease in hemipteran pest infestations on twigs, flush, and fruit 1 year after treatment (McCalla et al., 2023). Effective ant management is particularly critical for controlling Asian citrus psyllid in California. Research has shown that 55–66% of Asian citrus psyllid colonies are tended by Argentine ants (Tena et al., 2013; Schall and Hoddle, 2017), and when parasitoids of Asian citrus psyllid are encountered by Argentine ants on patches of Asian citrus psyllid nymphs, up to 95% of these natural enemies are evicted from colonies or killed (Schall and Hoddle, 2017).

The results presented here reinforce the importance of integrating ant management with cover crops as part of integrated pest management programs targeting Asian citrus psyllid and other hemipteran pests infesting citrus in southern California.

8 Promoting cover crop adoption and future trends in research

González-Chang et al. (2019) proposed that if sufficient local scientific knowledge is available and extended to growers, developing a straightforward, biodiversity-focused framework that farmers can readily implement could significantly promote the adoption of habitat management practices across large-scale farming systems. However, designing experiments to investigate the effects of cover crops on natural enemy and pest populations, especially at highly local levels with a focus on a particular cropping system, presents significant challenges.

For example, in the laboratory, buckwheat and mustard improved the longevity and reproductive fitness of *Hadronotus pennsylvanicus* (Ashmead), an egg parasitoid of the leaf-footed bug *Leptoglossus zonatus* (Dallas), which is primarily a pest of almonds and pistachios in California's Central Valley (Straser et al., 2024). However, when sown in a 1 m wide strip between rows in a pistachio orchard, the cover crops failed to affect the abundance of *H. pennsylvanicus* or parasitism of sentinel deployments of leaf-footed bug eggs (Straser et al., 2024). This discrepancy could stem from the limitations of sentinel egg masses in accurately representing natural biological control rates (Mitchell et al., 1999; Jones et al., 2014) or from an insufficient duration of field trials to detect measurable impacts on pest suppression (Straser et al., 2024). Because of competing confounding factors, there is uncertainty regarding the efficacy of cover crops for promoting enhanced natural enemy activity toward leaf-footed bugs in California pistachio crops. Additional field trials could resolve this issue.

Studies that successfully demonstrate enhanced parasitism or predation in cover crop plots compared to control plots lacking cover crops often attribute these results to either increased recruitment of natural enemies or improved fitness of natural enemies in cover crop plots. However, these differences may also result from the redistribution of natural enemies from control plots to cover crop plots. If redistribution occurs, the benefit in one area may be offset by corresponding losses in another, yielding no net advantage. Determining whether natural enemy redistribution occurs requires large-scale experiments with entire fields dedicated to cover crops and sufficient buffer zones to prevent cross-dispersion from different study plots.

The importance of using adequate buffer zones was highlighted in research by Irvin et al. (2018) that examined the dispersal of parasitoids and predators from a buckwheat cover crop sprayed with a solution of yellow dye, albumin, and casein proteins to mark natural enemies. Marked mymarid parasitoids and spiders were captured on sticky traps in control plots not sprayed with the mark which suggested that these predators were able to disperse further than the 36 m buffer zones separating buckwheat cover crop and control plots. In fact, some parasitoid species have been shown to disperse up to 16–20 m per day (Langhof et al., 2005; Lavandero et al., 2005; Wanner et al., 2007), or in some instances, up to 30 m (Freeman-Long et al. 1998). Predatory hoverflies, on the other hand, can disperse over much greater distances, traveling at least 200 m from flowering alyssum and buckwheat plots (Wratten et al., 2003; Gontijo et al., 2013).

Cover crop research investigating the use of cover crops for managing pests in agricultural systems may also require longer periods of study to detect benefits on pest control or primary crop yields. Gillespie et al. (2016) and Gurr et al. (2017) both stressed the need for more long-term studies. There have been many instances where provision of non-crop resources did not have a positive effect until the fourth or fifth year of the study because sufficient time is needed to build up the required ecological processes that support higher density natural enemy populations (Herz et al., 2019; Fountain, 2022).

For example, in a central England annual cropping system, it took 4 years for the positive effects of a wildflower cover crop on primary crop yield to appear, with benefits becoming increasingly pronounced over a 6-year period (Pywell et al., 2015). Similarly, in Canadian orchards, 5 years were necessary to achieve significant pest damage reduction following the introduction of flowering cover crops (Bostanian et al., 2004). Consequently, the gradual build-up of beneficial arthropod populations is likely a long-term process, especially when using perennial flower strips as cover crops in perennial cropping systems (Straser et al., 2024).

Growers may be more willing to adopt cover cropping practices if they demonstrate clear economic benefits. However, the effects of cover cropping on crop yields and detailed economic evaluations are rarely completed, with few studies, like Gurr et al. (2016), focusing on these aspects. Determining economic injury levels, or the lowest pest density that causes economic damage, can also be particularly challenging for some pests, and subsequent pest management decisions need to be made with consideration to control costs and the anticipated future value of the crop (Rincón et al., 2019; Shields et al., 2019). The challenge of accurate economic analyses is further compounded by the multifaceted ecosystem services provided by cover cropping, including pest and disease management, pollination, nutrient cycling, soil aeration, water supply, animal fodder, flood mitigation, runoff prevention, enhancement of wildlife, water purification, field trafficability, soil microbial biodiversity, seed dispersal, weed control, and carbon sequestration (Rapidel et al., 2015; Rayl et al., 2018; Castro et al., 2021; Garcia et al., 2018; Yousefi et al., 2024).

For example, the successful 'push-pull' system in East Africa (Khan et al., 2014) provides pest management, soil fertility, nitrogen fixation, and animal fodder, resulting in higher maize and sorghum yields while also enhancing milk production (Rayl et al., 2018). Some benefits of cover cropping extend beyond individual farms (Wratten et al., 2012), while others, such as aesthetics and contributions to human well-being, lack direct monetary value but can enhance mental health and well-being and may indirectly drive increased farm gate sales as a result (Roberts et al., 2015).

While cover cropping has the potential to deliver multiple ecosystem services, it also presents challenges or potential ecosystem disservices (i.e. increased pest or disease prevalence), as detailed in '6 Challenges associated with cover cropping.' Cover crops need to be properly researched and managed to provide ecosystem services while avoiding or minimizing disservices (Garcia et al., 2018). Factors such as primary crop species and varieties, climatic zone, soil properties, spatial arrangement and rotations for annual crops, sowing densities, strip arrangements, field architecture, cover crop species, mowing, tillage, irrigation, and fertilization all influence the ecosystem services achieved (Garcia et al., 2018, Yousefi et al., 2024). Some benefits develop quickly while others develop gradually and require longer timeframes to manifest (Schipanski et al., 2014).

Methods for evaluating and managing ecosystem services provided by cover crops in agroecosystems and for designing agroecosystems providing specific services have been proposed for annual crop (Schipanski et al., 2014) and perennial cropping systems (Rapidel et al., 2015; Garcia et al., 2018). These frameworks provide the means to quantify ecosystem services and facilitate the decision to transition to more multifunctional agricultural systems.

Future research should prioritize assessing whether cover crops effectively reduce crop damage to an extent that justifies their use in terms of economic benefits such as reduced pesticide costs or increased yields. The negative effect of cover crops on primary crop yield can occur as a result of water depletion, especially in dry agro-ecosystems (Yousefi et al., 2024). The use of technological advancements like automated sensors that use the 'Internet of Things' (IoT) to collect real-time data on soil moisture and temperature, and pest densities can greatly enhance the accuracy of pest control decisions and guide high-precision targeted controls to specific areas requiring management as opposed to crop-wide applications to control pests (Hoddle et al., 2022b). Similarly, data from IoT monitoring can be used to manage automated irrigation control systems, thereby optimizing irrigation schedules and delivery of the appropriate amount of water to cover crops without negatively affecting yield or quality of the primary crop (Yousefi et al., 2024).

Future studies should also investigate when and how best to apply insecticides to avoid negative effects on beneficial insects visiting cover crop flowers or natural enemies inhabiting cover crops. This includes identifying thresholds for the natural enemy-to-pest ratio required for effective biological control and determining when insecticide applications are most necessary (Herz et al., 2019). Furthermore, selecting cover crop species that can be harvested and sold (i.e. intercropping) presents a promising avenue for increasing crop yield per unit area while preserving the biodiversity benefits of cover cropping (Huss et al., 2022).

Developing management protocols that enhance multiple ecosystem services would provide greater value and likely boost grower adoption rates if economic benefits can be clearly demonstrated (Geertsema et al., 2016; Gurr et al., 2016). Therefore, multi-year studies are required, assessing multiple ecosystem services that are tailored to specific cropping systems and regional conditions.

Such research findings need to be communicated to growers in formats suitable for their educational levels and in a form that can be easily accessed and understood (Wyckhuys et al., 2018; Shields et al., 2019). Tools like videos, leaflets, television shows, web pages, postcards, or social media, as well as Apps on smart devices (i.e. phones or tablets), can aid in transferring knowledge and promoting sustainable pest management practices (Wyckhuys et al., 2018). Initiatives like 'farmer field schools' and 'farmer-to-farmer strategies' enhance communication between scientists and growers (Holt-Giménez, 2008; Amudavi et al., 2009; Altieri and Toledo, 2011). In these programs, growers test new technologies on their farms, and when they succeed, it encourages adoption by neighboring properties, leading to potential widespread regional adoption. Additionally, local and national agricultural incentive programs which offset initial investments to start up new farming practices can be used to promote cover crop adoption (Union of Concerned Scientists, 2013).

Globally, countries such as China, Bhutan, and parts of India have successfully integrated habitat management techniques like cover cropping into pest control programs (González-Chang et al., 2019). In December 2018, France banned the use of five bee-killing insecticides (Samuel, 2018), and recent EU policies aim to halt biodiversity loss and reduce pesticide use by 50% by 2030 (Stojanovic, 2021). These legislative changes drive innovation in agriculture, and cover crop research and adoption stand to benefit from these mandates. Many pesticides that are banned or are being phased out in the EU, Brazil, and China are unfortunately still widely used in the USA (Donley, 2019).

However, there are indications that stricter regulations are emerging in the USA as well, as seen recently in California. For example, in 2020, California joined Hawaii, Maryland, New York, and Oregon in banning the insecticide chlorpyrifos due to concerns over its neurodevelopmental effects, particularly in children (Price, 2019; Erickson, 2023). Additionally, the California Department of Pesticide Regulations is enacting regulations to reduce the use of neonicotinoid insecticides in agriculture and urban areas. As national and international policies increasingly regulate synthetic pesticides, sustainable practices, such as cover cropping, will likely become an increasingly important component of IPM strategies for growers.

9 Conclusion

In conclusion, cover cropping represents a promising biodiversity-based strategy for managing arthropod pests while addressing many of the environmental challenges associated with modern agriculture. By increasing habitat complexity and supporting communities of natural enemies, cover crops can suppress pest populations through both top-down effects by supporting natural enemies with resources like shelter, nectar, pollen, and alternative prey and through bottom-up effects by disrupting the pests' ability to find and exploit concentrated host crops. When carefully selected and properly managed through appropriate species choice, timing of sowing, placement, mowing, irrigation, and termination, cover crops can enhance biological control, improve ecosystem services, and complement other strategies within integrated pest management programs.

Despite these benefits, the success of cover cropping for pest management is highly context dependent. Differences in climate, cropping systems, pest complexes, and management practices can influence outcomes, and poorly implemented cover crops may lead to unintended trade-offs such as competition with the primary crop, increased pest pressure, frost risk, or changes in crop quality. The citrus orchard case study from southern California highlights both the potential and the complexity of implementing cover crops in commercial systems, demonstrating that effective pest suppression often requires integration with additional IPM tools, such as targeted ant management, to maximize the benefits of enhanced natural enemy populations.

Advancing the role of cover crops in pest management will require continued research focused on long-term, field-scale studies that evaluate ecological outcomes, crop yields, and economic viability. Integrating emerging technologies such as real-time monitoring systems, refining cover crop management strategies, and developing region-specific recommendations will help optimize their performance in diverse agricultural systems. Additionally, successful adoption depends on effective outreach and knowledge transfer to growers so that scientific findings can be translated into practical management guidelines.

As global agriculture seeks to reduce reliance on synthetic pesticides and transition toward more resilient and sustainable production systems, cover crops have the potential to play an increasingly important role. When integrated strategically into farming systems, cover cropping can contribute to healthier agroecosystems, improved pest management, and the long-term sustainability of food production.

10 Where to look for further information

10.1 Articles

Ministry of Agriculture and Lands (2019) identifies types of beneficial arthropods (predators, parasitoids, and pollinators) and provides photographs and descriptions of common examples of natural enemies.

González-Chang et al. (2019) contains a good overview of the benefits, limitations, and future prospects of habitat management, such as cover cropping, trap cropping, and intercropping, as strategies for pest control. They also provide a table summarizing 44 studies using 24 insectary plant cover crops and listing their target pests and effects on natural enemies.

Gillespie and Wratten (2017) and Singhal et al. (2020) give overviews of the ecosystem services and disservices that may result from cover cropping, discuss barriers to adoption and limitations of using cover crops, and offer advice on methods to manage or minimize ecosystem disservices.

Hokkanen (1991), Shelton and Badenes-Pérez (2006), Piñero and Manandhar (2015), Reddy (2017), Sarkar et al. (2018), and Badenes-Pérez (2019) provide detailed and updated lists of target insect species, their potential trap crops, and application strategies.

The following reviews and meta-analyses explore cover crop options, benefits, and management strategies in fruit crops (Fountain, 2022), apple orchards (Judt et al., 2023; Herz et al., 2019), vineyards (Wilson and Daane, 2017; Ingels et al., 1998; McGourty, 2004; Garcia et al., 2018), and vegetable and grain production (Lu et al., 2000).

10.2 Online publications

https://www.sare.org/publications/cover-crops-ecosystem-services/impact-of-cover-crops-on-natural-enemies-and-pests/: Explains how cover crops can enhance agricultural pest management by supporting populations of natural enemies that prey on or parasitize crop pests.

https://www.sare.org/wp-content/uploads/Manage-Insects-on-Your-Farm.pdf: Introduces strategies that focus on managing the entire farm ecosystem to maintain pest populations at acceptable levels using multiple complementary approaches including soil management and creating habitats that attract and sustain natural enemies of pests to naturally suppress pest populations. It includes real-world examples and case studies of farmers successfully implementing ecological pest management techniques, offering practical guidance for integrating these strategies into various farming systems.

https://attra.ncat.org/wp-content/uploads/2019/05/covercrop.pdf#:~:text=To%20impart%20a%20sense%20of%20the%20importance,weed%20suppression%2C%20and%20soil%20and%20water%20conservation.&text=In%20addition%20to%20nitrogen%20from%20legumes%2C%20cover,help%20recycle%20other%20nutrients%20on%20the%20farm.

Sullivan, P., 2003. Overview of Cover Crops and Green Manures, National Center for Appropriate Technology, ATTRA Sustainable Agriculture, Publication IP024; an overview of the principal uses, benefits, and limitations of cover crops, use in crop rotations, pest management, and economics of cover crops.

10.3 General websites

https://agricology.co.uk/resource/agrodiversity-toolbox/ Agroecology: provides information on more than 100 cover crops and contains links for cover crop selection tools for Europe and information on appropriate machinery, best management, and economics of cover crops.

https://www.midwestcovercrops.org/ Midwest Cover Crops: contains information on cover crop selection, planting, termination, and management. Cover crop recipes, field guides, and selector tools.

https://westerncovercrops.org/ Western Cover Crops Council: contains information and resources on cover cropping, plus a selector tool.

In most states, the Extension Service and Agricultural Experiment Stations offer information and publications on cover crops. Examples of extension sites are:

https://extension.oregonstate.edu/ Oregon State University: search 'cover crops' to find various resources and information about cover crops.

https://www.ucanr.edu/site/california-cover-crops-resources University of California Agriculture and Natural Resources: information on common and native cover crop species, cover crop management, benefits, and challenges. Includes links to selection tools and farmers' experiences with cover cropping.

https://www.canr.msu.edu/cover_crops/index Michigan State University: contains cover crop selection tool, information on growing and terminating a cover crop, and businesses that offer cover crop-related services to farmers across Michigan.

https://cropsandsoils.extension.wisc.edu/author/cover-crops-resources/ University of Wisconsin-Madison: cover crop recommendations for growing corn, soybean, and wheat, with information on common cover crops such as rye, barley, oats, radish, red clover, triticale, and field peas.

https://sarep.ucdavis.edu/covercrop UC SAREP Cover Crop Data Base: contains a thorough cover crop species database, with quantitative information on specific cover crops, and two slide shows covering the use of cover crops for annual and perennial farming systems. Also includes journal articles, conference proceedings, standard textbooks, unpublished data, and personal communications from researchers and farmers. The database includes over 400 cover crop images and valuable cover crop information such as pounds of nitrogen fixed, seeding rates, suggested timing for seeding, etc. Focuses on California and other similar climatic situations.

https://ahdb.org.uk/knowledge-library/cover-crops-for-integrated-pest-management: The Agriculture and Horticulture Development Board outlines the mechanisms of how cover crops contribute to pest management and lists cover crop resources.

https://mda.maryland.gov/resource_conservation/pages/cover_crop.aspx: Maryland Department of Agriculture Cover Crop Program offers financial assistance to farmers for planting cover crops in the fall.

https://eorganic.info/: eOrganic, the organic agriculture community of practice with extension, webinars, videos, articles, and news – search 'Cover Crops' in the search bar.

https://plants.usda.gov/cover-crop-plants: USDA Cover Crop Database.

https://covercropsguide.co.uk/ Guide to help farmers select, establish, and terminate cover crops with up-to-date information and advice on cover crop varieties and their management.

https://www.ars.usda.gov/plains-area/mandan-nd/ngprl/docs/cover-crop-chart/: A decision-support tool designed to assist growers in selecting and managing cover crops within crop and forage production. Includes details about the growth cycle, relative water use, plant architecture, seeding depth, forage quality, pollination characteristics, and nutrient cycling of 70 crop species that can be planted individually or in mixtures.

ATTRA contains information on specific cover crops via reprints, summaries of research, and other resources. This includes materials on living mulches, summer green manures, winter cover crops, and allelopathic cover crops, as well as on specific cover crops like hairy vetch and subterranean clover, and on the more obscure cover crops such as crotalaria, velvet bean, sesbania, and phacelia.

https://covercrop.org/: Cover Crop Guide for New York Growers, Cornell AgriTech.

https://www.drmcbug.com/: How to identify common natural enemies and pests.

https://futurefoodsolutions.co.uk/what-are-cover-crops-and-how-do-they-work/: Cover crop benefits, selection, management, termination, and timing.

https://www.covercropstrategies.com/articles/2442-things-that-can-go-wrong-with-cover-crops-prevention-and-management: Cover crop challenges and their solutions.

10.4 Cover crop tools

https://www.midwestcovercrops.org/selector-tools/.

https://westerncovercrops.org/cover-crop-selection-tool/.

https://www.nrcs.usda.gov/plant-materials/pacific-northwest-cover-crop-selection-tool.

https://southerncovercrops.org/cover-crop-resource-guide/row-crops/coastal-plain/cover-crop-selection-tool/.

https://sarep.ucdavis.edu/covercropsdb.

https://covercrop-selector.org/.

http://vm193-134.its.uni-kassel.de/toolbox/SC_DB_home.php: OSCAR Cover Crop Database.

https://umr-agronomie.shinyapps.io/EcoSystemix/: EcosysteMIX.

10.5 Books

Clark, A. (2007), *Managing Cover Crops Profitably, Third Edition, Handbook Series Book 9*, A. Clark (Ed), Sustainable Agriculture Research and Education: contains information on the characteristics, costs, seeding rate, and management of different cover crop species. Discusses the potential advantages and disadvantages of using cover crops and how to manage them effectively to minimize unwanted effects. It also provides comprehensive information on the major species used in the USA. Though geared primarily toward large acreage cropping systems, the information is readily adapted for smaller-scale systems. Available at: https://www.sare.org/wp-content/uploads/Managing-Cover-Crops-Profitably.pdf.

Lee-Mader, E., Stine, A., Fowler, J., Hopwood, J. and Vaughan M. (2015), *Cover Cropping for Pollinators and Beneficial Insects*, SARE Outreach: This 16-page bulletin contains a broad overview of pollinator and beneficial insect ecology, then describes cover crop selection and management and helpful and proven crop rotations. It also touches on the limitations of cover crops and pesticide harm reduction. Available at: Cover Cropping for Pollinators and Beneficial Insects – SARE.

Smith, R., Bugg, R. L. and Gaskell, M. (2011), *Cover Cropping for Vegetable Production: A Grower's Handbook*, Publication #3517, University of California Division of Agriculture and Natural Resources, Oakland, CA: describes primary cover crop species, including grasses, legumes, mustards, and other cool-season and warm-season options. Photos, seeding details, winter vigor descriptions, nitrogen fixation and scavenging, weed suppression, and nematode resistance are included for each species. Addresses the effects of cover cropping on water management, pest management, and farm economics.

Ingles, C. (1998), *Cover Cropping in Vineyards: A Growers Handbook*. Publication #3338. University of California Division of Agriculture and Natural Resources, Oakland, CA: offers cover cropping methods for enhancing vineyard performance and provides detailed information on how cover crops promote ecological stability.

Altieri, M. A. (2005), *Manage Insects on Your Farm: A Guide to Ecological Strategies*: promotes natural pest control by enhancing farm biodiversity and supporting beneficial insects and offers advice on using cover crops, no tillage farming, beetle banks, and farm corridors. Includes key to main beneficial insects and pests and how to design a functional biodiverse farm. Available at: https://www.sare.org/wp-content/uploads/Manage-Insects-on-Your-Farm.pdf

10.6 Videos

https://www.youtube.com/channel/UCRxVldHWHJAoU8xHNZ0efCQ: The ReMIX YouTube channel with videos on mixed cropping and other aspects of biological diversifications.

https://www.sare.org/What-We-Do/Impacts-from-the-Field/Cover-Crop-Innovators-Video-Series/.

Short videos of innovative farmers from around the USA, including both commodity crop and horticulture producers, illustrate why they are using cover crops, with examples of the benefits they are seeing from cover crops.

Brennan, Eric. Agriculture Research in High Value Organic Systems. www.youtube.com/user/EricBrennanOrganic Eric Brennan, a USDA Cooperative Extension Specialist in Monterey County, California, has created a series of informative and entertaining videos on cover crops including videos on use of cover crops for pest management such as 'A Biological Control Buffet in the Salad Bowl of America' and 'Efficient Intercropping for Biological Control of Aphids in Transplanted Organic Lettuce.'

https://www.youtube.com/playlist?list=PLdNOdHei9NV24njSkaCG6RpKpqqFDUGYP.

UC Santa Cruz Center for Agroecology. Short videos about when, how, and why to incorporate cover crops into your crop production.

https://www.youtube.com/watch?v=Lh-sv2MVWvg: Creative cover cropping for annual cropping systems. 1993. University of California Agriculture and Natural Resources: depicts opportunities and constraints of cover crop use and shows many types of cover crops used in various annual cropping systems for soil fertility and pest management.

https://www.youtube.com/watch?v=E3nLuK7D7LY: A Deep Look documentary about using flowering alyssum to attract and feed predatory hoverflies and enhance biological control of Asian citrus psyllid in southern California citrus.

11 References

Abad, J., et al. (2021), 'Cover crops in viticulture: a systematic review (2): implications on vineyard agronomic performance', *OENO One*, 2, 1–27.

Adamson, N., et al. (2016), *Choosing and managing cover crops to support beneficial insects for pest control and pollination. Southern Cover Crops.* 2016 Conference Fact Sheet, pp. 1–4. Available at: Choosing-and-Managing-Cover-Crops-Pest-Protection.pdf.

Almond Board of California (2020), *Cover Crop Best Management Practices*. Document #2021R0060. Almond Board of California. Available at: https://www.almonds.com/sites/default/files/2021-07/Cover%20Crops%20Best%20Management%20Practices%20BMPs_0.pdf.

Altieri, M. A and Schmidt, L. L. (1985), 'Cover crop manipulation in northern California orchards and vineyards: effects on arthropod communities', *Biol. Agric. Hortic.*, 3, 1–24.

Altieri, M. A. and Toledo, V. M. (2011), 'The agroecological revolution in Latin America: rescuing nature, ensuring food sovereignty and empowering peasants', *J. Peasant Stud.*, 38, 587–612.

Amudavi, D. M., et al. (2009), 'Evaluation of farmers' field days as a dissemination tool for push-pull technology in Western Kenya', *Crop Prot.*, 28, 225–23.

Anastasio, O. E., et al. (2021), 'Impacts of invasive ant-hemipteran interaction, edge effects and habitat complexities on the spatial distribution of ants in citrus orchards', *Agric. Ecosyst. Environ.*, 310, 107299.

Andersen, A. (1997), 'Densities of overwintering carabids and staphylinids (Carabidae and Staphylinidae) in cereal and grass fields and their boundaries', *J. Appl. Entomol.*, 121, 77–80.

Andow, D. A. and Risch, S. J. (1985), 'Predation in diversified agroecosystems: relations between a Coccinellid predator *Coleomegilla maculata* and its food', *J. Appl. Ecol.*, 22, 357–372.

Anjos, D. V., et al. (2021), 'Ants affect citrus pests and their natural enemies in contrasting ways', *Biol. Control*, 158, 104611.

Badenes-Pérez, F. R. (2019), 'Trap crops and insectary plants in the order Brassicales', *Ann. Entomol. Soc. Am.*, 112, 318–329.

Baggen, L. R. and Gurr, G. M. (1998), 'The Influence of food on *Copidosoma koehleri* (Hymenoptera: Encyrtidae), and the use of flowering plants as a habitat management tool to enhance biological control of potato moth, *Phthorimaea operculella* (Lepidoptera: Gelechiidae)', *Biol. Control*, 11, 9–17.

Baggen, L. R., et al. (1999), 'Flowering in tritrophic systems: Mechanisms allowing selective exploitation by insect natural enemies for conservation biological control', *Entomol. Exp. Appl.*, 91, 155–161.

Bakke, A. and Riege, L. (1982), 'The pheromone of the spruce bark beetle *Ips typographus* and its potential use in the suppression of beetle populations', In A. F. Kydonieus and M. Beroza (Eds), *Insect Suppression with Controlled Release Pheromone Systems, Vol. II*, CRC Press, pp. 3–15.

Balmer, O., et al. (2013), 'Noncrop flowering plants restore top-down herbivore control in agricultural fields', *Ecol. Evol.*, 3(8), 2634–2646.

Banks, J. E. and Ekbom, B. (1999), 'Modelling herbivore movement and colonization: pest management potential of intercropping and trap cropping', *Agric. For. Entomol.*, 1, 165–170.

Barnes, A., et al. (2009), 'Harnessing biodiversity to improve vineyard sustainability', *Outlooks Pest. Manag.*, 20, 250–255.

Begum, M., et al. (2006), 'Using selective food plants to maximize biological control of vineyard pests', *J. Appl. Ecol.*, 43, 547–554.

Berndt, L. A. and Wratten, S. D. (2005), 'Effects of alyssum flowers on the longevity, fecundity, and sex ratio of the leafroller parasitoid *Dolichogenidea tasmanica*', *Biol. Control*, 32, 65–69.

Berndt, L. A., et al. (2002), 'Effects of buckwheat flowers on leafroller (Lepidoptera: Tortricidae) parasitoids in a New Zealand vineyard', *Agric. For. Entomol.*, 4(1), 39–45.

Bickerton, M. W. and Hamilton, G. C. (2012), 'Effects of intercropping with flowering plants on predation of *Ostrinia nubilalis* (Lepidoptera: Crambidae) eggs by generalist predators in bell peppers', *Environ. Entomol.*, 41, 612–620.

Blubaugh, C. and Kaplan, I. (2015), 'Tillage compromises weed seed predator activity across developmental stages', *Biol. Control*, 81, 76–82.

Boinot, S., et al. (2019), 'Distribution of overwintering invertebrates in temperate agroforestry systems: implications for biodiversity conservation and biological control of crop pests', *Agric. Ecosyst. Environ.*, 285, 106630.

Bone, N. J., et al. (2009), 'Cover crops in Victorian apple orchards: effects on production, natural enemies and pests across a season', *Crop. Prot.*, 28, 675–683.

Bostanian, N. J., et al. (2004), 'Towards insecticide free apple orchards: flowering plants to attract beneficial arthropods', *Biocontrol Sci. Technol.*, 14, 25–37.

Bottenberg, H., et al. (1997), 'Yield and quality constraints of cabbage planted in rye mulch', *Biol. Agric. Hortic.*, 14, 323–342.

Bowers, C., et al. (2020), 'Cover crops improve early season natural enemy recruitment and pest management in cotton production', *Biol. Control*, 141, 10414.

Bowie, M. H., et al. (1995), 'Agronomy and phenology of "companion plants" of potential for enhancement of insect biological control', *N. Z. J. Crop Horticultural Sci.*, 23(4), 423–427.

Breil, K. (1999), 'Mulchrasen im obst- und weinbau: Arten, sorten, mischungen', *Erwerbsobstbau*, 41, 61–63.

Bruce, T. J. A., et al. (2005), 'Insect host location: a volatile situation', *Trends Plant Sci.*, 10(6), 269–274.

Buchanan, A., et al. (2018), 'Assessing annual and perennial flowering plants for biological control in asparagus', *Biol. Control*, 127, 1–8.

Buczkowski, G. and Bennett, G. W. (2008), 'Aggressive interactions between the introduced Argentine ant and the native odorous house ant: Implications for community displacement and biotic resistance', *Biol. Invasions*, 10(7), 1001–1011.

Bugg, R. L. (1991), 'Cover crops and control of arthropod pests in agriculture', In W. L. Hargrove (Ed), *Cover Crops for Clean Water*, Soil and Water Conservation Society, pp. 157–163.

Bugg, R. L. and Ellis, R. T. (1990), 'Insects associated with cover crops in Massachusetts', *Biol. Agr. Hort.*, 7, 47–68.

Bugg, R. L. and Waddington, C. (1994), 'Using cover crops to manage arthropod pests of orchards: a review', *Agric. Ecosyst. Environ.*, 50, 11–28.

Bugg, R. L., et al. (1989). 'Ichneumonidae (Hymenoptera) using extrafloral nectar of faba bean (*Vicia faba* L., Fabaceae) in Massachusetts', *Biol. Agr. Hort.*, 6(2), 107–114.

Bugg, R. L., et al. (1990), 'Insects associated with cool-season cover crops in southern Georgia: implications for pest control in the truck-farm and pecan agroecosystems', *Biol. Agr. Hort.*, 7, 17–45.

Bugg, R. L., et al. (1991), 'Cool-season cover crops relay intercropped with cantaloupe: influence on a generalist predator, *Geocoris punctipes* (Say) (Hemiptera: Lygaeidae)', *J. Econ. Entomol.*, 84, 408–416.

Bugg, R. L., et al. (2008), *Flower Flies (Syrphidae) and Other Biological Control Agents for Aphids in Vegetable Crops*. Publication 8285, May 2008, University of California Division of Agriculture and Natural Resources, University of California.

Carroll, B. (2017), *Use of Legumes in Pecan Orchards*. Oklahoma Cooperative Extension Fact Sheets HLA-6250 Division of Agricultural Sciences, Oklahoma State University Extension. Available at: https://extension.okstate.edu/fact-sheets/use-of-legumes-in-pecan-orchards.html.

Castro, J., et al. (2021), 'Structure of canopy and ground-dwelling arthropod communities in olive orchards is determined by the type of soil cover', *Eur. J. Entomol.*, 118, 159–170.

CDFA (2022), *Action Plan for Asian Citrus Psyllid and Huanglongbing (Citrus Greening) in California*. California Department of Food and Agriculture January 2022. Available at: https://www.cdfa.ca.gov/citruscommittee/docs/ActionPlan.pdf.

Chaney, W. E. (1998), 'Biological control of aphids in lettuce using in-field insectaries', In C. H. Pickett and R. L. Bugg (Eds), *Enhancing Biological Control: Habitat Management to Promote Natural Enemies of Arthropod Pests, 1st edition*, University of California Press, Vol. 126, pp. 73–83.

Chen, Y., et al. (2020), 'Alyssum (*Lobularia maritima*) selectively attracts and enhances the performance of *Cotesia vestalis*, a parasitoid of *Plutella xylostella*', *Sci. Rep.*, 10, 6447.

Claeys, W. L., et al. (2011), 'Exposure of several Belgian consumer groups to pesticide residues through fresh fruit and vegetable consumption', *Food Control*, 22, 508–516.

Clark, A. (2007), *Managing Cover Crops Profitably, Third Edition, Handbook Series Book 9*, A. Clark (Ed), Sustainable Agriculture Research and Education.

Colley, M. R. and Luna, J. M., (2000), 'Relative attractiveness of potential beneficial insectary plants to aphidophagous hoverflies (Diptera: Syrphidae)', *Environ. Entomol.*, 20(5), 1054–1059.

Collins, K. L., et al. (2003), 'Effects of different grass treatments used to create overwintering habitat for predatory arthropods on arable farmland', *Agric. Ecosyst. Environ.*, 96, 59–67.

Colugnati, G., et al. (2004), 'Gestione del terreno in viticoltura', *Vigne Vini*, 11, 53–83.

Cooper, M. L., et al. (2008), 'Liquid baits control Argentine ants sustainably in coastal vineyards', *Calif. Agric.*, 62, 177–183.

Cover Crop Resources (2024a), *Selecting Cover Crops*. Division of Extension, University of Wisconsin-Madison. Available at: https://cropsandsoils.extension.wisc.edu/articles/selecting-cover-crops/.

Cover Crop Resources (2024b), *Planting Methods for Cover Crops*. Division of Extension, University of Wisconsin-Madison. Available at: https://cropsandsoils.extension.wisc.edu/articles/planting-methods-for-cover-crops/.

Crews, T. E., et al. (2018), 'Is the future of agriculture perennial? Imperatives and opportunities to reinvent agriculture by shifting from annual monocultures to perennial polycultures', *Glob. Sustain.*, 1, e11.

Daane, K. M., et al. (2008), 'Testing baits to control Argentine ants (Hymenoptera: Formicidae) in vineyards', *J. Econ. Entomol.*, 101, 699–709.

Daane, K. M., et al. (2018), 'Native grass ground covers provide multiple ecosystem services in Californian vineyards', *J. Appl. Ecol.*, 55, 2473–2483.

Daane, K. M., et al. (2007), 'Impacts of Argentine ants on mealybugs and their natural enemies in California's coastal vineyards', *Ecol. Entomol.*, 32(6), 583–596.

Davies, A. P., et al. (2004), 'Is the potential of *Coccidoxenoides perminutus*, a mealybug parasitoid, limited by climatic or nutritional factors?', *Biol. Control*, 31, 181–188.

Degen, T., et al. (1999a), 'Pattern oviposition stimulants for carrot fly in leaves of various host plants', *J. Chem. Ecol.*, 25(1), 66–87.

Degen, T., et al. (1999b), 'Extracting oviposition stimulants for carrot fly from host-plant leaves', *J. Chem. Ecol.*, 25(1), 88–104.

Delmas, M. A. and Grant, L. E. (2014). 'Eco-labeling strategies and price-premium: the wine industry puzzle', *Bus. Soc.*, 53, 6–44.

Diehl, E., et al. (2012), 'Arable weeds in organically managed wheat fields foster carabid beetles by resource- and structure- mediated effects', *Arthropod. Plant. Interact.*, 6, 75–82.

Donley, N. (2019), 'The USA lags behind other agricultural nations in banning harmful pesticides', *Environ. Health*, 18, 44.

Duke, S. O., et al. (2000), 'Natural products as sources of herbicides: current status and future trends', *Weed Res.*, 40, 99–111.

Dutcher, J. D., et al (2010), 'Integration of insect and mite management with disease and weed control in pecan production', In A. Ciancio and K.G. Mukerji (Eds), *Integrated Management of Arthropod Pests and Insect Borne Diseases*, Springer, pp 133–162.

Erickson, B. E. (2023), 'Chlorpyrifos returns to the US market', *Chemical and Engineering News*, The American Chemical Society. Available at: https://cen.acs.org/environment/pesticides/Chlorpyrifos-returns-US-market/101/web/2023/11#:~:text=Some%20states%2C%20including%20California%2C%20Hawaii,Those%20bans%20remain%20in%20effect.

Fanadzo, M., et al. (2018), 'Application of conservation agriculture principles for the management of field crops pests', In S. Gaba, B. Smith and E. Lichtfouse (Eds), *Sustainable Agriculture Reviews 28*, Springer, pp. 125–152.

FAO (2021), *Citrus* Fruit Statistical Compendium *2020*, Citrus Fruit Fresh and Processed Statistical Bulletin 2020, Food and Agriculture Organization of the United Nations.

Fiedler, A. K. and Landis, D. A. (2007a), 'Attractiveness of Michigan native plants to arthropod natural enemies and herbivores', *Environ. Entomol.*, 36(4), 751–765.

Fiedler, A. K. and Landis, D. A. (2007b), 'Plant characteristics associated with natural enemy abundance at Michigan native plants', *Environ. Entomol.*, 36(4), 878–886.

Finch, S. (1976), 'Effect of secondary plant substances on host-plant selection by the cabbage root fly', *Coll. Internat. C.N.R.S.*, 265, 251–267.

Finch, S. and Collier, R. H. (2000), 'Host-plant selection by insects–a theory based on 'appropriate/inappropriate landings' by pest insects of cruciferous plants', *Entomol. Exp. Appl.*, 96, 91–102.

Finke, D. L. and Denno, R. F. (2002), 'Intraguild predation diminished in complex-structured vegetation: Implications for prey suppression', *Ecology*, 83(3), 643–652.

Fountain, M. T. (2022), 'Impacts of wildflower interventions on beneficial insects in fruit crops: a review', *Insects*, 13, 304.

Free, J. B. (1993), *Insect Pollination of Crops, 2nd edition*. Academic Press, pp. 684.

Freeman-Long, R., et al. (1998), 'Beneficial insects move from flowering plants to nearby crops', *Calif. Agric.*, 23–26.

Gagnon, A. È., et al. (2011), 'The ubiquity of intraguild predation among predatory arthropods', *PLoS One*, 6(11), e28061.

Garcia, L., et al. (2018), 'Management of service crops for the provision of ecosystem services in vineyards: a review', *Agric. Ecosyst. Environ.*, 251, 158–170.

Gardarin, A. (2023), 'Aphid biological control in arable crops via flower strips: the predominant role of food resources over diversity effects', *J. Appl. Ecol.*, 60, 2118–2131.

Geertsema, W., et al. (2016), 'Actionable knowledge for ecological intensification of agriculture', *Front. Ecol. Environ.*, 14, 209–216.

Géneau, C. E., et al. (2012), 'Selective flowers to enhance biological control of cabbage pests by parasitoids', *Basic Appl. Ecol.*, 13, 85–93.

Géneau, C. E., et al. (2013), 'Effects of extrafloral and floral nectar of *Centaurea cyanus* on the parasitoid wasp *Microplitis mediator*: olfactory attractiveness and parasitization rates', *Biol. Control*, 66, 16–20.

Giacalone, G., et al. (2021), 'Are cover crops affecting the quality and sustainability of fruit production?', *Agriculture*, 11, 1201.

Gillespie, M. A. K., et al. (2016), 'Beyond nectar provision: the other resource requirements of parasitoid biological control agents', *Entomol. Exp. Appl.*, 159, 207–221.

Gillespie, M. A. K. and Wratten, S. D. (2017), 'The role of ecosystem disservices in pest management', In C. Mosche and E. Wajnberg (Eds), *Environmental Pest Management: Challenges for Agronomists, Ecologists, Economists and Policymakers*, John Wiley and Sons Ltd., pp. 175–194.

Godfray, H. C. and Garnett, T. (2014), 'Food security and sustainable intensification', *Phil. Trans. R. Soc. B.*, 369, 20120273.

Goleva, I. and Zebitz, C. P. W. (2013), 'Suitability of different pollen as alternative food for the predatory mite *Amblyseius swirskii* (Acari, Phytoseiidae)', *Exp. Appl. Acarol.*, 61, 259–283.

Goller, E., et al. (1997), 'Faba beans as a cover crop in organically grown hops: influence on aphids and aphid antagonists', *Biol. Agric. Hortic.*, 15(1–4), 279–284.

Gontijo, L. M., et al. (2013), 'Flowers promote aphid suppression in apple orchards', *Biol. Control*, 66, 8–1.

González-Chang, M., et al. (2019), 'Habitat management for pest management: limitations and prospects', *Ann. Entomol. Soc. Am.*, 112(4), 302–317.

Gospodarek, J. (2021), 'Effect of *Sinapis alba* L. as an insectary plant on the occurrence of *Aphis fabae* Scop., Coccinellidae and Syrphidae in broad bean', *Agronomy*, 11, 2202.

Grafton-Cardwell, E. E. (2015), 'The status of citrus IPM in California', *Acta Hortic.*, 1065, 1083–1090.

Gratton, C. and Denno, R. F. (2003), 'Seasonal shift from bottom-up to topdown impact in phytophagous insect populations', *Oecologia*, 134, 487–495.

Guerra, B. and Steenwerth, K. (2012), 'Influence of floor management technique on grapevine growth, disease pressure, and juice and wine composition: a review', *Am. J. Enol. Vitic.*, 63(2), 149–164.

Gurr, G. M., et al. (2016), 'Multi-country evidence that crop diversification promotes ecological intensification of agriculture', *Nat. Plants*, 2, 16014.

Gurr, G. M., et al. (2017), 'Habitat management to suppress pest populations: progress and prospects', *Annu. Rev. Entomol.*, 62, 91–109.

Haenke, S., et al. (2014), 'Landscape configuration of crops and hedgerows drives local syrphid fly abundance', *J. Appl. Ecol.*, 51, 505–513.

He, X., et al. (2021), 'The effect of floral resources on predator longevity and fecundity: a systematic review and meta-analysis', *Biol. Control*, 153, 104476.

Hern, A., et al. (1996), 'A review of the pre-oviposition behavior of the small cabbage white butterfly, *Pieris rapae* (Lepidoptera, Pieridae)', *Ann. Appl. Biol.*, 128, 349–371.

Herrera, R. A., et al. (2022), 'Using flower strips to promote green lacewings to control cabbage insect pests', *J. Pest. Sci.*, 95, 669–683.

Herz, A., et al. (2019), 'Managing floral resources in apple orchards for pest control: ideas, experiences and future directions', *Insects*, 10, 247.

Hewlett, J. A., et al. (2019), 'The effects of sugarcane aphid density in sorghum on predation by lady beetles and lacewings', *Biol. Control*, 129, 171–177.

Hickman, J. and Wratten, S. D. (1996), 'Use of *Phacelia tanacetifolia* strips to enhance biological control of aphids by hoverfly larvae in cereal fields', *J. Econ. Entomol.*, 89, 832–840.

Hinds, J. and Barbercheck, M. E. (2020), 'Diversified floral provisioning enhances performance of the generalist predator, *Orius insidiosus* (Hemiptera: Anthocoridae)', *Biol. Control*, 149, 104313.

Hoddle, M. S., et al. (2022a), 'Successful biological control of Asian citrus psyllid, *Diaphorina citri*, in California', In R. G. van Driesche, et al. (Eds), *Contributions of Classical Biological Control to the US food Security, Forestry, and Biodiversity*, USDA Forest Service FHAAST-2019–05, pp. 127–143.

Hoddle, M. S., et al. (2022b), 'Using infra-red sensors and the Internet of things to automate Argentine ant counts', *CAPCA Advisor*, 15(5), 40–45.

Hogg, B. N., et al. (2011a), 'Attractiveness of common insectary and harvestable floral resources to beneficial insects', *Biol. Control*, 56, 76–84.

Hogg, B. N., et al. (2011b), 'Floral resources enhance aphid suppression by a hoverfly', *Entomol. Exp. Appl.*, 141(2), 138–144.

Hogg, B. N., et al. (2023), 'A comparison of candidate banker plants for management of pests in lettuce', *Environ. Entomol.*, 52(3), 379–390.

Hokkanen, H. M. T. (1989), 'Biological and agrotechnical control of the rape blossom beetle *Meligethes aeneus* (Coleoptera: Nitidulidae)', *Acta Entomol. Fenn.*, 53, 25–30.

Hokkanen, H. M. T. (1991), 'Trap cropping in pest management', *Annu. Rev. Entomol.*, 36, 119–138.

Holt-Giménez, E. (2008), 'Campesino a campesino. Voces de Latinoamerica', Movimiento campesino para la agricultura sustentable, Managua, Nicaragua: SIMAS 10.1017/CBO9781107415324.004.

Hooks, C. R. R. and Johnson, M. W. (2003), 'Impact of agricultural diversification on the insect community of cruciferous crops', *Crop Prot.*, 22, 223–238.

Horton, D. R., et al. (2003), 'Effects of mowing frequency on densities of natural enemies in three Pacific Northwest pear orchards', *Entomol. Exp. Appl.*, 106, 135–145.

Huang, D., et al. (2020), 'Volatile organic compounds (VOCs) regulate the spatial distribution of Lepidoptera insects in an orchard ecosystem', *Biol. Control*, 149, 104311.

Hubbard, P. T. (2023), *Planting in Sun or Shade*. Penn State Extension, College of Agricultural Sciences, The Pennsylvania State University. Available at: https://extension.psu.edu/planting-in-sun-or-shade#:~:text=Gardening%20in%20Sunny%20Areas,would%20be%20happy%20with%20more.

Hunter, M. D. and Price, P. W. (1992), 'Playing chutes and ladders: heterogeneity and the relative roles of bottom-up and top-down forces in natural communities', *Ecology*, 73, 724–732.

Huss, C. P., et al. (2022), 'Benefits and risks of intercropping for crop resilience and pest management', *J. Econ. Entomol.*, 115(5), 1350–1362.

Ingels, C. A., et al. (1998), *Cover Cropping in Vineyards: A Grower's Handbook*, University of California, Agriculture and Natural Resources.

Irvin, N. A. (1999), *Understorey Management for the Enhancementof Populations of a Leafroller (Lepidoptera: Tortricidae) Parasitoid* (Dolichogenidea tasmanica *(Cameron) in Canterbury, New Zealand Apple Orchards*. PhD Thesis, Lincoln University.

Irvin, N. A. and Hoddle, M. S. (2007), 'Evaluation of floral resources for enhancement of fitness of *Gonatocerus ashmeadi*, an egg parasitoid of the glassy-winged sharpshooter, *Homalodisca vitripennis*. *Biol. Control*, 40(1), 80–88.

Irvin, N. A. and Hoddle, M. S. (2021), 'The effects of floral nectar, extrafloral nectar and hemipteran honeydew on the fitness of *Tamarixia radiata* (Hymenoptera: Eulophidae), a parasitoid of *Diaphorina citri*', *Biol. Control*, 163, 104753.

Irvin, N. A., et al. (2006), 'The effects of floral understoreys on parasitism of leafrollers (Lepidoptera: Tortricidae) on apples in New Zealand', *Agric. Forest Entomol.*, 8, 25–34.

Irvin, N. A., et al. (2007), 'The effect of resource provisioning and sugar composition of foods on longevity of three *Gonatocerus* spp., egg parasitoids of *Homalodisca vitripennis*', *Biol. Control*, 40, 69–79.

Irvin, N. A., et al. (2014), 'Evaluating the potential of buckwheat and cahaba vetch as nectar producing cover crops for enhancing biological control of *Homalodisca vitripennis* in California vineyards', *Biol. Control*, 76, 10–18.

Irvin, N. A., et al. (2016), 'The effect of an irrigated buckwheat cover crop on grape vine productivity, and beneficial insect and grape pest abundance in Southern California', *Biol. Control*, 93, 72–83.

Irvin, N. A., et al. (2018), 'Measuring natural enemy dispersal from cover crops in a California vineyard', *Biol. Control*, 126, 15–25.

Irvin, N. A., et al. (2021), 'Evaluating the potential of flowering plants for enhancing predatory hoverflies (Syrphidae) for biological control of *Diaphorina citri* (Liviidae) in California', *Biol. Control*, 157, 104574.

Irvin, N. A., et al. (2022), 'Advances in cover cropping for enhancing natural enemies of Asian citrus psyllid in southern California citrus orchards', *CAPCA Advisor*, 15(5), 48–54.

Irvin, N. A., et al. (2024), 'Monitoring of coccinellid (Coleoptera) presence and syrphid (Diptera) species diversity and abundance in southern California citrus orchards: implications for conservation biological control of Asian citrus psyllid and other citrus pests', *Fla. Entomol.*, 107(1), 20240060.

Jacob, H. S. and Evans, E. W. (2000), 'Influence of carbohydrate foods and mating on longevity of the parasitoid *Bathyplectes curculionis* (Hymenoptera: Ichneumonidae)', *Environ. Entomol.*, 29, 1088–1095.

Jankowska, B. and Wojciechowicz-Żytko, E. (2016), 'Effect of intercropping carrot *(Daucus carota* L.) with two aromatic plants, coriander (*Coriandrum sativum* L.) and summer savory (*Satureja hortensis* L.), on the population density of select carrot pests', *Folia Hortic.*, 28, 13–18.

Jervis, M. A., et al. (1993), 'Flower-visiting by hymenopteran parasitoids', *J. Nat. Hist.*, 27, 67–105.

Johanowicz, D. and Mitchel, E. R. (2000), 'Effects of sweet alyssum flowers on the longevity of the parasitoid wasps *Cotesia marginiventris* (Hymenoptera: Braconidae) and *Diadegma insulare* (Hymenoptera: Ichneumonidae)', *Fla. Entomol.*, 83(1), 41–47.

Jones, A. L., et al. (2014), 'Sentinel eggs underestimate rates of parasitism of the exotic brown marmorated stink bug, *Halyomorpha halys*', *Biol. Control*, 78, 61–66.

Jones, I. M., et al (2017), 'The use of extrafloral nectar in pest management: overcoming context dependence', *J. Appl. Ecol.*, 54, 489–499.

Judt, C., et al. (2023), 'Floral resources and ground covers promote natural enemies but not pest insects in apple orchards: a global meta-analysis', *Sci. Total Environ.*, 903, 166139.

Kajak, A, and Lukasiewicz, J. (1994), 'Do semi-natural patches enrich crop fields with predatory epigean arthropods?' *Agric. Ecosyst. Environ.*, 49, 149–161.

Karp, D. S., et al. (2018), 'Crop pests and predators exhibit inconsistent responses to surrounding landscape composition', *Proc. Natl. Acad. Sci. U.S.A.*, 115, E7863–E7870.

Kaye, J. P. and Quemada, M. (2017), 'Using cover crops to mitigate and adapt to climate change: a review', *Agron. Sustain. Dev.*, 37, 4.

Kean, J., et al. (2003), 'The population consequences of natural enemy enhancement, and implications for conservation biological control', *Ecol. Lett.*, 6, 604–612.

Kennett, C. E., et al. (1979), 'Effect of wind-borne pollens on the population dynamics of *Amblyseius hibisci* (Acarina: Phytoseiidae)', *Entomophaga*, 2483–2498.

Kevan, P. G. and Baker, H. G. (1984), 'Insects on flowers', In C. B. Huffaker and R. L. Rabb (Eds), *Ecological Entomology*, John Wiley & Sons, pp. 607–631.

Khan, Z. R., et al. (2001), 'Habitat management strategies for the control of cereal stemborers and Striga in maize in Kenya', *Int. J. Trop. Insect Sci.*, 21, 375–380.

Khan, Z. R., et al. (2010), 'Exploiting phytochemicals for developing a 'push-pull' crop protection strategy for cereal farmers in Africa', *J. Exp. Bot.*, 61, 4185–4196.

Khan, Z. R., et al. (2014), 'Achieving food security for one million sub-Saharan African poor through push-pull innovation by 2020', *Philos. Trans. R. Soc. Lond. B. Biol. Sci.*, 369, 20120284.

Kistner, E. J., et al. (2017), 'Digital video surveillance of natural enemy activity on *Diaphorina citri* (Hemiptera: Liviidae) colonies infesting citrus in the southern California urban landscape', *Biol. Control*, 115, 141–151.

Koptur, S. (1994), 'Floral and extrafloral nectars of Costa Rican Inga trees: a comparison of their constituents and composition', *Biotropica*, 26, 276–284.

Koptur, S. (2005), 'Nectar as fuel for plant protectors', In F. L. Wäckers, et al. (Eds), *Plant-Provided Food for Carnivorous Insects: A Protective Mutualism and Its Applications*, Cambridge University Press, pp. 75–108.

Koptur, S. and Truong, N. (2005), 'Facultative ant-plant interactions: nectar sugar preferences of introduced pest ant species in South Florida', *Biotropica*, 30(2), 179–189.

Kost, C. and Heil, M. (2005), 'Increased availability of extrafloral nectar reduces herbivory in Lima bean plants (*Phaseolus lunatus*, Fabaceae)', *Basic Appl. Ecol.*, 6, 237–248.

Lach, L. (2005), 'Interference and exploitation competition of three nectar-thieving invasive ant species', *Insect. Soc.*, 52, 257–262.

Landis, D. A., et al. (2000), 'Habitat management to conserve natural enemies of arthropod pests in agriculture', *Annu. Rev. Entomol.*, 45, 175–201.

Langhof, M., et al. (2005), 'Measuring the field dispersal of *Aphidius colemani* (Hymenoptera: Braconidae)', *Agric. Ecosyst. Environ.*, 107, 137–143.

Lavandero, B., et al. (2005), 'Enhancing the effectiveness of the parasitoid *Diadegma semiclausum* (Helen): movement after use of nectar in the field', *Biol. Control*, 34, 152–158.

Lee, J. C. and Heimpel, G. E. (2002), 'Nectar availability and parasitoid sugar feeding', In G. Van Driesche (Ed), *Proceedings of the 1st International Symposium on Biological Control of Arthropods Jan. 14–18, 2002. Honolulu, Hawaii,* Forest Health Technology Enterprise Team, pp. 220–225.

Lee, J. C. and Heimpel, G. E. (2005), 'Impact of flowering buckwheat on Lepidopteran cabbage pests and their parasitoids at two spatial scales', *Biol. Control*, 34(3), 290–301.

Letourneau, D. K., et al. (2011), 'Does plant diversity benefit agroecosystems? A synthetic review', *Ecol. Appl.*, 21, 9–21.

Li, J., et al. (2019), 'Buckwheat strip crops increase parasitism of *Apolygus lucorum* in cotton', *BioControl*, 64, 645–654.

Liang, W. and Huang, M. (1994), 'Influence of citrus orchard ground cover plants on arthropod communities in China: a review', *Agric. Ecosyst. Environ.*, 50, 29–37.

Lu, Y. C., et al. (2000), 'Cover crops in sustainable food production', *Food Rev. Int.*, 16(2), 121–157.

Lu, Z., et al. (2015), 'Rice pest management by ecological engineering: a pioneering attempt in China', In K. L. Heong, et al. (Eds), *Rice Planthoppers*, Springer, pp. 161–178.

Lu, Z. X., et al. (2014), 'Mechanisms for flowering plants to benefit arthropod natural enemies of insect pests: prospects for enhanced use in agriculture', *Insect Sci.*, 21, 1–12.

Lucas, E. and Maisonhaute, J. E. (2019), 'Predator-prey interactions: intraguild predation', In J. C. Choe (Ed), *Encyclopedia of Animal Behavior, 2nd edition,* Academic Press, Elsevier Ltd, pp. 389–399.

Lundgren, J. G. (2009), 'Nutritional aspects of non-prey foods in the life histories of predaceous Coccinellidae', *Biol. Control*, 51, 294–305.

Lundin, O., et al. (2019), 'Identifying native plants for coordinated habitat management of arthropod pollinators, herbivores and natural enemies', *J. Appl. Ecol.*, 56, 665–676.

Luther, G. C., et al. (1996), 'Impact of cruciferous trap crops on lepidopteran pests of cabbage in Hawaii', *Environ. Entomol.*, 25, 39–47.

Lys, J. A., et al. (1994), 'Increase in activity density and species number of carabid beetles in cereals as a result of strip management', *Entomol. Exp. Appl.*, 73, 1–9.

Mathews, C. R., et al. (2017), 'Evaluating a polyculture trap crop for organic management of *Halyomorpha halys* and native stink bugs in peppers', *J. Pest. Sci.*, 90, 1245–1255.

McCalla, K. A., et al. (2020), 'Biodegradable alginate hydrogel bait delivery system effectively controls high-density populations of Argentine ant in commercial citrus', *J. Pest. Sci.*, 93, 1031–1042.

McCalla, K. A., et al. (2023), 'A low-toxicity baiting program precipitates collapse of Argentine ant and ant-associated hemipteran pest populations in commercial citrus', *Biol. Control*, 177, 105105.

McCauley, A., et al. (2017), *Soil pH and Organic Matter*, Montana State University Extension. Available at: https://apps.msuextension.org/publications/pub.html?sku=4449-8.

McGourty, G. (2004), *Cover Cropping Systems for Organically Farmed Vineyards*, Practical Winery & Vineyard, University of California Agriculture and Natural Resources, September/October 1–6. Available at: https://ucanr.edu/sites/Mendocino/files/17082.pdf.

McMurtry, J. A., et al. (2013), 'Revision of the lifestyles of phytoseiid mites (Acari: Phytoseiidae) and implications for biological control strategies', *Syst. Appl. Acarol.*, 18(4), 297–320.

Mena, G. T. and Gospodarek, J. (2024), 'White mustard, sweet alyssum, and coriander as insectary plants in agricultural systems: Impacts on ecosystem services and yield of crops', *Agriculture*, 14, 550.

Michaud, J. P. and Grant, A. K. (2005), 'Suitability of pollen sources for the development and reproduction of *Coleomegilla maculata* (Coleoptera: Coccinellidae) under simulated drought conditions', *Biol. Control*, 32, 363–370.

Milosavljević, I., et al. (2024), 'Spinosad-infused biodegradable hydrogel beads as a potential organic approach for argentine ant, *Linepithema humile* (Mayr) (hymenoptera: Formicidae), management in California citrus orchards', *J. Appl. Entomol.*, 148, 117–127.

Milosavljević, I., et al. (2021), 'Density dependent mortality, climate, and Argentine ants affect population dynamics of an invasive citrus pest, *Diaphorina citri*, and its specialist parasitoid, *Tamarixia radiata*, in Southern California, USA', *Biol. Control*, 159, 104627.

Ministry of Agriculture and Lands (2019), *Home & Garden Pest Management Guide*, Crown Publications, King's Printer for British Columbia, p. 472. Available at: https://www2.gov.bc.ca/assets/gov/farming-natural-resources-and-industry/agriculture-and-seafood/animal-and-crops/plant-health/gardenguide-6-beneficials.pdf.

Mitchell, E. R., et al. (1997), 'Diamondback moth (Lepidoptera: Plutellidae) infestation and parasitism by *Diadegma insulare* (Hymenoptera: Ichneumonidae) in collards and adjacent fields', *Florida Entomol.*, 80, 54–62.

Mitchell, E. R., et al. (2000), 'Management of diamondback moth (Lepidoptera: Plutellidae) in cabbage using collard as a trap crop', *Hortic. Sci.*, 35, 875–879.

Mitchell, P. L., et al. (1999), 'Natural mortality of leaf footed bug (Hemiptera: Heteroptera: Coreidae) eggs in cowpea', *J. Agric. Urban Entomol.*, 16(1), 25–36.

Mockford, A., et al. (2023), 'Developing perennial wildflower strips for use in Mediterranean orchard systems', *Ecol. Evol.* 13, e10285.

Mockford, A., et al. (2024), 'Wildflower strips enhance pest regulation services in citrus orchards', *Agric. Ecosyst. Environ.*, 370, 109069.

Monzó, C. and Stansly, P. A. (2020), 'Economic value of conservation biological control for management of the Asian citrus psyllid, vector of citrus Huanglongbing disease', *Pest Manag. Sci.*, 76, 1691–1698.

Monzó, C., et al. (2020), 'Cubiertas vegetales como estrategia de gestion de plagas en citricos', *Agricultura*, 1037, 40–44.

MSU Extension (2024), *Cover Crops: Annual Medics*, Michigan State University Extension. Available at: https://www.canr.msu.edu/cover_crops/species/annual-medics#:~:text=Annual%20medics%20(Medicago%20spp.),early%20maturing%2C%20unknown%20cold%20tolerance.

Muthiah, C. (2003), 'Integrated management of leafminer (*Aproaerema modicella*) in groundnut (*Arachis hypogaea*)', *Ind. J. Agric. Sci.*, 73, 466–468.

Myers, R. (2017), *Soil Health and Cover Crop Facts: Ten Ways Cover Crops Enhance Soil Health*, Soil Health Resource Series, USDA Natural Resources Conservation Service. Available at: https://www.sare.org/wp-content/uploads/10-ways-cover-crops-enhance-soil-health.pdf.

Myers, R. (2018), *Growing Canola for Oilseed or Cover Crop Use*, Division of Plant Sciences, University of Missouri Extension. Available at: https://extension.missouri.edu/publications/g4162.

Norris, R. F. and Kogan, M. (2000), 'Interactions between weeds, arthropod pests, and their natural enemies in managed ecosystems', *Weed Sci.*, 48, 94–158.

Oram, R., et al. (2005), 'Breeding Indian mustard [*Brassica juncea* (L.) Czern.] for cold-pressed, edible oil production – a review', *Aust. J. Agric. Res.*, 56, 581–596.

Orre, G. U. S., et al. (2010), 'Effects of an herbivore-induced plant volatile on arthropods from three trophic levels in brassicas', *Biol. Control*, 53, 62–67.

Pålsson, J., et al. (2022), 'Attract, reward and disrupt: responses of pests and natural enemies to combinations of habitat manipulation and semiochemicals in organic apple', *J. Pest Sci.*, 95, 19–631.

Pair, S. (1997), 'Evaluation of systemically treated squash trap plants and attracticidal baits for early-season control of striped and spotted cucumber beetles (Coleoptera: Chrysomelidae) and squash bug (Hemiptera: Coreidae) in cucurbit crops', *J. Econ. Entomol.*, 90(5), 1307–1314.

Park, Y., et al. (2024), 'Buckwheat as a companion plant in soybean fields: Implications for management of major pests', *Agric. For. Entomol.*, 26(4), 546–554.

Patt, J. M., et al. (1997), 'Foraging success of parasitoid wasps on flowers: interplay of insect morphology, floral architecture and searching behavior', *Entomol. Exp. Appl.*, 83, 21–30.

Patt, J. P. and Rohrig, E. (2017), 'Laboratory evaluations of the foraging success of *Tamarixia radiata* (Hymenoptera: Eulophidae) on flowers and extrafloral nectaries: Potential use of nectar plant for conservation biological control of Asian citrus psyllid (Hemiptera: Liviidae)', *Fla. Entomol.*, 100, 149–156.

Percival, M. S. (1955), 'The presentation of pollen in certain angiosperms and its collection by *Apis mellifera*', *New Phytol.*, 54(3), 353–368.

Petanidou, T. (2007), 'Ecological and evolutionary aspects of floral nectars in Mediterranean habitats', In S. W. Nicolson, et al. (Eds), *Nectaries and Nectar*, Springer, pp. 343–375.

Phatak, S. C. and Diaz-Perez, J. C. (2007), 'Managing pests with cover crops', In A. Clark (Ed), *Managing Cover Crops Profitably Third Edition, Handbook Series Book 9*, Sustainable Agriculture Research and Education, pp. 25–33.

Philips, C. R., et al. (2014), 'Herbert effect of buckwheat farmscapes on abundance and parasitism of *Pieris rapae* (L.) in Virginia collards', *J. Entomol. Sci.*, 49(3), 273–284.

Piñero, J. C. and Manandhar, R. (2015), 'Effects of increased crop diversity using trap crops, flowering plants, and living mulches on vegetable insect pests', *Trends Entomol.*, 11, 91–109.

Plata, A., et al. (2023), 'Exclusion of Mediterranean ant species enhances biological control of the invasive mealybug *Delottococcus aberiae* in citrus', *Pest Manag. Sci.*, 79, 2056–2065.

Platt, J. O., et al. (1999), 'Effect of buckwheat as a flowering border on populations of cucumber beetles and their natural enemies in cucumber and squash', *Crop Prot.*, 18(5), 305–313.

Prasifka, J. R., et al. (2006), 'Effects of living mulches on predator abundance and sentinel prey in a corn–soybean–forage rotation', *Environ. Entomol.*, 35(5), 1423–1431.

Price, A. (2019), *California Bans Poisonous Agricultural Pesticide: Chlorpyrifos Linked to Harmful Health Risks*, Berkeley Public Health, University of California. Available at:

https://publichealth.berkeley.edu/news-media/school-news/california-bans-poisonous-agricultural-pesticide.
Pywell, R. F., et al. (2015), 'Wildlife-friendly farming increases crop yield: Evidence for ecological intensification', *Proc. Biol. Sci.*, 282, 20151740.
Quintarelli, V., et al. (2022), 'Cover crops for sustainable cropping systems: a review', *Agriculture*, 12(12), 2076.
Rämert, B. and Ekbom, B. (1996), 'Intercropping as a management strategy against carrot rust fly (Diptera: Psilidae): a test of enemies and resource concentration hypotheses', *Environ. Entomol.*, 25, 1092–1100.
Rämert, B., et al. (2001), 'Assessment of trap crops for *Lygus* spp. in lettuce', *Int. J. Pest Manage.*, 47, 273–276.
Rapidel, B., et al. (2015), 'Analysis of ecosystem services trade-offs to design agroecosystems with perennial crops', *Agron. Sustain. Dev.*, 35, 1373–1390.
Rayl, R. J., et al. (2018), 'Conservation biological control of insect pests', In S. Gaba, et al. (Eds), *Sustainable Agriculture Reviews 28*, Springer International Publishing, pp. 103–124.
Rea, J. H., et al. (2002), 'Trap cropping to manage green vegetable bug *Nezara viridula* (L.) (Eteroptera Pentatomidae) in sweet corn in New Zealand', *Agric. For. Entomol.*, 4, 101–107.
Reddy, P. P. (2017), *Agro-Ecological Approaches to Pest Management for Sustainable Agriculture*, Springer Nature, p. 339.
Ribeiro, A. L. and Gontijo, L. M. (2017), 'Alyssum flowers promote biological control of collard pests', *BioControl*, 62, 185–196.
Rieux, R., et al. (1999), 'Role of hedgerows and ground cover management on arthropod populations in pear orchards', *Agric. Ecosyst. Environ.*, 73, 119–127.
Rincón, D. F., et al. (2019), 'Economic injury levels for the potato yellow vein disease and its vector, *Trialeurodes vaporariorum* (Hemiptera: Aleyrodidae), affecting potato crops in the Andes', *Crop Prot.*, 119, 52–58.
Rizvi, S. Z. M., et al. (2022), 'Prospects for habitat management to suppress vegetable pests in Australia', *Aust. Entomol.*, 6, 3–28.
Roberts, L., et al. (2015), *The Nature of Wellbeing: How* Nature's Ecosystem Services Contribute *to the Wellbeing of New Zealand and New Zealanders*, Department of Conservation, p. 145.
Rockstrom, J., et al. (2017), 'Sustainable intensification of agriculture for human prosperity and global sustainability', *Ambio.*, 46, 4–17.
Root, R. B. (1973), 'Organization of a plant-arthropod association in simple and diverse habitats: the fauna of collards (*Brassica oleracea*)', *Ecol. Monograph*, 43, 95–124.
Rusch, A., et al. (2016), 'Agricultural landscape simplification reduces natural pest control: a quantitative synthesis', *Agric. Ecosyst. Environ.*, 221, 198–204.
Russell, E. P. (1989), 'Enemies hypothesis: a review of the effect of vegetational diversity on predatory insects and parasitoids', *Environ. Entomol.*, 18, 590–599.
Samuel, H. (2018), 'France becomes first country in Europe to ban all five pesticides killing bees', *The Daily Telegraph*. Available at: https://www.telegraph.co.uk/news/2018/08/31/france-first-ban-five-pesticides-killing-bees/.
SARE Outreach (2015), *Cover Cropping for Pollinators and Beneficial Insects*, Sustainable Agriculture Research and Education. Available at: https://www.sare.org/resources/cover-cropping-for-pollinators-and-beneficial-insects/.

SAREP (2021), *Sesbania*, Cover Crops Database, University of Davis Sustainable Agriculture and Research Education Program, University of California Agriculture and Natural Resources. Available at: https://sarep.ucdavis.edu/covercrop/sesbania

Sarkar, S. C., et al. (2018), 'Application of trap cropping as companion plants for the management of agricultural pests: a review', *Insects*, 9(4), 128.

Sarrantonio, M. (2007), 'Selecting the best cover crops for your farm', In A. Clark (Ed), *Managing Cover Crops Profitably Third Edition, Handbook Series Book 9*, Sustainable Agriculture Research and Education, pp. 12–15.

Schaffers, A. P., et al. (2008), 'Arthropod assemblages are best predicted by plant species composition', *Ecology*, 89(3), 782–794.

Schall, K. A. and Hoddle, M. S. (2017), 'Disrupting the ultimate invasive pest partnership: how invasive ants impede biological control of ACP in southern California citrus', *Citrograph*, 8, 38–43.

Schall, K. A., et al. (2018), 'Harnessing hydrogels in the battle against invasive ants', *Citrograph*, 9, 30–35.

Schipanski, M. E., et al. (2014), 'A framework for evaluating ecosystem services provided by cover crops in agroecosystems', *Agric. Syst.*, 125, 12–22.

Schmidt, N. P., et al. (2007), 'Alfalfa living mulch advances biological control of soybean aphid', *Environ. Entomol.*, 36, 416–424.

Sengonça, C. and Frings, B. (1988), 'Einfluss von *Phacelia tanacetifolia* auf Schadlings und Nutzlings-populationen in Zuckerrubenfeldern', *Pedobiologia*, 32, 311–316.

Serée, L., et al. (2023), 'Within-field floral resources have the potential to increase parasitism rates in winter oilseed rape pests more than resources at field margins', *Agric. Ecosyst. Environ.*, 344, 108288.

Shafie, F. A. and Rennie, D. (2012), 'Consumer perceptions towards organic food', *Procedia Soc. Behav. Sci.*, 49, 360–367.

Shelton, A. M. and Badenes-Pérez, F. R. (2006), 'Concepts and applications of trap cropping in pest management', *Annu. Rev. Entomol.*, 51, 285–308.

Shields, M. W., et al. (2019), 'History, current situation and challenges for conservation biological control', *Biol. Control*, 131, 25–35.

Simpson, M., et al. (2011), 'Attract and reward: combining chemical ecology and habitat manipulation to enhance biological control in field crops', *J. Appl. Ecol.*, 48, 580–590.

Singhal, V., et al. (2020), 'Cover crop technology – a way towards conservation agriculture: a review', *Indian J. Agric. Sci.*, 90(12), 2275–2284.

Sluss, R. R. (1967), 'Population dynamics of the walnut aphid *Chromaphis juglandicola* (Kalt.) in Northern California', *Ecology*, 48, 41–48.

Song, B. and Han. Z. (2020), 'Assessment of the biocontrol effects of three aromatic plants on multiple trophic levels of the arthropod community in an agroforestry ecosystem', *Ecol. Entomol.*, 45, 831–839.

Song, J., et al. (2022), 'Effect of soybean volatiles on the behavior of the bean bug, *Riptortus pedestris*', *J. Chem. Ecol.*, 48, 207–218.

Sorribas, J. and Garcia-Mari, F. (2010), 'Comparative efficacy of different combinations of natural enemies for the biological control of California red scale in citrus groves', *Biol. Control*, 55(1), 42–48.

Snapp, S. S., et al. (2005), 'Evaluating cover crops for benefits, costs, and performance within cropping system niches', *Agron. J.*, 97, 322–332.

Stahl, L. (2022), *Minnesota Crop News: Can I Plant a Cover Crop If I Used a PRE Herbicide?* University of Minnesota Extension, Regents of the University of Minnesota. Available at: https://blog-crop-news.extension.umn.edu/2022/05/can-i-plant-cover-crop-if-i-used-pre.html.

Stephens, M. J., et al. (1998), 'Enhancing biological control of leafrollers (Lepidoptera: Tortricidae) by sowing buckwheat (*Fagopyrum esculentum*) in an orchard', *Biocontrol Sci. Technol.*, 8, 547–558.

Stojanovic, N. (2021), *European Green Deal and "Farm-to-Fork" Strategy for Fair, Healthy and Environmentally-Friendly Food System*, Collection of Papers from the Conference Organized on Occasion of the Day of the Faculty of Law, Vol. 2, pp. 130–149.

Straser, R. K., et al. (2024), 'Floral resources enhance fitness of the parasitoid *Hadronotus pennsylvanicus* (Hymenoptera: Scelionidae) but not biological control of its host *Leptoglossus zonatus* (Heteroptera: Coreidae)', *Environ. Entomol.*, 53(2), 212–222.

Straub, C. S., et al. (2014), 'Plant diversity increases herbivore movement and vulnerability to predation', *Basic Appl. Ecol.*, 15, 50–58.

Styrsky, J. D. and Eubanks, M. D. (2007), 'Ecological consequences of interactions between ants and honeydew-producing insects', *Proc. R. Soc. B.*, 274, 151–164.

Sundermeier, A. (2016), *Cover Crop Fundamentals*, Ohioline, Ohio State University Extension, Ohio State University Agricultural and Natural Resources, Publication AGF-142.

Teasdale, J. (2007), *Sustainable Production of Fresh-Market Tomatoes and Other Vegetables with Cover Crop Mulches*, United States Department of Agriculture, Agricultural Research Service, Farmers' Bulletin No. 2280 October 2007.

Tena, A., et al. (2013), 'Competition between honeydew producers in an ant-hemipteran interaction may enhance biological control of an invasive pest', *Bull. Entomol. Res.*, 103, 714–723.

Theunissen, J., et al. (1995), 'Effects of intercropping white cabbage with clovers on pest infestation and yield', *Entomol. Exp. Appl.*, 74, 7–16.

Thies, B. N. and Abawi, J. E. (2009), *Cornell Soil Health Assessment Training a Manual, 2nd edition,* Cornell University, School of Integrative Plant Science, pp. 79–98.

Thompson, J. N. and Pellmyr, O. (1991), 'Evolution of oviposition behavior and host preference in Lepidoptera', *Annu. Rev. Entomol.*, 36, 65–89.

Tian, J. C., et al. (2015), 'Bt crops benefit natural enemies to control non-target pests', *Sci. Rep.*, 5, 16636.

Tian, Y., et al. (2012), 'Influence of intercropping on dynamics of insect pests, natural enemies and the damage of *Ostrinia furnacalis* in sweet corn field', *Acta Phytophylacica Sin.*, 39, 1–6.

Tittonell, P. (2014), 'Ecological intensification of agriculture: sustainable by nature', *Curr. Opin. Environ. Sustain.*, 8, 53–61.

Tiwari, S., et al. (2018), 'Host plant selection by the wheat bug, *Nysius huttoni*', *J. Econ. Entomol.*, 111(2), 586–594.

Torres, J. B. and de Bueno, A. F. (2018), 'Conservation biological control using selective insecticides: a valuable tool for IPM', *Biol. Control*, 126, 53–64.

Tracey, E. (2014), 'The promise of biological control for sustainable agriculture: a stakeholder-based analysis', *J. Sci. Policy Govern.*, 5, 1–13.

Tscharntke, T., et al. (2016), 'When natural habitat fails to enhance biological pest control – five hypotheses', *Biol. Conserv.*, 204, 449–458.

Tschumi, M., et al. (2016), 'Tailored flower strips promote natural enemy biodiversity and pest control in potato crops', *J. Appl. Ecol.*, 53, 1169–1176.

Tylianakis, J. M., et al. (2004), 'Improved fitness of aphid parasitoids receiving resources subsidies', *Ecology*, 85, 658–666.

UCIPM (2020), *Agriculture Pest Management Guidelines: Citrus*, Statewide IPM Program, Agriculture and Natural Resources, Regents of the University of California, California. Available at: https://ipm.ucanr.edu/agriculture/citrus/.

Union of Concerned Scientists (2013), *Cover Crops: Public Investments Could Produce Big Payoffs.* Available at: https://www.ucsusa.org/sites/default/files/2019-09/cover-crop-fact-sheet-2012.pdf.

USDA (2023a), *USDA Plant Hardiness Zone Map*, United States Department of Agriculture. Available at: https://planthardiness.ars.usda.gov/.

USDA (2023b), *Citrus Fruits -2023 Summary.* United States Department of Agriculture National Agricultural Statistics Service. Available at: https://downloads.usda.library.cornell.edu/usda-esmis/files/j9602060k/4742bs21j/3n205h50s/cfrt0923.pdf.

Uyttenbroeck, R., et al. (2016), 'Pros and cons of flowers strips for farmers: a review', *Biotechnol. Agron. Soc.*, 20, 225–235.

van Rijn, P. C. J. and Wäckers, F. L. (2016), 'Nectar accessibility determines fitness, flower choice and abundance of hoverflies that provide natural pest control', *J. Appl. Ecol.*, 53, 925–933.

van Schalkwyk, J., et al. (2021), 'Maintaining high vegetation structural diversity in the landscape promotes arthropod diversity in dynamic production areas', *Landsc. Ecol.*, 36, 1773–1785.

Vattala H. D., et al. (2006), 'The influence of flower morphology and nectar quality on the longevity of a parasitoid biological control agent', *Biol. Control*, 39(2), 179–185.

Wäckers, F. L. (2001), 'A comparison of nectar- and honeydew sugars with respect to their utilization by the hymenoptera parasitoid *Cotesia glomerata*', *J. Insect Physiol.*, 47, 1077–1084.

Wäckers, F. L. and van Rijn, P. C. J. (2012), 'Pick and mix: selecting flowering plants to meet the requirements of target biological control insects', In E. Snyder and D. M. Y. Read (Eds), *Biodiversity and Insect Pests: Key Issues for Sustainable Management*, John Wiley & Sons, Ltd, pp. 139–165.

Wäckers, F. L., et al. (2007), 'Nectar and pollen feeding by insect herbivores and implications for multitrophic interactions', *Annu. Rev. Entomol.*, 52, 301–323.

Wan, N. F., et al. (2016), 'Ecological engineering of trap cropping promotes biocontrol services in peach orchard ecosystems', *Ecol. Eng.*, 90, 427–430.

Wang, R., et al. (2020), 'Habitat fragmentation changes top-down and bottom-up controls of food webs', *Ecology*, 101(8), e03062.

Wanner, H., et al. (2007), 'Tracking parasitoids with stable isotope Ca-44 in agroecosystems', *Agric. Ecosyst. Environ.*, 118, 143–148.

Warner, K. D. (2007), 'The quality of sustainability: agroecological partnerships and the geographic branding of California winegrapes', *J. Rural Stud.*, 23, 142–155.

Weiss, M. J., et al. (1994), 'Flea beetle (Coleoptera, Chrysomelidae) populations and crop yield in field pea and oilseed rape intercrops', *Environ. Entomol.*, 23, 654–658.

Wells, J., et al. (2020), 'Prototype *Lygus* spp. vacuum provides improved pest management in California strawberries', *Int. J. Fruit Sci.*, 20(1), 1019–1028.

Wilkinson, T. K. and Landis, D. A. (2005), 'Habitat diversification in biological control: the role of plant resources', In F. L. Wäckers, et al. (Eds), *Plant Provided Food and Plant-Carnivore Mutualism*, Cambridge University Press, pp. 305–325.

Wilson, H., et al. (2016), 'Host plant associations of *Anagrus* spp. (Hymenoptera: Mymaridae) and *Erythroneura elegantula* (Hemiptera: Cicadellidae) in northern California', *Ecol. Entomol.*, 45(3), 602–615.

Wilson, H., et al. (2017), 'Landscape diversity and crop vigor outweigh influence of local diversification on biological control of a vineyard pest', *Ecosphere*, 8(4), e01736.

Winkler, K., et al. (2010), 'Assessing risks and benefits of floral supplements in conservation biological control', *Biol. Control*, 55, 719–727.

Wingard, C. (1996), *Cover Crops in Integrated Vegetable Production Systems*. SARE Project Report #PG95-033. Southern Region SARE.

Wratten, S. D., et al. (2003), 'Field boundaries as barriers to movement of hover flies (Diptera: Syrphidae) in cultivated land', *Oecologia*, 134, 605–611.

Wratten, S. D., et al. (2012), 'Pollinator habitat enhancement: benefits to other ecosystem services', *Agric. Ecosyst. Environ.*, 159, 112–122.

Wyckhuys, K. A. G., et al. (2018), 'Maximizing farm-level uptake and diffusion of biological control innovations in today's digital era', *BioControl*, 63, 133–148.

Yancey, C. (1994), *Covers Challenge Cotton Chemicals*, The New Farm, pp. 20–23.

Yousefi, M., et al. (2024), 'Assessing the effectiveness of cover crops on ecosystem services: a review of the benefits, challenges, and trade-offs', *Int. J. Agric. Sustain.*, 22(1), 2335106.

Zannopoulos, S., et al. (2024), 'The potential of three summer legume cover crops to suppress weeds and provide ecosystem services–a review', *Agronomy*, 14, 1192.

Zhang, W., et al. (2007), 'Ecosystem services and dis-services to agriculture', *Ecol. Econ.*, 64, 253–260.

Zhao, J. Z., et al. (1992), 'Effects of neighboring nectar-producing plants on populations of pest Lepidoptera and their parasitoids in broccoli plantings', *Great Lakes Entomol.*, 24, 253–258.

Zhu, P., et al. (2013), 'Laboratory screening supports the selection of sesame (*Sesamum indicum*), to enhance *Anagrus* spp. parasitoids (Hymenoptera: Mymaridae) of rice planthoppers', *Biol. Control*, 64, 83–89.

Zhu, P., et al. (2018), 'Quantifying the respective and additive effects of nectar plant crop borders and withholding insecticides on biological control of pests in subtropical rice', *J. Pest Sci.*, 91, 575–584.

Chapter 5

Revisiting the contribution of cover crops to weed management in Conservation Agriculture

John M. Wallace and Laurel Wellman, The Pennsylvania State University, USA

1 Introduction

Increased adoption of integrated weed management (IWM) strategies is critically needed to reduce the evolution and spread of herbicide-resistant weed species in Conservation Agriculture (Mortensen et al., 2012). Cover cropping has the potential to be a foundational component of IWM in Conservation Agriculture by reducing weed fitness during active periods of growth and as surface residue after termination (Teasdale, 1996a). However, weed suppression is just one of several ecosystem services that winter annual cover crops regulate in agroecosystems (Schipanski et al., 2014).

Several contemporary trends may contribute to increased cover crop adoption in Conservation Agriculture. Cover cropping is increasingly incentivized as a means for improving soil health and climate resilience, which alters the assessment of agronomic and economic tradeoffs that influence adoption (Chami et al., 2023; Sawadgo, 2024). Continued advancement in precision agriculture technologies and development of decision support tools (DSTs) are also helping to optimize the agronomic benefits of cover crops

http://dx.doi.org/10.19103/AS.2025.0162.07

while minimizing liabilities associated with potential crop yield penalties. Finally, awareness and adoption of IWM is increasing due to the current pace of herbicide resistance evolution among weed species, which has far-reaching implications for the economics of weed control and sustainability of farming systems (Moss, 2019; Jussaume et al., 2022).

Adoption of IWM strategies remains elusive in modern production systems due to the common perception that it increases management complexity and cost (Jussaume and Dentzman, 2016). Integrating cover crops as a component of IWM certainly increases the complexity of management decision-making compared to herbicide-based weed control. The latter approach is a more prescriptive process often limited to the selection of herbicide products that result in acceptable weed control. In comparison, weed control by cover crops can vary in magnitude and sometimes direction depending on biotic, abiotic, and spatiotemporal conditions. Identifying the mechanisms, or ecological interactions, that are the source of variation in weed responses to cover cropping tactics is critical for advancing their utility (Catford et al., 2022). Cover crops also increase management complexity because they regulate multiple ecosystem services or disservices affecting cropping systems. Therefore, weed management performance targets should be weighed against other desired ecosystem services. Identifying the proper role of cover crops in IWM is a necessary first step toward negotiation of these tradeoffs.

In this chapter, we provide a unique contribution to the cover crop and weed ecology literature by focusing our synthesis on the integration of cover crop and herbicide-based weed control tactics in no-till grain crop production systems. Specifically, we aim to inform research that is oriented towards improving cover crop management decision-making by describing (1) scale-dependent processes that influence the ecology of cover crop–weed interactions; (2) the role of cover crops in IWM and proactive herbicide resistance management; and (3) cover cropping practices that aim to balance weed and cash crop management goals.

2 Ecology of cover crop-weed interactions

Cover crops can be managed to manipulate abiotic and biotic factors, hereafter referred to as ecological stressors, that lead to mortality or reduced fitness of weeds. A fundamental understanding of how these ecological stressors influence an individual's fitness is necessary to employ weed management strategies that integrate cover crop (ecological stressor) and herbicide-based (chemical stressor) tactics. In this section, we describe (1) cover crop traits and management tactics that affect the duration and intensity of stressors on an individual's fitness and (2) interspecific and intraspecific traits of weeds that influence susceptibility to these stressors.

Published by Burleigh Dodds Science Publishing Limited, 2026.

The regeneration niche. Niche theory is a useful framework for understanding stressors imposed by cover crops (Liebman and Davis, 2000). The 'regeneration' niche is defined as the abiotic, biotic, and physiochemical requirements necessary for germination, establishment, and growth of plant species (Grubb, 1977; Silvertown, 2004). Stressors imposed by cover crops affect weeds at each developmental stage involved in the regeneration niche. From this perspective, the management goal is to optimize the duration and intensity of cover crop stressors to preempt the regeneration niche of weeds, while minimizing negative impacts on the regeneration niche of cash crops.

Stressor duration, intensity, and heterogeneity. The duration of cover crop stressors can be delineated between the period of active cover crop growth and the period after termination in which cover crop surface residues impact the regeneration niche of weeds in Conservation Agriculture. The intensity of cover crop stressors can be thought of as the relative proportion of niche space that is occupied by cover crops, which is positively correlated with cover crop biomass and other measures of abundance. The spatial and temporal heterogeneity of stressors imposed by cover crops is an important and often overlooked determinant of weed suppression potential (Eslami and Davis, 2018). For example, cover crop biomass production can vary significantly at fine (< 1 m^2), patch (> 10 m^2), and field scales (Fig. 1). Weed suppression potential is optimized when cover crop management practices increase the duration and intensity of stressors, while reducing stressor heterogeneity. Management practices that influence these aims include cover crop species selection, sowing timing, establishment method, termination timing, and surface residue management.

Trait-based cover crop effects. The intensity of cover crop stressors on individual weed fitness is driven, in part, by differences in biomass potential among species (Mohler and Teasdale, 1993; Mirsky et al., 2013). Cover crops from all taxonomic groups can provide some level of weed suppression during the active period of growth or as surface residue, but winter cereal (e.g. cereal rye) and brassica species (e.g. canola) produce comparatively greater biomass and weed suppression levels compared to leguminous species due to higher growth rates in late fall and early spring (Finney et al., 2017).

Several cover crop traits mediate the effect of biomass on individual weed fitness. Establishment rate and early-season plant height are important traits for establishing a fitness advantage for cover crops, in which larger cover crop plants preempt light resources from emerging weeds via size-asymmetric competition (Schwinning and Weiner, 1998). Seed mass and germination requirements of cover crop species are primary indicators of both establishment rate and early-season growth rates (Nosratti et al., 2023). The growth habit of cover crops also mediates the effect of biomass on weeds during active periods of growth and as surface mulch. Teasdale and Mohler (2000) developed the mulch area index

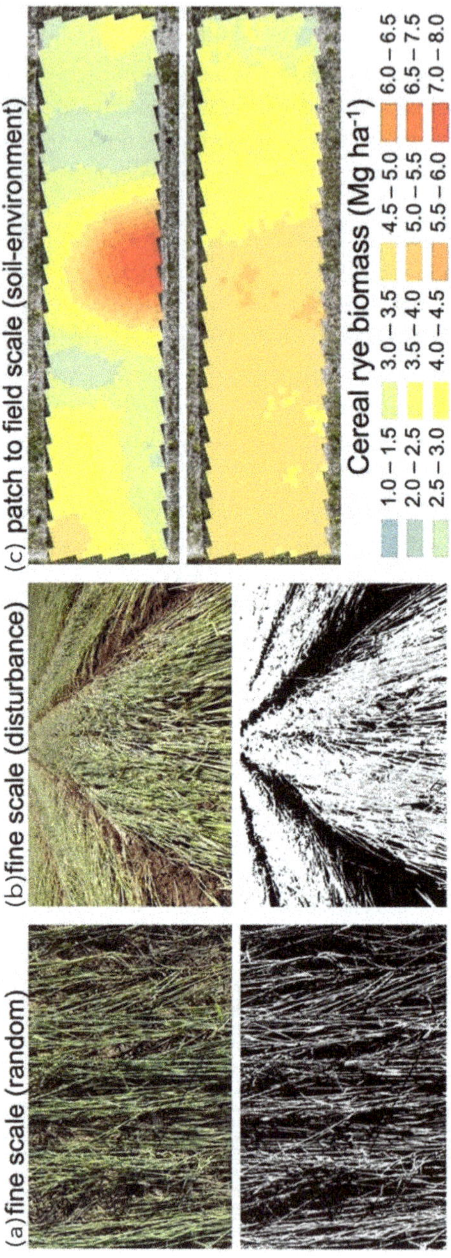

Figure 1 Examples of spatial heterogeneity of cover crop stressors at multiple scales. At fine-scales (< 1 m^2), (a) cover crop growth and residue management create random patterns (30% fractional green canopy cover; FGCC) and (b) planter-based soil disturbance creates regular patterns (58% FGCC). Heterogeneity of cover crop stressors is a function of varying soil-environment conditions at (c) patch (10 m^2) to field scales; demonstrated with geospatial maps of cover crop biomass (heading stage; Zadok's 5058) created by structure-for-motion imagery in an experiment conducted by Dobbs et al. (2024). Photo credits: a (L. Wellman); b (J. Wallace).

(MAI) to quantify the relationship between surface area and mass of cover crop species, which was then used to predict effects on weed fitness that can result from light resource preemption and physical impedance. They demonstrated that commonly used cover crops differ in the ratio of surface area to mass (hairy vetch > crimson clover > cereal rye), where greater surface area per unit of surface mass results in greater stressor intensity.

The duration of cover crop stressors is driven, in part, by biomass production and aboveground C:N ratio of plant tissue, which are traits that determine decomposition rates, or mass loss of surface residues. Species selection, including the presence or absence of leguminous species, and the phenological stage of cover crop species at the time of termination are the key factors that determine the C:N ratio of cover crop surface residues. Variation in cover crop decomposition rates results from complex interactions between intrinsic (soil, weather) and extrinsic (management) factors (Thapa et al., 2021). Decomposition rates decrease when cover crop management results in increased surface residue mass, as well as increased C:N, residue holo-cellulose concentration, and lignin:N. Relative humidity and number of rainy days are intrinsic factors that are positively correlated with cover crop decomposition rates (Thapa et al., 2021).

Trait-based weed responses. Weed life-history strategies influence responses to cover crop stressors. Cover crops have the greatest impact on weed species with winter- and summer-annual life cycles and have more limited impacts on species with biannual or perennial life cycles. Emergence periodicity, seed mass, and germination requirements are interrelated traits of annual weed species that are sources of variation in response to cover crop stressors.

Annual weed species with prolonged emergence periods, or peak emergence windows that correspond to advanced decomposition of surface residues, are less influenced by cover crop stressors due to greater regeneration niche availability. Annual weed species with greater seed mass are less susceptible to stressors created by cover crop surface residues (Mohler, 1996). Though weed species vary considerably in seed mass, they are often operationally defined as either small- or large-seeded using crop seed mass as a relative metric. Weed species with average seed mass that are at least two (10^2) to three (10^3) orders of magnitude smaller than corn and soybean are typically classified as small-seeded, whereas species that are less than an order of magnitude smaller than crops are classified as large-seeded.

Weed seed mass is associated with evolved life-history strategies that influence weed species responses to cover crop stressors at a population and community level. Classical plant ecology theory suggests that small-seeded species produce many seeds ('r-selected') to increase the probability of dispersal into a favorable environment for establishment (MacArthur and Wilson, 1967).

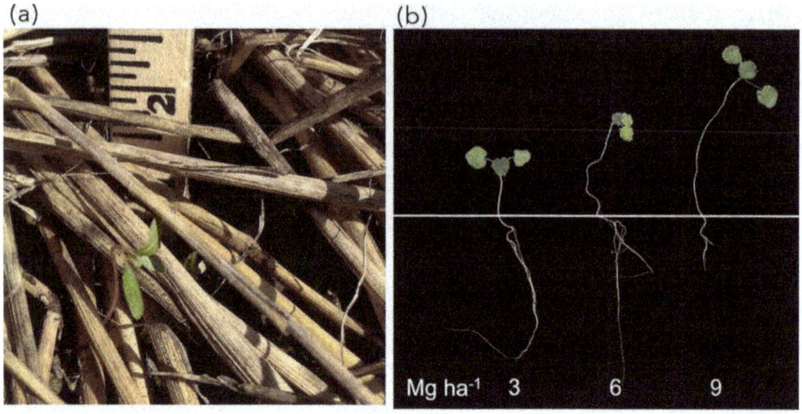

Figure 2 Effect of cereal rye surface residues on seedling emergence and growth of (a) waterhemp (*A. tuberculatus*), a small-seeded broadleaf species, in a Pennsylvania field experiment, and (b) velvetleaf (*A. theophrasti*), a large seeded broadleaf species, from a greenhouse experiment conducted by Ficks et al. (2022a) that measured the effect of cereal rye surface residues (3–9 Mg ha^{-1}) on seedling resource allocation to roots and shoots during the establishment phase. Photo credit: a (J. Wallace); b (T. Ficks).

Conversely, larger-seeded species produce fewer seeds per plant but are more tolerant to stressors and competition. Recent reviews have described the role of resource availability and disturbance regimes in r/K selection within agroecosystems (MacLaren et al., 2020), but explicit consideration of selection resulting from cover crop stressors has yet to be fully described (Fig. 2).

3 Cover crop stressor effects on individual weeds and weed communities

Cover crop stressors impact the fitness of individual weeds via effects on seedbank survival, germination and establishment rates, and seedling survival (i.e. vital rates). Cover crops indirectly affect weed seed mortality within soil by improving habitat for animals that consume weed seeds (i.e. seed predators) near the soil surface and microorganisms (bacteria, fungi) that contribute to seed decay within soil (*see* Smith et al., 2024). Seed predation rates can be significant but highly variable (Davis et al., 2008). A synthesis of the literature suggests that cropping system design, including use of continuous living cover and no-till practices, is the primary means for enhancing weed seed mortality rates through habitat improvement, while cover crop management practices within phases of crop rotations play a secondary role (Smith et al., 2024). Cover crop management decisions within a rotation sequence are likely to have a comparatively greater impact on germination, emergence, and seedling survivorship rates of annual weeds. In this section, we highlight how cover crop management practices influence individual weed fitness during these early developmental stages.

Germination. Weed seed germination occurs only after weed seeds have been released from either a primary or secondary dormancy state (Benech-Arnold et al., 2000). Dormancy release and germination are two different processes that are influenced by the same suite of interacting environmental controls, including light flux density, soil temperature amplitude, light quality, nitrate concentration, and biologically active gases (oxygen, carbon dioxide) in soil (Teasdale, 1996a). These environmental controls are indicators of soil disturbance or gap openings in established vegetation. Attenuation of germination stimuli is a mechanism by which cover crops influence weed fitness. This effect results in either induced secondary dormancy or delayed germination within a given growing season depending on the duration and intensity of cover crop stressors.

Cover crop effects on germination that result from changes in light flux density and light quality are strongly correlated with cover crop biomass production and residue management strategies. Teasdale and Mohler (1993) found that the effects of cereal rye and hairy vetch surface residues on photosynthetic photon flux density (PPFD) and red to far-red ratio (R:FR) are likely insufficient to prevent the germination process in phytochrome-sensitive seed located at the soil surface across a range of biomass levels (3–9 Mg ha^{-1}). The authors noted that although high levels of surface mulch did not exclude light from penetrating to the soil surface, the presence of surface mulch can prevent germination of seeds located at shallow soil depths (< 5 mm), where light penetration is marginal for germination under non-mulch conditions. Consequently, it is likely that soil disturbance and seed dispersal patterns that affect seed burial within Conservation Agriculture are important sources of variation that mediate the effect of cover crop surface residues on germination stimuli.

The magnitude of cover crop effects on germination that result from changes in soil temperature and moisture is likely to vary considerably among environmental and management gradients.

Teasdale and Mohler (1993) suggest that cover crop surface residue effects on maximum soil temperature are likely insufficient to prevent germination but may result in a delay in the onset of germination. Similarly, Teasdale and Daughtry (1993) reported a reduction in the amplitude of minimum and maximum soil temperatures beneath cover crop residues but suggested that prevention of germination was unlikely for seeds without a temperature amplitude requirement (~10 C; Taylorson, 1987). From a weed management perspective, a delay in germination in response to cover crop stressors has the potential to increase the efficacy of other tactics that follow in sequence, including crop competition and herbicide use.

The magnitude of cover crop effects on germination that result from changes in soil nitrate is likely to vary based on cover crop species selection, termination timing and method, and soil fertility. Spatial and temporal variation in soil nitrate concentrations is rarely quantified in relationship to weed recruitment. Consequently, evidence of inhibitory or stimulatory effects on germination is limited to observed differences in emergence rates of nitrophilic weed species in winter cereal or legume cover crops (Mohler et al., 2018). Compared to winter cereals, legume monocultures or mixtures often result in higher soil inorganic nitrogen pools due to less N immobilization prior to cover crop termination and higher mineralization rates after termination (Mohler and Teasdale, 1993; Pittman et al., 2020). Soil nitrate is known to stimulate germination in several economically important weeds, such as *Amaranthus* species and common lambsquarters (*Chenopodium album* L.), but germination rate and periodicity of these species are likely to be influenced by the interacting effects of soil nitrate pools and other abiotic factors (Little et al., 2021).

Seedling emergence. Cover crop stressors can also reduce the survival rate of a germinated seed, which is reliant on nonrenewable seed resources, to a photosynthetic autotrophic seedling (Forcella et al., 2000). Mortality after germination but prior to emergence (i.e. fatal germination) is an important seedbank loss pathway, which is often difficult to measure in field studies (Davis and Renner, 2007). Mechanisms by which cover crops influence fatal germination include resource limitation, physical impedance, allelopathy, and soilborne pathogens (Teasdale, 1996b).

Resource limitation and physical impedance are interacting factors that mediate elongation of coleoptiles and hypocotyls prior to seedling emergence. Teasdale and Mohler (1993) reported that the photosynthetic light compensation point (20–50 mol m^{-2} s^{-1}) can occur for an emerging seedling at approximately 4.5 Mg ha^{-1} of cover crop surface residue. Physical impedance of seedling elongation results from the occupation of space by cover crop surface residue above the location of seed, which creates a mechanical barrier that requires redirection of elongating seedlings and additional allocation of seed reserves to coleoptile or hypocotyl elongation (Teasdale, 1996b).

By controlling for the germination process, Ficks et al. (2022a) demonstrated that survivorship from germination to establishment is related to dicotyledon seed mass (Fig. 3a). Large-seeded species tend to have greater seed reserves and larger initial seedling size, which increases survival under low-light conditions that create carbon deficits during the establishment phase (Ganade and Westoby, 1999; Leishman and Westoby, 1994). Small-seeded species (r-selected) typically have higher relative growth rates, which can confer a fitness advantage over large-seeded species (K-selected) under non-limiting resource conditions in the post-establishment phase (Ganade and Westoby, 1999). However, establishment in cover crop surface residue is a barrier that

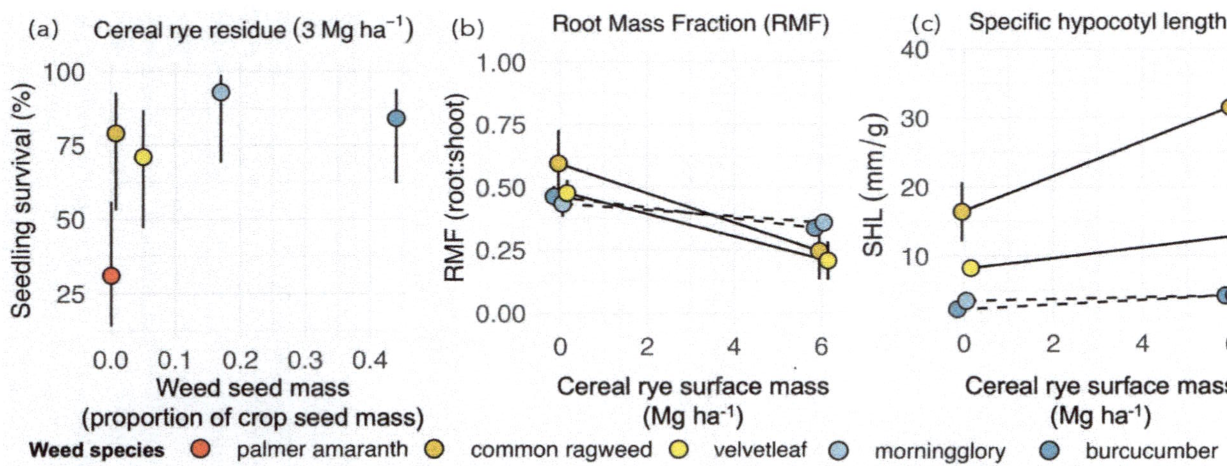

Figure 3 The relationship between cereal rye surface mass (Mg ha^{-1}) and weed seedling (a) survival rate to establishment; (b) root mass fraction (RMF), which quantifies the proportion of total seedling biomass allocated to roots; and (c) specific hypocotyl length (SHL), which quantifies shoot length per unit of shoot mass. Seed mass of weed species is one to three orders of magnitude smaller than crop (soybean) seed mass, including Palmer amaranth (10^3; *Amaranthus palmeri*), common ragweed (10^2; *Ambrosia artemisiifolia*), velvetleaf (10^2; *Abutilon theophrasti*), ivyleaf morningglory (10^1; *Ipomoea hederacea*), and burcucumber (10^1; *Sicyos angulatus*). Presentation of results is adapted from Ficks et al. (2022a).

small-seeded species must first navigate before high relative growth rates can create a fitness advantage.

Allelopathy, or chemically mediated interference of weed growth, has received considerable attention as a mechanism by which living cover crops and surface residues reduce seedling emergence rates. Allelopathy has been documented under controlled conditions for several taxonomic groups, including several winter cereal species (Putnam and DeFrank, 1983), medium red clover (Liebman and Davis, 2000), winter oilseed rape (Haramoto and Gallandt, 2005), and oilseed radish, black oat, buckwheat, and sunflower (Sturm et al., 2018).

Teasdale et al. (2012) reported that benzoxazinoid compounds found in cereal rye can inhibit weed growth under controlled conditions but do not reside in soil at concentrations necessary to inhibit weed growth under most management scenarios. Under no-till management, allelopathic compounds released by cereal rye, including phenolic acids and benzoxazinoids, (1) are more concentrated in cereal rye roots compared to shoots; (2) are found in soil at concentrations a magnitude of order below phytotoxic thresholds; (3) peak near cereal rye termination (3–7 d); (4) require soil moisture to be mobilized; and (5) are rapidly (< 2 d) degraded via microbial activity (Otte et al., 2020; Rice et al., 2022). These field-based observations are supported by a recent meta-analysis that suggests allelopathic effects of cover crops are variable and unpredictable under many management scenarios (Mahé et al., 2022).

Allelopathic effects of cover crops, when present, are also likely to differ among taxonomic groups and traits of weeds. For example, weed seed mass is positively correlated with emergence rate in medium red clover (Liebman and Sundberg, 2006) and oilseed rape (Kruidhof et al., 2011) residues containing allelopathic compounds. Later germinating weed cohorts are also more likely to survive to establishment in these residues (Kruidhof et al., 2011), which highlights that weed control potential is influenced by the temporal overlap between inhibitory concentrations of allelopathic compounds and the regeneration niche of annual weeds.

Current breeding efforts are targeting selection of increased allelopathic activity among cover crop cereal rye cultivars (Rebong et al., 2024). Separating allelopathic effects from other weed interference mechanisms under realistic management conditions is a significant challenge to such efforts. Breeding efforts require consideration of the relationship between allelopathic traits and traits that increase weed interference via other mechanisms. Breeding efforts will also need to consider the selection of allelopathic traits in the context of management to target the release of allelochemicals within the spatial and temporal regeneration niche of problematic weed species.

Seedling growth. Plants subjected to resource limitation during the establishment phase allocate more biomass to structures that increase acquisition of the limiting resource (Poorter et al., 2012). Therefore, non-lethal stressors imposed by cover crops are likely to affect resource allocation and acquisition patterns of emerging weed seedlings. Ficks et al. (2022a) demonstrated that dicot weed seedlings increased shoot biomass to facilitate hypocotyl elongation as cover crop residue mass increased, which came at a cost of reduced resource allocation to root biomass production (Fig. 3b). The authors found similar resource allocation tradeoffs in response to cover crop residue mass gradients for small- and large-seeded species, but small-seeded species elongated their hypocotyl with a lower tissue density compared to large-seeded species, likely due to comparatively lower seed reserves (Fig. 3c). Reduction in hypocotyl tissue density has the potential to decrease survivorship of seedlings when exposed to other abiotic or biotic stressors (Ganade and Westoby, 1999).

Soil nutrient availability can also be a limiting factor for establishing weed seedlings. Winter cereal cover crops that scavenge nitrogen prior to emergence of summer annual weeds can limit nitrogen acquisition during seedling growth (Wells et al., 2013; Poffenbarger et al., 2015a). High biomass production coupled with delaying termination until advanced growth stages of winter cereals can result in C:N ratios that exceed 20:1, which depletes soil inorganic nitrogen pools and delays decomposition and nitrogen mineralization from surface residues beyond early weed growth stages (Brockmueller et al., 2023; Keles et al., 2016). Menalled et al. (2021) demonstrated the importance of nitrogen limitation on weed growth by experimentally applying nitrogen after cereal rye termination at anthesis. The observed positive relationship between nitrogen addition and late season weed biomass found in this study demonstrates that nitrogen limitation is a mechanism that regulates cover crop stressor intensity.

3.1 Cover crop stressor effects on populations

The study of safe site and seed supply recruitment dynamics is a useful framework for understanding cover crop stressor effects on weeds at a population level. Harper (1977) defined a 'safe site' as a microsite that provides (1) necessary stimuli for terminating dormancy and germination, (2) resources required for seedling emergence, and (3) freedom from hazards that prevent seedling establishment, such as phytotoxic chemicals and pathogens within soil. Conceptually, a safe-site is an extension of the 'regeneration niche' to the population-level. Seed supply refers to the number of available propagules, either introduced via dispersal or stored within the weed seedbank. The size and structure of weed populations are emergent properties that result from interactions between the number of available safe sites and seed supply (Fig. 4). Most field-based studies measure cover cropping effects on weed

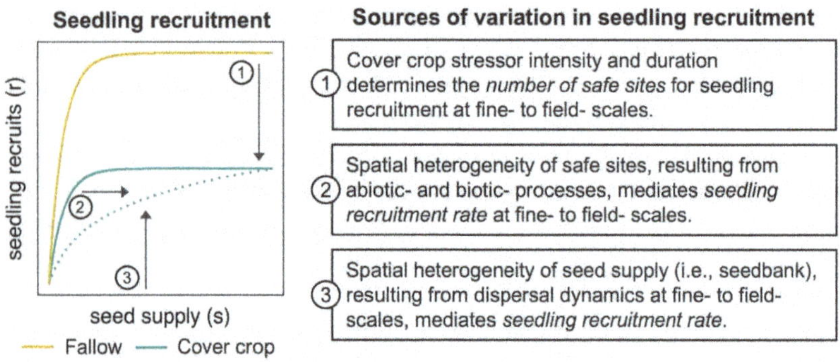

Figure 4 Hypothesized effect of cover crops on seedling recruitment processes based on model described in Duncan et al. (2009). Cover crop stressor intensity influences the (1) number of safe sites available for seedling recruitment. The rate at which safe sites are filled as seed supply increases is influenced by spatial heterogeneity of (2) seed dispersal dynamics and (3) cover crop stressors. Figure credit: L. Wellman.

populations (density or biomass) without consideration of seed supply, which varies considerably at fine- to field-level scales (Gallandt et al., 1999).

The direction and magnitude of cover crop stressor effects on recruitment dynamics are a function of the interaction between cover crop biomass, weed species traits, and extrinsic factors (Mohler and Teasdale, 1993; Nord et al., 2011; Nord et al., 2012). For example, cover crop surface residues below 2.5 Mg ha^{-1} can stimulate seedling recruitment of small-seeded *Amaranthus* species relative to bare ground in some years but decrease seedling recruitment in other years (Teasdale and Mohler, 2000; Nunes et al., 2024). Stimulatory effects of cover crop residues on weed recruitment can be attributed to improved soil moisture conditions for germination and establishment in years when soil moisture is a limiting factor (Mohler and Teasdale, 1993).

Sources of variation in population-level outcomes. Significant variation in weed population responses to cover crop stressors is evident among meta-analyses of field experiments conducted over a range of soil-environment conditions. Nichols et al. (2020) reported that cover crop residues exceeding 5 Mg ha^{-1} resulted in a 75% reduction in weed biomass in a meta-analysis of field studies (n = 15) within US Midwest corn and soybean production systems. However, the authors reported no effect of cover crop residues on weed density. In contrast, from a similar meta-analysis of field studies (n = 28) within the US Southeast, Weisberger et al. (2023) reported that cover crop residues exceeding 6.6 Mg ha^{-1} resulted in a 50% reduction in weed density but had no effect on weed biomass. In a global meta-analysis of field studies (n = 53), Osipitan et al. (2019) reported that increasing cover crop biomass reduced both weed density and biomass. In these studies, cover crop residues exceeding 5 Mg ha^{-1} and 6 Mg ha^{-1} resulted in a 50% reduction in weed biomass and weed density, respectively.

Several context-dependent factors are likely sources of variation in weed abundance across biomass gradients. Many foundational studies that were conducted in low-synthetic input or organic systems evaluated cover crop surface residue effects on season-long weed suppression in the absence of additional weed control interventions. In these scenarios, assessment of cover crop performance extends beyond the regeneration niche, which likely contributes to discrepancies between biomass and density responses. For example, Teasdale (1996b) suggested that cover crop residues influence weed biomass less than weed density because weeds that survive to establishment compensate for reduced density by producing greater biomass per plant due to less intraspecific competition. Weisberger et al. (2024) documented a reduction in Palmer amaranth density in response to cover crop residues but noted similar total weed biomass (per m^2) levels with and without cover crops.

These examples highlight the importance of considering the appropriate population-level measurement given the management goal, as well density-dependent and density-independent selection processes on weed populations. Looking forward, modeling relationships between cover crop biomass and weed seedling recruitment across a range of seedbank sizes (i.e. seed supply) will be useful for understanding density-dependent and density-independent recruitment processes that result from spatial and temporal heterogeneity of cover crop–weed interactions among seedbank gradients relevant to weed management (Duncan et al., 2009; Wandrag et al., 2019).

3.2 Cover crop stressor effects on weed communities

Community assembly theory provides a framework for understanding factors that shape weed community composition in Conservation Agriculture (Booth and Swanton, 2002; Smith and Mortensen, 2017). Within this framework, cover crops are considered a biotic filter situated below a hierarchy of environmental, cropping system, and management filters that structure species composition and abundance of weed communities. Smith and Mortensen (2017) suggest that the relative importance of biotic filters in structuring weed communities is greatest in cropping systems characterized by low disturbance frequency and intensity. Management filters that create high intensity and frequent disturbance events include herbicide-based and tillage-based weed control. Several studies have concluded that the strength of cover cropping effects on weed community structure depends on whether herbicides or tillage are tools in the weed control toolbox (Swanton et al., 1999; Moonen and Bàrberi, 2004; Adeux et al., 2021).

In the absence of other management filters, Smith et al. (2015) concluded that the filtering effects of cover crops are short-term, relatively stochastic, and a function of the relative abundance of species within the resident weed species pool. Within this context, there is general agreement on three trait-based responses of annual weed species to cover crops that shape community assembly. First, the filtering strength of cover crops will be greatest for annual weed species with life cycles that overlap with winter annual cover crop growth periods but emerge after cover crop sowing (Adeux et al., 2021). In comparison, annual weed species that have partially- or non-overlapping life cycles will be more influenced by other management filters. Next, annual dicots are more affected by cover crops than monocots (Swanton et al., 1999), likely due to the combination of phytochrome-sensitive germination traits and greater susceptibility of dicots to the physical impedance of hypocotyl elongation during the seedling establishment phase. And finally, small-seeded dicot species are more affected by cover crops than large-seeded dicot species (Teasdale, 1996b), likely due to reduced germination and survival under low-light conditions.

3.3 Summary

Given the heterogeneous environment at the scale of a non-dormant weed seed, Harper (1977) noted the impossibility of discovering which mechanism(s) directly control germination and seedling emergence. This observation highlights the difficulty, and perhaps futility, of research that aims to amplify a single weed suppression mechanism via a change in cover crop management tactics. Rather, the size and structure of weed populations are emergent properties of multiple stressors imposed by cover crops that influence processes at individual, population, and community levels (Fig. 5). Identifying cover crop management tactics that increase biomass production and reduce the heterogeneity of cover crop stressors at fine- to field-level scales should

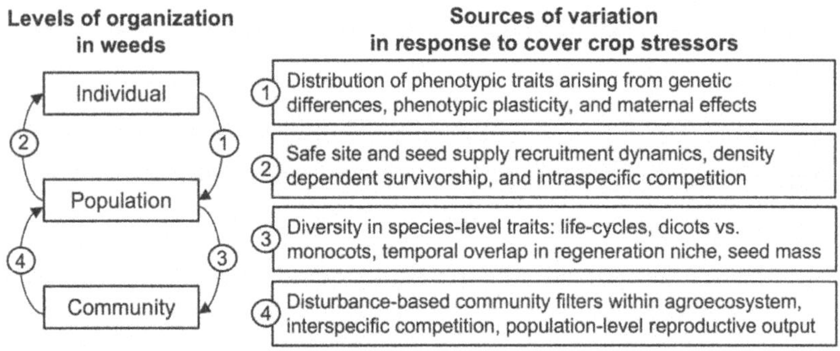

Figure 5 The levels of ecological organization influenced by cover crop stressors and the sources of variation that influence interactions among levels. Population-level responses to management practices involve interactions (denoted by arrows) among individuals within populations and among populations within weed communities. Figure credit: J. Wallace.

be the primary management goal. In the next section, we consider the role of cover crops when integrated with herbicide-based tactics.

4 The role of cover crops in integrated weed management

The design and practice of IWM are highly context dependent due to weed species traits, management constraints, and cropping system design (Buhler et al., 2000). Consequently, it is difficult to judge the performance of cover crops as an IWM tool without explicit consideration of context-dependent IWM goals. In this section, we link IWM principles to cover cropping performance targets for herbicide resistance management, describe herbicide–cover crop interactions through the lens of multi-stressor interactions, and synthesize field-based research focused on integration of cover crop and herbicide-based weed control.

4.1 IWM goals and performance target

As an extension of IPM principles, IWM aims to decrease the equilibrium density of weeds, reduce the impact of weeds on crop yield loss, and limit herbicide use to scenarios in which truncating weed population peaks to maintain below-threshold densities is necessary (Mortensen et al., 2000; Cardina et al., 2020). Adherence to these principles remains limited in modern weed management systems because their performance targets are difficult to measure and evaluate by practitioners operating in the field.

The 'many little hammers' concept is an extension of IWM principles that focuses on weed control at a population level and offers more measurable performance targets for integrating cover crops as an IWM tool (Liebman and Gallandt, 1997). This concept is based on employing multiple tactics (i.e. 'multi-tactic') that result in similar or greater weed control than a single and highly selective control tactic, such as herbicides or tillage. Management goals for integrating cover crops and herbicides as a multi-tactic strategy may include (1) acceptable weed control that prevents crop yield loss without increasing herbicide inputs; (2) reduced risk of herbicide-based weed control failures; and (3) reduced exposure of weed populations to selection and adaptation to herbicide-based tactics. Therefore, performance targets for cover cropping may include a measurable reduction in the (1) density of weed seedlings at the time of herbicide stressor events, (2) rate of weed seedling emergence relative to crop emergence; (3) early-season weed growth rates (i.e. size and biomass); (4) population-level reproductive output of targeted species; and (5) dominance of 'adapted' or 'resistant' weed species within weed communities.

4.2 The role of cover crops in herbicide resistance management

Weed species that have evolved resistance to multiple herbicide sites-of-action are now present and widespread in most crop production regions. Industry-led approaches to this challenge have focused on combining transgenic seed-trait technologies with rotation of herbicide mixtures that contain multiple, effective sites-of-action in preemergence and postemergence programs (Soteres and Peterson, 2015). Mortensen et al. have predicted (2012) and documented (2024) the limits of this approach and argue strongly for IWM as an alternative solution for reducing the agronomic and economic impacts of resistant weed species and the negative environmental consequences of increased herbicide use.

The design of IWM strategies as a proactive approach to slow herbicide resistance evolution will benefit from more explicit consideration of eco-evolutionary dynamics (Neve et al., 2014; Menalled et al., 2016). The study of eco-evolutionary dynamics rests on the principle that (1) the effect of evolutionary change on ecological change, and vice-versa, is strongest at the population level; and (2) understanding of phenotypic trait distributions within populations is critical because phenotypes interact with the environment, come under direct selection, and have direct effects on population growth (Hendry et al., 2011). From an eco-evolutionary perspective, IWM strategies should aim to manipulate selection processes to promote desired demographic outcomes (i.e. population decline) while reducing undesired evolutionary consequences.

Integrating cover crop and herbicide-based tactics can promote population decline by either altering the timing of selection or increasing the dimensionality (i.e. layering or stacking) of selection forces. For example, integrating cover crops with postemergence herbicides diversifies the timing of selection by imposing ecological stressors at developmental stages prior to when herbicide stressors are imposed (Fig. 6a). In comparison, integrating cover crop surface residues with soil-applied residual herbicides may increase the dimensionality of selection by layering both ecological and chemical stressors on individuals transitioning from the germination to establishment stage. In both scenarios, the desired demographic outcome is to maximize mortality, or control levels (%), to prevent population-reproductive output (i.e. zero seed rain).

Integrating cover crop and herbicide-based tactics will have the greatest impact on population decline when their interaction results in synergistic effects. Synergism occurs when the selection intensity of one stressor is amplified in the presence of another. For example, weed control levels and demographic outcomes will significantly improve if a postemergence herbicide that controls 80% of the emerged population when imposed in the absence of cover crop

Published by Burleigh Dodds Science Publishing Limited, 2026.

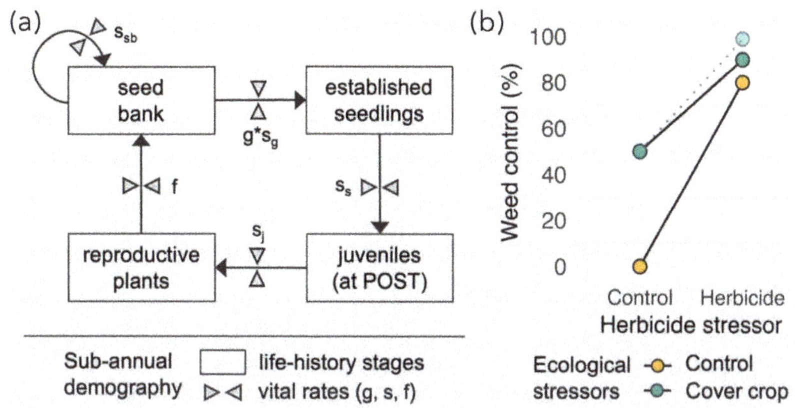

Figure 6 Diagram of (a) sub-annual developmental stages and vital rates influenced by multi-tactic strategies, including cover crop and postemergence (POST) herbicide stressors, on summer annual weed species within a cash crop growing season. Cover crops influence vital rates at early developmental stages, including: s_{sb} = survival of non-germinated seed within seedbank; g = germination rate; s_g = survival of germinated seed to establishment; and s_s = survival of established seedling to juvenile stage. Vital rates influenced by POST herbicide stressors include s_j = survival of juveniles and f = fecundity. Multi-stressor interactions on weed control (%) can be evaluated by (b) quantifying responses to single- (herbicide, cover crop) and multi-stressors (herbicide cover crop) in factorial experiments, which allows for assessment of synergistic effects (dashed line) based on predictions of their joint effects.

use controls 95% of the emerged population when used in sequence with cover crop surface residues (Fig. 6b).

Identifying management tactics that produce synergistic effects will require more explicit consideration of null mechanistic models that predict the joint effect of two stressors. We suggest that evaluation of IWM tactics would benefit from use of null mechanistic models designed in other disciplines. For example, Schäfer and Piggott (2018) offer alternative null models for the study of multi-stressor interactions in conservation biology that differ in assumptions related to (1) how a stressor affects an organism (i.e. mode-of-action); (2) correlation between stressor events (i.e. sequence or timing); and (3) differential sensitivity of organisms to a stressor (i.e. effect-size). In the context of IWM, it is reasonable to assume that the mode of action of cover crop stressors is dissimilar to herbicide stressors. However, the correlation between stressors will depend, in part, on the sequence or spatial extent of management tactics, and the effect size will depend on how phenotypic trait distributions within populations influence sensitivity to imposed stressors.

There is growing recognition that non-lethal herbicide stressors can select for resistance within the range of standing phenotypic variation of a weed population due to epigenetic inheritance mechanisms (Neve et al., 2014; Dyer, 2018). Such scenarios can rapidly shift the structure of populations towards

Published by Burleigh Dodds Science Publishing Limited, 2026.

a 'resistance phenotype' that survives labeled rates of herbicides. In these scenarios, stressors imposed by cover crops may act at the individual-level to reduce survivorship of resistant phenotypes to herbicide selection by reducing the mean lethal dose (Fig. 7a). Alternatively, stressors imposed by cover crops may act at the population-level by reducing survivorship of resistant phenotypes at developmental stages that precede herbicide selection events (Fig. 7b). In the next two sections, we explore these potential mechanisms in the context of cover crop interactions with preemergence and postemergence herbicides and their implications for designing IWM strategies.

4.3 Cover crop-soil residual herbicide interactions

The combined effects of cover crop surface residues and soil-applied residual herbicides may reduce survival of resistant phenotypes at the (1) individual level by lowering the effective, or lethal, herbicide dose needed to control emerging seedlings (Fig. 7a), or (2) population level by increasing the spatial extent of cover crop and herbicide stressors acting on emerging weed seedlings (Fig. 7b).

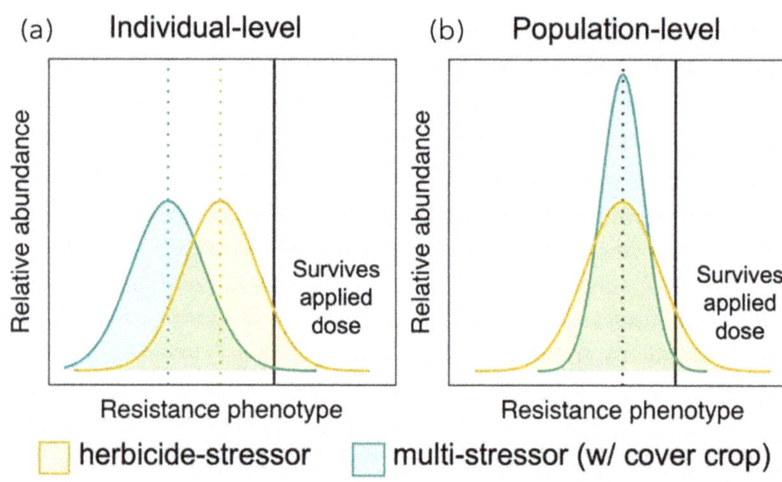

Figure 7 Hypotheses of cover crop effects on herbicide selection at the (a) individual and (b) population level based on low-dose herbicide resistance selection theory (Neve et al., 2014). This theory assumes the mean resistance phenotype is equivalent to the LD_{50} value (*dashed-line*) for a normal-distributed population (*bell-curve*) and herbicide selection (i.e. applied dose; *solid-line*) occurs within the range of standing variation. At the individual level, (a) cover crop stressors may reduce the mean LD_{50} of herbicide, leading to reduced survivorship (*right-tail*) of resistant phenotypes. At the population level, (b) cover crop stressors may remove resistant phenotypes from the population at developmental stages preceding herbicide selection, leading to reduced survival (*right-tail*) compared to imposing herbicide stressor only.

Published by Burleigh Dodds Science Publishing Limited, 2026.

To our knowledge, only a single study has documented synergistic effects of cover crop surface residues and soil-applied residual herbicides. Teasdale et al. (2005) found evidence of synergism from hairy vetch residues and S-metolachlor interactions on emergence of smooth pigweed (*A. hybridus*) in greenhouse studies. The combination of tactics reduced smooth pigweed emergence by 86%, which was greater than the prediction of their joint effects using a multiplicative null model. Multiplicative action assumes that stressors occur sequentially and all individuals within the population are exposed to both stressors. The authors concluded that the physical barrier to establishment created by hairy vetch residues resulted in exhaustion of seedling resources, which lowered the herbicide concentration needed to control emerging weed seedlings (i.e. lower mean LD_{50}) with very long chain fatty acid synthesis (VLCFAS; Group 15) inhibitors. Emergence of broadleaf weed species in cover crop surface residues requires partitioning of limited carbohydrate seed reserves to shoot (or hypocotyl) development during the seedling emergence phase. Thus, it is reasonable to assume that cover crop surface residue is uniquely suited as a synergist for VLCFAS inhibitors, which specifically inhibits seedling shoot elongation and is phytotoxic during the seedling emergence phase, in comparison to other sites of action, such as herbicides that disrupt photosynthesis.

Under many management scenarios, the combined action of cover crop surface residue and soil-applied residual herbicides can be compromised by cover crop interference of herbicide spray deposition to the soil surface (Teasdale et al., 2003). As a result, the potential for reducing the mean LD_{50} among phenotypes (Fig. 7a) may be negated by a reduction in the realized applied dose, or herbicide concentration in soil. For example, Whalen et al. (2020) found that concentration of soil-residual herbicides within the upper soil profile was significantly greater when applied at the jointing stage of cereal rye compared to more advanced heading stages. Nunes et al. (2023a) reported that increasing cereal rye surface mass results in reduced deposition of herbicides to the soil surface. Nunes et al. (2023b) also found that bioavailability of soil-residual herbicides remains significantly lower in systems using cover crop surface residue compared to no cover crop in the first 3 weeks after soybean planting, which overlaps with the regeneration niche of many summer annual weed species.

The effects of decaying crop residues on herbicide fate are well-studied in Conservation Agriculture, which provides a foundation for future research in cover cropping systems (Locke and Bryson, 1997; Alletto et al., 2010). In general, herbicide interception is likely to be a function of cover crop residue mass, height, and orientation to the soil surface (Banks and Robinson, 1982; Khalil et al., 2020). Herbicide retention on cover crop residues will likely be mediated by changes in physical (mass, surface area) and chemical composition (i.e. lignin content) of cover crop residues (Dao, 1991). And finally, precipitation timing and volume will mediate the amount of herbicide that is available to be washed off cover crop residues. Initial investigations by Smith (2025) found that several management scenarios can exacerbate the loss of herbicide bioavailability when cover crop surface residue levels exceed 4 Mg ha^{-1}. These scenarios include application of soil-residual herbicides into living cover crops, grass-legume mixtures, and roll-crimped cover crops. Smith (2026) also found a two- to fourfold increase in washoff potential of herbicides with intermediate polarity (log K_{ow}) compared to more lipophilic or hydrophilic herbicides when applied into cereal rye surface residues and subjected to simulated rainfall (1.9 cm). These results suggest that when higher levels of cover crop surface residue are achieved, repositioning the timing of soil-residual herbicide applications to later crop growth stages may help to improve the combined action of cover crop and herbicide stressors. Future research should also consider further quantifying the interactive effects between cover crop biochemical properties and herbicide properties on washoff processes to inform herbicide selection and application timing.

4.4 Cover crop-postemergence herbicide interactions

Stressors imposed by actively growing cover crops affect weed fitness prior to use of postemergence herbicides to terminate cover crops. Similarly, cover crop surface residue affects weed fitness prior to in-crop, postemergence applications. Consequently, cover cropping practices alter the timing of selection relative to postemergence herbicide use. The interactive effects of these stressors may reduce survival of resistant phenotypes at the individual level if reduced seedling fitness, which results from non-lethal cover crop stressors, lowers the effective herbicide dose needed to control resistant phenotypes at later growth stages (Fig. 7a). Alternatively, cover crop stressors may reduce survival of resistant phenotypes at the population level by increasing their mortality rates at earlier stages of development prior to herbicide selection (Fig. 7b).

Weed size and emergence timing (i.e. emergence cohort) are two phenotypic traits that are likely to influence the range of responses to postemergence herbicides within a population. Large individuals intercept less herbicide per biomass unit (Wauchope et al., 1997; Baker and Hunt, 1981), which, if translated to greater survivorship, would have a disproportionate effect on population growth (Heywood, 1986). Importantly, studies have documented that weed size at the time of herbicide stressor events can influence selection for phenotypes that evolve non-target site resistance mechanisms because large individuals have greater capacity to prevent the active ingredient from reaching the herbicide target site (Dinelli et al., 2006; Vila-Aiub et al., 2018).

The effects of cover crop stressors on weed size can be assessed in various ways. One method is to characterize the population structure by quantifying size inequality in weed populations. Size inequality occurs when relatively few individuals contribute most of the population's biomass due to interactions between phenotypes, heterogeneity of resources, and competition (Weiner, 1985). Postemergence herbicides may be more effective on populations with lower size inequality. Another approach with direct management implications is to quantify the number of days until a proportion of the emerged weed population exceeds a height threshold (e.g. > 10 cm) for postemergence applications (Wiggins et al., 2015; Perkins et al., 2021; Nunes et al., 2024). Many factors prevent timely postemergence applications, which can have a deleterious effect on weed demography and selection events that shape resistance evolution. Lengthening the number of days or weeks in which postemergence applications will still result in high levels of efficacy is a tangible management benefit of integrating cover crops and should be considered when weighing the costs and benefits of integrating cover crops into IWM strategies.

4.5 Examples of a 'two-hammer' approach

Many recent field-based studies have focused on the integration of cover crop and herbicide-based tactics for control of multiple herbicide-resistant species. Most have documented cover crop benefits related to reduced weed recruitment and lengthening of herbicide application windows, but few studies suggest that reducing herbicide inputs is attainable if minimizing reproductive output is the management goal.

For example, horseweed (*Conyza canadensis*) is a multiple herbicide-resistant weed species of economic importance in no-till production systems. It is a small-seeded and wind-dispersed facultative winter annual that is difficult to control at the time of pre-plant burndown applications, as well as in-crop postemergence applications in some regions. Wallace and colleagues (2019) demonstrated that cereal rye reduced horseweed population size (72 to 92%) and size inequality at the time of a pre-plant burndown application in the Northeast US. Field studies in other production regions have reported similar reductions in horseweed population size due to the use of cover crops. Pittman et al. (2019) concluded that a well-established cover crop stand can serve as a substitute for the use of fall-applied residual herbicides to control late-fall and spring emerging horseweed cohorts. In other studies, horseweed control levels from cover cropping tactics were insufficient to eliminate the use of postemergence fall herbicide applications or spring-applied pre-plant residual herbicides (Cholette et al., 2018; Essman et al., 2020; Schramski et al., 2021; Sherman et al., 2020).

The current pace of resistance evolution in waterhemp and Palmer amaranth has far-reaching implications for the sustainability of conservation tillage systems (Davis and Frisvold 2017). These species are currently managed with two- to three-pass herbicide programs that include combinations of soil-applied and foliar products to minimize population-level reproductive output (i.e. zero seed rain) in corn, soybean, and cotton production systems. They are dioecious species that have an extended emergence pattern, high relative growth rates and fecundity (Hartzler et al., 2004), and seeds are persistent for several years within soil seedbanks (Davis, 2008). The use of cereal rye and rye-legume mixtures can delay emergence and extend the number of days until individuals surpass height-based postemergence application thresholds (>10 cm) when used as a single tactic (Wiggins et al., 2015) or in combination with pre-plant residual herbicides (Perkins et al., 2021). While cover crops can result in meaningful reductions in the size of Palmer amaranth and waterhemp populations at early crop growth stages, comprehensive herbicide-based programs are still needed to minimize reproductive output and prevent population growth (Loux et al., 2017; Vollmer et al., 2020; Montgomery et al., 2018).

4.6 Summary

Adoption of cover cropping as a foundational component of IWM will only increase if practitioners understand both short- and long-term benefits of ecologically based practices. Cover crops become a meaningful IWM tool when their use reduces the risk of weed control failures in response to herbicide stressors, which can result from unforeseen events (e.g. weather, misapplication) or from selection of resistant phenotypes. Targeting fewer and smaller weeds with postemergence herbicides and maintaining high-efficacy rates without increasing herbicide inputs are two performance targets that can be observed by the practitioner and measured by the researcher.

Deeper understanding of eco-evolutionary processes is necessary to design IWM strategies that limit or slow herbicide resistance evolution. Though modeling sub-annual population dynamics can help identify IWM strategies that reduce population-level reproductive output (Liebman et al., 2022), future efforts may benefit from integrating trait-based and demographic approaches, which has been described in other disciplines (Salguero-Gómez et al., 2018), to explore relationships between IWM strategies and selection for herbicide-resistant phenotypes.

5 Integrated crop management decision making

Cover crop management decisions routinely result in tradeoffs among ecosystem services. Design of management strategies that aim to negotiate tradeoffs among ecosystem services requires understanding of weed ecology, soil quality, and crop productivity relationships (reviewed in Gallandt et al., 1999;

Smith et al., 2011; Nichols et al., 2015). Here, we review current understanding of weed, soil fertility, and crop management tradeoffs associated with key cover crop management decisions in no-till production systems.

5.1 Cover crop species selection

Considerable attention has been given to the hypothesis that diverse cover crop mixtures should provide higher levels of weed suppression than monocultures due to increased productivity, resource utilization, and nutrient retention in agroecosystems (Florence et al., 2019; MacLaren et al., 2019). However, cover crop biomass production, rather than species diversity, is the primary driver of weed suppression during the period of active cover crop growth (Smith et al., 2020). In many studies, the most weed suppressive monoculture provides similar or greater weed suppression than cover crop mixtures that contain the most weed suppressive species (Finney and Kaye, 2017; Baraibar et al., 2018a). In corn production systems, using cereal rye monocultures to maximize weed suppression potential can negatively affect nitrogen supply relative to corn demand (Finney et al., 2017). This tradeoff can be overcome, in part, with use of grass-legume mixtures (Poffenbarger et al., 2015b) or with use of split-applications of synthetic nitrogen that align with peak nitrogen demand during the corn-growing season (Quinn et al., 2023).

5.2 Cover crop establishment

Fall growing degree days (GDDs) are strongly correlated with spring cover crop biomass production and a primary driver of biomass-related ecosystem services for fall-sown cover crops, such as cereal rye (Feyereisen et al., 2006; Mirsky et al., 2011; Mirsky et al., 2017b) and hairy vetch (Mischler et al., 2010; Mirsky et al., 2017a). For example, sowing cereal rye 2 to 3 weeks after optimal establishment dates can reduce spring biomass potential by 1 Mg ha^{-1} to 3 Mg ha^{-1} (Farsad et al., 2011; Mirsky et al., 2011), which limits weed suppression potential of cover crop surface residues. Cash crop life cycles and rotation sequences in some regions are significant constraints to lengthening establishment windows for cover crops.

In northern US corn production regions, post-harvest establishment of cover crops during optimal sowing windows is limited to sequences that follow small grains or annual forages in rotation. Within these sowing windows, Baraibar and colleagues (2018b) found that cover crop and weed biomass were positively correlated when measured in the fall and negatively correlated in the spring. This relationship reflects a management tradeoff, where early fall sowing dates more directly overlap with the regeneration niche of winter annual weeds, whereas later fall sowing dates permit an opportunity to control emerged winter annual weeds prior to sowing the cover crop.

There is considerable interest in relay intercropping practices as a means for establishing cover crops within grain crop rotations that have narrow post-harvest establishment windows. The timing of cover crop establishment using relaying-intercropping practices depends on methods and equipment, ranging from interseeding at early cash crop growth stages using modified grain drills (Curran et al., 2018; Noland et al., 2018) to later crop growth stages using high-clearance broadcast seeders (Hively et al., 2015), fixed-wing aircraft (Baker and Griffis, 2009; Wilson et al., 2014), or UAVs (i.e. drones). Relay-established cover crops typically do not overlap with the regeneration niche of summer annual weeds within the cash growing season, but successful establishment has been shown to suppress fall- and spring-emerging weed species (Curran et al., 2018; Wallace et al., 2021) when compared to fallow management. However, performance of cover crops that are established using relay intercropping methods is often characterized by high spatial heterogeneity, which reduces the weed suppression benefits of this practice.

5.3 Delayed cover crop termination

Planting green is an emerging agronomic practice where cover crop termination is delayed until at or after corn or soybean planting to produce continuous living roots, improve planting conditions, and build soil health (Reed et al., 2019). Planting green is also a means for reaching biomass thresholds (> 5 Mg ha^{-1}) that are correlated with meaningful levels (50–75%) of weed suppression (Nichols et al., 2020; Weisberger et al., 2023). This practice presents a significant opportunity for increasing cover crop stressors on weed populations but also presents crop management tradeoffs that must be negotiated by growers. The role of planting green as an IWM tactic should therefore be placed in context of adaptive management decision-making to balance crop and weed management goals.

Delaying termination of winter cereal cover crops until at or after soybean planting can significantly improve control of multiple herbicide-resistant species when integrated with herbicide-based tactics, while producing comparable soybean yields to no cover crop or early-terminated cover crop scenarios (Ficks et al., 2023; Grint et al., 2022; Nunes et al., 2024). Adaptive management decisions when planting green in soybean production systems include consideration of the relative benefits of early soybean planting dates to maximize yield potential compared to weed suppression and soil health benefits associated with greater cover crop biomass accumulation.

Planting green represents a change in long-held management recommendations for cover crop termination in corn production systems (Duiker and Curran, 2005), which requires re-evaluation of both pest and fertility management recommendations to optimize the benefits and minimize the risks associated with this practice. In most regions, cover crop establishment before corn phases is limited to winter cereals, such as cereal rye. Agronomic tradeoffs related to delaying cereal rye termination prior to corn production include increased nitrogen immobilization (Patel et al., 2019; Quinn et al., 2023), reduced soil-water availability (Raimbault et al., 1990), increased incidence of seedling disease (Acharya et al., 2022), reduced stand establishment due to residue interference (Champagne et al., 2021; Wallace et al., 2023), and increased potential for interference via allelopathy (Koehler-Cole et al., 2020). Planting green into a leguminous cover crop carries fewer production risks in corn production, but this practice remains limited by narrow establishment and growing season windows for leguminous cover crops.

5.4 Cultural practices for cover crops

Increased sowing density and more uniform spatial arrangement of cover crops are two methods for increasing light resource preemption and competition for resources with weeds that have overlapping life cycles (Schwinning and Weiner, 1998; Weiner et al., 2001). In the context of cover cropping, most studies have measured the effect of increased winter cereal sowing rates on ground cover, biomass, and weed suppression. Increasing cereal rye seeding rates can increase early spring ground cover (Ryan et al., 2011, Haramoto, 2019; Schramski et al., 2021) but has marginal impacts on total biomass potential (Bish et al., 2021; Ryan et al., 2011; Haramoto, 2019). Therefore, it's likely that higher sowing density will be a more important cultural tactic for suppressing winter annual weed species that have an overlapping regeneration niche with fall-sown cover crops.

Future research may consider how interactions between cover crop sowing density and spatial arrangement can mediate the preemption of resources and the regeneration niche of weed species. For example, plants are known to over-proliferate roots in response to neighbors when sown at high densities (Gersani et al., 2001; Cabal, 2022). Winter cereal seeding rates can mediate root distribution and belowground biomass allocation (Hecht et al., 2016). High cereal rye seeding rates can also increase tissue C:N, which can be attributed to increased depletion of soil N due to below-ground competition (Brockmueller et al., 2023). Future cover crop breeding efforts may benefit from evaluating weed suppressive traits in the context of these cultural management practices (Weiner, 2022).

5.5 Residue management

Achieving consistent and precise crop seed placement when planting into cover crop surface residue has limited adoption of high-residue cover crop management practices, particularly in corn (Mirsky et al., 2012; Mirsky et al., 2013). However, significant technological advancements have been made in the past decade since the introduction of roll-crimping (Kornecki et al., 2009) as a mechanical practice that creates a surface residue mulch at the time of planting. No-till planters can now be equipped with row-cleaners, double-disk openers, electric metering systems, automated hydraulic downforce, and closing wheel assemblies that are designed for higher residue systems and make use of precision technologies (Brockmueller et al., 2024).

Manipulating cover crop residue to improve cash crop establishment involves a crop–weed management tradeoff. For example, the combination of row-cleaning and roll-crimping cereal rye results in significantly higher intra-row weed recruitment, but lower inter-row weed recruitment compared to leaving cereal rye standing and not employing row cleaners (Fig. 8; Wallace et al., 2023). Row-cleaning and roll-crimping can also reduce etiolation of corn seedlings compared to those emerging in standing residue (Wallace et al., 2023). Etiolation of corn seedlings is a shade avoidance response (i.e. ecological stressor) common in monocot species (Page et al., 2010; Rajcan et al., 2004), which is created by low R:FR light conditions within standing cover crops and can impact corn productivity. Etiolation of soybean seedlings is also common, but fewer cases of soybean yield responses to standing cover crops have been observed relative to roll-crimping practices (Sias et al., 2024).

6 Case study

The following case study presents a body of research that focused on defining achievable performance targets for cover crop-based IWM in no-till corn and soybean production systems within the Northeastern US. In response to increasing prevalence of glyphosate-resistant weed species, Bunchek et al. (2020) designed experiments to quantify the role of cover crops when integrated with alternative herbicide programs, including a one-pass postemergence (POST) program compared to a two-pass program (preemergence, PRE and POST). Fall-sown cereal rye (135 kg ha^{-1}) was compared to winter fallow management in soybean experiments, and cereal rye–hairy vetch (101 kg ha^{-1} + 22 kg ha^{-1}) was sown in corn experiments. Cover crops were terminated approximately 10 days prior to cash crop planting, producing 2–8.5 Mg ha^{-1} of dry matter biomass before corn and 3–-5.5 Mg ha^{-1} before soybean. Cover crop residue management included the use of integrated roll-crimpers and double-disk row cleaners (Fig. 8) in both crops to create uniform surface residue cover and improve crop seed placement. The primary weed species of interest were glyphosate-resistant horseweed, a facultative winter annual, and smooth pigweed (*A. hybridus*), a small-seeded summer annual species.

Corn experimental results. Fall-sown cover crops significantly reduced horseweed density at the pre-plant burndown application timing (Fig. 9a) and smooth pigweed at the POST application timing (Fig. 9d). In addition, the

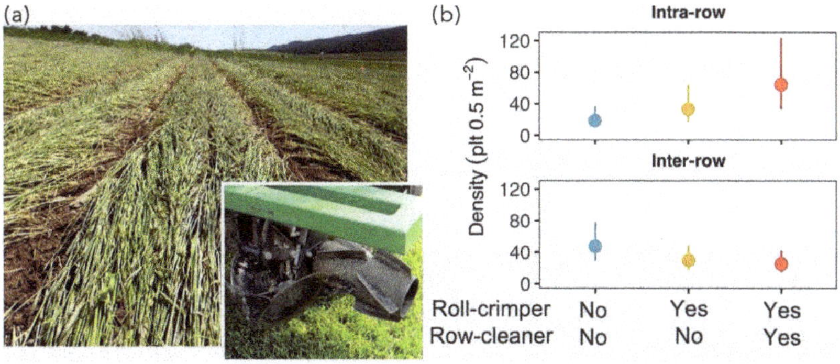

Figure 8 Example of (a) cover crop surface residues following use of integrated roll-crimper and row-cleaners; and (b) effect of this residue management strategy (*red*) on early-season intra- and inter-row weed recruitment compared to planting into standing cereal rye without residue management tools (*blue*), or roll-crimping only (*yellow*), in a study conducted by Wallace et al. (2023).

population structure of both species was characterized by lower size inequality, due to fewer large individuals in the presence of cover crops (Fig. 9b, e). The combined effects on population size and structure were considered to increase the potential for synergistic effects when integrated with POST herbicides. Similar findings have been reported in studies of cover crop effects on multiple herbicide-resistant species, such as Palmer amaranth (Vollmer et al., 2020) and waterhemp (Nunes et al., 2024).

Cover crops and two-pass herbicide (PRE/POST) programs reduced weed abundance at the time of crop maturity by approximately 60% and 90%, respectively, when imposed as a single tactic. The highest level of weed suppression occurred when cover crops were combined with the two-pass herbicide program despite cover crop interference of PRE-herbicide deposition (Fig. 9c, f). Corn yields were not influenced by the cereal rye–hairy vetch mixtures. Given the lack of yield response, additional inputs associated with cover crop seeding resulted in lower net returns when compared to treatments employing only herbicides. Weed suppression benefits associated with cereal rye can be achieved at lower seeding rates (< 56 kg ha^{-1}), and thus input costs, than used in this study (Ficks et al., 2022b).

These results reflect a tradeoff that should be considered when designing IWM practices. Integrating cover crops with more intensive herbicide programs (two-pass PRE/POST) increases control levels and promotes greater population decline. However, similar corn and soybean yields were observed when cover crops were integrated with reduced herbicide programs (POST only). The presence of multiple herbicide-resistant species that rapidly evolve resistance likely influences management decision-making related to herbicide inputs and the risk of weed escapes. Consideration of cover crop input costs should also

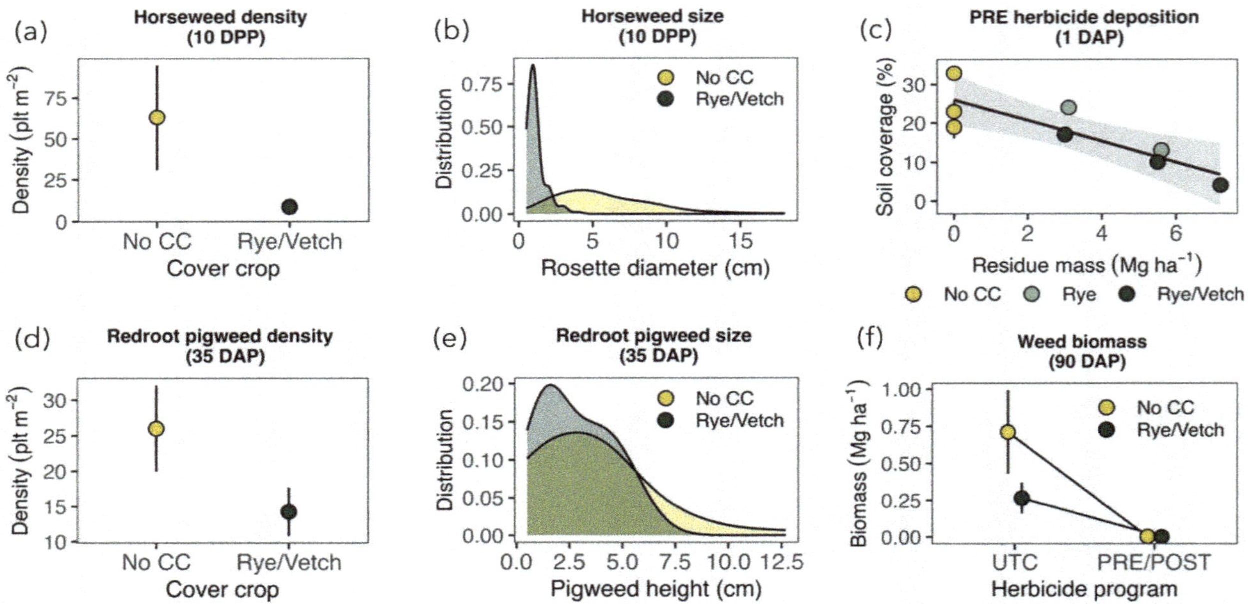

Figure 9 Effect of cereal rye/hairy vetch mixture relative to no cover crop (No CC) on horseweed (a) density and (b) distribution of rosette sizes at time of cover crop termination (10 day pre-plant; DPP); (c) deposition of PRE herbicide applied 1 day after plant (DAP), including a cereal rye monoculture treatment; smooth pigweed (d) density and (e) distribution of pigweed heights at the time of POST application (35 DAP); and (f) total late season weed biomass among multi-tactic strategies, including pre-plant burndown only (UTC) compared to a two-pass approach (PRE/POST). Adapted from Bunchek et al. (2020).

weigh other ecosystem services and disservices. As a component of IWM and Conservation Agriculture, cover crops should be valued for their short-term benefits and as an investment in the long-term success of the farm.

7 Conclusion and future trends

A central theme of this chapter is that use of cover crops in IWM strategies requires consideration of scale-dependent ecological interactions and tradeoffs among crop- and weed-management goals. Such complexity stands in contrast to more prescriptive, herbicide-based strategies that are common in modern agriculture. Several current and future advances can help reduce the complexity of cover crop management decisions and increase adoption of IWM strategies described in this chapter.

Development of decision support tools (DSTs) that can accurately describe or predict variation in cover crop stressor intensity, duration, and spatial heterogeneity is in a nascent stage and will likely improve management decision-making. For example, photogrammetric (i.e. structure-from-motion) and proximal remote sensing technologies can now be mounted to agricultural equipment to provide real-time estimates of cover crop biomass levels and heterogeneity within agricultural production fields at scales that are relevant to weed management decision-making (Dobbs et al., 2022, 2024). The expansion of cover crop research has also provided opportunities for utilizing 'big-data' (Huddell et al., 2024) to develop and evaluate process-based, machine-learning, and deep-learning models that can inform future DSTs. For example, process-based models that utilize site-specific data to predict cover crop decomposition and inform nitrogen management decisions (Thapa et al., 2021, 2022) may also have utility in describing the stressor duration of cover crop residues on the regeneration niche of weed species.

As a foundational component of IWM in Conservation Agriculture, cover crops will play an increasingly important role in reducing the risk of weed control failures and population growth that are associated with evolution and spread of herbicide-resistant species. The last decade has produced a large body of field-based research demonstrating weed control benefits associated with cover crops. The next decade of research should aim to quantify the effect of cover crops and other non-chemical tactics on processes that shape herbicide resistance evolution. To accomplish this goal, weed scientists will need to adopt an eco-evolutionary framework for studying IWM strategies (Menalled et al., 2016), leverage epidemiological approaches that have recently been incorporated into our discipline (Comont and Neve, 2021), and construct thoughtful long-term cropping systems experiments that evaluate proactive herbicide resistance management practices in the context of other ecosystem services and disservices.

Finally, cover crops ought to play a significant role in mitigating risks of surface- and sub-surface runoff of soil-applied herbicides, as well as extreme weather impacts on herbicide efficacy under future climate change scenarios (Landau et al., 2021). Increased understanding of short- and long-term cover cropping effects on hydrologic processes within agricultural fields (Basche and Delonge, 2019) should be leveraged to improve herbicide stewardship, weed control efficacy, and herbicide resistance management.

8 Where to look for further information

Our contribution to the cover crop–weed ecology literature is intentionally placed in context of early-season weed management decision-making in corn and soybean production systems that practice no-tillage, employ herbicide-based weed control tactics, and reside in regions where crop production is not limited by precipitation. Several contributions to the literature have been made that address areas beyond our scope: Smith et al. (2024) provide a broad synthesis of how cover crop management can influence weed populations and communities over longer temporal scales. Kumar et al. (2020) review the role of cover crops in IWM for the US semiarid Great Plains region, where more limited rainfall influences IWM and agronomic tradeoffs. Toler et al. (2019) synthesize cover crop research in the context of IWM in cotton production systems of the Southeast US, where the climate is well-suited for intensive cover crop management. Finally, cover crops play a central role in organic grain crop production systems that aim to reduce the intensity and frequency of tillage. In fact, many foundational studies on weed suppression mechanisms related to cover crop surface residues, as well as systems-level analyses of crop–weed management tradeoffs, were conducted in organic systems and provide a blueprint for cover crop-based IWM (Mirsky et al., 2013; Vincent-Caboud et al., 2019).

9 Acknowledgments

The authors would like to thank Drs William Curran and Matt Liebman for suggestions on an earlier draft of this manuscript, as well as the editors for their instructive feedback.

10 References

Acharya, J., et al., 2022. Effect of planting into a green winter cereal rye cover crop on growth and development, seedling disease, and yield of corn. *Plant Disease* 106, 114–120. https://doi.org/10.1094/PDIS-04-21-0836-RE.

Adeux, G., et al., 2021. Cover crops promote crop productivity but do not enhance weed management in tillage-based cropping systems. *European Journal of Agronomy* 123, 126221. https://doi.org/10.1016/j.eja.2020.126221.

Alletto, L., et al., 2010. Tillage management effects on pesticide fate in soils: a review. *Agronomy for Sustainable Development* 30, 367–400. https://doi.org/10.1051/agro/2009018.

Baker, E.A. and Hunt, G.M., 1981. Developing changes in leaf epicuticular waxes in relation to foliar penetration. *New Phytologist* 88, 731–747. https://doi.org/10.1111/j.1469-8137.1981.tb01750.x.

Baker, J.M. and Griffis, T.J., 2009. Evaluating the potential use of winter cover crops in corn–soybean systems for sustainable co-production of food and fuel. *Agricultural and Forest Meteorology* 149, 2120–2132. https://doi.org/10.1016/j.agrformet.2009.05.017.

Banks, P.A. and Robinson, E.L., 1982. The Influence of straw mulch on the soil reception and persistence of metribuzin. *Weed Science* 30, 164–168. https://doi.org/10.1017/S0043174500062263.

Baraibar, B., et al., 2018a. Weed suppression in cover crop monocultures and mixtures. *Weed Science* 66, 121–133. https://doi.org/10.1017/wsc.2017.59.

Baraibar, B., et al., 2018b. Growing degree days and cover crop type explain weed biomass in winter cover crops. *Agronomy for Sustainable Development* 38, 65. https://doi.org/10.1007/s13593-018-0543-1.

Basche, A.D. and Delonge, M.S., 2019. Comparing infiltration rates in soils managed with conventional and alternative farming methods: a meta-analysis. *PLoS One* 14(19):,e0215702. https:doi.org/10.1371/journal.pone.0215702.

Benech-Arnold, R.L., et al., 2000. Environmental control of dormancy in weed seed banks in soil. *Field Crops Research* 67, 105–122. https://doi.org/10.1016/S0378-4290(00)00087-3.

Bish, M., et al., 2021. Effects of cereal rye seeding rate on waterhemp (*Amaranthus tuberculatus*) emergence and soybean growth and yield. *Weed Technology* 1–7. https://doi.org/10.1017/wet.2021.28.

Booth, B.D. and Swanton, C.J., 2002. Assembly theory applied to weed communities. *Weed Science* 50, 2–13. https://doi.org/10.1614/0043-1745(2002)050[0002:AIATAT]2.0.CO;2.

Brockmueller, B., et al., 2024. Row unit down force and coulter effects vary by environmental conditions in organic no-till soybeans. *Agronomy Journal* 116, 1478–1491. https://doi.org/10.1002/agj2.21552.

Brockmueller, B., et al., 2023. Winter rye cover crop seeding rate and termination timing effects on cover crop biomass and quality. *Communications in Soil Science and Plant Analysis* 54, 2360–2375. https://doi.org/10.1080/00103624.2023.2221299.

Buhler, D.D., et al., 2000. Theoretical and practical challenges to an IPM approach to weed management. *Weed Science* 48, 274–280. https://doi.org/10.1614/0043-1745(2000)048[0274:TAPCTA]2.0.CO;2.

Bunchek, J.M., et al., 2020. Alternative performance targets for integrating cover crops as a proactive herbicide-resistance management tool. *Weed Science* 68, 534–544. https://doi.org/10.1017/wsc.2020.49.

Cabal, C., 2022. Root tragedy of the commons: Revisiting the mechanisms of a misunderstood theory. *Frontiers in Plant Science* 13, 960942. https://doi.org/10.3389/fpls.2022.960942.

Cardina, J., et al., 2020. Development of Weed IPM: Levels of integration for weed management, in: Buhler, D.D. (Ed.), Expanding the context of weed management. CRC Press, pp. 239–267. https://doi.org/10.1201/9781003075202-13.

Cassigneul, A., et al., 2015. Nature and decomposition degree of cover crops influence pesticide sorption: Quantification and modelling. *Chemosphere* 119, 1007–1014. https://doi.org/10.1016/j.chemosphere.2014.08.082.

Catford, J.A., et al., 2022. Addressing context dependence in ecology. *Trends in Ecology & Evolution* 37, 158–170. https://doi.org/10.1016/j.tree.2021.09.007.

Chami, B., et al., 2023. Incentive programs promote cover crop adoption in the northeastern United States. *Agricultural & Environmental Letters* 8, e20114. https://doi.org/10.1002/ael2.20114.

Champagne, R.J., et al., 2021. Rotational no-till and tillage-based organic corn produce management tradeoffs in the Northeast. *Agronomy Journal* 113, 5348–5361. https://doi.org/10.1002/agj2.20823.

Cholette, T.B., et al., 2018. Suppression of Glyphosate-resistant Canada Fleabane (*Conyza canadensis*) in corn with cover crops seeded after wheat harvest the previous year. *Weed Technology* 32, 244–250. https://doi.org/10.1017/wet.2018.19.

Comont, D. and Neve, P., 2021. Adopting epidemiological approaches for herbicide resistance monitoring and management. *Weed Research* 61, 81–87. https://doi.org/10.1111/wre.12420.

Curran, W.S., et al., 2018. Evaluation of cover crops drill interseeded into corn across the Mid-Atlantic region. *Agronomy Journal* 110, 435–443. https://doi.org/10.2134/agronj2017.07.0395.

Dao, T.H., 1991. Field decay of wheat straw and its effects on metribuzin and S-Ethyl metribuzin sorption and elution from crop residues. *Journal of Environmental Quality* 20, 203–208. https://doi.org/10.2134/jeq1991.00472425002000010032x.

Davis, A.S. and Frisvold, G.B., 2017. Are herbicides a once in a century method of weed control? *Pest Management Science* 73, 2209–2220. https://doi.org/10.1002/ps.4643.

Davis, A.S. and Renner, K.A., 2007. Influence of seed depth and pathogens on fatal germination of Velvetleaf (*Abutilon theophrasti*) and Giant foxtail (*Setaria faberi*). *Weed Science* 55, 30–35. https://doi.org/10.1614/W-06-099.1.

Davis, A.S., et al., 2008. Chemical and physical defense of weed seeds in relation to soil seedbank persistence. *Weed Science* 56, 676–684. https://doi.org/10.1614/WS-07-196.1.

Dinelli, G., et al., 2006. Physiological and molecular insight on the mechanisms of resistance to glyphosate in *Conyza canadensis* (L.) Cronq. biotypes. *Pesticide Biochemistry and Physiology* 86, 30–41. https://doi.org/10.1016/j.pestbp.2006.01.004.

Dobbs, A.M., et al., 2022. New directions in weed management and research using 3D imaging. *Weed Science* 70, 641–647. https://doi.org/10.1017/wsc.2022.56.

Dobbs, A.M., et al., 2024. Mapping predicted biomass in cereal rye using 3D imaging and geostatistics. *Weed Science* 72, 553–561. https://doi.org/10.1017/wsc.2024.62

Duiker, S.W. and Curran, W.S., 2005. Rye cover crop management for corn production in the Northern Mid-Atlantic region. *Agronomy Journal* 97, 1413–1418. https://doi.org/10.2134/agronj2004.0317.

Duncan, R.P., et al., 2009. Safe sites, seed supply, and the recruitment function in plant populations. *Ecology* 90, 2129–2138. https://doi.org/10.1890/08-1436.1.

Dyer, W.E., 2018. Stress-induced evolution of herbicide resistance and related pleiotropic effects. *Pest Management Science* 74, 1759–1768. https://doi.org/10.1002/ps.5043.

Eslami, S.V. and Davis, A.S. 2018, Weed interference with no-till soyabeans influenced by fine-scale covariation between soil properties and cover crop performance. *Weed Research* 58, 463–474.

Essman, A.I., et al., 2020. The effects of integrating a cereal rye cover crop with herbicides on glyphosate-resistant horseweed (*Conyza canadensis*) in no-till soybean. *Weed Science* 68, 527–533. https://doi.org/10.1017/wsc.2020.47.

Farsad, A., et al., 2011. Spatial modeling of critical planting date for winter rye cover crop to enhance nutrient recovery. *Agronomy Journal* 103, 1252–1257. https://doi.org/10.2134/agronj2010.0433.

Feyereisen, G.W., et al., 2006. Potential for a rye cover crop to reduce nitrate loss in Southwestern Minnesota. *Agronomy Journal* 98, 1416–1426. https://doi.org/10.2134/agronj2005.0134.

Ficks, T.S., et al., 2023. Delayed cover-crop termination and reduced herbicide inputs produce trade-offs in soybean phase of US Northeast forage-grain rotation. *Weed Technology* 37, 132–140. https://doi.org/10.1017/wet.2023.18.

Ficks, T.S., et al., 2022a. Weed seed mass affects seedling establishment and resource allocation patterns across a cereal rye surface residue gradient. *Weed Research* 62, 49–58. https://doi.org/10.1111/wre.12521.

Ficks, T.S., et al., 2022b. Cereal rye seeding rate does not affect magnitude of weed suppression when planting green within Mid-Atlantic United States. *Weed Technology* 36, 838–843. https://doi.org/10.1017/wet.2022.91.

Finney, D.M. and Kaye, J.P., 2017. Functional diversity in cover crop polycultures increases multifunctionality of an agricultural system. *Journal of Applied Ecology* 54, 509–517. https://doi.org/10.1111/1365-2664.12765.

Finney, D.M., et al., 2017. Ecosystem services and disservices are bundled in simple and diverse cover cropping Systems. *Agricultural & Environmental Letters* 2, 170033. https://doi.org/10.2134/ael2017.09.0033.

Finney, D.M., et al., 2016. Biomass production and carbon/nitrogen ratio influence ecosystem services from cover crop mixtures. *Agronomy Journal* 108, 39–52. https://doi.org/10.2134/agronj15.0182.

Florence, A.M., et al., 2019. Cover crop mixture diversity, biomass productivity, weed suppression, and stability. *PLoS One* 14, e0206195. https://doi.org/10.1371/journal.pone.0206195.

Forcella, F., et al., 2000. Modeling seedling emergence. *Field Crops Research* 67, 123–139. https://doi.org/10.1016/S0378-4290(00)00088-5.

Gallandt, E.R., et al., 1999. Improving Soil Quality: Implications for weed management. *Journal of Crop Production* 2, 95–121. https://doi.org/10.1300/J144v02n01_06.

Ganade, G. and Westoby, M., 1999. Seed mass and the evolution of early-seedling etiolation. *The American Naturalist* 154, 469–480.

Gersani, M., et al., 2001. Tragedy of the commons as a result of root competition: game theory and root competition. *Journal of Ecology* 89, 660–669. https://doi.org/10.1046/j.0022-0477.2001.00609.x.

Grint, K.R., et al., 2022. Cover crops and preemergence herbicides: an integrated approach for weed management in corn-soybean systems in the US Midwest. *Frontiers in Agronomy* 4, 888349. https://doi.org/10.3389/fagro.2022.888349.

Grubb, P.J., 1977. The maintenance of species richness in plant communities: the importance of the regeneration niche. *Biological Reviews* 52, 107–145. https://doi.org/10.1111/j.1469-185X.1977.tb01347.x.

Haramoto, E.R., 2019. Species, seeding rate, and planting method influence cover crop services prior to soybean. *Agronomy Journal* 111, 1068–1078. https://doi.org/10.2134/agronj2018.09.0560.

Haramoto, E.R. and Gallandt, E.R., 2005. Brassica cover cropping: I. Effects on weed and crop establishment. *Weed Science* 53, 695–701. https://doi.org/10.1614/WS-04-162R.1.

Harper, J., 1977. *Population biology of plants*. Academic Press.

Hartzler, R.G., et al., 2004. Effect of common waterhemp (*Amaranthus rudis*) emergence date on growth and fecundity in soybean. *Weed Science* 52, 242–245. https://doi.org/10.1614/WS-03-004R.

Hecht, V.L., et al., 2016. Sowing density: a neglected factor fundamentally affecting root distribution and biomass allocation of field grown spring barley (*Hordeum vulgare* L.). *Frontiers in Plant Science* 7, 944. https://doi.org/10.3389/fpls.2016.00944.

Hendry, A.P., et al., 2011. Evolutionary principles and their practical application. *Evolutionary Applications* 4, 159–183. https://doi.org/10.1111/j.1752-4571.2010.00165.x.

Heywood, J.S., 1986. The effect of plant size variation on genetic drift in populations of annuals. *The American Naturalist* 127, 851–861. https://doi.org/10.1086/284529.

Hively, W.D., et al., 2015. Remote sensing to monitor cover crop adoption in southeastern Pennsylvania. *Journal of Soil and Water Conservation* 70, 340–352. https://doi.org/10.2489/jswc.70.6.340.

Huddell, A.M., et al., 2024. U.S. cereal rye winter cover crop growth database. *Scientific Data* 11, 200. https://doi.org/10.1038/s41597-024-02996-9.

Jussaume, R. and Dentzman, K., 2016. Farmers' perspectives on management options for herbicide-resistant weeds. *Choices* 31, 1–7.

Jussaume, R.A., et al., 2022. Factors that influence on-farm decision-making: evidence from weed management. Society & Natural Resources 35, 527–546. https://doi.org/10.1080/08941920.2021.2001123.

Keles, G., et al., 2016. Forage yields and feeding value of small grain winter cereals for lambs: winter cereals for lambs. *Journal of the Science of Food and Agriculture*. 96, 4168–4177. https://doi.org/10.1002/jsfa.7619.

Khalil, Y., et al., 2020. Pyroxasulfone efficacy for annual ryegrass control is affected by wheat residue height, amount and orientation. *Pest Management Science* 76, 861–867. https://doi.org/10.1002/ps.5590.

Koehler-Cole, K., et al., 2020. Is allelopathy from winter cover crops affecting row crops? *Agricultural & Environmental Letters* 5, e20015. https://doi.org/10.1002/ael2.20015.

Kornecki, T.S., et al., 2009. New roller crimper concepts for mechanical termination of cover crops in conservation agriculture. *Renewable Agriculture and Food Systems* 24, 165–173. https://doi.org/10.1017/S1742170509002580.

Kruidhof, H.M., et al., 2011. Selective weed suppression by cover crop residues: effects of seed mass and timing of species' sensitivity. *Weed Research* 51, 177–186. https://doi.org/10.1111/j.1365-3180.2010.00825.x.

Kumar, V., et al., 2020. Integrating cover crops for weed management in the semiarid U.S. Great Plains: opportunities and challenges. *Weed Science* 68, 311–323. doi:10.1017/wsc.2020.29.

Landau, C.A., et al., 2021. Diminishing weed control exacerbates maize yield loss to adverse weather. *Global Change Biology* 27, 6156–6165. https://doi.org/10.1111/gcb.15857.

Leishman, M.R. and Westoby, M., 1994. The role of large seed size in shaded conditions: ēxperimental evidence. *Functional Ecology* 8, 205. https://doi.org/10.2307/2389903.

Liebman, M. and Davis, A.S., 2000. Integration of soil, crop and weed management in low-external-input farming systems. *Weed Research* 40, 27–47. https://doi.org/10.1046/j.1365-3180.2000.00164.x.

Liebman, M. and Sundberg, D.N. 2006. Seed mass affects the susceptibility of weed and crop species to phytotoxins extracted from red clover shoots. *Weed Science* 54, 340–345.

Liebman, M., et al., 2022. How can cover crops contribute to weed management? A modelling approach illustrated with rye (*Secale cereale*) and *Amaranthus tuberculatus*. *Weed Research* 62, 1–11. https://doi.org/10.1111/wre.12508.

Liebman, M. and Gallandt, E., 1997. Many little hammers: ecological management of crop-weed interactions, in: Jackson, L.E. (Ed.), *Ecology in agriculture, physiological ecology*. Academic Press.

Little, N.G., et al. 2021. Effects of fertility amendments on weed growth and weed-crop competition: a review. *Weed Science* 69: 132–146. https://doi.org/10.1017/wsc.2021.1.

Locke, M.A. and Bryson, C.T., 1997. Herbicide-soil interactions in reduced tillage and plant residue management systems. *Weed Science* 45, 307–320. https://doi.org/10.1017/S0043174500092882.

Loux, M.M., et al., 2017. Influence of cover crops on management of *Amaranthus* species in glyphosate- and glufosinate-resistant soybean. *Weed Technology* 31, 487–495. https://doi.org/10.1017/wet.2017.30.

MacArthur, R. and Wilson, E., 1967. *The theory of island biogeography*. Princeton Univeristy Press.

MacLaren, C., et al., 2020. An ecological future for weed science to sustain crop production and the environment. A review. *Agronomy for Sustainable Development* 40, 24. https://doi.org/10.1007/s13593-020-00631-6.

MacLaren, C., et al., 2019. Cover crop biomass production is more important than diversity for weed suppression. *Crop Science* 59, 733–748. https://doi.org/10.2135/cropsci2018.05.0329.

Mahé, I., et al., 2022. Deciphering field-based evidence for crop allelopathy in weed regulation. A review. *Agronomy for Sustainable Development* 42, 50. https://doi.org/10.1007/s13593-021-00749-1.

Menalled, F., et al., 2016. The eco-evolutionary imperative: revisiting weed management in the midst of an herbicide resistance crisis. *Sustainability* 8, 1297. https://doi.org/10.3390/su8121297.

Menalled, U.D., et al., 2021. High seeding rates and low soil nitrogen environments optimize weed suppression and profitability in organic no-till planted soybean. *Frontiers in Agronomy* 3, 678567. https://doi.org/10.3389/fagro.2021.678567.

Mirsky, S.B., et al., 2017a. Hairy vetch biomass across the Eastern United States: ēffects of latitude, seeding rate and date, and termination timing. *Agronomy Journal* 109, 1510–1519. https://doi.org/10.2134/agronj2016.09.0556.

Mirsky, S.B., et al., 2011. Timing of Cover-crop management effects on weed suppression in no-till planted soybean using a roller-crimper. *Weed Science* 59, 380–389. https://doi.org/10.1614/WS-D-10-00101.1.

Mirsky, S.B., et al., 2012. Conservation tillage issues: Cover crop-based organic rotational no-till grain production in the mid-Atlantic region, USA. *Renewable Agriculture and Food Systems* 27, 31–40. https://doi.org/10.1017/S1742170511000457.

Mirsky, S.B., et al., 2013. Overcoming weed management challenges in cover crop–based organic rotational no-Till soybean production in the Eastern United States. *Weed Technology* 27, 193–203. https://doi.org/10.1614/WT-D-12-00078.1.

Mirsky, S.B., et al., 2017b. Characterizing cereal rye biomass and allometric relationships across a range of fall available nitrogen rates in the Eastern United States. *Agronomy Journal* 109, 1520–1531. https://doi.org/10.2134/agronj2016.09.0557.

Mischler, R., et al. 2010. Hairy Vetch Management for No-Till Organic Corn Production. *Agronomy Journal* 102: 355–362. https://doi.org/10.2134/agronj2009.0183.

Mohler, C.L., 1996. Ecological bases for the cultural control of annual weeds. *Journal of Production Agriculture* 9, 468–474. https://doi.org/10.2134/jpa1996.0468.

Mohler, C.L., et al., 2018. Effects of incorporated rye and hairy vetch cover crop residue on the persistence of weed seeds in the soil. *Weed Science* 66, 379–385. https://doi.org/10.1017/wsc.2017.80.

Mohler, C.L. and Teasdale, J.R., 1993. Response of weed emergence to rate of *Vicia villosa* Roth and *Secale cereale* L. residue. *Weed Research* 33, 487–499. https://doi.org/10.1111/j.1365-3180.1993.tb01965.x.

Montgomery, G.B., et al., 2018. Dicamba-tolerant soybean combined cover crop to control Palmer amaranth. *Weed Technology* 32, 109–115. https://doi.org/10.1017/wet.2017.96.

Moonen, A.C. and Bàrberi, P., 2004. Size and composition of the weed seedbank after 7 years of different cover-crop-maize management systems. *Weed Research* 44, 163–177. https://doi.org/10.1111/j.1365-3180.2004.00388.x.

Mortensen, D.A., 2000. The role of ecology in the development of weed management systems: an outlook. *Weed Research* 40, 49–62. https://doi.org/10.1046/j.1365-3180.2000.00174.x.

Mortensen, D.A., et al., 2012. Navigating a critical juncture for sustainable weed management. *BioScience* 62, 75–84. https://doi.org/10.1525/bio.2012.62.1.12.

Mortensen, D.A., et al., 2024. Another step on the transgene-facilitated herbicide treadmill. *Pest Management Science* 80, 4145–4149. https://doi.org/10.1002/ps.8105.

Moss, S., 2019. Integrated weed management (IWM): why are farmers reluctant to adopt non-chemical alternatives to herbicides? *Pest Management Science* 75, 1205–1211. https://doi.org/10.1002/ps.5267.

Neve, P., et al., 2014. Expanding the eco-evolutionary context of herbicide resistance research. *Pest Management Science* 70, 1385–1393. https://doi.org/10.1002/ps.3757.

Nichols, V., et al., 2020. Cover crops and weed suppression in the U.S. Midwest: a meta-analysis and modeling study. Agricultural & Environmental Letters 5(1), e20022. https://doi.org/10.1002/ael2.20022.

Nichols, V., et al., 2015. Weed dynamics and conservation agriculture principles: a review. *Field Crops Research* 183, 56–68. https://doi.org/10.1016/j.fcr.2015.07.012.

Noland, R.L., et al., 2018. Establishment and function of cover crops interseeded into corn. *Crop Science* 58, 863–873. https://doi.org/10.2135/cropsci2017.06.0375.

Nord, E.A., et al., 2011. Integrating multiple tactics for managing weeds in high residue no-till soybean. *Agronomy Journal* 103, 1542–1551. https://doi.org/10.2134/agronj2011.0024.

Nord, E.A., et al., 2012. Effects of management type and timing on weed suppression in soybean no-till planted into rolled-crimped cereal cye. *Weed Science* 60, 624–633. https://doi.org/10.1614/WS-D-12-00024.1.

Nosratti, I., et al., 2023. Knowledge of cover crop seed traits and treatments to enhance weed suppression: a narrative review. Agronomy 13, 1683. https://doi.org/10.3390/agronomy13071683.

Nunes, J., et al., 2023a. Planting into a living cover crop alters preemergence herbicide dynamics and can reduce soybean yield. *Weed Technology* 37, 226–235. https://doi.org/10.1017/wet.2023.41.

Nunes, J., et al., 2023b. Impact of cereal rye cover crop on the fate of preemergence herbicides flumioxazin and pyroxasulfone and control of *Amaranthus* spp. in soybean. *Weed Science* 71, 493–505. https://doi.org/10.1017/wsc.2023.46.

Nunes, J., et al., 2024. Planting soybean green: how cereal rye biomass and preemergence herbicides impact *Amaranthus* spp. management and soybean yield. *Weed Science* 72, 615–629. https://doi.org/10.1017/wsc.2024.47.

Osipitan, O.A., et al., 2019. Impact of cover crop management on level of weed suppression: a meta-analysis. *Crop Science* 59, 833–842. https://doi.org/10.2135/cropsci2018.09.0589.

Otte, B.A., et al., 2020. Phenolic acids released to soil during cereal rye cover crop decomposition. *Chemoecology* 30, 25–34. https://doi.org/10.1007/s00049-019-00295-z.

Page, E.R., et al., 2010. Shade avoidance: an integral component of crop–weed competition. *Weed Research* 50, 281–288. https://doi.org/10.1111/j.1365-3180.2010.00781.x.

Patel, S., et al., 2019. Can management practices enhance corn productivity in a rye cover crop system? *Agronomy Journal* 111, 3161–3171. https://doi.org/10.2134/agronj2019.03.0158.

Perkins, C.M., et al., 2021. Efficacy of residual herbicides influenced by cover-crop residue for control of *Amaranthus palmeri* and *A. tuberculatus* in soybean. *Weed Technology* 35, 77–81. https://doi.org/10.1017/wet.2020.77.

Pittman, K.B., et al., 2019. Horseweed (*Conyza canadensis*) Suppression from cover crop mixtures and fall-applied residual herbicides. *Weed Technology* 33, 303–311. https://doi.org/10.1017/wet.2018.111.

Pittman, K.B., et al., 2020. Cover crop residue components and their effect on summer annual weed suppression in corn and soybean. *Weed Science* 68, 301–310. https://doi.org/10.1017/wsc.2020.16.

Poffenbarger, H.J., et al., 2015a. Nitrogen competition between corn and weeds in soils under organic and conventional management. *Weed Science* 63(2), 461–476. http://dx.doi.org/10.1614/WS-D-14-00099.1.

Poffenbarger, H.J., et al., 2015b. Legume proportion, poultry litter, and tillage effects on cover crop decomposition. *Agronomy Journal* 107, 2083–2096. https://doi.org/10.2134/agronj15.0065.

Poorter, H., et al., 2012. Biomass allocation to leaves, stems and roots: meta-analyses of interspecific variation and environmental control. *New Phytologist* 193, 30–50. https://doi.org/10.1111/j.1469-8137.2011.03952.x.

Putnam, A.R. and DeFrank, J., 1983. Use of phytotoxic plant residues for selective weed control. *Crop Protection* 2, 173–181. https://doi.org/10.1016/0261-2194(83)90042-X.

Quinn, D.J., et al., 2023. Corn optimum nitrogen fertilizer rate and application timing when following a rye cover crop. *Field Crops Research* 291, 108794. https://doi.org/10.1016/j.fcr.2022.108794.

Raimbault, B.A., et al. 1990. Corn Response to Rye Cover Crop Management and Spring Tillage Systems. *Agronomy Journal*, 82, 1088–1093. https://doi.org/10.2134/agronj1990.00021962008200060012x.
Rajcan, I., et al., 2004. Red–far-red ratio of reflected light: a hypothesis of why early-season weed control is important in corn. *Weed Science* 52, 774–778. https://doi.org/10.1614/WS-03-158R.
Rebong, D., et al., 2024. Breeding allelopathy in cereal rye for weed suppression. *Weed Science* 72, 30–40. https://doi.org/10.1017/wsc.2023.64.
Reed, H.K., et al., 2019. Planting green effects on corn and soybean production. *Agronomy Journal* 111, 2314–2325. https://doi.org/10.2134/agronj2018.11.0711.
Rice, C.P., et al., 2022. Benzoxazinoids in roots and shoots of cereal rye (*Secale cereale*) and their fates in soil after cover crop termination. *Chemoecology* 32, 117–128. https://doi.org/10.1007/s00049-022-00371-x.
Ryan, M.R., et al., 2011. Effects of seeding rate and poultry litter on weed suppression from a rolled cereal rye cover crop. *Weed Science* 59, 438–444. https://doi.org/10.1614/ws-d-10-00180.1.
Salguero-Gòmez, R., et al., 2018. Delivering the promises of trait-based approaches to the needs of demographic approaches, and vice versa. *Functional Ecology* 32, 1424–1435.
Sawadgo, W.P.M., 2024. Are crop insurance discount programs for cover crops effective? Evidence from Iowa. *Agricultural & Environmental Letters* 9, e20125. https://doi.org/10.1002/ael2.20125.
Schäfer, R.B. and Piggott, J.J., 2018. Advancing understanding and prediction in multiple stressor research through a mechanistic basis for null models. *Global Change Biology* 24, 1817–1826. https://doi.org/10.1111/gcb.14073.
Schipanski, M.E., et al., 2014. A framework for evaluating ecosystem services provided by cover crops in agroecosystems. *Agricultural Systems* 125, 12–22. https://doi.org/10.1016/j.agsy.2013.11.004.
Schramski, J.A., et al., 2021. Effects of fall-planted cereal cover-crop termination time on glyphosate-resistant horseweed (*Conyza canadensis*) suppression. *Weed Technology* 35, 223–233. https://doi.org/10.1017/wet.2020.103.
Schwinning, S. and Weiner, J., 1998. Mechanisms determining the degree of size asymmetry in competition among plants. *Oecologia* 113, 447–455. https://doi.org/10.1007/s004420050397.
Sherman, A.D., et al., 2020. Integrating fall and spring herbicides with a cereal rye cover crop for horseweed (*Conyza canadensis*) management prior to soybean. *Weed Technology* 34, 64–72. https://doi.org/10.1017/wet.2019.116.
Sias, C., et al., 2024. Cereal rye cover crop termination management for Palmer amaranth (*Amaranthus palmeri*) suppression in soybean. *Weed Technology* 38, e94. https://doi.org/10.1017/wet.2024.73.
Silvertown, J., 2004. Plant coexistence and the niche. *Trends in Ecology & Evolution* 19, 605–611. https://doi.org/10.1016/j.tree.2004.09.003.
Smith, C., et al., 2025. Herbicide deposition and wash-off potential is affected by cover crop management tactics used in planting green systems. *Weed Science* 73(1):e93. doi:10.1017/wsc.2025.10070.
Smith, C., et al., 2026. Wash-off potential from living and aged cover crop residues differs among soil-residual herbicide sorption properties. *Weed Science* 74(1):e14. doi:10.1017/wsc.2026.10090.

Smith, R.G., et al, 2015. Cover-crop species as distinct biotic filters in weed community assembly. *Weed Science* 63, 282–295. https://doi.org/10.1614/WS-D-14-00071.1.

Smith, R.G., et al., 2024. Mechanisms of weed suppression by cover crops, intercrops, and mulches, in: Korres, N.E., et al. (Eds.), *Ecologically-based weed management*. Wiley, pp. 172–195. https://doi.org/10.1002/9781119709763.ch10.

Smith, R.G. and Mortensen, D.A., 2017. A Disturbance-based rramework for understanding weed community assembly in aagroecosystems: challenges and opportunities for agroecological weed management, in: Wezel, A. (Ed.), *Agroecological practices for sustainable agriculture*. World Scientific (Europe), pp. 127–154. https://doi.org/10.1142/9781786343062_0005.

Smith, R.G., et al., 2011. Direct and indirect impacts of weed management practices on soil quality, in: Hatfield, J.L., Sauer, T.J. (Eds.), *Soil Management: Building a stable base for Agriculture*. Soil Science Society of America, pp. 275–286. https://doi.org/10.2136/2011.soilmanagement.c18.

Smith, R.G., et al., 2020. Are cover crop mixtures better at suppressing weeds than cover crop monocultures? *Weed Science* 68, 186–194. https://doi.org/10.1017/wsc.2020.12.

Soteres, J.K. and Peterson, M.A., 2015. Industry views of monitoring and mitigation of herbicide resistance. *Weed Science* 63, 972–975. https://doi.org/10.1614/WS-D-15-00101.1.

Sturm, D.J., et al., 2018. Contribution of allelopathic effects to the overall weed suppression by different cover crops. *Weed Research* 58, 331–337. https://doi.org/10.1111/wre.12316.

Swanton, C.J., et al., 1999. Effect of tillage systems, N, and cover crop on the composition of weed flora. *Weed Science* 47, 454–461. https://doi.org/10.1017/S0043174500092079

Taylorson, R.B. 1987. Environmental and chemical manipulation of weed seed dormancy. *Weed Science* 3, 135–154.

Teasdale, J.R., 1996a. Contribution of cover crops to weed management in sustainable agricultural systems. *Journal of Production Agriculture* 9, 475–479. https://doi.org/10.2134/jpa1996.0475.

Teasdale, J.R., 1996b. Cover crops, smother plants, and weed management, in: Hatfield, J.L., et al. (Eds.), *Integrated weed and soil management*. Ann Arbor Press, pp. 247–270.

Teasdale, J.R. and Daughtry, C.S.T., 1993. Weed suppression by live and desiccated hairy vetch (*Vicia villosa*). *Weed Science* 41, 207–212. https://doi.org/10.1017/S0043174500076074.

Teasdale, J.R. and Mohler, C.L., 2000. The quantitative relationship between weed emergence and the physical properties of mulches. *Weed Science* 48, 385–392. https://doi.org/10.1614/0043-1745(2000)048[0385:TQRBWE]2.0.CO;2.

Teasdale, J.R. and Mohler, C.L., 1993. Light transmittance, soil temperature, and soil moisture under residue of hairy vetch and rye. *Agronomy Journal* 85, 673–680. https://doi.org/10.2134/agronj1993.00021962008500030029x.

Teasdale, J.R., et al., 2005. Synergism between cover crop residue and herbicide activity on emergence and early growth of weeds. *Weed Science* 53, 521–527. https://doi.org/10.1614/WS-04-212R.

Teasdale, J.R., et al., 2012. Expression of allelopathy in the soil environment: soil concentration and activity of benzoxazinoid compounds released by rye cover crop residue. *Plant Ecology* 213, 1893–1905.

Teasdale, J.R., et al., 2003. Influence of hairy vetch residue on atrazine and metolachlor soil solution concentration and weed emergence. *Weed Science* 51, 628–634. https://doi.org/10.1614/0043-1745(2003)051[0628:IOHVRO]2.0.CO;2.

Thapa, R., et al., 2021. Cover crop residue moisture content controls diurnal variations in surface residue decomposition. *Agricultural and Forest Meteorology* 308–309, 108537. https://doi.org/10.1016/j.agrformet.2021.108537.

Thapa, R., et al., 2022. Cover crop residue decomposition in no-till cropping systems: insights from multi-state on-farm litter bag studies. *Agriculture, Ecosystems & Environment* 326, 107823. https://doi.org/10.1016/j.agee.2021.107823.

Toler, H.D., et al., 2019. Global meta-analysis of cotton yield and weed suppression from cover crops. *Crop Science* 59, 1248-1260. https://doi.org/10.2135/cropsci2018.10.0603.

Vila-Aiub, M., et al., 2018. The role of plant size in the selection of glyphosate resistance in *Sorghum halepense*: plant size and herbicide resistance. *Pest Management Science* 74, 2460–2467. https://doi.org/10.1002/ps.4923.

Vincent-Caboud, L., et al., 2019. Using mulch from cover crops to facilitate organic no-till soybean and maize production: a review. *Agronomy for Sustainable Development* 39, 45. https://doi.org/10.1007/s13593-019-0590-2.

Vollmer, K.M., et al., 2020. Influence of cereal rye management on weed control in soybean. *Frontiers in Agronomy* 2, 600568. https://doi.org/10.3389/fagro.2020.600568.

Wallace, J.M., et al., 2019. Cover crop effects on horseweed (*Erigeron canadensis*) density and size inequality at the time of herbicide exposure. *Weed Science* 67, 327–338. https://doi.org/10.1017/wsc.2019.3.

Wallace, J.M., et al., 2021. Drill and broadcast establishment methods influence interseeded cover crop performance in organic corn. *Renewable Agriculture and Food Systems* 36, 77–85. https://doi.org/10.1017/S174217052000006X.

Wallace, J.M., et al., 2023. Cereal rye residue management tactics influence interrow and intrarow weed recruitment dynamics in field corn when planting green. *Weed Technology* 37, 422–430. https://doi.org/10.1017/wet.2023.56.

Wandrag, E.M., et al., 2019. Quantifying niche availability, niche overlap and competition for recruitment sites in plant populations without explicit knowledge of niche axes. *Journal of Ecology* 107, 1791–1803. https://doi.org/10.1111/1365-2745.13137.

Wauchope, R.D., et al., 1997. A measurement of the total mass of spray and irrigation mixtures intercepted by small whole plants. *Weed Technology* 11, 466–472. https://doi.org/10.1017/S0890037X00045267.

Weiner, J., 1985. Size hierarchies in experimental populations of annual plants. *Ecology* 66, 743–752. https://doi.org/10.2307/1940535.

Weiner, J., 2022. Weed suppression by cereals: Beyond 'competitive ability'. *Weed Research* 63, 133–138. https://doi.org/10.1111/wre.12572.

Weiner, J., et al., 2001. Suppression of weeds by spring wheat *Triticum aestivum* increases with crop density and spatial uniformity. *Journal of Applied Ecology* 38, 784–790. https://doi.org/10.1046/j.1365-2664.2001.00634.x.

Weisberger, D.A., et al., 2023. Do cover crops suppress weeds in the U.S. Southeast? A meta-analysis. *Weed Science* 71, 244–254. https://doi.org/10.1017/wsc.2023.21.

Weisberger, D.A., et al., 2024. Demographics of Palmer amaranth (*Amaranthus palmeri*) in annual and perennial cover crops. *Weed Science* 72, 96–107. https://doi.org/10.1017/wsc.2023.66.

Wells, M.S., et al., 2013. The reduction of plant-available nitrogen by cover crop mulches and subsequent effects on soybean performance and weed interference. *Agronomy Journal* 105, 539–545. https://doi.org/10.2134/agronj2012.0396.

Wiggins, M.S., et al., 2015. Integrating cover crops and POST herbicides for Glyphosate-resistant Palmer Amaranth (*Amaranthus palmeri*) control in Corn. *Weed Technology* 29, 412–418. https://doi.org/10.1614/WT-D-14-00145.1.

Wilson, M.L., et al., 2014. Aerially seeding cover crops in the northern US corn belt: limitations, future research needs, and alternative practices. *Journal of Soil and Water Conservation* 69, 67A–72A. https://doi.org/10.2489/jswc.69.3.67A.

Chapter 6

The role of cover crops in forage production

Heather Darby, Claire Benning and John Bruce, University of Vermont, USA; Kurt Cotanch, Barn Swallow Consulting, USA; and Ivy Krezinski, Amber Machia, Lindsey Ruhl, Jeffrey Sanders, Elizabeth M. Seyler, Laura Sullivan and Sara Ziegler, University of Vermont, USA

1 Introduction

Cover cropping can provide ecosystem services to farms and communities. Cover crops are broadly defined as crops that provide soil protection and improvement during periods when a cash crop is not present in a field (Clark, 2008; SSSA, 2008). The ecosystem services commonly promoted through cover cropping include regulating services related to soil health, water quality, climate, erosion, and carbon sequestration, as well as supporting services focused on wildlife habitat and maintenance of biological diversity (Blanco-Canqui et al., 2015). One provisioning ecosystem service often overlooked is the material output of forage that is possible from cover crops (Blanco-Canqui et al., 2023). Farmers usually select cover crops for nitrogen fixation, weed suppression, breaking disease cycles, reducing soil erosion, and minimizing nutrient runoff and loss. But more farmers are beginning to focus their cover crop goals on both soil conservation and livestock forage production (Bowman et al., 2024; Blanco-Canqui et al., 2023; Popovici et al., 2023). This is especially true for dairy and feedlot operations: in the USA, 30% of them reported adopting cover crops in the 2017 USDA Census

http://dx.doi.org/10.19103/AS.2025.0162.08

of Agriculture (Bowman et al., 2024). Of these farms, 72% of dairy and 89% of cow-calf operations reported harvesting or grazing at least some of their cover crop acreage for forage (Bowman et al., 2024). Growing cover crops for forage is not only an option for livestock farms but also for cash grain farms that can offer contract-grazing opportunities to local livestock farms seeking winter forage options. An increasing desire to reintegrate crop and livestock production is encouraging these types of beneficial partnerships (Bowman et al., 2024; Roesch-McNally et al., 2018; Sulc and Franzluebbers, 2014). In fact, the expansion of cover cropping across the United States will likely come from increased grazing arrangements between livestock and cash-crop farms (Bowman et al., 2024).

Most livestock operations grow perennial forages as a staple for the ruminant diet and blend them with a mix of concentrates grown on or imported onto the farm to form a complete ration (Beever and Mould, 2000). The cash crops most often grown to complement perennial forages in the diet include corn (*Zea mays*, L.) for silage, corn for grain, and soybeans (*Glycine max* (L.) Merr.). These crops may be rotated with perennial forages or grown in continuous culture, depending on the farm. During the annual crop years, many farmers will grow corn for silage (Jordan and Fourdraine, 1993). This crop fits well into the rotation, both from the livestock nutrition perspective and because it requires some of the same harvest equipment and storage infrastructure as do perennial forages. When corn silage is harvested, the entire plant is removed, chopped, and ensiled in a bunker or silo structure. Because the entire plant is removed from the field, there is essentially no residue left covering the soil from early fall until early summer of the next year. This system is predisposed to a variety of soil health and water quality concerns (Jokela et al., 2009; Wilhelm et al., 2010). The promotion of cover crops in corn silage cropping systems has been paramount to improving soil health and resiliency and water quality in many regions where dairy and beef operations are located (Bowman et al., 2024). In fact, Bowman et al. (2024) reported that 20% of U.S. corn silage acres included fall cover crops; no other crop includes such a high percentage.

In addition to a mixture of perennial and annual crops, livestock farms produce manure that is applied to the crop fields as a nutrient source. Depending on state regulations, manure is often applied throughout the year, regardless of whether there is a crop growing in the field. Concentrated animal feeding operations (CAFOs), however, must follow an approved nutrient management plan, per federal regulations, when applying manure (Copeland, 2010). Farmers need to match soil characteristics, landscape restrictions, and crop nutrient removal with total annual quantities of manure and fertilizer nutrients applied. The goal is to reduce the chance of nutrient losses to the surrounding environment. Hence, cover crops have been promoted to reduce the risk of manure nutrient losses, especially in the fall and winter months (Kleinman et al., 2005; Thilakarathna et al., 2015).

The adoption of cover crops on livestock operations poses many challenges that vary by location. Climate, landscape, soil type, and individual state regulations can all influence a farm's interest in and ability to integrate cover crops into the livestock operation. Many livestock operations are located in cooler climates, and corn silage is grown because it fits within the short season in these areas. Corn silage is a full-season crop generally harvested in late September to mid-October in the northern US. This leaves little time to spread fall manure, conduct fall tillage, and get a cover crop established before winter sets in, when temperatures are too low for cover crop germination and growth. In the spring (March–May), management of the cover crop is difficult because of the short window for field work, manure application, and corn planting. Incorporating the cover crop into these livestock systems also increases labor, management, and input costs. These are significant barriers for farmers who are already operating on tight margins and tight timelines. The increasingly erratic and unpredictable climate adds one more unknown to the equation.

State and federal cost-share programs are often well utilized by U.S. livestock operations because they provide funding for manure storage facilities, barnyards for heavy-use areas, grazing infrastructure, and many other livestock-related practices and infrastructure (Reimer and Prokopy, 2014). These operations also participate in subsidized conservation practices such as crop rotation, forage seedings, cover cropping, and nutrient and pasture management, to name a few. Livestock operators take advantage of many of these federal programs (Reimer and Prokopy, 2014; Chami et al., 2023). They are strictly regulated by state and federal governments, which often require that they adopt practices such as cover cropping, crop rotation, and buffers to manage manure nutrients on the farmstead and in the fields (Hively et al., 2015; VGA, 2015).

Because of the short growing season in the northern US, it remains difficult for farmers to adopt cover cropping. However, as innovative livestock farmers started growing cover crops, they realized that this lush green cover growing early in the spring looked like an excellent source of forage. In some cases, farmers who were short on feed inventory considered the cover crop an emergency forage for grazing or harvested feed (Ketterings et al., 2015a). In other cases, when the cover crop started to grow more rapidly than the farmers felt they could handle prior to cash crop planting, they turned to harvesting cover crop for stored feed just to reduce the amount of cover crop biomass to manage prior to main crop establishment (Darby et al., 2018). In all of these examples, how farmers view the cover crop and its associated benefits started to shift on livestock operations–cover crops for many are now part of a double-crop system. Several studies indicate that harvesting or grazing cover crops to feed livestock makes cover cropping profitable (Schomberg et al., 2014; Sadeghpour et al., 2022; Wallender et al., 2021; Tobin et al., 2020), although others indicate that the revenue from grazing does not cover the cost of cover cropping (Sestak et al., 2023). Economic analysis is limited, and the economic impacts of cover cropping are variable, depending on several factors. Studies on a farm-by-farm basis will likely be necessary to determine the feasibility for any single operation.

Nonetheless, more and more livestock farmers are viewing cover crops as multi-benefit, both for improving soil health, meeting water quality regulations, and protecting soil from erosion during the winter months while also serving as a valuable and nutritious feedstuff for their herds. Harvesting the cover crop as a feedstuff has increased the adoption of cover crops on livestock farms (Bowman et al., 2024). Considering cover crops a valuable feedstuff has also likely improved overall cover crop management to encourage timelier fall establishment, planting methods and rates that improve establishment, and fertility management to encourage rapid growth (Ketterings et al., 2015b). Farmers have also had to make other adjustments to their cash crop production practices to ensure that the cover crop can produce a high-yielding and quality forage crop (Darby et al., 2018; Ketterings et al., 2015b). This chapter focuses on the management of cover crops for forage. We review stored feed and grazing systems and highlight examples of farm adoption.

2 Species selection of cover crops grown for forage

Cover crop species selection is dependent on a variety of factors specific to farm location, environment, cropping system, and goals. Nationally, farmers' primary cover cropping goals relate to soil, weed, and nutrient management; a much smaller number grow cover crops primarily for livestock forage production (Anon, 2023). The main goals driving species selection for cover crops intended to be harvested or grazed as forage include early vigor, hardiness, optimizing maturation timing, high nutrient quality, palatability, grazing adaptability, and high yield potential. These goals limit suitable cover crop species to mainly cool-season forages, including annual ryegrass (*Lolium multiflorum,* L.), cereal grains (wheat (*Triticum aestivum,* L.), barley (*Hordeum vulgare,* L.), rye (*Secale cereale,* L.), triticale(*Secale cereale x Triticum aestivum)*, and oats (*Avena sativa,* L.)), brassicas (*Brassica spp.*, L such as turnips, radishes, kales, and hybrids), and legumes (clovers (*Trifolium* spp.), peas (*Pisum sativum* L.), and vetch (*Vicia villosa* (Roth)).

Cover crop mixtures are used to achieve multiple agroecosystem benefits simultaneously. Utilizing mixtures, especially in beef, sheep, and some dairy systems, has been a strategic means to gain more environmental benefits from the cover crop while also producing a high-biomass forage crop. However, it can be difficult to select cover crop species that perform well in a mixture and meet a plethora of agroecosystem goals. Studies on cover crop mixtures that meet farm-level goals for forage production are limited. In both monoculture and mixture cover cropping for forage, species selection is based on vigor, seasonality, hardiness, maturation, and timing, as well as maximum yield and quality potential.

2.1 Vigor, seasonality and hardiness

Cover crop species with high fall and spring vigor will ensure maximum growth in the short growing window between cash crops. The fastest biomass accumulation in the fall is usually obtained from spring cereal grains – such as oats – planted in the late summer months, because they stay vegetative until the onset of winter temperatures (Darby et al., 2022a). If the intent is to fall harvest or graze, spring cereal grains will out-yield winter cereal grains; however, the spring cereal grains are not winter hardy and will be killed at temperatures below −17.8°C (Gusta and O'Connor, 1987). To maximize the benefit of oats or other small grains, early planting (typically in August) is critical. This limits their use in many double-cropping scenarios and regions. However, if planting can be accomplished approximately 60 days prior to a killing frost, these spring grains can produce sizeable yields of high-quality forage suitable for a range of livestock (Darby et al., 2022b). Similarly, many brassica species, while more cold-tolerant than oats, will also winterkill in colder climates and benefit from late summer seeding. These cover crops have been shown both to produce higher yields and to retain quality better than annual ryegrass (Dillard et al., 2020) and small grain monocultures (Omokanye et al., 2019). Though cover crop mixtures do not always enhance yields or ecosystem services over single species cover crops, mixing species may be desirable when selecting a cover crop for forage (Sanderson et al., 2018; Farney et al., 2018). Though brassicas are quick to establish and have high yield and forage-quality potential, they are best utilized in mixtures with other species due to their very low fiber content, high protein content, and high digestibility, which can cause health issues in animals fed exclusively brassicas (Dillard et al., 2020; Wiedenhoeft and Barton, 1994).

If the goal is to produce spring forage from the cover crop, winter cereals are the best choice. Though winter cereals will stay vegetative until spring in northern climates, they are able to survive low temperatures, thus overwintering and resuming growth in the spring. As such, the primary cover crop species grown throughout the US is winter rye: it is most adapted to cold temperatures and a wide variety of soil conditions, and it can grow and develop under cool soil and air temperatures (Myers and Watts, 2015). Other cereal grains such as winter wheat, triticale, and barley can also be suitable for double cropping; however, these species are generally less vigorous and not as widely adapted as rye, making them less suitable in many cases (Landry et al., 2019; Jemison et al., 2012).

Little information exists on winter grain varieties and their performance as double-cropped forages, but Darby and Bruce (2024) found significant differences in fall and spring vigor, maturation timing, biomass production, and forage quality between twelve winter rye varieties. All were commercially available varieties (hybrids and open-pollinated), however, the varieties were promoted for different end-uses, including some specific for cover crop, feed grain, food grain, and forage production. In this study, varieties differed in their fall vigor as measured by ground cover, ranging from 29.8% to 73.5%. Those varieties with the highest fall ground cover also had the highest spring cover. Interestingly, varieties bred specifically for cover cropping did not always provide the highest ground cover (Aroostook, 39.4%), while varieties often destined for baking or distilling produced some of the best ground cover (Danko, 73.5%). Progas, a hybrid rye forage-specific variety, did have some of the highest forage yields but also had the lowest fall and spring cover. More research is needed on the best dual-purpose rye varieties for optimizing both conservation goals and forage production.

2.2 Maturation timing and quality

Harvest timing is the most critical determinant of forage quality in both perennial and annual forages. The maturation timing of a double-crop forage species/variety will directly impact the farm's ability to capture its yield and quality benefits. The nutritional needs of the livestock being fed will determine the stage of maturity targeted for harvest. For dairy and beef operations, sufficient forage quality is achieved if harvest occurs from flag leaf through boot stages of cereal cover crops (Landry et al., 2019). Winter rye reaches these maturity stages 7 to 10 days earlier than winter wheat or triticale (Darby et al., 2017a; Landry et al., 2019). While there is limited information on using rye and other winter grains as spring forages, Darby and Bruce (2024) found that the timing of boot-stage maturity in rye varieties ranges from May 9 to 18 in northern Vermont. This again indicates the need for more variety-specific information to help farmers select the best dual-purpose cover crop. In corn cropping systems, winter rye's earlier maturation timing allows for cover crop harvesting and corn planting to occur within optimal growing season windows. Jemison et al. (2012) found that corn silage yields were significantly reduced when the double-cropped winter grain silage harvest was delayed into the soft dough stage. Delaying winter grain harvest until the soft dough stage delayed corn planting and resulted in the corn crop not reaching proper silage maturity prior to a killing frost. Cover crops such as wheat and triticale are later maturing and may be best suited as a double-crop in areas with longer growing seasons.

2.3 Yield potential

Cover crops must produce enough dry matter yield to warrant the cost of harvesting for livestock feed. Forage yields are determined by the cover crop species/variety and a wide range of other field management methods. Although management may play a more significant role in determining yield, species and variety selection is still a critical component.

Grazing systems that incorporate annual cover crops as forage often benefit from the availability of higher-quality feed during times when perennial forage is dormant or too mature (Blanco-Canqui et al., 2023). Across both southern and northern climates in the US, cool-season species have outperformed warm-season species in terms of biomass production (Suazo et al., 2024; Wyffels et al., 2021). Although warm-season cover crop species may have the potential to produce large quantities of biomass, they are often terminated to meet the schedule of the cash crop and likely do not reach their full yield potential. Mixtures containing oats often have higher yields than mixtures without and biomass production level often supports economical mechanical harvest in addition to grazing (Bowman et al., 2024; Wyffels et al., 2021). Harvested forage yields increased after the earliest harvest date in South Dakota (Hansen et al., 2013), a possible option for maximizing yield. Depending on the end-use, stored feed or grazing, species selection will need to be considered so that yield and quality can be optimized from cover-cropped fields.

As previously mentioned, fall-planted spring grains may not be suitable for some climates or cropping systems because they don't survive cold winter temperatures. However, if planted approximately 60 days prior to a killing frost, spring grains can produce up to 3,923 Dry Matter (DM) kg ha^{-1} and can be mechanically harvested and ensiled or grazed (Darby and Ziegler, 2024; Maloney et al., 1999). Similarly, another study found that fall-planted brassica cover crops produce 3,363 to 6,725 DM kg ha^{-1} in monoculture and 5,604 DM kg ha^{-1} in mixtures with cereal grains and other species, depending on location and climate (Table 1) (Darby et al., 2017b; Dillard et al., 2020; Villalobos and Brummer, 2015). Winter grains can produce up to 2,242 DM kg ha^{-1} in the fall when they are in the vegetative growth stage and 8,967 DM kg ha^{-1} in the spring when harvested in the boot stage, depending on species, variety, and location (Jemison et al., 2012; Maloney et al., 1999; Darby et al., 2017a; Darby and Bruce, 2024).

Seed companies have started to release cereal varieties specifically bred or selected for forage production, and now offer forage-specific varieties of oats, rye, and triticale, often boasting of high yields and nutrition/quality

Table 1 Mean maximum yields and standard deviations for cover crop forages by region. Harvest timing varied by cropping systems. Harvest timing varied by cropping systems. Across all experiments, samples were harvested once to measure cover crop growth in the respective season. Standard deviation of zero indicates only one reported value.

Location	Category	Species	Dry matter yield (metric ton ha^{-1})	Sources
Midwest	Single species	Grasses	4.44 (2.33)	Black et al., 2023; Dhakal et al., 2022; Fae et al., 2009; Maloney et al., 1999
	Mixture		4.93 (1.59)	
Great Plains	Single species	Grasses	5.22 (4.01)	Blanco-Canqui et al., 2020; Hansen et al., 2013; Simon et al., 2021, 2024; Titlow et al., 2014
		Legumes	2.47 (0)	
		Brassicas	3.36 (0)	
	Mixture		3.30 (1.95)	
Southwest	Mixture		6.12 (0.27)	Acharya et al., 2022
Northeast	Single species	Grasses	5.45 (3.74)	Darby et al., 2017a, 2017b; Dillard et al., 2020; Jemison Jr et al., 2012
		Legumes	1.37 (0.18)	
		Brassicas	1.50 (1.26)	
	Mixture		4.46 (0.99)	
Canada	Single species	Grasses	7.62 (0)	Okokanye et al., 2020, 2021; Wyffels et al., 2021
		Forbs	5.65 (3.16)	
		Legumes	3.27 (0.85)	
		Brassicas	5.29 (1.21)	
	Mixture		3.92 (3.43)	

performance. Darby and Bruce (2024) compared 12 varieties of winter rye and found similar yields and forage quality between many of the varieties, regardless of their targeted end use (i.e. forage specific or grain specific). Similarly, Darby and Ziegler (2024) evaluated yield and quality of oat varieties marketed for forage or grain production and found that an early maturing grain oat variety produced similar quantities of forage as a forage oat variety. This, again, indicates that variety evaluation is critical for farmers wanting to optimize conservation and forage benefits.

2.4 Palatability and grazing adaptation

Cover crops differ in their suitability for grazing. Though grasses, legumes, and brassicas may produce forage suitable for grazing livestock, animal preferences can influence their use and therefore the benefit gained from the double crop. Farney (2019) found that cattle exhibited a strong preference for oat and barley and a strong aversion to the brassica species offered. Robinson (2020) found that sheep and goats preferred ryegrass, cereal rye, and berseem clover

over Austrian winter pea and radish. Omokanye (2021) found that beef cattle prefer buckwheat (*Fagopyrum esculentum,* Moench), daikon radish, inka brand marrowstem kale, and forage collards based on their DM yield and nitrate levels (Omokanye et al., 2021). Interestingly, several review articles have indicated that grazing animals can adapt and overcome taste/sensory aversions or build preferences with repeated exposure and training (Gregorini et al., 2017; Provenza, 1996). Hence, more diverse cover crop mixtures can likely be grazed with proper focus on achieving

Grazing livestock may also exhibit preferences for cover crop species based on their phenological growth habits. For instance, cattle graze by wrapping their tongues around a forage mass to rip it off, but sheep use their lips and teeth to bite forage. Cattle may find grazing taller small grains, brassicas, and legumes easier and more efficient than species with low, prostrate growth habits such as vetch and winter pea.

The presence of anti-quality compounds and other animal health risks are also an important consideration. Brassica species contain glucosinolates and other compounds that can cause bloat and anemia, interfere with thyroid function, and have other health impacts (McCartney et al., 2009). In addition, some legume species may also cause bloat under certain conditions.

2.5 Other factors to consider when selecting cover crops for forage

Cover crop species selection will depend in part on the timing and planting method, which may limit the best options available. Due to short growing seasons in most livestock-focused regions, post-harvest establishment in the late summer and early fall is common practice. Cereal rye is favoredin corn and soybean systems because it establishes easily under a variety of conditions and overwinters, providing a forage resource the following spring (Cates et al. 2018; Baker and Griffis, 2009). The planting date of cover crops will greatly impact their overall establishment and biomass production. Several studies have evaluated the response of cereal rye to a range of planting dates and the impact on establishment, weed suppression, and biomass production (Mirsky et al., 2011; Balkcom et al., 2023). For forage production, farmers are interested in leafy cover crops that can accumulate biomass quickly in the spring to produce ample forage prior to the cash crop planting window. Tillering helps to increase fall ground cover and is an important contributor to yield (Hucl and Baker, 1991). Since fall tiller development is related to accumulated growing degree days (GDDs) after planting, timely planting of the winter cover crop will allow for greater tiller production. Baron et al. (1999) reported that earlier seeded cereal grains produced earlier spring pasture. Winter rye produced more biomass earlier than winter rye and winter triticale. This is because rye is a high-tillering crop, often producing five tillers per plant (Gan and Mcleod, 1997). The more fall GDDs after planting, the higher the spring pasture yields. These findings underscore the need to properly match the species selected with the environment, farm location, management practices, and cover crop goals.

While fall planting of cereal rye or other cereal grains following the cash crop harvest is common, other methods allow for more diverse species selection. For example, interseeding cover crops into standing cash crops circumvents the short growing season challenges of late fall planting (den Hollander et al., 2007; Sarrantonio and Gallandt, 2003). While interseeding cover crops earlier in the growing season provides more options for species selection, interseeded cover crops may not reliably produce enough biomass for grazing or harvest due to poor establishment under the canopy of the main crop. Success with interseeding a wide variety of cool and warm season cover crop species into corn and soybean systems has been highly variable and thus remains more an experimental practice than commonplace (Anon, 2023; Brooker et al., 2020; Fisher et al., 2011).

While the goals of cover cropping for forage may differ from conservation goals, the aim of species and variety selection should be to optimize conservation efforts while producing high-yielding and quality feed sources for livestock, thus increasing the ecosystem services overall.

3 Nutrient management of cover crops grown for forage

Cover crops are generally grown for their environmental benefits, including reducing erosion, enhancing soil physical properties (Liesch et al., 2011), improving water quality via uptake of excess nutrients, reducing fertilizer inputs by supplying scavenged nutrients to subsequent cash crops, and, in the case of legumes, fixing nitrogen. Harvesting cover crops as forage can provide the added benefit of increasing total annual forage yields (Ketterings, et al., 2015a; Jemison et al., 2012; Binder et al., 2021; Fouli et al., 2012). Cover crops are not usually fertilized, since one objective is to scavenge residual soil nutrients. If forage harvest is the objective, however, then the cover crop should be managed as if it were a cash crop. This includes supplying nutrients for optimal forage yield and quality. This requires knowledge of soil-available nutrients and the additional nutrients needed to meet the crop's yield and quality goals. Forage yields of cover crops will vary by species, planting date, and harvest date, in addition to weather conditions and residual soil fertility. Furthermore, the removal of the cover crop biomass as a forage crop will alter the fertility requirements for the subsequent crop.

3.1 Cover crop nutrient application timing

Nutrient applications to cover crops for forage should be based on a soil test. There are two opportunities to apply fertilizer to cover crops: (1) prior to cover crop seeding and (2) during cover crop growth. For overwintering species, the second option could occur in the fall prior to dormancy or in the spring at the break of dormancy.

Late summer seeding of spring cereal grains such as oats may see yield increases of 0% to 40% with nitrogen fertilization at seeding (Zhang et al., 2006; Coblentz et al., 2014; Darby and Ziegler, 2024), depending on residual fertility, nitrogen application rates, and growing conditions. In some cases, despite minimal impacts on yield, nitrogen fertilization may increase protein content by 5% to 7%. Depending on the available fertility source and cost of application, these gains may be economically viable. However, avoid overfertilizing fall annual cover crops, because this may increase the risk of leaching or runoff losses (the environmental goal), and they may accumulate nitrates, which can be toxic to livestock (the forage goal).

For overwintering species that will be harvested the following spring, nutrient applications may occur in the fall and/or spring. For example, in livestock operations with corn silage double-cropping systems, it is common to apply manure after corn harvest prior to or after cover crop seeding. These nutrients are at risk of loss to the environment via leaching, runoff, and volatilization/nitrification, but cereal grains planted in the fall may recover some of them. Split nutrient applications are typically recommended to avoid such losses.

Timing and rates of fertilizer applications are important to cover crop response. A survey of New York State double croppers found that 79% applied nitrogen at spring dormancy break to achieve higher yields (Ketterings et al., 2015b). Based on data collected in 62 field trials across New York, the most economic rate of nitrogen applied at dormancy break averaged approximately 78.5 kg ha^{-1} (Lyons et al., 2019). Low rates of nitrogen (13.5 kg ha^{-1}) applied in the fall at planting may have no impact on rye cover crop biomass (Nichoals and Gailans, 2022). Balkcom et al. (2018) found that nitrogen fertilizer (ammonium nitrate) applications increased rye cover crop biomass by 13% compared to similar rates of poultry litter nitrogen applications, and that rye biomass responded more favorably to higher rates of nitrogen (67.3 kg ha^{-1} and 100.9 kg ha^{-1}, compared to 0 kg ha^{-1} and 33.6 kg ha^{-1}) applied in the fall than in the spring. However, harvest timing should also be considered when determining fertility rates. For example, apply less fertilizer if harvest will happen earlier in the season, and more if harvesting later in the season, because favorable weather conditions can cause additional growth in as little as 2weeks (Krueger et al., 2012; Kantar, et al., 2011).

3.2 Adjustments to cash crop nutrient management

In traditional cover cropping systems where biomass is left in the field, nutrient credits from or tied up by the cover crop is accounted for in fertility application decisions for the subsequent cash crop. Nutrient recommendations take cover crops into consideration and are adjusted to protect cash crop yields. However, when harvesting a cover crop for forage, the above-ground biomass and its nutrients are completely removed from the field, not returned to the soil.

Such practices and others related to using cover crops for forage may decrease cash crop yields. Corn yield may be reduced by: nutrients removed with the cover crop harvest, delayed planting to accommodate spring cover crop management, elevated drought risk according to Fouli et al. (2012), absence of tillage after rye, depletion of soil water, and reduced nitrogen availability according to Krueger et al. (2012). West et al., (2020) conducted a five-season study and found that harvest rye cover crop resulted in a 13% decrease in corn silage yield, but total production of corn silage and harvested rye equaled or exceeded corn silage yields without rye. Another two-year study conducted by Bracey et al. 2022 found that forage harvest did not negatively affect the subsequent corn or soy yield or quality.

A cover crop harvested for forage removes nutrients that would have been available to the primary cash crop. This can be addressed in the farm nutrient management plan. Cover crop nitrogen removal can often be mitigated by adjusting spring manure rates or pre-plant/starter fertilizer rates or with adequate sidedressing of corn crops (West et al., 2020). Corn nitrogen requirements at sidedress time can be determined with a Pre-Sidedress Nitrate Test (PSNT).

3.3 Environmental impacts

Some environmental benefits of cover crops, such as increased organic matter, are derived from the return of above-ground biomass to the soil. However, while not the case for cover crops harvested as forage, they offer reduced feed costs and many other benefits described above. In addition, the belowground biomass is not harvested and the roots help to improve and/or maintain aggregate stability even when the aboveground portion is removed (Liesch et al., 2011; Bonetti et al., 2019). Cereal rye and triticale planted as a cover crop and harvested as forage can scavenge residual nutrients and also reduce soil erosion by the remaining stubble (Ketterings, et al., 2015a). In a five-season study, there was a 40% reduction in soil residual nitrogen when a rye cover crop was harvested as forage compared to no cover crop or a cover crop terminated with herbicide in a no-till system. (West et al., 2020). Particularly important in heavily manured systems, double-crop systems can prevent soil nitrate accumulation, reduce nitrate concentrations in the soil solution, and increase soil organic carbon (Krueger et al., 2012). Cover crops can scavenge nitrogen from manure that may otherwise be lost to the environment by erosion, leaching, or as a greenhouse gas (West et al., 2020).

4 Harvest, storage and feeding of cover crops grown for forage

Cover crops grown for forage will produce significant biomass in the spring. For farms that do not plan to remove the cover crop for forage, take other measures earlier in the growing season to manage cover crop growth and prevent compromising the cash crop. Harvesting the cover crop for forage is one of the most effective ways to remove that biomass in preparation for cash crop planting and provides additional benefit to the farm while preserving many of the cover crop ecosystem services.

Cover crops can be terminated through tillage, herbicide application, or other mechanical means such as roll and crimping. Cover crop biomass can also be removed from the field through livestock grazing or through mowing, chopping/baling, or complete removal from the field.

4.1 Grazing cover crops

For farms with sufficient grazing infrastructure, grazing cover crops can be an inexpensive way to recoup the cover crop forage value and nutrients. A cover crop can help to fill voids in the perennial forage pasture system (Fig. 1). For example, cool season pasture perennials such as orchardgrass (*Dactylis glomerata* L.) and various clover species (*Trifolium* spp.) are prone to slow

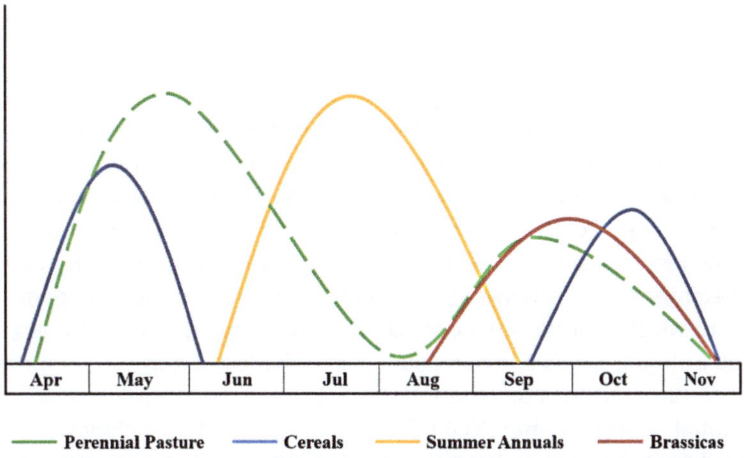

Figure 1 Dry matter production throughout the grazing season in northern U.S. climates characterized by cold winters and hot summers.

recovery periods and reductions in overall productivity when conditions get hot and dry, a phenomenon colloquially as 'summer slump.' Additionally, farms continue to face increasingly extreme and unpredictable weather conditions that further challenge the productivity of our traditional forage production systems. Unfavorable weather conditions can impede perennial forage stand establishment, survival, productivity, and quality. Furthermore, as feed prices increase and milk prices decrease, farmers are seeking strategies to increase production of homegrown forages to supplement pasture and minimize the cost of purchasing feed.

In mild climates, animals with low nutrient requirements (such as beef and sheep) can graze cover crops through the winter months. In northern climates, fall cover crop grazing is possible in rotations where late summer cover crop planting allows for sufficient growth prior to the onset of winter. In corn and soybean double-crop scenarios in the north, options are limited to spring grazing of overwintered species. Summer cover crops can be used to renovate pastures or hay fields. Section 6 of this chapter provides a case study highlighting this strategy.

The optimum time to graze the cover crop will depend on several factors, including the nutrient requirements of the livestock, the growth stage of the cover crop, and climatic conditions. Most cover crops are short-lived annuals with little regrowth potential once they are grazed. In some cases, yields can be maximized by grazing multiple times; in these cases, grazing should be timed to make sure the growing point is not damaged. Depending on the conditions and management practices, some cover crops, such as cereals, can regrow quickly and be ready for a second grazing cycle within 7 to 14 days (Darby et al., 2010). The second grazing will be more variable in both yield and quality because often only the primary shoot will grow back, making the stand less dense. If less lush feed is required, grazing could be delayed, allowing the cover crop to become more mature. The forage yield and quality change considerably depending on the growth stage of the cover crop, so managing grazing to target the quality of feed required by the class of livestock is critical. When grazed at the vegetative stage of growth, small grain species are highly digestible and often high in protein. The bulky characteristics of forage come from fiber. Forage feeding values are negatively associated with fiber since the less digestible portions of plants are contained in the fiber fraction. The detergent fiber analysis system separates forages into two parts: cell contents, which include sugars, starches, proteins, non-protein nitrogen, fats, and other highly digestible compounds; and the less digestible components found in the fiber fraction. The total fiber content of forage is contained in the neutral detergent fiber (NDF). Chemically, this fraction includes cellulose, hemicellulose, and lignin. Because of these chemical components and their association with the bulkiness of feeds, NDF is closely related to feed intake and rumen fill in cows and is considered a good measure of feed quality. In general, legume forages are considered good quality if they have less than 40% NDF, while a legume with over 50% quality would be poor quality. In the case of grasses, NDF that is less than 50% is considered high quality while over 60% is viewed as low quality (Van Saun, 2013).

In recent years, the need to determine rates of digestion in the rumen of the cow has led to the development of NDFd. This in vitro digestibility calculation is very important when looking at how fast feed is being digested and passed through the cow's rumen. Higher rates of digestion lead to higher dry matter intakes and higher milk production or rates of gain. Similar types of feeds can have varying NDFd values based on growing conditions and a variety of other factors. The range of NDFd in legumes and grasses is large (25% to 75% of NDF) but good to excellent quality forage has 60% and higher NDFd (Hoffman et al., 2001).

Ultimately, the class of livestock and stage of growth will determine the quality of feed required to meet nutritional requirements (Table 2). As an example, animals that are lactating will require more digestible nutrients and protein than an older animal in maintenance phase.

Darby et al. (2010) evaluated the yield and quality of different species of cereal grains grazed at the vegetative stage or at the boot stage. Forage quality was highest in the first grazing with fiber digestibility levels over 80% and crude protein (CP) concentrations near 20% (Table 3). These forages, in a balanced ration, could support the needs of lactating animals. Forage quality differences were observed between species, especially as they became more mature. When harvested in the boot stage (Feekes 10), there was no significant difference among yields between the cereal grains (Table 4). However, barley clearly outperformed both wheat and triticale in overall quality. The growth

Table 2 Forage quality requirements for animals at different stages of life. Adapted from Ball et al., 2001.

Cattle stage of life	Crude Protein†	Neutral Detergent Fiber
	% of DM	
Maintenance	>8	<70
Early growth stages	16 to 18	<30
50% of mature weight	12	55
Late pregnancy	10 to 12	<50
Nursing	12 to 14	<55
High-producing lactating	16 to 18	<30
Fattening	8 to 10	<25
Maintenance	>8	<70

†CP, crude protein; NDF, neutral detergent fiber.

Table 3 Forage yield and quality of cereal grains grazed at various growth stages in northern VT. Adapted from Darby et al., 2010.

		Forage quality characteristics		
	DM† yield	CP	NDF	NDFd
Harvest stage	kg/ha	% DM		% NDF
Graze @ 30.5cm	1008	19.3	37.3	83.4
Graze regrowth @ 30.5cm	1267	20.4	47.3	65.6
Graze @ boot stage (Feekes 10)	3284	13.6	51.9	66.5
LSD‡ (p = 0.10)	420	1.32	1.27	1.66

†DM, dry matter; CP, crude protein; NDF, neutral detergent fiber; NDFd, digestible neutral detergent fiber after 30 h.
‡LSD, least significant difference.

Table 4 Dry matter yield and quality of barley, triticale, and wheat grazed at 30.5cm and at the boot stage (Feekes 10) in northern VT. Adapted from Darby et al., 2010.

		Forage quality characteristics		
	DM yield	CP	NDF	dNDF
Small grain	lbs/ac	% DM		% NDF
Graze @ 30.5 cm				
Barley	267	21.5	32.5	85.7
Triticale	1,887	17.3	38.8	83.3
Wheat	638	18.1	38.7	84.9
LSD (0.10)	887	NS	2.28	NS
Graze @ Boot stage (Feekes 10)				
Barley	2,968	14.1	47.0	73.2
Triticale	2,791	14.9	53.2	63.2
Wheat	3,906	11.8	52.3	65.1
LSD (0.10)	NS	1.65	1.84	1.62

†DM, dry matter; CP, crude protein; NDF, neutral detergent fiber; dNDF, digestible neutral detergent fiber after 30 hours.
‡LSD, least significant difference.
§NS, no significant difference between treatments.

stage at which the farmer decides to have livestock begin grazing will likely be determined by the quality of feed required to support their livestock needs.

Winter barley and winter wheat were ready for grazing two weeks prior to perennial cool-season pasture in Vermont, resulting in more than a ton of high-quality dry matter prior to normal turnout of grazing animals (Fig. 2). This can have positive economic returns for pasture-based farms that may have limited high-quality stored feed.

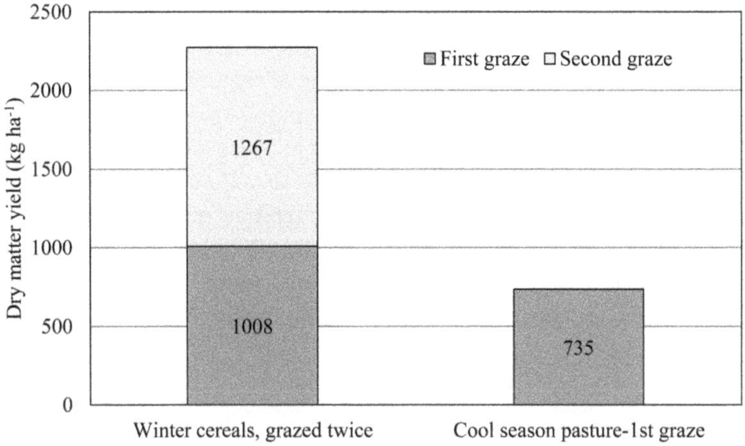

Figure 2 Dry matter yield of winter cereal grains harvested for grazing by 7 May compared to dry matter yields of cool-season pasture grazed on 14 May in northern Vermont.

Grazing cover crops can impact meat, milk, and fiber production. Grazing nontraditional crops may have negative dietary effects that should be considered when selecting the cover crop species. There are limited studies evaluating animal performance when grazing cover crops. Rushing et al. (2023) reported differences in beef steer average daily weight gains when grazing oats, oats and crimson clover (*Trifolium incarnatum* L.), or oats, crimson clover and radish; highest gains were observed when cattle grazed oats or just oats and clover. Cox et al. (2017) also found that average daily gains were greater when beef steers grazed an oat and brassica cover crop mixture compared to corn stalk residue. Phillips et al. (2021) evaluated the impact on meat quality from dairy steers grazing rye or wheat pasture in Minnesota. They demonstrated improvements in the fatty acid profile, and consumers preferred the taste and mouth feel of beef finished on winter wheat compared to winter rye. More livestock-specific research is needed to better understand system-wide benefits and challenges of grazing different classes of livestock on cover crop pasture.

4.2 Harvesting cover crops for stored forage

Cover crops can be harvested and baled or chopped, hauled from the field, and allowed to ferment in storage for later use. Harvest timing will depend on a wide variety of factors, but in most cases, the goal is to optimize yield and quality and not impede the cash crop planting. Cover crop yield and quality will be primarily determined by the stage of maturity at harvest. As plants mature, fiber content increases and negatively impacts energy values. At the seed-head stage of maturity, digestible fiber and protein levels drop rapidly. As undigestible fiber increases and energy drops, the feed becomes slow to digest. Higher levels of undigestible fiber can limit intake by taxing gut fill and slowing rumen throughput. Lower intake will negatively impact productivity.

Harvest can take place at different stages of maturity: earlier harvests are often higher in quality, and later harvests are often higher in yield. Sometimes, a matter of days can have a big impact on yield. For example, it took only 2 to 3 days for cereal rye, triticale, and winter wheat to mature from flag leaf to boot stage and resulted in a 20% increase in yield (Liebert et al., 2023). Harvesting at the boot stage creates a trade-off between lower yield and higher quality; harvesting at later stages of maturity produces higher yields, but quality is reduced (Table 5). The boot stage is just before seed head emergence, when the head can be felt near the top of the leaf whorl within the sheath of the flag leaf (Feekes growth stage 8). Also, boot stage is short, lasting only a few days. To optimize both yield and quality, cereal grain cover crops should be harvested on the earlier side of this stage, well before the flag leaf splits and heads begin to show.

Species and variety can also influence maturation. Darby et al. (2023a) found that winter rye matured 7 to 10 days earlier than wheat and triticale. They also found that winter rye, wheat, and barley varieties can vary in heading timing by as much as 10 days under the same conditions (Darby et al., 2023; Darby and Wilcox Warren, 2023). Timing is important because most winter cereal grains are in the boot stage (Feekes 10) near the same time as optimal cash crop planting dates – or even later, depending on the species/variety. So, farmers often must balance forage harvest and cash crop planting.

Harvesting cover crops for forage in double-cropping scenarios poses a risk of high ash content in the resulting stored feed due to soil contamination. Normal ash levels are <7% of dry matter (DM). Ash values >7% DM indicate possible soil contamination (Schmidt, October 8, 2022b), the primary source of undesirable pathogenic soil organisms such as enterobacteria and clostridia, which may become dominant during silage fermentation. These organisms can

Table 5 Dry matter yield and forage quality characteristics of cereals at various harvest stages. Adapted from Darby et al., 2021.

			Forage quality characteristics		
Harvest stage	DM† at harvest	DM yield	CP	NDF	NDFd
	%	kg ha^{-1}	% DM		% NDF
Boot (Feekes 10)	14.8‡	2,275	19.2	44.4	64.1
Milk (Feekes 10.54–11.1)	27.5	6,366	10.2	59.5	51.1
Soft Dough (Feekes 11.2)	26.3	6,535	10.0	54.9	47.0
LSD‡ (p = 0.10)	1.32	369	0.54	1.57	1.70

†DM, dry matter; CP, crude protein; NDF, neutral detergent fiber; dNDF, digestible neutral detergent fiber after 30 hours.
‡LSD, least significant difference.

result in high levels of butyric acid and degraded protein that can be harmful to animals. To mitigate this risk, cover crops should not be mowed lower than 7.6 cm. In addition, the use of ground-driven rakes should be avoided, and merger, rake, and pick-up reel tines should not scrape soil but rather comb the forage lying on top of stubble.

Harvesting at the proper moisture content for silage is critical. Often, climate conditions for adequate wilting and dry down are difficult when crops are in the boot stage: Cool temperatures, low sunshine, and low winds can limit the dry-down rate. Crops harvested at <30% DM are at higher risk of nutrient loss through leachate and of poor silage fermentation, which results in higher butyric acid and more protein degradation. Harvesting at 33% to 36% DM is better for limiting nutrient loss and avoiding fermentation failures (Hoffman, 2019). Cover crop species grown for forage have large, hollow plant stems that can trap air pockets in the silage mass and prolong the aerobic phase of silage fermentation. A shorter chop length of 0.6 to 1.0 cm may allow for improved packing density and exclusion of air from the silage mass. A prolonged aerobic fermentation period results in excessive heating and binding of plant proteins to sugars. As the plant matures beyond seed-head formation, undigestible fiber increases and becomes tougher and less compactable, making it more difficult to pack tightly (and exclude air). Silage packing density is critical for lactic acid formation, proper pH drop, and for reaching the anaerobic phase of silage fermentation. Use of silage inoculants is recommended to aid in fermentation. Silage fill rate and packing weights need to be adequate to reach the goal packing density of >721 kg. wet weight per cubic meter (Schmidt, 2022b). Segregate cover crops harvested for ensiled feed rather than comingling them with other forages in the feed storage system. This separation will allow for improved feeding management and feed out targeted to specific animal groups.

4.3 Feeding cover crop forage as ensiled feed

Cover crops harvested for ensiled feed are best suited for lactating animals and growing youngstock (growing and pregnant heifers). High levels of highly digestible fiber (dNDF) and 12%–15% CP allow for adequate intake of energy and protein to support milk and milk component production, as well as growth in young stock. Typical ration inclusion levels of 2–9 kg DM is common. However, these forage cover crop species can often have significant amounts of silage-volatile fatty acids (VFAs), including lactic, acetic, and butyric acids, especially if harvested at <30%DM. Total ration acetate levels of >1.5% of ration DM have been shown to inhibit dry matter intake (DMI). High levels of butyric acid can predispose lactating animals to ketosis, a metabolic condition in which the animal is mobilizing body fat for energy, stressing the animal's organs. Feeding

<50 g butyric or keeping the total ration of butyric acid to <1.5% of DM is safe. Subclinical ketosis can occur with 50–100 g of butyric acid, and clinical ketosis can occur at 200 g. Butyric acid is volatile, and the aeration of the silage can help reduce or blow off some of the acid. For silages with >0.50% butyric, air out the silage before including it in rations.

Generally, ensiled cover crop species should not be fed to dry cows. Potential high levels of butyric acid and other risks resulting from predisposition to poor silage fermentation may decrease intake and put animals in the late stage of pregnancy at risk for additional health problems during transition. Also, these species tend to be high in potassium (K), which elevates the ration dietary cation anion difference (DCAD) at a time when reduced DCAD is advised to avoid metabolic complications. For milking cows, the higher K levels may be a benefit by helping to improve rumen fiber degradation and milk production.

When cover crops grown for forage have adequate CP and energy at harvest and are preserved correctly, they can be well suited to young livestock. Less-than-ideal fermentation may result in palatability issues and can reduce intake and rate of gains.

There are many calculations and metrics used to evaluate and compare forage nutrient values. Today's forage labs can accurately estimate fiber digestibility, which ration-balancing software programs use to calculate forage nutrient value. Working with a nutritionist is highly recommended to evaluate forage quality and develop rations that are optimal for growth and milk production and avoid potential health risks.

4.4 Other considerations when harvesting cover crops for forage

There are several considerations for a cover crop program that removes the cover crop for forage versus terminating the cover crop using other methods (i.e. mechanically or chemically). Mechanical termination involves the physical incorporation of the cover crop into the soil vs. the removal of the cover crop from the field as feed. First, the additional machinery (livestock or mechanical) used to harvest the cover crop may introduce or exacerbate soil compaction, which can negatively impact soil health and the subsequent cash crop. Establishing a good stand can help reduce excessive soil moisture via the evapotranspiration of the cover crop to help avoid wet soil conditions at the time of harvest.

Second, one of the significant benefits of cover cropping is that nutrients are stored in the cover crop and then released back into the soil after conventional termination. This cycle gets disrupted when the above-ground cover crop is removed from the field as animal feedstock. However, the nutrients may eventually be returned in the form of livestock manure. Cover crop and residue management affect soil nitrate nitrogen availability, and therefore nitrogen uptake and yield of the subsequent crop. For example, nitrogen is often depleted following a grass cover crop. This will also affect farm economics, depending on the need for additional fertilizer. This needs to be considered in management decisions (Kuo and Jellum, 2002). See section 3 of this chapter for a discussion of nutrient management.

Third, removing the cover crop biomass reduces the organic matter, carbon, and nutrient contributions that would typically improve soil health. This represents a significant reduction in the amount of total biomass that would otherwise have been contributed to the soil ecosystem. The below-ground biomass remains, however, and studies have indicated that 30% of the total biomass of the cover crop (depending on species) remains in the soil as roots, and this portion continues to provide benefits to the soil properties (Blanco-Canqui et al., 2023; Ruis et al., 2019). In addition, there is some evidence that some of the above-ground benefits of cover cropping can be maintained with management adjustments. Harvesting cover crops at low cutting heights reduces the amount of residue left and can potentially eliminate the soil benefits of cover crops. However, with proper cutting heights (e.g., about 10 cm) sufficient residue can be left to maintain soil services (Blanco-Canqui et al., 2013). Furthermore, by the time the cover crop is harvested, many of the benefits of protecting the soil through the fall, winter, and early spring, have already been captured, and the biomass is being terminated at a similar time as in most common cover crop systems. These are important considerations for the management of the cover crop system in order to maximize all the potential benefits of the cover crop.

Strategies to harvest, store, and feed forage made from cover crops will be highly dependent on the goals, resources, and infrastructure that each farm has available for managing these crops. With adequate planning and management, cover crops can produce high yields and quality feed for any number of livestock classes. Often, this feed adds significant value to the farm because it can be produced and used outside of the main grazing or cropping season. Careful attention should be paid during harvest to minimize any unintended impacts to soil health, such as compaction or erosion. Gaining forage from the cover crop should not come at the expense of the farm's conservation efforts.

5 Examples of use: case studies from Vermont

5.1 Double-cropping cereal rye on dairy farms

As dairy farms continue to seek ways to increase milk production and components, some have turned to double-cropping to supplement feedstocks. When implemented correctly, this practice provides the best environmental and economic benefits of cover cropping in a dairy corn silage system. But producing a high-quality feed product using winter cereal grains such as rye or triticale is possible only with a properly executed plan; it will not happen by accident. Careful planning of the planting, fertilizing, harvesting, and storing of the ensiled feed is critical to success.

Key to establishing a good cover crop in the fall are seed selection, field selection, and planting timing. Triticale and cereal rye are common species choices for double cropping. Cereal rye is much less expensive (roughly half the price) and, when harvested properly produces excellent feed. Triticale, though more expensive, often matures later than cereal rye, providing a wider harvest window for excellent quality feed.

For double cropping, it is best to select fields that are relatively well drained and have few stones. Winter grains rely on tillering (producing new shoots) in the fall to produce more biomass in the spring. This outcome is maximized when the crop is planted in early fall. In northern Vermont, the second to third week of September is ideal for planting, depending on the farm's fertility program. If adding fall manure before planting, the cover crop should be planted later in the window to prevent overgrowth and smothering during the winter months. If no manure is added, planting earlier in the window allows for more growth to occur with a reduced risk of smothering.

Even when planting is timely, the recommended seeding rate for double cropping is substantially higher than for a cover crop because the goal of double cropping is to harvest high biomass. For example, cereal rye should be planted at 135 to 157 kilograms per hectare. The later the crop is planted, the higher the seed rate should be. However, farmers cannot substitute the seed rate for the effects of tillering. There is a point in the fall (around the fourth week of September) when cereal rye cannot tiller enough to produce a high-volume harvestable crop in the spring. Double cropping in October is seldom successful due to the cereal rye's low tillering rate. For all cover crops, the seed should be planted with a grain drill or broadcast and then lightly harrowed in. It is advisable to roll the soil after seeding to ensure excellent soil-to-seed contact and to smooth the field for spring harvest and subsequent no-till planting of the summer cash crop. The main objective when rolling the soil surface is smoothing the surface of the soil, knowing that a mowing machine will be the next pass over the field.

When double-cropping and applying manure in the fall, adjust the spring fertilizer rates. Fertilizer should be added to the field when the plant comes out of dormancy, which for cereal rye is when the temperature rises above 3.3°C. In northern Vermont, that is in April. Add roughly 27 kg of nitrogen to cereal rye for harvest at this time. This will increase yields and boost protein levels in the plant. Be sure to consider the nutrient needs of the subsequent crop. A high-yielding (1,800+ kg of DM) cover crop will pull many nutrients out of the soil. It is important to replace them, especially the nitrogen, for a successful cash crop after the cover crop.

Ideally, small grains should be harvested when they are in the boot stage, prior to seed head emergence, to recover as much yield and quality as possible. Once the seed head has emerged, the quality will be too low to feed to lactating animals and, if left to mature further, will be of little value to the dairy farm. However, in northern Vermont, weather drives all field-level timing decisions. Careful attention should be paid to plant maturity and weather conditions in the spring to properly time harvest. When using winter cereal grains, it may be best to harvest at the first reasonable weather window. Though this practice may seem to sacrifice quantity, if harvest is delayed and the weather turns wet, the cover crops will mature rapidly, and the forage quality will plummet. High-quality forage for livestock far outweighs a higher quantity but low-quality dry matter.

Species and varieties may vary in maturation timing, but in general small grains reach the boot stage around mid-to-late May in Vermont, USA.. Therefore, harvest of small grains for forage typically aligns with the first harvest of perennial forage crops, leaving enough time to plant the annual crop (corn for silage in most cases). Typically, the cover crop is mowed when there is at least a three-day window of dry weather approaching. Because the weather during this time is typically cool and moist, drying can be slow. In addition, the amount of wet (80+% moisture) biomass can be substantial. To combat these conditions, mow the crop into wide swaths. To avoid soil contamination and subsequent high ash content in the forage, avoid using ground-driven rakes. The feed is typically collected at 28%–35% dry matter. It should be stored separately from other forages and fed within 90 days of ensiling to prevent cow health issues that can result from high moisture content.

Double-crop feed is comparable to or better than excellent-quality first-cut grass silage. Protein levels of >15% are common, and fiber digestibility is in the 60s to low 70s (Table 6). These qualities make an excellent supplemental feed source for high-producing dairy cows. Research has shown that a 1% increase in forage fiber digestibility corresponds to a 0.18 kg/cow/day increase in DMI and, subsequently, a 0.25 kg/cow/day increase in milk production. Increasing farm inventory of exceptional-quality forage will pay off by maintaining high milk production throughout the season, especially during hot spells in the summer. Obtaining this forage from a cover crop is a win-win for the dairy farm's land, animals, and economics.

The costs of double cropping vary widely depending on location, timing of the forage seeding, field conditions, access to equipment, and harvest goals. Seed costs will vary depending on the winter grain species chosen. Typical prices can vary from $0.49 to $0.99 per kilogram of seed. Typical seeding rates range from 135 to 157 kilogram per hectare, depending on the planting date. Additional fieldwork will likely be needed depending on which fields are selected for planting. There could be stones to pick, and fields are often

Table 6 Example forage analysis of double crop triticale in Vermont.

Nutrient	DM	Range (%)	Nutrient	DM	Range (%)
Protein quality			Moisture (%)	74.4	65–82
Crude protein (%)	17.0	10–23	Dry matter (%)	25.6	18–35
Soluble CP (%CP)	81.3	65–88	Ash (%)	8.80	8–12
ADICP (%)	0.09	0–1.11			
Ammonia (%CP)	7.91	<20	Minerals		
cTotal_RDP (%)	14.0	0–40.83	Calcium (%)	0.30	0.19–0.48
Total RDP %CP (%)	82.5	76–84	Phosphorus (%)	0.43	0.31–0.48
Carbohydrates			Magnesium (%)	0.14	0.03–0.37
Total sugar (%)	4.86	0–8	Potassium (%)	3.71	2.45–4.12
Adjusted starch (%)	0.01	0–1	Sodium (%)	0.02	0–0.04
Agel (%)	0.01	0–1	Chloride (%)	0.33	0–1.24
NFC (%)	19.1	12–25	Sulfur (%)	0.22	0.17–0.24
cADF (%)	29.5	23–37	DCAD (mEQ/100g)	72.9	
cNDF (%)	51.1	43–60			
dNDF (%)	30.7	24–31			
NDFD (%NDF)	60.0	48–59			

Table 7 Average costs for double cropping management above traditional cover cropping.

Expense type	Cost per hectare
Seed	+$44
Planting	+$0
Post-seeding rolling	+$37
Mowing	+$44
Merging	+$37
Fertilizer	+$99
Total	+$262

rolled after planting to improve seed-to-soil contact and smooth the field surface. These additional time and work are added expenses. Table 7 outlines approximate expenses for double cropping beyond those of managing a regular cover crop.

Additional expenses include: $44.00 for the additional seed (average for cereal rye), $37.00 per hectare to roll the soil, $44.00 per hectare to mow the crop, $37.00 per hectare to merge the crop, and $161.00 per hectare to harvest the crop (prices based on local costs). Additional fertilizer, especially nitrogen, may be needed for the subsequent crop, which could cost up to $99 per hectare at 2024 nitrogen prices. The post-harvest tillage operation and herbicide application are not an additional cost because farmers likely would have had to do so regardless of whether or not they double-cropped. In total, double cropping costs an additional $262 per hectare over cover cropping. Given a haylage market value of $66 per metric ton, the break-even yield is only 0.99 dry matter metric tons per hectare (or 986 kg ha^{-1}). While dry matter yields will vary from year to year and with planting date, a reasonably conservative estimate for northern Vermont is 3,363 dry matter kilograms per hectare, which would equate to a net value of $183 per hectare. The degree of benefit received with this system is highly dependent on how well it is managed from planting through feed out. Understanding the risks and strategies to mitigate those risks will set up the farm for success.

This practice has become routine on several farms in northern Vermont. Although results can vary from year to year and additional planning and management are required, farms that have adopted this practice stick with it because of the valuable benefits double cropping provides their farm and the environment.

5.2 Grazing cover crops on organic grass-fed dairy farms

On a grass-fed dairy farm where no grain feeding is allowed, having a variety of high-quality forage crops available throughout the year is critical to supporting productivity and farm economics. Although a more typical double-cropping scenario with a corn silage crop is not possible on these farms, some of them are finding ways to incorporate a wide range of cover crops. On Severy Farm in Cornwall, Vermont, annual cover crops are often grown in every season. The farmer manages a herd of about 60 milking cows that calve year-round and have been grass-fed for four years. The farm produces approximately half of the herd's dry matter needs and purchases the rest.

Since transitioning to grass-fed, the farm has been increasing its land base to produce more forage to replace the dry matter in the herd's diet that was once purchased in the form of grain. The farmers currently manage about 115 hectares of land, and approximately 40 hectares are permanent pasture. The farm's soil varies from heavy clay to loamy sand. In the heat of summer, growth of the perennial cool-season pastures often slows, and there is a very long recovery period before they can be grazed again. This is when summer annual cover crop species, such as sorghum sudangrass (*Sorghum bicolor* (L.) Moench/*Sorghum bicolor var. sudanense)* and pearl millet (*Pennisetum americanum* L.), are useful. Each year, the farm chooses an area of 1.2 to 2.4 hectares of perennial forage to reseed. The herd grazes those acres early in the spring, then the field is tilled and the summer annual cover crop is planted, usually in June. By mid-summer when the heat usually sets in, the field is ready to graze. The cows usually have access to the annual cover crop pasture for just a few hours at a time – just long enough to stimulate them to eat more, fill up, and go back to the perennial pasture. The farmers will strip-graze the field, giving the cows a small new section of the annuals daily. They are careful to move the cows through the annuals quickly so they don't graze the plants too short, which would inhibit regrowth. After the cows have strip grazed the entire field once, it is ready for grazing again in as little as three weeks, depending on weather conditions. Typically, the herd will graze the annuals twice in a season.

Sorghum sudangrass, pearl millet, and Japanese millet (*Echinochloa esculenta* (A. Braun) H.Scholz) are usually planted. These species can provide very good weed suppression and leave a clean, well-primed seedbed for a perennial pasture mix or a fall annual cover crop. The annual cover crop is often a mixture of oats and peas that can be either harvested and stored as baleage or grazed to further extend the grazing season. Because these cover crops will winterkill, the ground is well-prepared for seeding a perennial pasture mixture in the following spring. Or, winter grains such as winter barley or triticale are planted. These will not be grazed in the fall but are typically harvested as baleage the following spring. The quality of these small-grain forages can be very high and make great supplemental forage for a grass-fed herd, for which getting ample metabolizable energy is a constant struggle. The metabolizable energy content of both the oat/pea (2.60 Mcal/kg) and triticale baleage (2.47 Mcal/kg) is equal to or higher than the earliest lush spring pasture on the farm at 2.54 Mcal/kg. Similar to the seedbed after summer cover crops, the post-harvest seedbed after winter grains is ideally suited to no-tilling in a perennial pasture mix. Additionally, if planted at the right time in the spring, small grains tiller prolifically, filling in the spaces between the rows to provide good weed suppression. These examples show the flexibility that can be gained by integrating cover crops for forage.

The Choiniere Family Farm, an organic grass-fed dairy located in Highgate, Vermont, has been experimenting with diverse cover crop mixtures that can be grown during the hot summer months. A herd of about 80 milking cows calves year-round and has been grass-fed for 9 years on about 200 hectares of land with approximately 28 in permanent pasture. Hay fields near the milking facility are often added to the grazing rotation later in the season. The farmers produce all the forage their herd needs on their land. In the heat of summer, growth of the perennial pastures often slows considerably, and there is a very long recovery period before they can be grazed again. This is when summer cover crop mixtures that include heat-loving species such as sorghum Sudangrass, pearl millet, cowpeas (*Vigna unguiculata* (L.) Walp, and sunflower (*Helianthus annuus* L.) are planted. The mixtures typically include six to 10 different species, each contributing different benefits. Legumes such as vetch and cowpea help to fix nitrogen, deep taprooted species such as daikon radish and sunflower break up compaction, and dense fibrous-rooted grass such as annual ryegrass increase organic matter and soil aggregation.

The field is grazed early in the spring, then it is tilled, and the summer annual is planted, usually in June. By mid-summer, when it begins to get hot, the field is ready to graze. The cows usually have access to the annual pasture for just a couple of hours at a time; just enough to stimulate them to eat more, fill up, and go back to the perennial pasture. In this strip grazing system the cows are turned onto a small new section of the annuals every day.

Many of the species in these diverse mixtures will not regrow, so they are only grazed once. The cows do a good job of trampling in the residue, and the ground is typically in pretty good condition for seeding a perennial pasture mix. Even if only grazed once during the summer, this crop allows them to introduce a diversity of species that otherwise would not be on the farm, promoting soil biodiversity, enhancing soil condition, and reducing demand on the perennial pastures.

These case studies demonstrate various ways to incorporate cover crops on a diverse set of livestock operations. Dairy farms in Vermont have continued to innovate with cover crops to enhance the profitability of their operations. Capturing and sharing farmer experiences will allow other agricultural producers to see the successful implementation of this strategy.

6 Conclusion

Good stands of cover crops provide ecosystem services of significant environmental and economic benefit to agricultural systems (Sarrantonio and Gallandt, 2003). When the cover crop biomass is harvested for forage, the ecosystem service benefits may be more substantial, but there are important considerations and, as with any crop, risks. Harvesting annuals for forage is not a new practice. However, planting annuals as a dual-purpose crop to provide ecosystem services as well as high-quality and yielding forage is a new approach to managing cover crops. The use of cover crops across the nation remains low: only 5% of the total crop acreage was planted in cover crops in 2017. Allowing farmers to harvest cover crops for forage may increase the direct economic value to a farm and, in turn, expand the adoption of cover crops across the nation. Many factors will influence whether a cover crop harvested for forage provides an economic benefit to the farm, including proximity to livestock, cost of growing a high-yield, high-quality cover crop, and overall demand for forage in any given location.

Farmers know how to grow high-yielding cash crops, and similar efforts are needed if cover crops are going to produce maximum output. Cover crops grown for forage must be treated like a cash crop, including using proper seeding methods, seeding rates, and fertility recommendations. However, growing crops in late fall and early spring has more inherent risk than during the regular growing season. The amount of growing degree days, moisture, ice, and heaving of the soil all affect yield and are often beyond the grower's control. Creating an economic benefit for a farm by harvesting cover crops for forage may, in turn, produce more effective cover crops that provide more consistent environmental benefit to farms and communities. There is still much to learn about how best to optimize the conservation benefits as well as the forage benefits from cover crops. Sacrificing cover crop environmental benefits for forage production is not an acceptable compromise. Current research suggests that both benefits can be achieved; however, more research is needed throughout the nation to capture region-specific situations and garner local farmer support.

7 Where to look for more information

The following articles, guides, and factsheets provide basic information on how to adapt cover crops for forage production.

- Darby, H., et al. (2018), *Under cover: Integrating cover crops into silage corn systems*. Northwest Crops & Soils Program. 584. https://scholarworks.uvm.edu/nwcsp/584 (accessed 18 November 2024).
- Darby, H., et al. (2021), *Guide to using annual forages in the Northeast*. Northwest Crops & Soils Program. 582. https://scholarworks.uvm.edu/nwcsp/582 (accessed 18 November 2024).
- Hoorman, J. J., et al. (n.d.), *Growing and managing forage cover crops*. SAG-XX-15, Ohio State University Extension. https://midwestcovercrops.org/wp-content/uploads/2016/10/OH_2015_Using-Cover-Crops-as-Forage-051815.pdf (accessed 21 November 2024).
- Sarrantonio, M. and Gallandt, E. (2003), 'The role of cover crops in North American cropping systems', *J. Crop Prod.*, 8(1–2), 53–74.
- Suazo, S. A., et al. (2024), '394 Cover crops for efficient grazing systems', *J. Anim. Sci.*, 102(3), 189.
- Williams, A., et al. (2018), *Grazing cover crops: A how-to guide*, Pasture Center at the Wallace Center. https://pastureproject.org/publications/grazing-cover-crops-how-to-guide/ (accessed 21 November 2024).

These articles evaluate the impact of harvesting cover crops for forage on soil quality and conservation.

- Blanco-Canqui, H., et al. (2020), 'Does cover crop grazing damage soils and reduce crop yields?', *Agrosyst. Geosci. Environ.*, 3(1), 1–11.
- Drewnoski, M., et al. (2018), 'Forages and pastures symposium: cover crops in livestock production: whole-system approach. Can cover crops pull double duty: conservation and profitable forage production in the Midwestern United States?', *J. Anim. Sci.*, 96(8), 3503–3512.

- Franzluebbers, A. and Stuedemann, J. (2015), 'Does grazing of cover crops impact biologically active soil carbon and nitrogen fractions under inversion or no tillage management?', *J. Soil Water Conserv.*, 70(6), 365–373.
- Poffenbarger, H. (2010), 'Ruminant grazing of cover crops: Effects on soil properties and agricultural production', *J. Nat. Resour. Life Sci. Educ.*, 38(1), 49–52.

8 References

Anon. (2023), *2022–2023 National cover crop survey report*, Conservation Technology Information Center, https://www.ctic.org/data/Cover_Crops_Research_and_Demonstration_Cover_Crop_Survey (accessed 18 November 2024).

Archarya, P., et al. (2022), 'Net greenhouse gas balance with cover crops in semi-arid irrigated cropping systems', *Scientific Reports*, 12, 12386.

Baker, J. M. and Griffis, T. J. (2009), 'Evaluating the potential use of winter cover crops in corn–soybean systems for sustainable so-production of food and fuel', *Agric. For. Meteorol.*, 149(12), 2120–2132.

Balkcom, K. S., et al. (2018), 'Fertilizer management for a rye cover crop to enhance biomass production', *Agron. J.*, 110(4), 1233–1242.

Balkcom, K. S., et al. (2023), 'Rye planting date impacts biomass production more than seeding rate and nitrogen fertilizer', *Agron. J.*, 115(5), 2351–2368.

Ball, D. M., et al. (2001), 'Understanding forage quality', *American Farm Bureau Federation Publication*, 1(1), 1–15.

Baron, V. S., et al. (1999), 'Fall seeding date and species effects on spring forage yield of winter cereals', *J. Prod. Agric.*, 12(1), 110–115.

Beever, D. E. and Mould, F. L. (2000), 'Forage evaluation for efficient ruminant livestock production'. In D. I. Givens, et al. (Eds), *Forage Evaluation in Ruminant Nutrition*, CAB International, pp. 15–42.

Binder, J. M., et al. (2021), 'Winter annual management to increase nutrient recovery and forage production on dairies', *Agrosyst. Geosci. Environ.*, 4(2), 1–12.

Black, K. L., et al. (2023), Interseeding Wide-Row Corn with Forage Cover Crops: Investigating System Potential for Expanded Economic Opportunities in Corn Production Systems. *Agronomy, 13*(2), Article 2. https://doi.org/10.3390/agronomy13020307Blanco-Canqui, H., et al. (2013), 'Replacing fallow with cover crops in a semiarid soil: Effects on soil properties', *Soil Sci. Soc. of Am. J.*, 77(3), 1026–1034.

Blanco-Canqui, H., et al. (2015), 'Cover crops and ecosystem services: Insights from studies in temperate soils', *Agron. J.*, 107(6), 2449–2474.

Blanco-Canqui, H., et al. (2020), 'Does cover crop grazing damage soils and reduce crop yields?', *Agrosyst. Geosci. Environ.*, 3(1), 1–11.

Blanco-Canqui, H., et al. (2023), 'Grazing cover crops: How does it influence soils and crops?', *Agron. J.*, 115(6), 2801–2828.

Bonetti, J. A., et al. (2019), 'Impact of a long-term crop-livestock system on the physical and hydraulic properties of an Oxisol', *Soil Till. Res.*, 186, 280–291.

Bowman, M., et al. (2024), *Cover crops on livestock operations: Potential for expansion in the United States*. Report No. AP-120, US Department of Agriculture, Economic Research Service.

Bracey, W. M., et al. (2022), 'Forage yield, quality, and impact on subsequent cash crop of cover crops in an integrated forage/row crop system', *Agronomy*, 12(5), 1–19.

Brooker, A. P., et al. (2020), 'Interseeding cover crops in corn: Establishment, biomass, and competitiveness in on-farm trials', *Agron. J.*, 112(5), 3733–3743.

Brown, A. N., et al. (2018), 'Nutritional composition and in vitro digestibility of grass and legume winter (cover) crops', *J. Dairy Sci.*, 101(3), 2037–2047.

Cates, A. M., et al. (2018), 'What do we know about cover crop efficacy in the North Central United States?', *J. Soil Water Conserv.*, 73(6), 153A–157A.

Cela, S., et al. (2014), 'Characterization of nitrogen, phosphorus, and potassium mass balances of dairy farms in New York State', *J. Dairy Sci.*, 97, 7614–7632.

Chami, B., et al. (2023), 'Incentive programs promote cover crop adoption in the northeastern United States', *Agric. Environ. Lett.*, 8(2), 1–6.

Clark, A. (ed.) (2008), *Managing cover crops profitably*. Sustainable Agriculture Research and Education. https://www.sare.org/resources/managing-cover-crops-profitably-3rd-edition/

Coblentz, W. K., et al. (2014), Fall-grown oat to extend the fall grazing season for replacement dairy heifers. *Journal of dairy science*, *97*(3), 1645-1660.

Copeland, C. (2010), *Animal waste and water quality: EPA regulation of concentrated animal feeding operations (CAFOs)*. CRS Report RL 31851, Congressional Research Service.

Cox-O'Neill, J. L., et al. (2017), 'The effects of backgrounding system on growing and finishing performance and carcass characteristics of beef steers', *J. Animal Sci.*, 95(12), 5309–5319.

Darby, H., et al. (2010), *Winter cereals as a multipurpose crop*. Northwest Crops and Soils Program. 42. https://scholarworks.uvm.edu/nwcsp/42 (accessed 18 November 2024).

Darby, H., et al. (2017a), *Cool season annual forage mixtures trial*. Northwest Crops & Soils Program. 10. https://scholarworks.uvm.edu/nwcsp/10 (accessed 18 November 2024).

Darby, H., et al. (2017b), *Maximizing forage yields in corn silage systems with winter grains*. Northwest Crops & Soils Program. 7. https://scholarworks.uvm.edu/nwcsp/7 (accessed 18 November 2024).

Darby, H., et al. (2018), *Under cover: integrating cover crops into silage corn systems*. Northwest Crops & Soils Program. 584. https://scholarworks.uvm.edu/nwcsp/584 (accessed 18 November 2024).

Darby, H. et al. (2018), Cool Season Annual Forage Mixtures Trial. Northwest Crops and Soils Program. https://scholarworks.uvm.edu/items/41dd1cee-2060-4e44-a814-9817af7ac1e8

Darby, H., et al. (2021), *Guide to using annual forages in the Northeast*. Northwest Crops & Soils Program. 582. https://scholarworks.uvm.edu/nwcsp/582 (accessed 18 November 2024).

Darby, H., et al. (2022a), *2020 Cool season annual forages trial*. Northwest Crops & Soils Program. 434. https://scholarworks.uvm.edu/nwcsp/434 (accessed 21 November 2024).

Darby, H., et al. (2022b), *2022 Rye variety trial*. Northwest Crops & Soils Program. 509. https://scholarworks.uvm.edu/nwcsp/509 (accessed 18 November 2024).

Darby, H. and Wilcox Warren, S. (2023), *2022 Spring wheat crosses trial*. Northwest Crops & Soils Program. 505. https://scholarworks.uvm.edu/nwcsp/505 (accessed 18 November 2024).

Darby, H., Brown, A. and Emick, H. (2023), *2022 Triticale variety trial*. Northwest Crops & Soils Program. 504. https://scholarworks.uvm.edu/nwcsp/504 (accessed 18 November 2024).

Darby, H. and Bruce, J. (2024), *2023 Winter rye cover crop and forage variety trial*. Northwest Crops & Soils Program. 585. https://scholarworks.uvm.edu/nwcsp/585 (accessed 18 November 2024).

Darby, H. and Ziegler, S. (2024), *Fall annual forages trial*. Northwest Crops & Soils Program. 545. https://scholarworks.uvm.edu/nwcsp/545/ (accessed 18 November 2024).

den Hollander, N. G., et al. (2007), 'Clover as a cover crop for weed suppression in an intercropping design: II. Competitive ability of several clover species', *Eur. J. Agron.*, 26(2), 104–112.

Dhakal, D. et al. (2022), Grazing cover crops in a no-till corn and soybean rotation. *Agronomy Journal*, *114*(2), 1255–1268.

Dillard, S. L., et al. (2020), 'Assessment of forage brassica species for dairy and beef-cattle fall grazing systems', *Appl. Anim. Sci.*, 36(2), 157–166.

Drewnoski, M., et al. (2018), 'Forages and pastures symposium: cover crops in livestock production: whole-system approach. Can cover crops pull double duty: conservation and profitable forage production in the Midwestern United States?', *J. Anim. Sci.*, 96(8), 3503–3512.

Faé, G. S., Sulc, R. M., Barker, D. J., Dick, R. P., Eastridge, M. L., & Lorenz, N. (2009), Integrating Winter Annual Forages into a No-Till Corn Silage System. *Agronomy Journal*, *101*(5), 1286–1296. https://doi.org/10.2134/agronj2009.0144

Farney, J. K., et al. (2018), 'Composition, forage production, and costs are variable in three-way cover crop mixes as fall forage', *Crop Forage Turfgrass Manag.*, 4(1), 1–7.

Farney, J. K. (2019), 'Cattle preference for annual forages', *Kansas Agricultural Experiment Station Research Reports*, 5(2), 7732. https://doi.org/10.4148/2378-5977.7732.

Fisher, K. A., et al. (2011), 'Is broadcasting seed an effective winter cover crop planting method?', *Agron. J.*, 103(2), 472–478.

Fouli, Y., et al. (2012), 'Double cropping effects on forage yield and the field water balance', *Agric. Water Manage.*, 115, 104–117.

Franzluebbers, A. J. and Stuedemann, J. A. (2014), 'Crop and cattle production responses to tillage and cover crop management in an integrated crop–livestock system in the southeastern USA', *European J. Agron.*, 57, 62–70.

Franzluebbers, A. and Stuedemann, J. (2015), 'Does grazing of cover crops impact biologically active soil carbon and nitrogen fractions under inversion or no tillage management?', *J. Soil Water Conserv.*, 70(6), 365–373.

Gailans, S. and Nichols, G. (2022), *Fertilizing cover crops*. Practical Farmers of Iowa Cooperators' Program Farmer-Led Research. https://practicalfarmers.org/research/fertilizing-cover-crops/ (accessed 18 November 2024).

Gan, Y. and McLeod, J. G. (1997), 'Winter rye leaf emergence and its relationship to the phyllochron', *Crop Sci.*, 37(1), 162–166.

Gregorini, P., et al. (2017), 'Grazing management: setting the table, designing the menu and influencing the diner', *Anim. Prod. Sci.*, 57(7), 1248–1268.

Gusta, L. V. and O' Connor, B. J. (1987), 'Frost tolerance of wheat, oats, barley, canola and mustard and the role of ice-nucleating bacteria', *Can. J. Plant Sci.*, 67(4), 1155–1165.

Hansen, M., et al. (2013), Suitability of cover crop monocultures for late-season forage in South Dakota. *Canadian Journal of Plant Science*, *93*(4), 589-597.

Hively, W. D., et al. (2015), 'Remote sensing to monitor cover crop adoption in southeastern Pennsylvania', *J. Soil Water Conserv.*, 70(6), 340–352.

Hoffman, P. (2019), Wet small grain silage: how do I feed these? https://www.vitaplus.com/forage-foundations/wet-small-grain-silages-how-do-i-feed-these/ (accessed 18 November 2024).

Hoffman, P., et al. (2001), 'Understanding NDF digestibility of forages', *Focus Forage*, 3, 1–3.

Hoorman, J. J., et al. (n.d.), *Growing and managing forage cover crops*. SAG-XX-15, Ohio State University Extension. https://midwestcovercrops.org/wp-content/uploads/2016/10/OH_2015_Using-Cover-Crops-as-Forage-051815.pdf (accessed 21 November 2024).

Hucl, P. and Baker, R. J. (1991), 'Effectiveness of early-generation selection for tillering capacity in spring wheat', *Crop Sci.*, 31(4), 938–942.

Jemison Jr, J. M., et al. (2012), 'Winter grain–short season corn double crop forage production for New England', *Agron. J.*, 104(2), 256–264.

Jokela, W. E., et al. (2009), 'Cover crop and liquid manure effects on soil quality indicators in a corn silage system', *Agron. J.*, 101(4), 727–737.

Jordan, E. R. and Fourdraine, R. H. (1993), 'Characterization of the management practices of the top milk producing herds in the country', *J. Dairy Sci.*, 76(10), 3247–3256.

Kantar, M., et al. (2011), 'Growth stage influences forage yield and quality of winter rye', *Forage Grazinglands*, 9(1), 1–7.

Ketterings, Q. M., et al. (2015a), 'Integrating cover crops for nitrogen management in corn systems on Northeastern US dairies', *Agron. J.*, 107(4), 1365–1376.

Ketterings, Q. M., et al. (2015b), 'Winter cereals as double crops in corn rotations on New York dairy farms', *J. Agric. Sci.*, 7(2),18–25.

Kleinman, P. J., et al. (2005), 'Effect of cover crops established at time of corn planting on phosphorus runoff from soils before and after dairy manure application', *J. Soil Water Conserv.*, 60(6), 311–322.

Krueger, E. S., et al. (2012), 'Rye–corn silage double-cropping reduces corn yield but improves environmental impacts', *Agron. J.*, 104(4), 888–896.

Kuo, S. and Jellum, E. J. (2002), 'Influence of winter cover crop and residue management on soil nitrogen availability and corn', *Agron. J.* 94(3), 501–508.

Landry, E., et al. (2019), 'Winter cereal cover crops for spring forage in temperate climates', *Agron. J.*, 111(1), 217–223.

Liebert, J., et al. (2023), Winter cereal species, cultivar, and harvest timing affect trade-offs between forage quality and yield. *Frontiers in Sustainable Food Systems, 7*, 1067506.

Liesch, A. M., et al. (2011), 'Soil structure and physical properties under rye-corn silage double-cropping systems', *Soil Sci. Soc. Am. J.*, 75(4), 1307–1314.

Lyons, S. E., et al. (2019), Double-cropping with forage sorghum and forage triticale in New York. *Agronomy Journal*, *111*(6), 3374-3382

Maloney, T. S., et al. (1999), 'Small grains for fall and spring forage', *J. Prod. Agric.*, 12(3), 371–494.

McCartney, D., et al. (2009), 'Potential of warm-season annual forages and Brassica crops for grazing: a Canadian review', *Can. J. Anim. Sci.*, 89(4), 431–440.

Mirsky, S. B., et al. (2011), 'Timing of cover-crop management effects on weed suppression in no-till planted soybean using a roller-crimper', *Weed Sci.*, 59(3), 380–389.

Myers, R. and Watts, C. (2015), 'Progress and perspectives with cover crops: Interpreting three years of farmer surveys on cover crops', *J. Soil Water Conserv.*, 70(6), 125A–129A.
Nichols, G. A., et al. (2022), Fertilizing Cover Crops. Practical Farmers of Iowa Cooperators' Program. https://practicalfarmers.org/wp-content/uploads/2022/11/22_FC_CC_FertilizingCoverCrops_FINAL-1.pdf. (accessed February 2026).
Omokanye, A., et al. (2019), 'Forage production, economic performance indicators and beef cattle nutritional suitability of multispecies annual crop mixtures in Northwestern Alberta, Canada', *J. Appl. Anim. Res.*, 47(1), 303–313.
Omokanye, A. et al. (2020), Forage potential of corn intercrops for beef cattle diets in northwestern Alberta. Crop, Forage & Turfgrass Management, 6(1), e20056.
Omokanye, A., et al. (2021), 'Alternative forage feeds for beef cattle in Northwestern Alberta, Canada: forage yield and nutritive value of forage brassicas and forbs', *J. Appl. Anim. Res.*, 49(1), 203–210.
Phillips, H. N., et al. (2021), 'Biomass yield and nutritive value of rye (Secale cereale L.) and wheat (Triticum aestivum L.) forages while grazed by cattle', *Crops*, 1(2), 42–53.
Poffenbarger, H. (2010), 'Ruminant grazing of cover crops: Effects on soil properties and agricultural production', *J. Nat. Resour. Life Sci. Educ.*, 38(1), 49–52.
Popovici, R., et al. (2023), 'The social factors influencing cover crop adoption in the Midwest: a controlled comparison', *Environ. Manage.*, 72(3), 614–629.
Provenza, F. D. (1996), 'Acquired aversions as the basis for varied diets of ruminants foraging on rangelands', *J. Ani. Sci.*, 74(8), 2010–2020.
Ranck, E. J., et al. (2020), 'Economic and environmental effects of double cropping winter annuals and corn using the Integrated Farm System Model', *J. Dairy Sci.*, 103(4), 3804–3815.
Reimer, A. P. and Prokopy, L. S. (2014), 'Farmer participation in US Farm Bill conservation programs', *Environ. Manage.*, 53, 318–332.
Robinson, J. L. J. (2020), *Sheep and goat preference of five common cover crops*. Master of Science (Agriculture), Sam Houston State University. Available at: https://www.proquest.com/docview/2488262522?pq-origsite=gscholar&fromopenview=true&sourcetype=Dissertations%20&%20Theses (accessed 21 November 2024).
Ruis, S. J., et al. (2019), 'Cover crop biomass production in temperate agroecozones'. *Agron. J.*, 111(4), 1535–1551.
Rushing, J. B., et al. (2023), 'The effects of grazing cover crops on animal performance and soybean production', *Crop Forage Turfgrass Manag.*, 9(1), e20203.
Sadeghpour, A., et al. (2022), 'Profitability of dual-purpose rye cover crop as influenced by harvesting date', *Agron. J.*, 114(1), 627–640.
Sanderson, M., et al. (2018), 'Cover crop mixtures grown for annual forage in a semi-arid environment', *Agron. J.*, 110(2), 525–534.
Sarrantonio, M. and Gallandt, E. (2003), 'The role of cover crops in North American cropping systems', *J. Crop Prod.*, 8(1–2), 53–74.
Schmidt, R. (2022a), Small-grains silage: Big opportunities for your forage program. https://www.agproud.com/articles/55352-small-grains-silage-big-opportunities-for-your-forage-program (accessed 18 November 2024).
Schmidt, R. (2022b), Small-grain silages: evaluating and feeding. https://www.agproud.com/articles/56190-small-grain-silages-evaluating-and-feeding (accessed 21 November 2024).
Schomberg, H. H., et al. (2014), 'Grazing winter rye cover crop in a cotton no-till system: yield and economics', *Agron. J.*, 106(3), 1041–1050.

Sestak, J. R., et al. (2023), 'Profitability of grazing no-till established winter cereal pasture and summer cover crops', *Agron. J.*, 115(3), 1295–1307.

Simon, L. M. et al. (2021), Forage productivity and soil properties in dual-purpose cover crop systems. Agron. J. 113(6), 5569-5583.

Simon, L. M. et al. (2024), Cover crop grazing effects on soil properties in no-tillage dryland cropping systems in the central Great Plains. Agric., Ecosys., and Environ., 374, 109140.

Soil Science Society of America (2008), *Glossary of soil science terms 2008*. Soil Science Society of America.

Suazo, S. A., et al. (2024), '394 Cover crops for efficient grazing systems', *J. Anim. Sci.*, 102(3), 189.

Sulc, R. M. and Franzluebbers, A. J. (2014), Exploring integrated crop–livestock systems in different ecoregions of the United States. *European journal of agronomy*, *57*, 21-30.

Thilakarathna, M. S., et al. (2015), 'Management of manure nitrogen using cover crops', *Agron. J.*, 107(4), 1595–1607.

Titlow, A. et al. (2014), Using dryland annual froage mixtures as a forage option for grazing beef cattle. Forage & Grazinglands, 12(1) FG-2013-0041-RS.

Tobin, C., et al. (2020), 'Demonstrating short-term impacts of grazing and cover crops on soil health and economic benefits in an integrated crop-livestock system in South Dakota', *Open J. Soil Sci.*, 10(03), 109–136.

Van Saun, R. J. (2013), *Determining forage quality: understanding feed analysis*. Penn State Extension.

Vermont General Assembly (2015), 'Act 64: an act relating to improving the quality of state waters', *State of Vermont*.

West, J. R., et al. (2020), 'Sustainable intensification of corn silage cropping systems with winter rye', *Agron. Sustain. Dev.*, 40, 1–12.

Wiedenhoeft, M. H. and Barton, B. A. (1994), 'Management and environment effects on brassica forage quality', *Agron. J.*, 86(2), 227–232.

Wilhelm, W. W., et al. (2010), 'Balancing limiting factors and economic drivers for sustainable Midwestern US agricultural residue feedstock supplies', *Ind. Biotechnol.*, 6(5), 271–287.

Williams, A., et al. (2018), *Grazing cover crops: a how-to guide*. Pasture Center at the Wallace Center. https://pastureproject.org/publications/grazing-cover-crops-how-to-guide/ (accessed 21 November 2024).

Wyffels, S. A., et al. (2021), 'Introducing cover crops as a fallow replacement in the Northern Great Plains: I. Evaluation of cover crop mixes as a forage source for grazing cattle', *Renew. Agric. Food Sys.*, 37(4), 292–302.

Villalobos, L. A. and Brummer, J. E. (2015), 'Forage brassicas stockpiled for fall grazing: yield and nutritive value', *Crop Forage Turfgrass Manag.*, 1(1), 1–6.

Zhang, M., et al. (2006), 'Solid and liquid cattle manure application in a subarctic soil: Bromegrass and oat production and soil properties', *Agron. J.*, 98(6), 1551–1558.

Chapter 7

Assessing the effects of cover crops on cash crop yields

Inderjot Chahal, University of Guelph, Canada; Yvonne Lawley, University of Manitoba, Canada; Alex L. Woodley, North Carolina State University, USA; Maryse Bourgault, University of Saskatchewan, Canada; and Laura L. Van Eerd, University of Guelph, Canada

1 Introduction

Cover crops (CCs) provide an opportunity to reduce the environmental externalities of farming (Blanco-Canqui et al. 2015; Daryanto et al. 2018). However, CCs have been found to have positive, negative, or no effect on following crop yields (Table 1, Alvarez et al. 2017; Chahal and Van Eerd 2021,

http://dx.doi.org/10.19103/AS.2025.0162.09

Table 1 Summary of the literature on the cover crop effects on subsequent crop yield. Data presented are from meta-analyses and review articles.

	Reference	Study years	Location/Region	Type of study	Cover crop type	Main crop	Yield response
1	Miguez and Bollero 2005	1965–2004	USA and Canada	Meta-analysis	Legume	Corn	+24%
					Grass	Corn	ns
					Mixture	Corn	+21%
2	Marcillo and Miguez 2017	1965–2014	USA and Canada	Meta-analysis	Legume	Corn	+30 to 33%
					Grass	Corn	ns
					Mixture	Corn	+30%
3	Bourgeois et al. 2022	1910–2015	USA, Canada, Europe	Meta-analysis	Legume	Corn	+31%
						Small grain cereal	+33%
						Soybean	ns
					Non-legume broadleaf	Corn	ns
						Small grain cereal	ns
						Soybean	ns
					Mixture	Corn	+29%
						Small grain cereal	+26%
						Soybean	ns
					Grass	Corn	-2%
						Small grain cereal	ns
						Soybean	ns

4	Chahal and Van Eerd 2023	2015–2021	USA and Canada	Meta-analysis	Grass	Grain crop	ns
						Vegetable	ns
					Legume	Grain crop	ns
						Vegetable	+20%
					Non-legume broadleaf	Grain crop	ns
						Vegetable	ns
					Mixture	Grain crop	ns
						Vegetable	ns
5	Tonitto et al. 2006	1970–2002	Temperate	Meta-analysis	Legume	Corn	−12% vs. fertilized[z]
6	Norris and Congreves 2018	1996–2017	Temperate	Meta-analysis	Cover crop	Vegetable	ns
7	Allam et al. 2023	2019–2022	Europe	Meta-analysis	Legume	Cereal	+12%
					Non-legume	Cereal	+9%
					Mixture	Cereal	+7%
8	Joshi et al. 2023	Until 2022	Global	Meta-analysis	Legume	Corn	+35%
					Non-legume	Corn	ns
					Mixture	Corn	ns
9	Quemada et al. 2013	1910–2012	Global	Meta-analysis	Legume	Cash crop[y]	+25%
					Non-legume	Cash crop	ns
10	Thapa et al. 2018	1931–2017	Global	Meta-analysis	Legume	Cash crop	ns
					Non-legume	Cash crop	ns
11	Alvarez et al. 2017	2006–2016	Argentina Pampas	Meta-analysis	Legume	Corn	+7%

(Continued)

Table 1 *Continued*

	Reference	Study years	Location/Region	Type of study	Cover crop type	Main crop	Yield response
					Non-legume	Corn	–8%
					Mixture	Corn	ns
12	Daryanto et al. 2018	1985–2017	Global	Quantitative synthesis	Legume	Cash crop	+27%
					Non-legume	Cash crop	+6%
13	Toler et al. 2019	Until 2014	Global	Meta-analysis	Legume	Cotton	+11%
					Non-legume	Cotton	–10%
14	Florence and McGuire 2020	1995–2018	Global	Systematic review	Single CC species vs CC mixture	Cash crop	ns
15	Wang et al. 2021	1980–2020	Global	Meta-analysis	Legume	Cash crop	ns
					Non-legume	Cash crop	ns
					Mixture	Cash crop	ns
16	Abdalla et al. 2019	2017–2018	Global	Review	Legume	Cereal	–4%
					Non-legume	Cereal	ns
					Mixture	Cereal	+13%
17	Garba et al. 2022	1994–2021	Dryland	Meta-analysis	Legume	Cash crop	–4.9%
					Non-legume	Cash crop	–10.1%
					Mixture	Cash crop	–5.6%
18	Zhao et al. 2022	Until 2020	Global	Systematic review	Legume	Cereal	+21%
					Non-legume	Cereal	ns

19	Peng et al. 2024	Until 2023	Global	Synthesis	Legume	Cash crop	+22%
					Non-legume	Cash crop	+10%
20	Vendig et al. 2023	Until 2020	Global	Meta-analysis	Legume	Cash crop	+20%
					Non-legume	Cash crop	+9.8%
21	Osipitan et al. 2018	1990–2017	Global	Meta-analysis	Non-legume	Grain crop	ns
					Mixture	Grain crop	ns
					Non-legume	Vegetable	+2.9%
					Mixture	Vegetable	+5.9%
22	Snapp et al. 2005	Until 2004	USA	Review	Legume	Cash crop	+
23	Lavergne et al. 2021	2018–2020	North America and Europe	Review	Legume	Cash crop	+
					Mixture	Cash crop	-
24	Deines et al. 2022	2000–2021	USA	Meta-analysis	Cover crop vs no cover crop	Corn	–5.5%
						Soybean	–3.5%

+, -, and ns represent positive, negative, or non-significant effect of cover crops on the cash crop yield, respectively.
[z] relative to conventionally fertilized corn production systems.
[y] 'Cash crop' term was used when type of main crop was not specified.

2023; Daryanto et al. 2018; Deines et al. 2022; Nielsen et al. 2016; Sanchez et al. 2019; Tonitto et al. 2006; Qin et al. 2021). The conflicting results among the studies can be attributed to soil, agronomic, and climatic factors such as soil texture and fertility, CC species, main crop type, CC and main crop management, precipitation, and temperature, to name a few (Blanco-Canqui et al. 2015; Malone et al. 2022; Nielsen et al. 2016). A better understanding of contributing factors of CC effects on main crop yield is therefore needed. In this chapter, we will review the agronomic and pedoclimatic variables under which CCs either enhance or decrease crop yield. We will also highlight some of the potential mechanisms of CC-induced impacts on the subsequent crop yield along with five case studies from the USA and Canada that highlight regional differences of CC effects on crop yields.

In a particular agroecological environment, the above and belowground CC biomass quantity and composition (e.g. C:N ratio) depend on CC species and management (e.g. CC planting and termination time) and the pedoclimatic conditions. Cover crop biomass, in turn, affects the CC effect on the following crop's yield (Kaspar and Bakker 2015; Ruis et al. 2019). The selection of CC species, therefore, is largely dependent on the soil, climatic conditions, and agronomic management practices. A meta-analysis by Fan et al. (2021) confirmed that CC species and CC growth were key factors influencing CC effects on main crop yield. Additionally, the impact of CC on yield depends on the main crop grown and how it is managed (e.g. tillage) with soil and climatic conditions modulating overall effects.

2 Cover crop factors affecting yields: species, biomass accumulation and length of use

Legume CC species tend to increase the yield of non-legume crops following them (Table 1, Fig. 1, Bourgeois et al. 2022; Chahal and Van Eerd 2023; Daryanto et al. 2018; Marcillo and Miguez 2017; Peng et al. 2024; Tonitto et al. 2006). The yield gain with legume CCs is correlated with aboveground CC biomass and N content (Bourgeois et al. 2022; Chahal and Van Eerd 2023; Marcillo and Miguez 2017; Valkama et al. 2015). Bourgeois et al. (2022) reported that corn yield response to CC increased when the CC aboveground biomass N content increased from 50 kg N ha^{-1} to 200 kg N ha^{-1}. Relative to fall termination, spring termination of legume CC in regions where precipitation is not limited increased the main crop yield by allowing for a longer period of CC growth and increased accumulation of CC biomass and N. Likewise, several meta-analyses conducted in temperate climates (globally and USA) demonstrated that including legume CC in mixtures also contributed to a yield increase (Daryanto et al. 2018; Marcillo and Miguez 2017; Miguez and Bollero 2005).

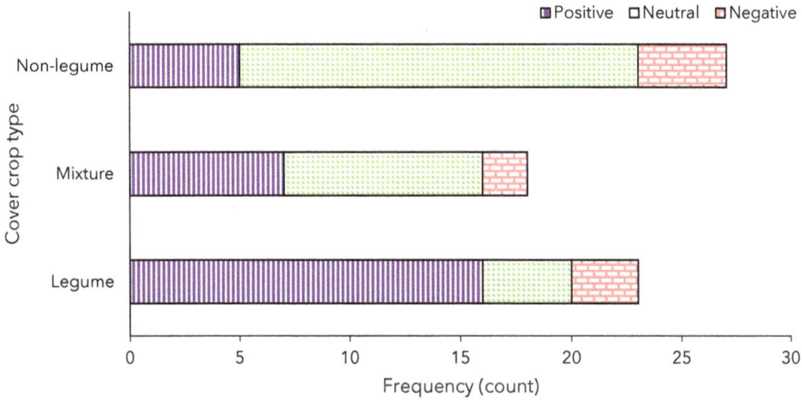

Figure 1 Frequency distribution of meta-analysis summarized cover crop effect on the main crop yield separated by cover crop type. Data presented is based on the crop yield effect reported in Table 1.

In contrast, grass and non-legume broadleaf CC species typically decrease or have no effect on the following non-legume crop yield (Blanco-Canqui et al. 2017, 2020; Daryanto et al. 2018; Marcillo and Miguez 2017; Miguez and Bollero, 2005; Ruis et al. 2017; Table 1). The decline in crop yield with non-legume CC species is mainly dependent on the CC termination time and duration of growth. Typically, large amounts of aboveground CC biomass correlate with increasing C:N ratios. Grass CC species have a wide C:N ratio at the reproductive growth stage which generally leads to N immobilization that decreases the main non-legume crop yield, especially when no supplemental N is applied (Marcillo and Miguez 2017; White et al. 2016).

2.1 Cover crop growth and biomass accumulation

In addition to species, CC growth and biomass accumulation have a large impact on the main crop yield. For instance, in humid regions with mild winters, successive CC is a better option to produce sufficient CC biomass and obtain main crop yield benefits than intercropped CC (Bourgeois et al. 2022). One example of successive CC planting in a temperate humid climate could be tomato-corn rotation, where tomatoes are harvested in early September, CC planted in mid-September, CC terminated in early May, and corn planted in mid-May (Chahal and Van Eerd 2021). Intercropping is another strategy of CC planting to obtain adequate CC biomass. It is a promising strategy, especially in regions with shorter growing seasons (such as Canadian Prairies), where successive CC planting might not be a feasible option to obtain sufficient CC biomass. An example of intercropping legume CC could be winter wheat undersown with red clover in a temperate humid climate (Gaudin et al. 2013).

Intercropping legume CC allows for a longer period of CC growth, high biomass production, and high N accumulation; thus, potentially resulting in yield benefits for the non-legume crop following it. Timely planting of CC allows enough time for CC growth and biomass accumulation before the frost and/or the onset of the winter season, specifically for the winter killed CC species like oats and radishes. High CC biomass, when returned to the soil, contributes to providing numerous benefits such as enhancing soil C storage, aggregate stability, and reducing nutrient losses via leaching and erosion. All these contributions are directly or indirectly linked to enhancing the main crop yield in the long-term (see Section 3).

Cover crop termination time also influences the amount of biomass produced by the CC, N fixed by the legume CC species, and the N release through decomposition of CC residue during the spring season. The amount of CC biomass production decreases with an early termination whereas late termination promotes an increase in biomass accumulation. The timing of CC termination also impacts the C:N ratio. For instance, early termination of grass CC species leads to less biomass production with a low C:N ratio, resulting in quick decomposition of CC residue. When CC are terminated late, it generally indicates that CC have reached the reproductive stage and have a high C:N ratio. Late termination of grass CC species, therefore, leads to N immobilization, reducing N availability for the main crop, and a potential decline in main crop yield. However, the decision regarding when to terminate the CC species is highly site specific and is dependent upon the type of the main crop grown. Generally, the winter CC species are terminated 2 to 4 weeks prior to the main crop planting in spring. This allows for sufficient time for the CC biomass to desiccate and decompose, and facilitates the next crop's planting; thus, allowing for a better seed-to-soil contact for seed germination and emergence with implications on increasing the main crop yield.

Cover crop root biomass is also a critical component to elucidate the CC effects on main crop yield. A data synthesis by Blanco-Canqui et al. (2020) demonstrated that the average CC root biomass (consisting of a wide range of CC species) in the upper 30 cm of the soil was 1.33 Mg ha^{-1}, which was about 42% of the amount of the aboveground CC biomass. Nonetheless, very few studies report and measure the root CC biomass, highlighting a major gap in the literature.

2.2 Short-term versus long-term use of CCs

It is generally thought that the positive effects of CC on main crop yield will increase over time due to soil health improvement. A study by Blanco-Canqui et al. (2015) and Chahal et al. (2020) observed that CC effects on the main crop yield are small and variable in the short-term (<5 years) and are generally observed in the medium- and long-term (>10 years). In the long-term, positive influences of CC on the soil organic matter, soil microbial biomass, soil N availability, and soil physical properties are some of the key mechanisms through which CCs contribute to yield increases in the next crop (Blanco-Canqui and Ruis 2020; Chahal et al. 2020; Haruna et al. 2020; Kim et al. 2020). However, studies documenting the effect of CCs on main crop yields over the long term are scarce (Langelier et al. 2021). Long-term CC use was therefore studied in a trial at Ridgetown, ON, Canada (see next Section).

3 Case studies 1 and 2: intercropping and long-term use of cover crops

3.1 *Case study 1: Intercropping cover crops in wheat and canola on the Canadian Prairies*

As shown above, high biomass accumulation of legume CCs or mixtures typically increases yields of the following crops. But a major challenge on the Canadian Prairies is the short period of growth available to plant and grow CCs after the harvest of the main cash crops wheat and canola, leading to low CC biomass. Oftentimes, producers are still harvesting crops when the first fall frost occurs. Therefore, the Universities of Alberta, Saskatchewan and Manitoba investigated whether CCs could be established in the main crops before harvest (Odiketa et al. 2024). One option is to broadcast CC seeds between rows of wheat or canola in late June, and another option is to broadcast or plant CCs at the same time as wheat or canola. If CC could be well established earlier in the season, they would then benefit immediately from the opening of the canopy of the cash crop at harvest and could then accumulate more biomass, and consequently provide environmental benefits more effectively. However, because of the semi-arid climate prevalent in the Canadian Prairies, there are also risks of the CC competing against the cash crop for soil moisture and reducing yields. Preliminary data suggested that the success of broadcasting CC seeds depended largely on whether they received timely rainfall shortly after being broadcasted. In addition, planting CC with the wheat crop (i.e. at the same time and in the same row) led to poor establishment of the CC, presumably because of the greater competitiveness of the wheat crop. In the prairies, row spacing for spring wheat tends to range from 10–12 inches.

In this experiment, row spacing for wheat was kept the same but additional rows for the intercropped CC were added five to six inches from the wheat seed row. Intercropping CC in alternate rows, however, led to the successful establishment of all four CC species tested, and indicated an opportunity to optimize alternate-row intercropping (Figs. 2 and 3).

Figure 2 Canola with intercropped clover cover crops seeded in alternating rows at the rosette stage (left), bolting stage (middle), and after harvest (right). Photo credit: Janelle Gawiak.

Figure 3 Wheat intercropped with clover, alfalfa, and phacelia at the tillering stage (left), flag leaf (middle), and at heading (right). Photo credit: Janelle Gawiak (left, middle) and Maryse Bourgault (right).

3.2 Case study 2: Long-term cover cropping in a humid temperate climate

In this experiment, the same annual CCs were grown since 2007 on site A and repeated at an adjacent site since 2008 (site B). This novel experiment is in a rotation that includes both horticultural (processing vegetables) and grain crops (corn, soybean, winter wheat) and is an ideal system to make direct comparisons of the various CC species (oat, cereal rye, radish, mix of radish and rye, and no-CC control) on soil, plants, and the environment (Chahal and Van Eerd 2021). Cover crops are planted after the main crop harvest, except after winter wheat or grain corn; hence, CC had been grown 12 times in 18 years by 2024. Typical management practices are used for main crops and CC. Most recently CCs were sown in August 2021 followed by corn, soybean, and winter

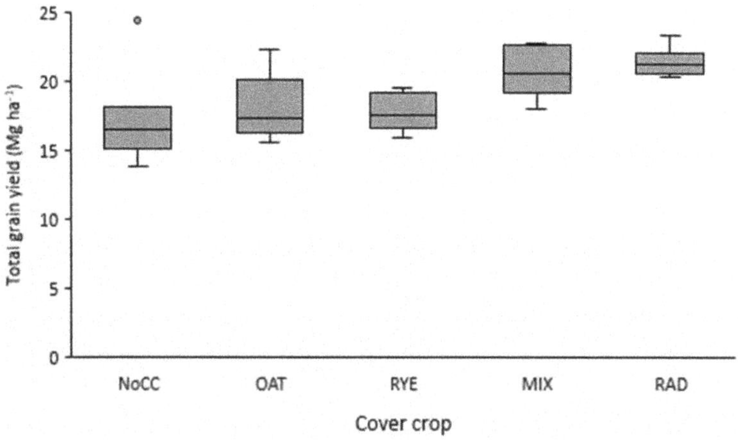

Figure 4 Effect of long-term cover cropping on total grain yield (corn, soybean, winter wheat) grown in 2022 to 2024, respectively, in a temperate humid climate (Ridgetown, ON, Canada). Boxes with different letters indicate statistical significance at $P < 0.05$, n = 40. NoCC = no cover crop control, OAT = oat, RYE = fall cereal rye, MIX = mixture of cereal rye and radish, RAD = radish.

wheat in 2022, 2023 and 2024, respectively. No fertilizer N was applied to corn (to study N contribution of CC) and soybean (as provincially recommended), but 105 kg N ha^{-1} was spring applied to the winter wheat crop. Both 2022 and 2023 main crop growing seasons were relatively dry (i.e. 13.3% and 13.8% less than the 30-year norm, respectively) and 2024 was wetter and cooler than normal.

To quantify the overall productivity of the rotation from 2022 to 2024, the sum of corn, soybean, and winter wheat was compared. Greater total yields with radish and a mix of radish and rye than no-CC control and grass CC (Fig. 4) were largely due to higher corn yields with the CCs and a slight, non-significant yield increase for soybean and wheat. Long-term CCing did not affect the measured seed quality attributes (i.e. %moisture, test weight) for winter wheat and soybean. Greater soil organic C content in the surface 15 cm of the soil with long-term CC (31.8 ± 1.42 Mg ha^{-1}) than the no-CC control (29.1 ± 1.64 Mg ha^{-1}; $P = 0.0563$) in 2022 is considered to be the major reason for increased grain yields with CCs (Peng and Van Eerd 2024).

4 Crop management factors affecting yields: crop type and tillage

4.1 Main crop type

Cover crop effects on the subsequent crop yield are dependent on the type of main crop. Vegetable crops are generally more responsive to CC than grain crops (Chahal and Van Eerd 2023; Chahal et al. 2020; Chehade et al. 2019; Marcillo and Miguez 2017). Depending on the vegetable crop, they are harvested between August to September in temperate humid climates and planted later in the spring than grain crops. Therefore, compared to grain crops, vegetable production systems provide a greater opportunity for timely CC planting, better establishment, and greater biomass (specifically for legume CC), which perhaps is a major factor demonstrating the success of CC on vegetable crop yields. Among the grain crops, several studies have reported that the legume CC increased the yield of small grain cereals and corn (Bourgeois et al. 2022; Daryanto et al. 2018; Marcillo and Miguez 2017) but not of soybeans (Bourgeois et al. 2022; Chahal and Van Eerd 2023; Dozier et al. 2017; Peterson et al. 2019). The lack of CC response on soybean yield was attributed to N fixing ability of soybean.

Studies by Daryanto et al. (2018), Miguez and Bollero (2005), Peng et al. (2024), and Tonitto et al. (2006) demonstrated that the application of nitrogenous fertilizers to main crops reduced the positive effects of legume CC on the subsequent crop yield. The decline in main crop yield with grass CC was reduced when N fertilizer inputs were increased to the main crop. Grass CC negatively impacts the next crop yield mainly due to N immobilization, so with N fertilizer application to the main crop, N availability is no longer a limitation for the main crop yield, and the other variables such as temperature, soil moisture, weed infestation etc. become the major yield-limiting factors (Peng et al. 2024). Additionally, N application through fertilizers negates the N fertility-induced benefits of legume CC; hence, the positive effect of legume CC is smaller when the main crop is fertilized (Bourgeois et al. 2022). Overall, the interactive effects of N fertility and CC on the main crop yield are largely dependent on the type of main and CC species; hence, a site-specific approach of N fertilizer inputs based on the CC type is crucial to optimize the CC benefits on the main crop yield.

4.2 Tillage system

Tillage is another moderating variable influencing the CC benefits on the main crop yield. Reviews by Blanco-Canqui et al. (2015) and Marcillo and Miguez (2017) have reported that crop yield following CC increased in no-tillage systems. The benefits of CC on crop yield were more pronounced in no-tillage mainly due to the changes in soil physical, chemical, and biological characteristics associated with greater surface residue in no-tillage than plow tillage systems (Blanco-Canqui et al. 2015). Additionally, soil moisture conservation, due to the CC residues left on the soil surface, particularly in no-tillage systems, might be beneficial for the next crop yield, specifically in dry regions or during the dry summer season. Cover crop residues tend to have a narrower C:N ratio (i.e.

14:1) than grain crop residues (e.g. corn approx. 80:1), so the addition of these residues, particularly on the soil surface in no-tillage systems, may accelerate the decomposition of main crop residues. Plow tillage disrupts soil aggregates and accelerates the decomposition by increasing soil-to-residue contact, resulting in a rapid release of N (specifically upon mineralization of leguminous CC residue), which may or may not be in synchrony with crop N uptake and crop yield. Plow tillage is a physical method of termination, whereas no-tillage systems rely on chemical or frost termination. Regardless of the system, improper CC termination decreases main crop yields through competition, and the risk of incomplete termination is greater in no-tillage systems.

5 Soil and climatic conditions modulating cover crop effects on yields

5.1 Soil texture

Crop yield enhancement with CC tends to be greater in coarse than medium and fine textured soils (Chahal and Van Eerd 2023; Daryanto et al. 2018; Peng et al. 2024). As reported by Cordeiro et al. (2021) and Rawls et al. (2004), adopting CC in fine textured soil led to yield penalties. Yield advantages with CC in coarse textured soil are not entirely clear but might be attributed to better crop establishment, soil aeration, less compaction, and quick drying and warming of soil during the spring season under coarse than fine and medium textured soils. Enhanced soil moisture conservation associated with CC residues decreasing evaporation and increasing moisture holding capacity might also be a contributing factor for increasing main crop yield specifically in coarse textured soils. Regardless, caution is needed while developing strategies to implement CC in fine textured soils.

5.2 Climate

Temperature and precipitation are strong modulators of CC-induced effects on the next crop yield. However, the CC performance under different climatic conditions is dependent on the CC species. Legume CC generally increased main crop yield (Bourgeois et al. 2022). Climate conditions (specifically precipitation) had a greater influence on the main crop yield following grass CC species than legumes and mixtures of CC in a temperate humid climate (Bourgeois et al. 2022). For instance, under dry conditions (<500 mm), grass CC species decreased grain corn yield (Blanco-Canqui et al. 2015; Bourgeois et al. 2022). The corn yield decline with grass CC under dry conditions (<500 mm) is not entirely clear but might be attributed to the slow degradation and leaching of allelopathic compounds from the grass CC residue (mainly cereal rye) (Bourgeois et al. 2022). The immobilization of N due to grass CC was also found to be greater under low precipitation conditions (Blanco-Canqui et al. 2015). While the magnitude of yield decline following grass CC decreased with an increase in precipitation, the yield lag due to grass CC even under high precipitation conditions in temperate humid climate might be attributed to increased biomass production possibly interfering with planting or leading to N immobilization (if terminated late in the spring). Likewise, in CC mixtures of legume and grass, the yield benefits of legume CC are counteracted due to an increased N immobilization and reduced N inputs from legume biomass to the main crop associated with a greater growth of grass CC species in the mix under high rainfall conditions in temperate humid climate (Bourgeois et al. 2022). Thus, small grain cereal and corn yields are not affected or decreased with grass CC, regardless of rainfall. Nevertheless, future research to elucidate the processes driving the CC effects on crop yield under different moisture regimes is needed.

6 Case study 3: cover crops as a fallow replacement

In the USA, CCs are being promoted as an environmental practice to replace fallow in cereal-based production systems. In the semi-arid Northern Great Plains, CCs are not necessarily grown for the full season but often terminated in early July to conserve soil moisture for the winter wheat that will be planted in September. A 9-year phased experiment was established in 2012 at the Northern Agricultural Research Center of Montana State University to evaluate the productivity of various CC mixtures and their potential impact on the subsequent wheat yields, with CC-spring wheat or winter wheat rotations. This was conducted on two adjacent fields, and CCs were further terminated by haying (i.e. cut and bailed), grazing by livestock, or chemical termination with herbicides. Results showed that CC biomass was greater in the mixtures of cool-season species (such as turnip, radish, vetch, oat, and pea) compared to warm-season species (such as sorghum-sudan grass, millet, and sunflower) or mixtures containing both cool and warm-season species (Wyffels et al. 2021). Yields of wheat following CC were decreased in 3 out of 8 years, by as much as 1.4 Mg ha^{-1}, or 60% in one year (Bourgault et al. 2022), which was attributed to CC management (i.e. CC species), and highlights some of the production risks associated with growing CC in environments with limited precipitation.

7 Potential mechanisms in cover crop effects on yields: soil conditions and water availability

Cover crops can have no effect, increase or decrease the next crop yield (Table 1). The variability in the CC-induced crop yield effects could be attributed to numerous factors such as the CC and crop agronomic and pedoclimatic variables. The key mechanisms through which CCs influence subsequent crop yield are complex. Maximizing yield in any cropping system is not easy and including CC increases complexity and management. Some underlying processes governing the CC influence on main crop yield are summarized in Table 2. For each mechanism, it is important to note that CC can have a positive or negative influence on crop productivity and the direction of influence varies with both CC and main crop management as well as growing conditions (e.g. weather, and soil type). Obtaining a good CC establishment and growth is crucial for achieving the main crop yield benefits. Results of three North American CC surveys (CTIC and SARE 2023; Morrison and Lawley 2021a,b) revealed that farmers consider CC failure or poor establishment as the main agronomic challenge in the USA, the Canadian Prairies, and Ontario (Fig. 5), which perhaps contributed to no differences in main crop yield.

The ability of CCs to significantly alter soil physical, chemical, and biological characteristics has been well investigated in the literature (Blanco-Canqui et al. 2015; Daryanto et al. 2018). Cover crops can help in improving soil structure, soil porosity, aggregation, and reducing compaction (Blanco-Canqui et al. 2015). All these benefits, generally observed in the long term, help to provide better soil conditions for crop growth, increase soil tilth, and increase crop yield (Blanco-Canqui et al. 2015). Nonetheless, too much crop residue, particularly with overwintering CC species could reduce the seed-to-soil contact leading to poor planting conditions, impacting establishment and subsequent yield decline.

Soil chemical and biological properties such as soil organic matter and soil microbial activity improve with CCs (Blanco-Canqui et al. 2015; Vendig et al. 2023). The above and belowground biomass C inputs from CCs are major contributors to increasing soil organic matter in the long-term. Root exudates and belowground CC C inputs also increase soil biological health by increasing soil microbial biomass, for example. Soil organic matter provides energy for the soil microbial activity. Soil microbes eventually mineralize soil organic matter, drive the cycling of nutrients such as N, P, and K, and release nutrients for plant uptake (Vendig et al. 2023); hence, it contributes to the yield increase of the main crop.

Influence of CC on soil N fertility is another mechanism through which CCs either increase or decrease the next crop yield. Cover crops can scavenge the fall residual N, convert it into organic forms, and gradually release N during decomposition in the spring season. Overwintering legume CC such as red clover and hairy vetch have a greater potential in imparting next crop's yield gains than the winter killed legume CC. Overwintering legume CCs have a longer period of growth and can accumulate greater aboveground biomass N than winter killed legume CC species. When the CC residue undergoes mineralization in the spring season, there is an enhanced potential for soil N availability to the main crop. Depending on the synchronization of CC residue

Table 2 Potential mechanisms explaining the impact of cover crops (CC) on the main crop yield.

Mechanism	Yield gains (+) or losses (−)		Common CC type[z]	Considerations	Reference(s)
Water	Conservation	+	Reproductive grass	Long-lasting residues on soil surface act as mulch reducing evaporation and soil temperature	Blanco-Canqui et al. 2015
		–	Overwintering	Spring surface residues limit evaporation and soil warming; thus, delaying planting	Sharma et al. 2018
	Deficits	–	Region specific	Semi-arid regions- winter precipitation replenishes soil moisture but cover crops deplete water	Nielsen et al. 2015
Weed control	Competition	+	Brassica (radish)	Quick emergence and growth	O'Reilly et al. 2011
	Physical barrier or mulch	+	Cereal rye or mix of cereal rye and hairy vetch	Roller crimper = anthesis	Camargo Silva and Bagavathiannan 2023
	Allelopathy	+/–	Cereal rye	Could injure crop (e.g corn)	
Weed problem	Cover crop could become a weed	–	Overwintering species (Cereal rye, red clover)	Improper termination	Baraibar et al. 2018
			Other broadleaf (Buckwheat)	Quick growing annuals set seed or long cover crop growing season	
			Hairy vetch	Hard seed does not germinate until next season	

Insect	Suppression	+	Potentially enhance soil food web supporting high population of predators		Acharya et al. 2021; Beaumelle et al. 2021
	Support pollinators	+n	Reproductive cover crop (buckwheat, sunflowers clovers, brassicas)	Long growing season to reach flowering	Beaumelle et al. 2021
	Host/attract pests	–	Cereal rye	Army worm	Dunbar et al. 2016
Disease	Suppression	+	Brassica (Radish) and grass cover crop	Trap for plant pathogens; Cover crops enhance soil health and/or fertility, selecting different cover crop species introduces a break in disease cycle	Harbach et al. 2021; Davis et al. 2010; Stone 2012; Trueman et al. 2023
	Green bridging	–	Cereal rye	Corn	Acharya et al. 2017
Nematode	Suppression	+	Oat and cereal rye	Highly species specific (nematode, cover crop, and main crop)	Acharya et al. 2021
	Host	–	Legume		Chen et al. 2006
Soil physical characteristics	Planting conditions	+	Buckwheat	Increase soil tilth	Garcia-Gonzalez et al. 2018
		–	Overwintering cover crop	Too much residue reduces seed to soil contact, cannot close trench	Kichler et al. 2023
	Compaction reduction	+	Radish		Chen and Weil 2010

(*Continued*)

Table 2 *Continued*

Mechanism	Yield gains (+) or losses (−)		Common CC type[z]	Considerations	Reference(s)
N fertility	Mineralization	+	Overwintering legumes (Red clover, hairy vetch)		Finney et al. 2016
		+	Grass (Cereal rye)	Manage as green manure (incorporate living vegetative stage)	
	Immobilization	−	Reproductive stage grasses (Cereal rye)	Less likely with no-till	Acharya et al. 2017
Soil chemical characteristics	Brings nutrients to surface	+n		Deep rooted CC (red clover, radish)	Cober et al. 2019; Lozier et al. 2017
		−n		Nutrients prone to erosion and run off	
Soil biological characteristics	Support earthworm population	+n	Alfalfa and red clover	Quality of cover crop residue controls the earthworm population dynamics	Blanco-Canqui et al. 2011
	Support soil microbial and enzyme activity	+n	Legume and grass mixture cover crop	Influences the nutrient cycling and availability	Fernandez et al. 2016

[z] represents the most likely cover crop species, functional group, life cycle or winter survival (common examples are in parenthesis) to observe an effect on crop yield. n not likely to influence yields directly.

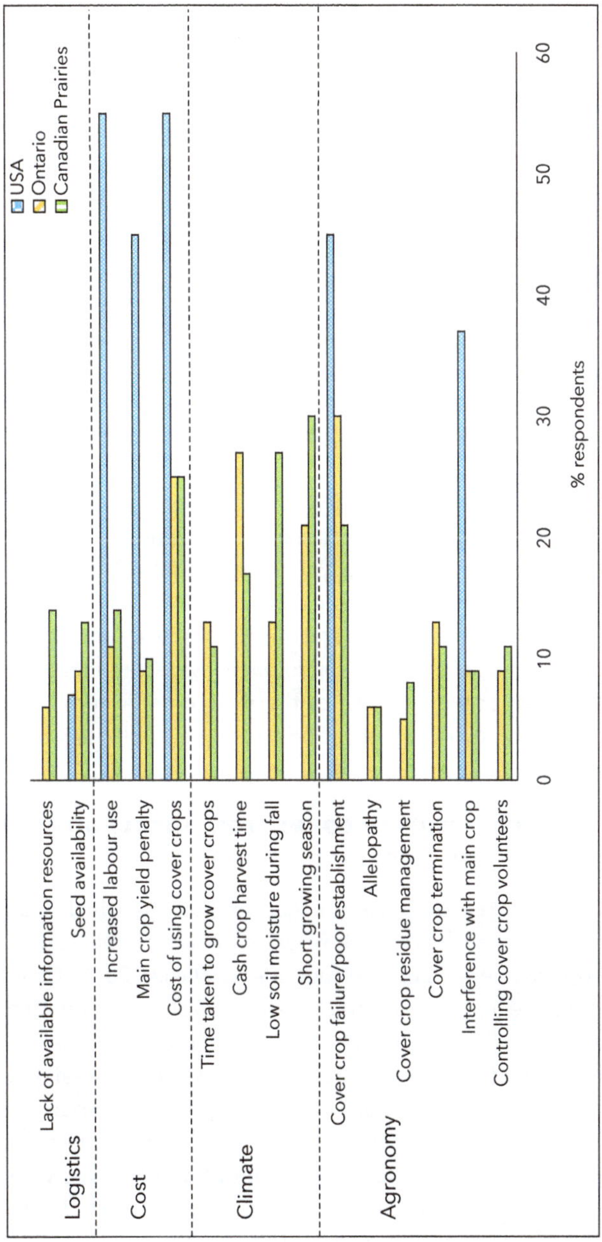

Figure 5 Grower identified challenges related to cover crop adoption based on the survey respondents in Canadian Prairies (n = 281), Ontario (n = 520), and USA (n = 82). Adapted from Morrison and Lawley 2021a 2021b, CTIC and SARE (2020), respectively.

mineralization and crop N uptake, CC can reduce the reliance on external N inputs and can contribute to crop yield increase. In contrast, delayed spring termination of overwintering grass CC could result in N immobilization and reduce the soil N availability to the next crop. The main crop N stress eventually results in a crop yield decline, especially without N fertilizer additions (Abdalla et al. 2019; Deines et al. 2022).

Water availability is a crucial factor governing the main crop yield response to CC (Peng et al. 2024). Overwintering CC species (mainly grass) residue acts as a mulch on the soil surface which helps to increase water infiltration and reduce evaporation and soil temperature; thereby increasing crop water availability, leading to main crop yield gain. On the contrary, CCs may reduce available soil water content for the next crop in the season because of increased evapotranspiration while the CC is growing (Adil et al. 2022; Deines et al. 2022). However, if soil moisture is excessive in spring, CCs can help with the timely planting of the next crop in the spring due to the removal of soil moisture via evapotranspiration. However, if dead CC residue covers the soil surface, this reduces evaporation and soil drying, leading to a delayed planting of the main crop. Additionally, delayed soil drying due to spring CC residue reduces oxygen supply during the early growing season, especially in fine textured soils in temperate climates (Deines et al. 2022; Peng et al. 2024). The residue of dead CC residue on the soil surface helps to conserve soil water, but in wet springs, the additional soil water enhances the soil microbial activity and aggravates the oxygen depletion in the soil. The soil microbial activity, stimulated by the CCs, consumes the oxygen at the expense of the main crop, in turn, stressing the main crop in the early growing season, and causing a yield loss.

8 Potential mechanisms in cover crop effects on yields: weed, disease and pest management

Weed control with CC is a potential mechanism through which CC indirectly impacts the nutrient availability to the next crop with implications for yield enhancement (Bhaskar et al. 2021; Lemessa and Wakjira 2015; Osipitan et al. 2019). Cover crop species such as radish have a quick emergence and growth and contribute to weed control mainly via competition with weed species for soil nutrients and water (O'Reilly et al. 2011). The mixture of cereal rye and hairy vetch contributes to reducing weeds via providing a physical barrier or acting as a mulch and reducing weed seed emergence. Cereal rye can also control weeds through the release of allelopathic compounds. Improved weed control with CCs may increase the next crop yield. Conversely, the CC itself could also become a weed and cause a decline in main crop yield if not completely terminated, being of particular concern in organically managed systems that do not allow herbicides. For instance, improper spring termination, particularly of overwintering CC species (cereal rye, red clover) competing with the main crop. Cover crop seeds germinating during the main crop season are weeds and most often are a result of hard seed (hairy vetch), or quick-growing annual CC species which set seed (buckwheat). Under these scenarios, CC can decrease the next crop yield.

Disease, nematode, and insect suppression with CC have been demonstrated as a pathway to increase subsequent crop yield. Studies by Acharya et al. (2021) and Harbach et al. (2021) have reported that brassica and grass CCs can act as a trap for plant pathogens, such as soybean cyst nematode, thereby leading to reduced pest populations, improved plant health, and ultimately increasing subsequent crop yield. Oat and cereal rye CC have helped to reduce the infestation of root-knot nematodes (Wang et al. 2004). Long-term CC usage reduced disease pressure in tomatoes, potentially via increasing soil health (Trueman et al. 2023). Integrating CC, such as brassica (radish) and Sudan grass, in crop rotations also helped to break the disease cycle of verticillium wilt in potatoes (Davis et al. 2010) and root rot in corn (Stone 2012). Cover crops that provide floral resources, such as buckwheat, sunflower, and brassica (camelina) help to attract and support the bees and other beneficial insects but the direct link to crop yields is tenuous.

Conversely, some CC species could act as a 'green bridge' host for insect pests and diseases negatively impacting the next season's main crop yield. For example, stripe rust survived on grass CC species like triticale in central Alberta, Canada and negatively affected the following winter wheat crop (Xi et al. 2015). Cereal rye has also acted as a green bridge for several diseases and insect pests (e.g. armyworm) infecting the following corn crop (Dunbar et al. 2016). Cover crops, mainly legume CC, could also act as a host for nematodes and increase their population, ultimately reducing main crop yield (Ibrahim Garba et al. 2024). For example, alfalfa, hairy vetch, and clovers are very susceptible to root-knot nematode infestation which could increase the nematode's population and decrease the yield of susceptible crops that follow these crops in rotation.

9 Case study 4: incorporating cover crops in organic sweet potato systems using surface residues and incorporation

Sweet potato (*Ipomoea batatas*) is an economically important crop for North Carolina, which grows 40-50% of the US sweet potatoes. Sweet potatoes are often grown in the coastal plain region of the Southeastern US which is characterized by coarse textured soil with low soil organic matter content (Ultisols). The crop is grown with full field tillage followed by formation of hilled ridges in which vines are planted. These hilled systems help improve drainage and aeration and increase sweet potato root growth and yield while in-season cultivation is used for weed control. There is a growing interest in improving the sustainability of sweet potato production by including CC for enhanced soil organic matter, nutrient supply and weed control. However, historically CC have not been used due to perceived risks that a CC will cause a 'greenbridge' for wireworms (Coleoptera: Elateridae), which can drastically reduce marketable yields. To address how CC might be included before sweet potato production two approaches were taken: (1) surface residue CC and (2) incorporated CC.

9.1 Surface residue cover crops

The potential benefits of roller-crimping CC and leaving the residue on the surface include: erosion control, weed suppression and reduced evaporation. However, their inclusion in hilled systems prevents in-season cultivation, important for weed control on both conventional and organically managed farms. A 3-year organic sweet potato study conducted in 2001 (Treadwell et al. 2007) evaluated the use of a cereal rye-hairy vetch mix grown either on flat ground and incorporated in the spring or sown on fall-formed hills and terminated and left on the surface (reduced tillage treatment). These treatments were compared to conventionally and organically managed CC-free controls. Cover crop biomass was significant ranging from 7410 to 13 800 kg ha^{-1} with an estimated 70% of the biomass derived from cereal rye. In two of 3 years, there were no differences in marketable yield between treatments. In 1 year, however, yields in the reduced tillage treatment were 60% lower compared to the average of the remaining treatments. In all treatments, excluding the reduced tillage, in-season cultivation is possible until the vines run into the alleys providing critical weed control. In contrast, once committed to the surface CC residue for weed control there is no ability to cultivate if weeds become dominant, highlighting the need to understand CC residue biomass requirements to ensure season-long weed control. A similar study confirmed that low biomass CC (1400–3000 kg ha^{-1}) resulted in increased weed infestation and despite hand weeding resulted in a near total loss of marketable yields (1091 kg ha^{-1}) compared to in-season cultivated plots (15 497 kg ha^{-1}). In the southeast, experiments in corn and soybean have demonstrated that for cereal rye to be an effective weed control, CC biomass must be above 8000 kg ha^{-1} (Mirsky et al. 2013).

9.2 Incorporated cover crops

In this system, CC are drilled into flat ground before being terminated in the spring and incorporated into hills when formed (Fig. 6). Importantly, in-season cultivation is possible for weed management. In a 4 site-year study (Kile 2022; Pellegrino et al. 2021) conducted in the coastal plain of North Carolina, crimson clover (*Trifolium incarnatum*), winter pea (*Pisum sativum*) and cereal rye were

Figure 6 Roller-crimping cereal rye in a hilled system at the Center for Environmental Farming Systems in Goldsboro, NC, USA.

evaluated for their impact on nitrogen dynamics, wireworm prevalence and marketable yield in organically managed 'Covington' sweet potato. These treatments were compared to a range of nitrogen rate treatments from 0% to 150% (96 kg N ha^{-1}) of the recommended rate of nitrogen to sweet potato to determine the fertilizer equivalency value of the CC. Cover crop biomass was variable ranging from 1350 to 8274 kg ha^{-1} for the legume CC with 27 to 100 kg N ha^{-1} in their biomass. Marketable yields were the same between the CC treatments at 23.7 Mg ha^{-1} and comparable to the low nitrogen application rates (0 to 60 kg ha^{-1}), despite the nitrogen content in the CC biomass often approaching nitrogen requirements.

Two key findings from this study indicate the importance of timing of CC termination. Sweet potatoes were planted in early June and CC termination occurred in the middle of May. This is often a month later than normal winter CC termination in the southeast. By mid-May crimson clover was very mature with C:N ratios as high as 50:1 and winter pea 30:1. This late termination will result in significantly different nitrogen release compared to a low C:N ratio in early spring terminated CC. In addition, the high heat and precipitation in May caused most of the nitrogen from the CC to be mineralized before the sweet potatoes were planted. Thus, in the southeast, the existing research on CC in corn and soybean systems may not be representative of how best to optimize the use of CC in late-planted crops, calling for further research. Lastly, there was no evidence of increased wireworm damage from CC to the sweet potato (Pellegrino et al. 2021), which had been an important barrier to CC adoption in these systems.

10 Case study 5: on-farm research and adaptation of cover crop management

On-farm research compliments field experiments conducted at experiment stations to drive cropping system innovation and adoption of CC. Testing CC with farmers in commercial fields with large scale equipment, regional crop rotations, and real word logistics for seeding, crop protection, and harvest can also help to identify adoption barriers and feed new questions into research and policy development. Cover crop adoption in the Canadian Prairies is in the early adoption phase. Key barriers to CC adoption identified by Morrison and Lawley (2021b) in their survey of early adopters included the short growing season and limited fall moisture to establish CC.

On-farm experiments in five commercial farm fields in the eastern Canadian Prairies (Manitoba) were conducted over a 3-year period from 2019–2021 to identify where CC fit into annual grain crop rotations. Out of 17 site years, CC were established in 13 site-years. The inclusion of cereal crops, such as spring wheat and oat, in the rotation was important for having a successful window to establish CC, while longer season crops, such as canola, soybean, and edible bean, were harvested too late for establishing CC in Manitoba. With an average of 660 GDD accumulated between CC seeding and the first fall frost, the average CC biomass across the 13 sites years was 780 kg ha^{-1} and provided an average ground cover of 30%. Limited biomass production due to drought conditions during the study leads farmers to adapt their CC management. These adaptive strategies included switching from CC species that winter killed to those that would overwinter and grow additional biomass in the spring, intercropping CC with cash crops to establish them when more soil moisture was available to germinate CC seeds (see Case study 1), and utilizing voluntary emerging seeds from the previous crop and crop regrowth to compliment CC as a strategy to increase ground cover.

11 Future research trends

Additional research on the following topics might help to discern the CC potential to enhance cash crop yields or mitigate yield losses. Research at multiple scales will be key to driving the adaptation and adoption of CC across regions.

1. Basic CC agronomic research is needed. Development of region and cropping system-specific CC management strategies to optimize CC benefits on crop yields. Research studies on introducing CC into alternative production systems (Norris and Congreves 2018), CC mixtures and the amount of biomass accumulation (Keene et al. 2017) and altering the crop production systems to allow for a longer CC growing season such as by planting short-duration main crop cultivars or early planting of main crop (Ruis et al. 2019) are needed. Studies reporting the belowground CC biomass and its influence on the next crop yield are limited, suggesting the need for future research.
2. Most of the CC research has been conducted in grain production systems (such as corn and soybean systems) with less emphasis on vegetable cropping systems. Vegetable systems have degraded soil health due to intensive tillage and reduced crop residue return to maintain organic matter content. Therefore, there is a large potential for future research targeting the implementation of CC into vegetable systems and understanding the CC-induced effects on vegetable crop yields.
3. Research into breeding and selecting CC cultivars that perform different functions and/or to fit better into cropping systems is needed. For instance, CC to provide specific ecosystem services (e.g. disease suppression) and associated yield benefits is needed. Breeding CC to express greater root exudates and rhizodeposits would benefit the soil microbiome (Tosi et al. 2021), thus, contributing to increased nutrient availability to the crop, possibly increasing crop yield. Shade-tolerant CC cultivars would allow for successful interseeding of CC in corn without yield penalties (Belfry and Van Eerd 2016).
4. Implementing CC along with additional best management practices with an aim of understanding the synergistic interaction between CC and sustainable soil and crop management strategies. For instance, field research combining CC and 4R nutrient stewardship practices to optimize yield gains and fertilizer use efficiency is needed.
5. Long-term research is needed to fully understand and discern the CC effects on the next crop yield. Cover crop effects on soil health and ecosystem services are not evident in the short term (≤5 years). Nevertheless, very few long-term CC experiments exist in North America (Norris et al. 2022); thus, highlighting the need for further research.
6. Beyond yield benefits, the economics of CC is a major challenge identified by farmers in the USA and Canada (Fig. 5). Comprehensive economic analysis is needed to better understand the trade-offs with CC and to identify the best CC management strategies to maximize the yield benefits.

7. The use of technology and decision support tools to increase yield gains from CC. For instance, CC mixes that contain legumes and grasses create a heterogenous N supply as species composition varies within the field. Using machine learning to differentiate CC species and biomass to determine N supply variations within the field and subsequently map out precision N application rates will be essential to reduce inorganic N inputs. In addition, choosing the appropriate CC for a specific soil, climate and cropping system is a significant contributor to variability in CC success in improving crop yields. Developed decision support tools such as the Cover Crop Species Selection Tool (https://covercrop-selector.org/) allow growers to find the best CC species based on goals, soils, main crops, and climate. Social science-based research coupled with agronomic research needs to be able to quantify this continual improvement of site-specific tools for growers and its relationship with more consistent yield improvement.

12 Conclusion

In this chapter, we have provided a comprehensive assessment of the various agronomic, soil, and environmental factors that influence CC impacts on the subsequent crop yield. We demonstrate that the effects of CC on the next crop yield are variable and largely controlled by the CC species type, regional climatic conditions, and the main crop production system. Legume CC species tend to increase the following crop yield, while grasses and non-legume broadleaf CC species either decrease or have no effect on the main crop yield.

As demonstrated by the case studies, CC increased crop yield in a long-term CC experiment in a temperate, humid climate (Ridgetown, Ontario, Case study 2), which is characterized by sufficient total annual precipitation. In contrast, CCs reduce or have a higher risk of reducing next crop yield in semi-dry environments like the Canadian Prairies (Case studies 1, 3, and 5). Furthermore, the short growing season observed in the Prairies (USA and Canada) is another limiting factor regarding CC adoption and its success on the next crop yield. Under these conditions, strategies such as intercropping or broadcasting CC might be beneficial in ensuring successful CC establishment, growth, and biomass production.

Another barrier to CC adoption is the potential of CC to act as a host of insect pests and diseases and negatively impact the subsequent crop yield. However, the CC research in organic sweet potato systems in the southeastern USA did not find any evidence of increased insect damage with CC (Case study 4). Our review of CC-induced effects on the next crop yield also demonstrated that CC adoption is well suited in vegetable production systems in temperate humid climates. Vegetable crops are harvested at the end of August to early

September, allowing for sufficient time for CC establishment, growth, and biomass accumulation till the next spring. While CC have the potential to improve the next crop yield in the long-term, future research targeting the strategies and policies to overcome the barriers limiting CC adoption under diverse climatic and agronomic conditions is needed.

13 Where to look for further information

Readers are suggested to explore some of the review articles and meta-analyses (listed below) discussing the impact of cover crops on the following crop yield.

- Blanco-Canqui, H., Shaver, T. M., Lindquist, J. L., Shapiro, C. A., Elmore, R. W., Francis, C. A., and Hergert, G. W. 2015. Cover crops and ecosystem services: insights from studies in temperate soils. *Agron. J.* 107: 2449–2474.
- Chahal, I., and Van Eerd, L. L. 2023. Do cover crops increase subsequent crop yield in temperate climates? A meta-analysis. *Sustainability*. 15: 6517.
- Peng, Y., Wang, L., Jacinthe, P.-A., and Ren, W. 2024. Global synthesis of cover cropimpacts on main crop yield. *Field Crops Res*. 310: 109343.

Additionally, the references listed in Table 1 of this chapter are a great source to learn more about the mechanisms and potential impact of cover crops on main crop yield.

14 References

Abdalla, M., et al. 2019. A critical review of the impacts of cover crops on nitrogen leaching, net greenhouse gas balance and crop productivity. *Glob. Chang. Biol.* 25: 2530–2543.

Acharya, J., et al. 2017. Time interval between cover crop termination and planting influences corn seedling disease, plant growth, and yield. *Plant Dis.* 101: 591–600.

Acharya, K., et al. 2021. Effects of cover crops on population reduction of soybean cyst nematode (*Heterodera glycines*). *Plant Dis.* 105: 764–769.

Adil, M., et al. 2022. Effects of fallow management practices on soil water, crop yield and water use efficiency in winter wheat monoculture system: a meta-analysis. *Front. Plant Sci.* 13. https://doi.org/10.3389/fpls.2022.825309.

Allam, M., et al. 2023. A meta-analysis approach to estimate the effect of cover crops on the grain yield of succeeding cereal crops within European cropping systems. *Agron*. 13:1714.

Alvarez, R., et al. 2017. Cover crop effects on soils and subsequent crops in the pampas: a meta-analysis. Soil Till. Res. 170: 53–65.

Baraibar, B., et al. 2018. Growing degree days and cover crop type explain weed biomass in winter cover crops. *Agron. Sustain. Dev.* 38: 1–9.

Beaumelle, L., et al. 2021. Benefits of increased cover crop diversity for predators and biological pest control depend on the landscape context. *Ecol. Solut. Evid.* 2: e12086.

Belfry, K. D. and Van Eerd, L. L. 2016. Establishment and impact of cover crops intersown into corn. *Crop Sci.* 56:1245–1256. doi:10.2135/cropsci2015.06.0351.

Bhaskar, V., et al. 2021. Integrated management of living mulches for weed control: a review. *Weed Tech.* 35: 856–868.

Blanco-Canqui, H., et al. 2011. Addition of cover crops enhances no-till potential for improving soil physical properties. *Soil Sci. Soc. Am. J.* 75: 1471–1482.

Blanco-Canqui, H. and Ruis, S.J. 2020. Cover crop impacts on soil physical properties: a review. *Soil Sci. Soc. Am. J.* 84: 1527–1576.

Blanco-Canqui, H., et al. 2015. Cover crops and ecosystem services: insights from studies in temperate soils. *Agron. J.* 107: 2449–2474.

Blanco-Canqui, H., et al. 2017. Aerial interseeded cover crop and corn residue harvest: soil and crop impacts. *Agron. J.* 109: 1344–1351.

Blanco-Canqui, H., et al. 2020. Harvesting cover crops for biofuel and livestock production: another ecosystem service? *Agron. J.* 112: 2373–2400.

Bourgault, M., et al. 2022. Introducing cover crops as fallow replacement in the Northern Great Plains: II. Impact on following wheat crops. *Renew. Agric. Food Syst.* 37: 303. https://doi.org/10.1017/S1742170521000508.

Bourgeois, B., et al. 2022. Interactive effects between cover crop management and the environment modulate benefits to cash crop yields: a meta-analysis. *Can. J. Plant* Sci. 102: 656–678. https://doi.org/10.1139/CJPS-2021-0177.

Camargo Silva, G. and Bagavathiannan, M. 2023. Mechanisms of weed suppression by cereal rye cover crop: a review. *Agron. J.* 115: 1571–1585.

Chahal, I. and Van Eerd, L.L. 2021 Cover crops increase tomato productivity and reduce nitrogen losses in a temperate humid climate. *Nutr. Cycl. Agroecosyst.* 119: 195–211.

Chahal, I. and Van Eerd, L.L. 2023. Do cover crops increase subsequent crop yield in temperate climates? A meta-analysis. *Sustainability*. 15: 6517.

Chahal, I., et al. 2020. Cumulative impact of cover crops on soil carbon sequestration and profitability in a temperate humid climate. *Sci. Rep.* 10: 13381.

Chehade, L.A., et al. 2019. Evaluation of the agronomic performance of organic processing tomato as affected by different cover crop residues management. *Agron.* 9: 504.

Chen, G. and Weil, R. R. 2010. Penetration of cover crop roots through compacted soils. *Plant Soil*. 331: 31–43.

Chen, S., et al. 2006. Effect of cover crops alfalfa, red clover, perennial ryegrass, and rye on soybean cyst nematode population and soybean and corn yields in Minnesota. *J. Nematol.* 38: 267.

Cober, J.R., et al. 2019. Winter phosphorus release from cover crops and linkages with runoff chemistry. *J. Env. Quality*. Special Section: agricultural water quality in cold environment. 48: 907–914. doi:10.2134/jeq2018.08.0307.

Cordeiro, C., et al. 2021. Cover crop increases soybean yield cropped after degraded pasture in sandy soil. *Rev. Bras. De. Eng. Agr. Ambient.* 25: 514–521.

CTIC and SARE. 2020. 2019–2020 cover crop survey report. Conservation technology information center (CTIC), Sustainable Agriculture Research & Education (SARE). Available at: https://www.ctic.org/data/Cover_Crops_Research_and_Demonstration_Cover_Crop_Survey#:~:text=2019%2D2020%20Cover%20Crop

%20Survey&text=Among%20farmers%20who%20planted%20green,and%202.625%20in%20spring%20wheat.

CTIC and SARE. 2023. 2022–2023 cover crop survey. Conservation technology information center (CTIC), Sustainable Agriculture Research & Education (SARE). Available at: https://www.ctic.org/files/2022-2023%20National%20Cover%20Crop%20Survey%20Report-%20FNL%281%29.pdf.

Daryanto, S., et al.. 2018. Quantitative synthesis on the ecosystem services of cover crops. *Earth-Sci. Rev.* 185: 357–373.

Davis, J.R., et al. 2010. Ecological relationships of Verticillium wilt suppression of potato by green manures. *Am. J. Potato Res.* 87: 315–326.

Deines, J.M., et al.. 2022. Recent cover crop adoption is associated with small maize and soybean yield losses in the United States. *Glob. Change Biol.* 29: 794–807.

Dozier, I.A., et al. 2017. Tillage and cover cropping effects on soil properties and crop production in Illinois. *Agron. J.* 109: 1261–1270.

Dunbar, M.W., et al. 2016. Increased risk of insect injury to corn following rye cover crop. *J. Econ. Entomol.* 109:1691–1697.

Fan, F., et al. 2021. Cover crops promote primary crop yield in China: a meta-regression of factors affecting yield gain. *Field Crops Res.* 271: 108237.

Fernandez, A.L., et al. 2016. Structure of bacterial communities in soil following cover crop and organic fertilizer incorporation. *Appl. Microbiol. Biotechnol.* 100: 9331–9341.

Finney, D.M., et al. 2016. Biomass production and carbon/nitrogen ratio influence ecosystem services from cover crop mixtures. *Agron. J.* 108: 39–52.

Florence, A.M. and McGuire, A.M. 2020. Do diverse cover crop mixtures perform better than monocultures? A systematic review. *Agron. J.* 112: 3513–3534.

Garba, I.I., et al. 2022. Cover crop legacy impacts on soil water and nitrogen dynamics, and on subsequent crop yields in drylands: a meta-analysis. *Agron. Sustain. Dev.* 42: 34.

Garcia-Gonzalez, I., et al. 2018. Cover crops to mitigate soil degradation and enhance soil functionality in irrigated land. *Geoderma* 322: 81–88.

Gaudin, A.C.M., et al. 2013. Improving resilience of northern field crop systems using inter-seeded red clover: a review. *Agron.* 3: 148–180.

Harbach, C.J., et al. 2021. A mechanistic approach to assessing the potential for cover crops to serve as trap crops for the soybean cyst nematode. *Plant Dis.* 105: 1136–1142.

Haruna, S.I., et al. 2020. Improving soil physical properties through the use of cover crops: a review. *Agrosyst. Geosci. Environ.* 3: e20105.

Ibrahim Garba, I., et al. 2024. Cover crop functional types alter soil nematode community composition and structure in dryland crop-fallow rotations. *Appl. Soil Ecol.* 194: 105196.

Joshi, D.R., et al. 2023. A global meta-analysis of cover crop response on soil carbon storage within a corn production system. *Agron. J.* 115: 1543–1556.

Kaspar, T.C. and Bakker, M.G. 2015. Biomass production of 12 winter cereal cover crop cultivars and their effect on subsequent no-till corn yield. *J. Soil Water Conserv.* 70: 353.

Keene, C.L., et al. 2017. Cover crop termination timing is critical in organic rotational no-till systems. *Agron. J.* 109: 272–282.

Kichler, C.M., et al. 2023. Cover crop termination methods and custom residue manager effects on collard production. *Agron.* 13: 2595.

Kile, L. 2022. *The influence of winter cover crops and nitrogen fertilizer rates on nitrogen uptake and yield response in organic sweet potato (Ipomoea batatas) production* (pp. 1–41). The Influence of Winter Cover Crops and Fertilizer Application on Nutrient Uptake and Yield Response in Sweet Potato (*Ipomoea batatas*) Production Thesis. North Carolina State University.

Kim, N., et al. 2020. Do cover crops benefit soil microbiome? A meta-analysis of current research. *Soil Biol. Biochem.* 142: 107701.

Langelier, M., et al. 2021. Nitrogen-15 labelling and tracing techniques reveal cover crops transfer more fertilizer N to the soil reserve than to the subsequent crop. *Agr. Ecosyst. Environ.* 313: 107359.

Lavergne, S., et al. 2021. Using fall-seeded cover crop mixtures to enhance agroecosystem services: a review. *Agrosyst. Geosci. Environ.* 4: e20161.

Lemessa, F. and Wakjira, M. 2015. Cover crops as a means of ecological weed management in agroecosystems. *J. Crop Sci. Biotechnol.* 18: 123–135.

Lozier, T.M., et al. 2017. Release of phosphorus from crop residue and cover crops over the non-growing season in a cool temperate region. *Agric. Water Manage.* 189: 39–51.

Malone, L.C., et al. 2022. Cover crops in a Wisconsin annual cropping system: feasibility and yield effects. *Agron. J.* 114: 1052–1067.

Marcillo, G. and Miguez, F. 2017. Corn yield response to winter cover crops: an updated meta-analysis. *J. Soil Water Conserv.* 72: 226–239.

Miguez, F.E. and Bollero, G.A. 2005. Review of corn yield response under winter cover cropping systems using meta-analytic methods. *Crop Sci.* 45: 2318–2329.

Mirsky, S.B., et al. 2013. Overcoming weed management challenges in cover crop–based organic rotational no-till soybean production in the eastern United States. *Weed Tech.* 27: 193–203. https://doi.org/10.1614/WT-D-12-00078.1.

Morrison, C.L. and Lawley, Y. 2021a. 2020 Ontario Cover Crop Feedback Report, Department of Plant Science, University of Manitoba. https://gfo.ca/agronomy/soil-leadership/.

Morrison, C.L. and Lawley, Y. 2021b. 2020 Prairie Cover Crop Survey Report. Department of Plant Science, University of Manitoba. https://umanitoba.ca/agricultural-food-sciences/make/make-ag-foodresources#crops.

Nielsen, D.C., et al. 2016. Cover crop effect on subsequent wheat yield in the Central Great Plains. *Agron. J.* 108: 243–256.

Norris, C.E. and Congreves, K.A. 2018. Alternative management practices improve soil health indices in intensive vegetable cropping systems: a review. *Front. Environ. Sci.* 6: 50.

Norris, C.E., et al. 2022. The story of long-term research sites and soil health in Canadian agriculture. *Can. J. Soil Sci.* Special Issue Soil Health. https://doi.org/10.1139/CJSS-2021-0174.

Odiketa, J., et al. 2024. *Making cover crops work with grain cropping systems in Western Canada: impact on subsequent cash crops.* ASA, CSSA, SSSA International Annual Meeting, San Antonio, TX.

O'Reilly, K.A., et al. 2011. Weed populations, sweet corn yield, and economics following fall cover crops. *Weed Tech.* 25: 374–384.

Osipitan, O.A., et al. 2018. Cover crop for early season weed suppression in crops: systematic review and meta-analysis. *Agron. J.* 110: 2211–2221.

Osipitan, O.A., et al. 2019. Impact of cover crop management on level of weed suppression: a meta-analysis. *Crop Sci.* 59: 833–842.

Pellegrino, A.M., et al. 2021. Understanding the relationship between wireworm (Coleoptera: Elateridae) damage, varietal resistance, and cover crop use in organic sweet potato. *J. Econ. Entomol.* 114: 2127–2134. https://doi.org/10.1093/jee/toab118.
Peng, Y. and Van Eerd, L.L. 2024. *Surface soil sampling underestimates soil carbon and nitrogen storage of long-term cover cropping*. Regional Geoderma. October e00885. In press.
Peng, Y., et al. 2024. Global synthesis of cover crop impacts on main crop yield. *Field Crops Res.* 310: 109343.
Peterson, A.T., et al. 2019. Intersowing cover crops into standing soybean in the US upper Midwest. *Agron.* 9: 264.
Qin, Z., et al. 2021. Assessing the impacts of cover crops on maize and soybean yield in the U.S. Midwestern agroecosystems. *Field Crops Res.* 273: 108264.
Quemada, M., et al. 2013. Metaanalysis of strategies to control nitrate leaching in irrigated agricultural systems and their effects on crop yield. *Agric. Ecosyst. Environ.* 174: 1–10.
Rawls, W., et al. 2004. Effect of soil organic carbon on soil hydraulic properties. *Dev. Soil Sci.* 30: 95–114.
Ruis, S.J., et al. 2019. Cover crop biomass production in temperate ecozones. *Agron. J.* 111: 1–17.
Ruis, S.J. and Blanco-Canqui, H. 2017. Cover crops could offset crop residue removal effects on soil carbon and other properties: a review. *Agron. J.* 109: 1785–1805.
Sanchez, I., et al. 2019. Cover crops impacts on Louisiana corn production and soil properties. *Agrosyst. Geosci. Environ.* 2: 190015.
Sharma, P., et al. 2018. The role of cover crops towards sustainable soil health and agriculture: a review paper. *Am. J. Plant Sci.* 9: 1935.
Snapp, S.S., et al. 2005. Evaluating cover crops for benefits, costs and performance within cropping system niches. *Agron. J.* 97: 322–332.
Stone, A. 2012. *Cover crops for disease suppression*. eXtension Foundation webinar.
Thapa, R., et al. 2018. Cover crops reduce nitrate leaching in agroecosystems: a global meta-analysis. *J. Environ. Qual.* 47: 1400–1411.
Tonitto, C., et al. 2006. Replacing bare fallows with cover crops in fertilizer-intensive cropping systems: a meta-analysis of crop yield and N dynamics. *Agric. Ecosyst. Environ.* 112: 58–72. https://doi.org/10.1016/j.agee.2005.07.003.
Toler, H.D., et al. 2019. Global meta-analysis of cotton yield and weed suppression from cover crops. *Crop Sci.* 59: 1248–1261.
Tosi, M., et al. 2021. Long-term N inputs shape microbial communities more strongly than current-year inputs in soils under 10-year continuous corn cropping. *Soil Biol. Biochem.* 160: 108361.
Treadwell, D.D., et al. 2007. Cover crop management affects weeds and yield in organically managed sweet potato systems. *Weed Tech.* 21: 1039–1048. https://doi.org/10.1614/WT-07-005.1.
Trueman, C.L., et al. 2023. Long-term cover cropping suppresses foliar and fruit disease in processing tomatoes. *Plant Dis.* 107: 2160–2168.
Valkama, E., et al. 2015. Meta-analysis of the effects of undersown catch crops on N leaching loss and grain yields in the Nordic countries. *Agric. Ecosyst. Environ.* 203: 93–101.
Vendig, I., et al. 2023. Quantifying direct yield benefits of soil carbon increases from cover cropping. *Nat. Sustain.* 6: 1125–1134.

Wang, K.H., et al. 2004. Effect of winter cover crops on nematode population levels in north Florida. *J. Nematol.* 36: 517–523.

Wang, J., et al. 2021. A meta-analysis on cover crop impact on soil water storage, succeeding crop yield, and water-use efficiency. *Agric. Water Manag*. 256: 107085.

White, C.M., et al. 2016. A model-data fusion approach for predicting cover cop nitrogen supply to corn. *Agron. J.* 108: 2527–2540.

Wyffels, S.A., et al. 2021. Introducing cover crops as a fallow replacement in the Northern Great Plains:I. Evaluation of cover crop mixes as a forage source for grazing cattle. *Renew. Agric. Food Syst*. 37: 292–302. https://doi.org/10.1017/S1742170521000417.

Xi, K., et al. 2015. Understanding the development and management of stripe rust in central Alberta. *Can. J. Plant Path*. 37: 21–39.

Zhao, J., et al. 2022. Global systematic review with meta-analysis reveals yield advantage of legume-based rotations and its drivers. *Nat. Commun*. 13: 4926.

Part 2

Managing cover crops

Chapter 8

Cover crop decision support tools

Victoria J. Ackroyd, University of Maryland, USA; Heather Darby, University of Vermont, USA; Mikah Pinegar and S. Chris Reberg-Horton, North Carolina State University, USA; and Steven B. Mirsky, USDA-ARS, USA

1 Introduction

Benefits of cover crops for sustainable soil management are documented in this book and the scientific literature (Abdalla et al., 2019; Clark, 2008; Plastina et al., 2020). Cover crops do, however, increase the management complexity of agroecosystems (Bowman et al., 2022; Dunn et al., 2016; Myers and Watts, 2015; Plastina et al., 2020). Cover crops interact with pests (i.e. weeds, insects), diseases, and nutrient cycles in a myriad of ways and can negatively impact cash crop growth and yield if not managed appropriately (Abdalla et al., 2019; Bourgeois et al., 2022; Qin et al., 2021). Management requires resources, including time, money, and an understanding of the field dynamics involved. Farmers need information that is accurate, reliable, dependable, and specific to their farm/environment to help them make decisions and build knowledge. However, that information must be packaged in a streamlined format that is easy to digest and implement.

http://dx.doi.org/10.19103/AS.2025.0162.11
Published by Burleigh Dodds Science Publishing Limited, 2026.

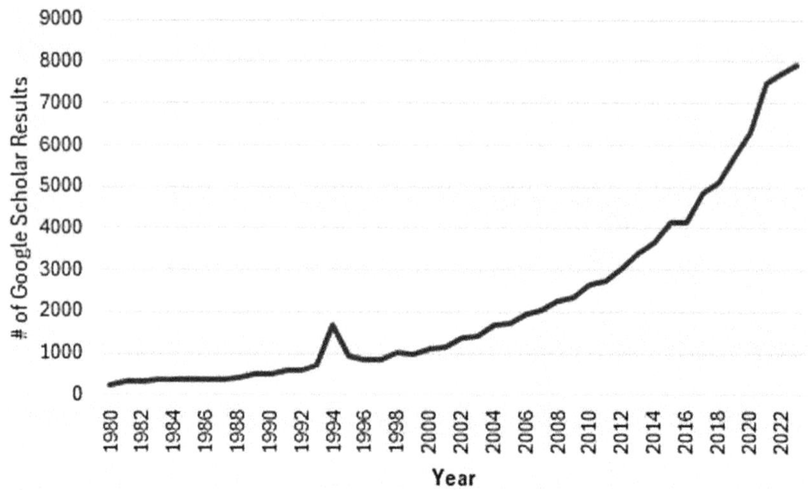

Figure 1 Number of Google Scholar results generated for each year from 1980 to 2023 in an Advanced Search with the exact phrase 'cover crop*' in October 2024.

For illustrative purposes, consider using the 'Advanced Search' feature of Google Scholar for the term 'cover crop*' (the * causes the search to return results with any variant including 'crop', e.g. 'crops', 'cropped', 'cropping', etc.) that returns articles dated within a single year. A search conducted in October of 2024 for the period 1980–2023 turned up 253 records in 1980, increasing to 7,930 results in 2023 (Fig. 1). This example drives home the sheer amount of cover crop information generated each year. Lack of information has understandably been rejected as the sole driver of farmer behavior (Crane, 2014), but how much of this information is relevant to a farmer in a particular U.S. location, considering his/her concerns and constraints? An answer would require extensive further sifting beyond the scope of this illustration. To that point, growers and policy-makers must have trustworthy information that is useful and usable (Bawden and Robinson, 2009).

2 Current cover crop decision support tools and their limitations

To meet the need for 'useful and usable' cover crop management information in a readily accessible format, crop advisers, including university extension agents, USDA NRCS conservation planners, crop consultants, and other technical assistance providers, have developed cover crop decision support tools. These tools may be as low-tech as crop rotation plans sketched out on paper, or as high-tech as online tools that interact with multiple databases to populate input and generate recommendations. Tools developed by both the private sector (e.g. Green Cover's Smartmix calculator; https://smartmix.greencoverseed

.com/) and the public sector (e.g. the Midwestern Cover Crops Council's Cover Crop Species Selector Tool; https://www.midwestcovercrops.org/selector-tools/) have made their way to the farming community over the last 5 years.

However, decision support tools have historically suffered from several issues centered on data quality and user friendliness. It is extremely difficult to create a tool that incorporates a dataset that is relevant across the entire US, takes into account how data is used to manage complex and dynamic systems, has a mechanism for updating the data as new information becomes available, and can readily exchange information with other tools (i.e. is interoperable). Agricultural data underlying tools has been collected at varying levels of granularity, in disparate formats and structures, and using varied protocols (Raturi, 2017). Decision support tools have tended to be of use to a limited geographic region (e.g. a cover crop seeding rate calculator based on information gathered at one site in one state) or, at the other end of the spectrum, to be implemented for an entire region or all of the US, but lacking in site specificity.

Furthermore, it is resource-intensive to create easy-to-use, human-centered tools that display complex information in a way that can be easily understood by users. This process requires large amounts of user feedback and design iterations at multiple points in tool design and development. Finally, there has been a need for better national coordination to minimize the number of tools that may be redundant in terms of their goal (e.g. providing cover crop seeding rate recommendations). Collaborative efforts more efficiently utilize resources, data, and marketing while decreasing the sheer number of tools with redundant purposes but limited applicability.

3 The Precision Sustainable Agriculture (PSA) network

To address the concerns discussed above, the Precision Sustainable Agriculture (PSA) network and its predecessors (https://www.precisionsustainableag.org/about-us), in conjunction with NRCS and the regional cover crop councils, developed cover crop decision support tools that are no- or low-cost, user-friendly, scalable, and adaptable nationwide. The PSA network is a transdisciplinary community of farmers, scientists, engineers, designers, educators, commodity group members, government agents, and private industry leaders that works together to develop decision support tools.

The PSA network is a grassroots effort that emerged from an on-farm research network in 2014 and has evolved into a cutting-edge technology research and development team. The PSA network examines how sustainable practices implemented in conjunction with precision farming technologies (e.g. variable rate fertilizer and herbicide sprayers, field mapping sensors) interact with climate and soils to affect crop yield and impact the environment. This is accomplished through the development of Internet of Things (IoT) plant and soil sensing technologies, cyberinfrastructure, and the community of practice established to coordinate on-farm and on-station research networks. Data collected from these networks support national web-based decision support tools for farm management and long-term planning.

One of the primary objectives of the PSA network is to create cutting-edge web-based cover crop decision support tools for farmers, technical service providers, researchers, and policy makers. These tools to aid growers in making decisions that match their farming objectives, enabling them to consider the economic, environmental, and production implications of cover cropping were designed by:

a. Conducting stakeholder assessments to understand decision support needs,
b. Evaluating the potential of existing data and tools to meet these needs, and
c. Partnering with the developers of existing data and tools to modify them to meet identified needs or generate new data and tools for cover crop selection and management support.

The next section explores these issues in more detail.

4 Developing new cover crop decision support tools by the PSA network: key issues

Providing farmers with timely, relevant, and specific information is key to supporting the cover crop decision-making process. Design of technology to address the challenges discussed in the chapter introduction must involve a systems approach that considers institutional knowledge and how, where, and in what form data are collected, stored, analyzed, and used; stakeholder goals; and the environmental context in which tools are used - while keeping user experience at the forefront of the design process. To ensure that the network creates usable and useful technologies, we involve stakeholders throughout the design, development, and deployment process. Given the variety of human factors that influence agricultural practices, the social and economic contexts of technology must be considered. User-centered design offers techniques to directly engage cover crop stakeholders in the participatory design of technology (Niles and Mueller, 2016; Norman and Draper, 1986).

User-testing with farmers and other stakeholders is conducted frequently, especially during the initial development phase of a tool. Farmers and stakeholders who participate in user-testing are recruited early in the process.

We recruit farmers from the PSA on-farm trial network and via extension collaborators to test the tools during the design, refinement, and deployment phases. Other tool development activities include user testing to measure usability (including A/B design testing; King et al., 2017) and satisfaction, and train-the-trainer activities with crop consultants and extension agents. This feedback is critical in order to produce a final product that is user-friendly and delivers the information in a format that is useful for our stakeholders. All phases of tool development, testing, and deployment are iterative.

The tool development process differs depending on the nature of the underlying data: expert opinion vs. process model based or a combination of both. Expert opinion-based tools rely heavily on information provided by groups of experts in a given US state or region, who themselves draw on the scientific literature and research results, their 'boots on the ground' experience, and practical knowledge. Expert opinion-based tools often contain extensive qualitative information wherein, for example, cover crops are ranked comparatively against each other for suitability fulfilling a given cropping system goal (e.g. 'prevents fall soil erosion') or possessing a trait of interest (e.g. 'heat tolerance'). While some information is available in the literature for the potential of a given cover crop to prevent fall soil erosion or tolerate heat, judgment calls must still be made to provide ratings and rankings due to the enormous body of data involved. No single research study has evaluated such traits across all 40+ species in the Northeast species selector tool, for example. Decision support tools can contain tens of thousands of data points, ranging from the mundane and relatively easily sourced (e.g. cover crop species name; PSA uses itis.gov as its authority) to the complex and sometimes controversial (e.g. 'penetrates plow pan').

The limitations inherent in relying on expert opinion make critical the need to gather input from a diverse and representative group of experts that includes farmers, extension agents and researchers, government agency personnel, and representatives from the agricultural industry and NGOs. See later for a use case example of an expert opinion-based tool. The PSA network recognizes the limitations of expert opinion-based tools, and in the long term seeks to develop tools that are primarily process model based.

Cover crop decision support tools driven by models calibrated for farm-specific conditions that integrate on-site sensor data would enable the provision of recommendations that account for real-time environmental conditions and other site-specific factors (O'Grady and O'Hare, 2017). However, there has historically been limited linkage between modeling and decision support tool development efforts in agriculture. Efforts that integrate geospatial data, such as land suitability tools (Elsheikh et al., 2013), have typically failed to consider how the tools will be used for actual decision making. Model-based decision support tools have thus suffered from usability issues to the detriment of efforts to encourage the adoption of the tools and practices they support. Effective tools must enable data sharing and interoperability (Kruize et al., 2016) to enable high-quality recommendations (Rose et al., 2016) and use.

Irrespective of the nature of the data and the techniques used to generate recommendations for a decision support tool, care must be taken to ensure that the tool's interface meets user expectations and needs. Tool interfaces and the recommendations they provide as outputs should have a 'clean, simple appearance', be easy to browse, be laid out in a logical progression that matches the stakeholder's thought process, and be mobile-friendly (Cabrera and Holt, 2020). Information overload can increase anxiety (Bawden and Robinson, 2009; Eppler and Mengis, 2004). Ease of use and interpretation is particularly crucial since research suggests that online support tools are a key resource for farmers seeking to make changes to their cropping system management (Cabrera and Holt, 2020). The PSA cover crop decision support team takes all of these factors into account when developing, implementing, and testing support tools.

One of the guiding principles of the PSA network is the desire to avoid reinventing the wheel. To that end, PSA members frequently collaborate to update and refine existing decision support tools, which they then offer online with a user-friendly website interface. Such proof-of-concept tools have frequently taken the form of Excel spreadsheets running macros and were typically developed by cover crop councils (e.g. the PSA cover crop species selector tool), USDA NRCS personnel (e.g. the seeding rate calculator and mix maker), or individual team members or labs (e.g. the nitrogen release and biomass decomposition calculator) (see https://www.precisionsustainableag.org/decision-support-tools for a full list). With the permission and assistance of their initial creators, the PSA team then methodically expands the tool range of inference region by region. The team works with regional experts, drawing on their field experience, the scientific literature, and/or output from models, PSA field data, and other sources such as remote-sensing technologies, to refine the data for that region. Tools range from heavily expert opinion based to virtually entirely process model based. Below are two examples of tools, one occupying each end of the spectrum.

5 The PSA Cover Crop Species Selector Tool

Collaborators in early iterations of the PSA network initiated their tool development program in 2015 when they recognized the need for a cohesive, user-friendly, site-specific cover crop species selector to assist farmers and crop consultants in quickly and easily sifting through cover crop options to identify the species best suited to their cropping system goals and constraints. At the time, team members noted that cover crop selection and management for optimal benefits is knowledge-intensive; farmers repeatedly cite site- and system-specific information and management complexity as barriers to adoption (Dunn et al., 2016; Myers and Watts, 2015). The PSA team was well-positioned to develop a solution given that many collaborators were (often founding) members of their respective regional cover crop councils.

The resultant PSA Cover Crop Species Selector (Fig. 2; https://covercrop-selector.org/) is an example of an expert opinion-based tool. This decision support tool was modified from an online species selector that was originally developed by the Midwest Cover Crops Council (MCCC; https://www.midwestcovercrops.org/selector-tools/; Baas et al., 2011) in 2008. The PSA Cover Crop Species Selector provides recommendations to assist in the selection of cover crops. Users must, at a minimum, input location. Goals range from 'good (for) grazing' to 'increase soil aggregation' to 'prevent fall (soil) erosion' to 'suppresses winter annual weeds'. Based on user location, the tool auto-populates soil properties such as drainage [from the NRCS Soil Survey Geographic database (SSURGO)] and annual weather metrics such as average precipitation (from an API custom-built by the PSA team to access weather data collected by the US government) (https://covercrop-ncalc.org/#/about). Users may optionally input information about their cash crop. The species selector then generates recommendations, which the user may further refine using a series of filters (Fig. 3).

The PSA team retained the essential nature of the MCCC data generation and modification process. This process consists of a preliminary review of the scope of existing resources by a small (5–15) group of leaders and a data manager (usually the council's program manager). The leadership team then

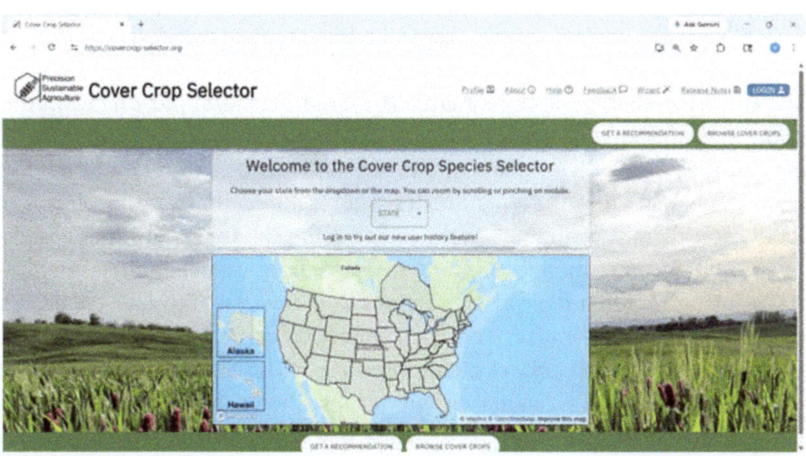

Figure 2 Screenshot of the PSA Cover Crop Species Selector landing page as of March 2026.

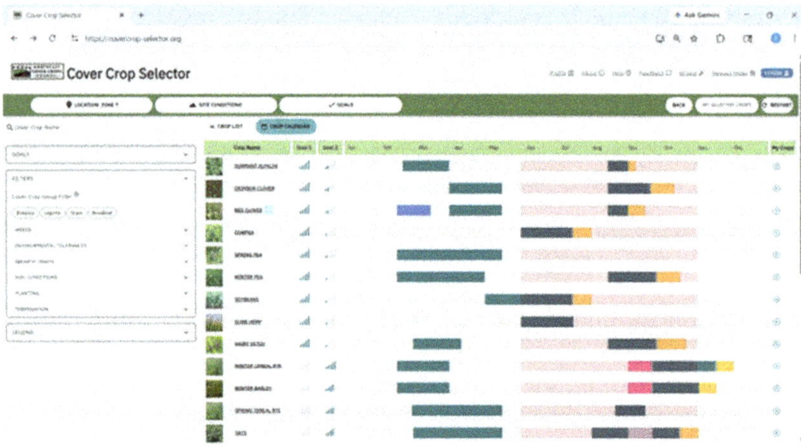

Figure 3 Cover crop species recommendation output as of March 2026 by the PSA Cover Crop Species Selector for a field in Maryland (USDA plant hardiness zone 7) with a cash crop growing window of June 1-October 30 (pink bars) and the desired goals of 'nitrogen fixation' (Goal 1) and 'nitrogen scavenging' (Goal 2).

worked with stakeholders ranging from farmers to representatives from the agricultural industry, NGOs, government agencies such as USDA NRCS, and academia to select the top-level parameters that should be in the tool (e.g. cover crop characteristics to rate and rank) and the levels at which data should be reviewed and modified. This data division was crucial to adapting the tool to each US region and its unique needs and constraints.

In the Midwest, data were generally reviewed on a state-by-state basis. The PSA team extended this 'data verification' first to the Northeast, then the South, and finally the Western US. In the Northeast, data was refined and modified by USDA plant hardiness zone (https://planthardiness.ars.usda.gov/), starting with Zone 6 (which began with the data from Ohio in the MCCC tool). In the South, data were reviewed by a combination of hardiness zone and physiographic region. In the West, data verification was conducted at levels ranging from All-West (i.e. a cover crop was given one rating or value for a given trait that does not vary by location in the West) to the ecoregion to the state level. It is, in part, these different groupings that allows for the differentiation that enables site-specific recommendations. Comments sections for each dataset allow for the further enumeration of edge cases and caveats. The data verification process was highly iterative. The data manager carried concerns between groups conducting the process, supporting a larger discussion within each region that unified the individual expert groups. The program managers of the four regional cover crop councils estimate that over 400 experts contributed to cover crop species selector design, data modification, implementation, and testing.

The high-level process of selection of the parameters to be included in the decision support tool prior to the commencement of modification of individual data points also supported tool regionalization based on distinctive needs, constraints, and stakeholder concerns. Each of the four regional versions of the tool has unique aspects. For example, the Western version of the tool incorporates extensive and detailed information about cover crop termination timing that will actually require additional programming and the release of an updated version of the tool once fully implemented. The Midwestern tool accounts for average frost dates at the county level when generating planting window recommendations. Data verification for the Southern species selector tool aggregated and organized extensive cultivar-level information from USDA NRCS Plant Materials Centers (PMC) Cover Crop Performance and Adaptation Trials (https://www.nrcs.usda.gov/plant-materials/cover-crop-performance-and-adaptation-trials) that will likewise not be implemented until the release of a future version of the tool.

The data groupings within each species selector region are not the only ways in which the PSA tool differs from the MCCC original upon which it is based. The MCCC originally had no easy method to import new data into the online interface, making data updates difficult. The PSA team overcame this barrier by storing and modifying data in Airtable (https://airtable.com/), which combines elements of a spreadsheet and a database, rather than in a basic Microsoft Excel spreadsheet. The PSA team completely redesigned the selector tool user interface, creating a 'wizard' that walks users through data input rather than employing a single static page for data entry. The recommendations output was augmented with the addition of filters, which allow users to further refine their cover crop options. A 'My Selected Crops' feature enables users to select cover crops of interest for further, detailed comparison.

For all of their differences, the original MCCC and PSA cover crop selector tools share in common the goal of reducing information overload to support farmers in cover crop decision making. The MCCC maintains its original (albeit updated) tool independently on its website; however, in the interest of simplicity and completeness, the data from that tool are also available in the PSA tool (i.e. users can generate species selection recommendations for the Midwestern states using the PSA tool). The MCCC and PSA species selectors are both meant to be a learning tool and were consciously designed as such. Both selectors gray out, rather than remove from view, non-suitable cover crop options. Both selector tools can generate an information sheet for each cover crop in the tool summarizing all of the available data for said cover crop for a given state. Both selector tools provide links to resources for further reading. Both selector tools aim to increase cover crop adoption and support the maximization of ecosystem services provided by those cover crops through effective and efficient management choices.

6 Measuring cover crop nitrogen release and biomass decomposition: The PSA Cover Crop Nitrogen Calculator

Once the PSA team committed to the development of a suite of cover crop decision support tools, it quickly recognized the need for a tool that estimates cover crop nitrogen release and biomass decomposition. The topic of cover crop N provision (in the case of legumes) and N immobilization and release in the soil has proven vexing (Thapa et al., 2022a). Farmers often seek to add N through cover crops but need to know how much N a cover crop has likely provided (i.e. the N credit) and the rate at which it will be released (Thapa et al., 2022a). On the other hand, concerns about potential negative effects on a following cash crop may be based, at least partly, on the risk of N immobilization in the soil (Qin et al., 2021). The amount of biomass produced and the composition (i.e. C and N content) of the cover crop biomass governs these dynamics (Thapa et al., 2022a, b) and dictates N management in the following cash crop. The PSA network was ideally positioned to meet the critical need for a cover crop and N management tool, given the modeling experience of team members and the opportunity presented by its extensive set of on-farm and on-station research sites to conduct multi-state field studies examining cover crop N release and biomass decomposition.

The resultant PSA Cover Crop Nitrogen Calculator (CC-NCALC; https://covercrop-ncalc.org/; Fig. 4), a process model-based tool, lies at the other end of the spectrum from the PSA Cover Crop Species Selector. This calculator originated in the lab of Dr. Miguel Cabrera at the University of Georgia; its

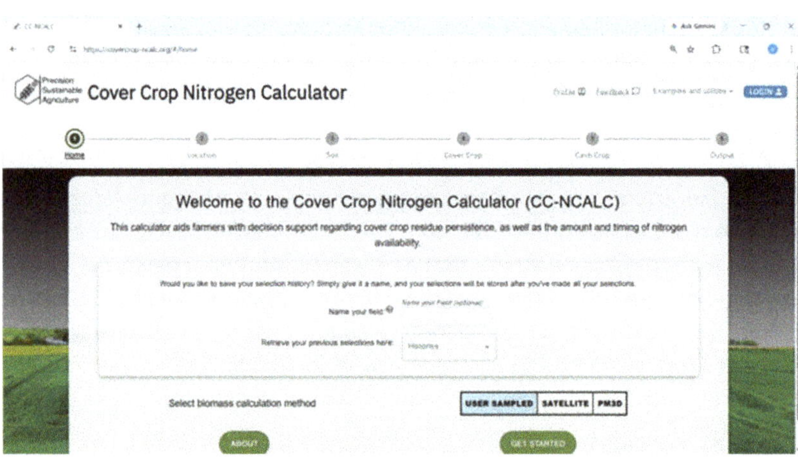

Figure 4 Screenshot of the PSA Cover Crop Nitrogen Calculator (CC-NCALC) landing page as of March 2026.

Published by Burleigh Dodds Science Publishing Limited, 2026.

underpinnings have been reported in the literature (Quemada and Cabrera, 1995; Quemada et al., 1997; Zubillaga et al., 2007).

CC-NCALC estimates (a) the amount of N released from decomposing cover crop residue over time, (b) the rate of biomass decomposition over time, (c) the amount of N taken up by corn, and (d) the recommended N fertilizer application rate for a following cash crop, accounting for a cover crop N credit (https://covercrop-ncalc.org/#/about) (Fig. 5). The calculator currently works for no-till systems.

Based on user location, the calculator auto-populates soil properties [from the NRCS Soil Survey Geographic database (SSURGO)], soil moisture and temperature [from Iteris (https://docs.clearag.com/documentation/Soil_Conditions/Soil_Conditions/latest)], and hourly weather conditions (from an API custom-built by the PSA team to access weather data collected by the US government) (https://covercrop-ncalc.org/#/about). It requires the user to input, at a minimum, the field location. Additional information needed to generate output can be provided manually by the user ('User Sampled'), using the 'Satellite' button to populate cover crop data from remote imaging, or with the PlantMap3D camera system ('PM3D'). No matter the data entry method, the model in the tool needs cover crop biomass and N content to run. Users can also input further information about the cash crop of interest. The model underlying the calculator then provides information about N release from the cover crop after its termination and the amount of cover crop residue (i.e. mulch, biomass) remaining, as well as fertilizer application quantity recommendations; Fig. 5). The decision support tool, the modeling underlying the tool, and the

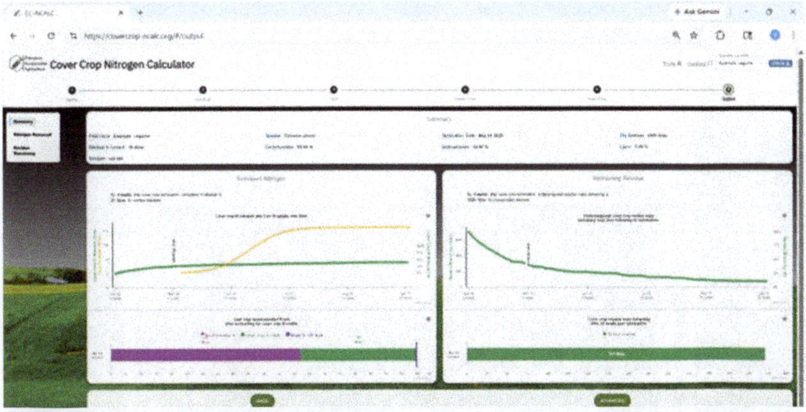

Figure 5 PSA Cover Crop Nitrogen Calculator (CC-NCALC) output as of March 2026 for nitrogen release (left) and biomass residue remaining (right) using the 'Example: Legume' feature of the calculator to populate the data for a cover crop terminated May 1, 2025 at a site in Beltsville, MD with a following cash crop of corn planted June 1, 2025 with a 150 bu/A yield goal and a Target Nitrogen Application Rate of 100 lbs/A.

field studies that populated the data in the model are extensively documented in the literature (Dann et al., 2021; Gaskin et al., 2020; Meeks et al., 2023; Thapa et al., 2021, 2022a, 2022b; Woodruff et al., 2018).

The model at the heart of CC-NCALC is a modified version of CERES-N (Crop Environment REsource Synthesis-N submodule) (Thapa et al., 2022a). The original tool developer, Dr. Cabrera, worked with collaborators within the PSA network to (a) further refine the model, (b) expand the geographic range of inference for the calculator, and (c) update the tool user interface. The PSA team adjusted the CERES-N model to account for dynamics related to residue decomposition by adding or altering model variables related to the interplay between residue water potential and temperature, N limitations, and the amount of residue left in contact with the soil (Thapa et al., 2022a). The calculator was then calibrated and validated with 99 site-years of data from the 2017 to 2019 PSA on-farm litter bag biomass decomposition experiment (Thapa et al., 2022b), which expanded the region to which the calculator is relevant. Finally, the calculator website interface was given an extensive overhaul to simplify and improve the user experience.

PSA tool developers understood from the outset that a major limitation of the calculator is the need to provide an estimate of cover crop biomass. While there are certainly ways for a farmer or crop consultant to collect this information on a small scale, it presents a significant barrier to calculator adoption. In the short term, the PSA development team created 'Example data'

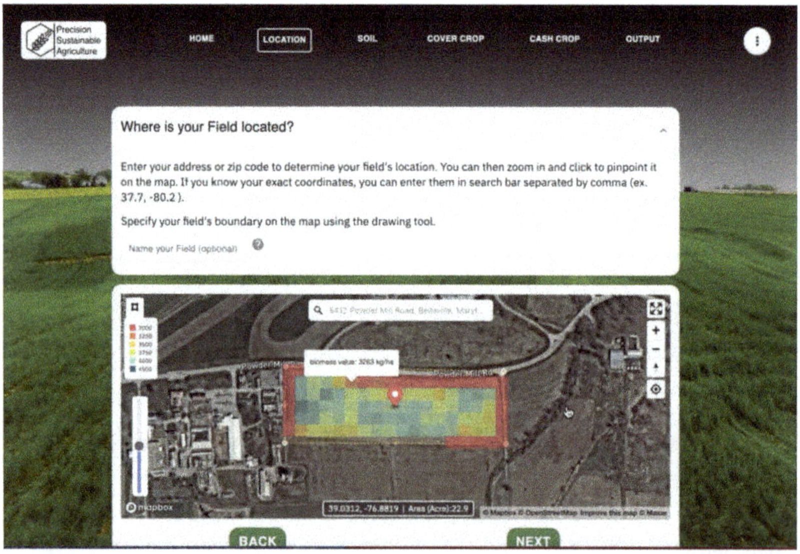

Figure 6 Estimated plant biomass (lbs/A) for a site near Beltsville, MD, based on satellite imagery. Areas in red represent hard surfaces such as roads and parking lots, while shades of green indicate research fields.

Published by Burleigh Dodds Science Publishing Limited, 2026.

sets, one for a grass and one for a legume cover crop scenario, that remove the need to populate anything other than a user's location. This feature allows for the demonstration of the tool even when a user is uncomfortable providing estimated biomass. In the long-run, the CC-NCALC user experience is being simplified and improved through a novel and innovative method that updates CC-NCALC with remote sensing techniques and uses satellite- or camera-based (PlantMap3D) imagery to estimate cover crop biomass (Fig. 6; https://www.precisionsustainableag.org/remote-sensing-1). As with the PSA Cover Crop Species Selector, the goal of the CC-NCALC tool is to support the maximization of ecosystem services provided by cover crops through efficient and effective management choices.

7 Future challenges

The primary challenge faced by cover crop decision support tool creators is the same challenge that drove the development of such tools in the first place: data. No matter the source of the data populating a tool (i.e. expert opinion based, process modeling based, or a mixture), there will remain the need to systematically recognize and address knowledge gaps to further refine the site specificity of the underlying data driving recommendations.

Knowledge gaps related to data that inform expert opinions, such as the comparative rankings of a group of cover crops for a goal like 'prevents fall soil erosion' under a given set of circumstances in a given geographic region, should be relatively simple to fill through basic field experiments. However, the existence of an extensive body of historical literature may lead to a perception among funders that there is no need to conduct further research, leading to difficulties accessing resources with which to carry out the work. Up until the digital age, research agronomists have been hindered in their ability to generate recommendations based on study results due to the limited field of inference inherent in experiments that are typically conducted at one or a few sites for 1–3 years. Simply put, numerous studies have indeed been conducted to investigate the impact of cover crops on the many facets of crop system dynamics (e.g. soil health, pests, disease, nutrient cycling, yield). However, this body of knowledge has not lent itself to the generation of site-specific recommendations, even when subjected to statistical methods such as meta-analysis.

While the rise of new statistical analyses using artificial intelligence (AI) methods, such as the use of evidence-based reasoning, multistrategy machine learning, automated search, and mixed-initiative interaction, may eventually render the existing literature more useful for applied purposes than it has been, there is the further challenge of climate change. Data will have to be continuously updated and replaced as the climate changes, and tools will have to evolve. Climate change will no doubt pose unforeseen challenges that will force alterations in crop system management to maintain the viability and sustainability of agroecosystems, which will need to be reflected in decision support tools.

The challenge, then, is how to better capture the complexity of the interactions among cover and cash crop genetics, the environment (e.g. soil and weather conditions), cash and cover crop management, and the social aspects tied up in crop system management choices in order to develop decision support tools of value to farmers and technical service providers. The PSA network has been working to do just that through its extensive collection of on-farm and on-station research sites that are developing, refining, implementing, and delivering to the agricultural industry cutting-edge tools, such as automated field-mapping camera systems (i.e. PlantMap3D), that can also feed data into the models underlying decision support tools.

However, this solution to the problem of a lack of usable data to generate decision tool recommendations creates challenges of its own. The amount of data collected by the PSA network is enormous. For that data to be of use, manpower must be available to utilize this data to generate, test, and calibrate the models running process model-based tools or to otherwise analyze and then import it into the more expert opinion-based tools. This is also true of data generated by other research teams.

Further, research results are not the only type of data that goes into tools. Other points of information, such as current incentive program requirements, are also germane when generating cover crop selection and management recommendations. However, the need for dedicated manpower to keep such information up to date has, to date, precluded its inclusion in PSA-developed decision support tools. While it is paramount that the information and data driving recommendations be as up to date as possible, the need to maintain the codebase running online decision support tools should not be ignored and will also remain a challenge. It takes skilled manpower to keep an online decision support tool running in the face of the ever-changing internet environment.

8 Conclusion

To maximize the agroecosystem services provided by cover crops, farmers and their advisers need detailed information. Public and private groups have sought to distill the large body of knowledge on cover crops into decision support tools that aim to relieve information overload and promote cover crop adoption by providing cover crop selection and management recommendations. However, such tools are often limited to particular geographical regions and have suffered from a lack of site specificity, failure to integrate existing data across agricultural systems and scales, a lack of adaptivity and interoperability, and suboptimal user interface design. The Precision Sustainable Agriculture network develops tools that are site specific, inter-operable, relevant across the US, and user-friendly. The PSA team employs a human-centered design approach that identifies stakeholder needs and the ways stakeholders will use the tools.

Tools can generally be divided into expert opinion based and process model based. Examples given in this chapter include an expert opinion-based species selector and a process model-based cover crop nitrogen calculator. Both were developed using an iterative design, implementation, and testing process.

Future challenges facing cover crop decision support tool development relate to the amount of data held within tools, the need to keep tools updated, and the difficulty in elucidating interactions among cash and cover crop genetics, the environment, and management to provide the best possible recommendations. Cover crop decision support tools are important to support farmers in sustainably producing fuel, food, feed, and fiber as the world population climbs towards an estimated 10 billion people in 2050 under changing climate conditions.

9 Where to look for further information

The cover crop decision support tool domain is a fast-growing and rapidly-changing field. For the most current information, contact teams that are actively developing decision support tools, such as the Precision Sustainable Agriculture network (https://www.precisionsustainableag.org/). The relevant regional cover crop council(s) and their conferences and instructional events represent a wellspring of up-to-date information:

- Midwest Cover Crops Council (https://www.midwestcovercrops.org/).
- Northeast Cover Crops Council (https://northeastcovercrops.com/).
- Southern Cover Crops Council (https://southerncovercrops.org/).
- Western Cover Crops Council (https://westerncovercrops.org/).

Each council has a program manager or other point of contact and a list of board members who can direct interested parties to current tool development, testing, and deployment efforts. Since public sector decision support tools for sustainable agriculture are ipso facto publicly funded, a search of federal grant tracking and reporting tools is an excellent way to discover more about current tool creation activities and teams. Suggested websites include:

- US Department of Agriculture's Data Gateway (https://www.nifa.usda.gov/data/data-gateway).

- US Department of Agriculture Natural Resources Conservation Service project search (https://cig.sc.egov.usda.gov/cig-projects).
- Sustainable Agriculture Research and Education program grant management website (https://projects.sare.org/search-projects/).

10 References

Abdalla, M., et al., (2019). A critical review of the impacts of cover crops on nitrogen leaching, net greenhouse gas balance and crop productivity. *Glob. Change Biol.*, *25*(8), 2530–2543.

Baas, D. G., et al., (2011). A state-specific online cover crop decision tool for Midwest farmers. *J. Ext.*, *49*(6), Article 35. https://tigerprints.clemson.edu/joe/vol49/iss6/35

Bawden, D. and Robinson, L., (2009). The dark side of information: overload, anxiety and other paradoxes and pathologies. *J. Inf. Sci.*, *35*(2), 180–191.

Bourgeois, B., et al., (2022). Interactive effects between cover crop management and the environment modulate benefits to cash crop yields: a meta-analysis. *Can. J. Plant Sci.*, *102*(3), 656–678.

Bowman, M., et al., (2022). Farmers employ diverse cover crop management strategies to meet soil health goals. *Agric. Environ. Lett.*, *7*(1), e20070.

Cabrera, E. and Holt, J., (2020). An evaluation of website usability for a cover crops resource in the Southern Region. *J. Applied Commun.*, 104, 1. https://doi.org/10.4148/1051-0834.2305.

Clark, A., ed., (2008). *Managing cover crops profitably*. USDA SARE. https://www.sare.org/resources/managing-cover-crops-profitably-3rd-edition/.

Crane, T. A., (2014). Bringing science and technology studies into agricultural anthropology: technology development as cultural encounter between farmers and researchers. *Cult. Agric. Food Environ.*, *36*(1), 45–55.

Dann, C. E., et al., (2021). Modeling water potential of cover crop residues on the soil surface. *Ecol. Model.*, *459*, 109708.

Dunn, M., et al., (2016). Perceptions and use of cover crops among early adopters: findings from a national survey. *J. Soil Water Conserv.*, *71*(1), 29–40.

Elsheikh, R., et al., (2013). Agriculture Land Suitability Evaluator (ALSE): a decision and planning support tool for tropical and subtropical crops. *Comput. Electron. Agric.*, *93*, 98–110.

Eppler, M. J. and Mengis, J., (2004). The concept of information overload - A review of literature from organization science, accounting, marketing, MIS, and related disciplines. *Inf. Soc. Internat. J.*, *20*, 1–20.

Gaskin, J. W., et al., (2020). Using the cover crop N calculator for adaptive nitrogen fertilizer management: a proof of concept. *Renew. Ag. Food Syst.*, *35*(5), pp.550–560.

King, R., et al., (2017). *Designing with data: improving the user experience with A/B testing*. O'Reilly Media, Inc.

Kruize, J. W., et al., (2016). A reference architecture for farm software ecosystems. *Comput. Electon. Agric.*, *125*, 12–28.

Meeks, C., et al., (2023). Biochemical composition of cover crop residues determines water retention and rewetting characteristics. *Agron. J.*, *115*(6), 3173–3187.

Myers, R. and Watts, C., (2015). Progress and perspectives with cover crops: interpreting three years of farmer surveys on cover crops. *J. Soil Water Conserv., 70*(6), 125A–129A.

Niles, M. T. and Mueller, N. D., (2016). Farmer perceptions of climate change: associations with observed temperature and precipitation trends, irrigation, and climate beliefs. *Glob. Environ. Change., 39*, 133–142.

Norman, D. A. and Draper, S., (1986). *User-centered system design: new perspectives on human–computer interaction*. L. Erlbaum Associates Inc.

O'Grady, M. J. and O'Hare, G. M. P., (2017). Modelling the smart farm. *Inf. Process. Agric. 4*, 179–187.

Plastina, A., et al., (2020). Cover crops use in Midwestern US agriculture: perceived benefits and net returns. *Renew. Agric. Food Systs., 35*(1), 38–48.

Qin, Z., et al., (2021). Assessing the impacts of cover crops on maize and soybean yield in the US Midwestern agroecosystems. *Field Crops Res., 273*, 108264.

Quemada, M. and Cabrera, M. L., (1995). CERES-N model predictions of nitrogen mineralized from cover crop residues. *Soil Sci. Soc. Am. J., 59*(4), 1059–1065.

Quemada, M., et al., (1997). Nitrogen release from surface-applied cover crop residues: evaluating the CERES-N submodel. *Agron. J., 89*(5), 723–729.

Raturi, A., (2017). *Modeling sustainable agriculture* (Doctoral dissertation, UC Irvine).

Rose, D. C., et al., (2016). Decision support tools for agriculture: towards effective design and delivery. *Agric. Systs., 149*, 165–174.

Thapa, R., et al., (2021). Effects of moisture and temperature on C and N mineralization from surface-applied cover crop residues. *Biol. Fert. Soil., 57*, 485–498.

Thapa, R., et al., (2022a). Modeling surface residue decomposition and N release using the Cover Crop Nitrogen Calculator (CC-NCALC), *Nut. Cycling Agroecosyst., 124*(1), 81–99.

Thapa, R., et al., (2022b). Cover crop residue decomposition in no-till cropping systems: insights from multi-state on-farm litter bag studies. *Agric. Ecosyst. Environ., 326*, 07823.

Woodruff, L. K., et al., (2018). A web-based model of N mineralization from cover crop residue decomposition. *Soil Sci. Soc. Am. J., 82*(4), 983–993.

Zubillaga, M. M., et al., (2007). Modeling field-scale N mineralization in Coastal Plain soils (USA). *Ecol. Model., 207*(2–4), 243–250.

Chapter 9

Cover crop establishment

Andy Clark, formerly University of Maryland and USDA Sustainable Agriculture Research and Education (SARE), USA; and Sjoerd W. Duiker, The Pennsylvania State University

1 Introduction

Cover crops are an integral component of sustainable agroecosystems. Cover crop benefits are optimized within Conservation Agriculture systems that include (1) minimal soil disturbance; (2) permanent living vegetation or dead mulch on the soil surface; and (3) crop diversity, either in time or space. A fourth principle that is now widely considered as an additional opportunity to improve soil with Conservation Agriculture is livestock integration – especially grazing. Adding cover crops to existing cropping systems increases management complexity, with unique challenges for both establishment and termination. This is particularly the case in Conservation Agriculture because of the goal to maintain permanent crop residue cover at the soil surface. Management decisions about cover crops must consider both establishment and termination and must be balanced against other field operations needed for successful

http://dx.doi.org/10.19103/AS.2025.0162.23

cash crop production. In this chapter, we discuss cover crop establishment, with a primary focus on grain production systems in temperate humid climates.

Cover crop management is site-specific and differs depending on the subsequent cash crop. Weather, soil type, previous crop, and growing season all affect management decisions. Before planting a cover crop, identify specific goals for that cover crop (Clark, 2007). Is it soil erosion protection, soil health improvement, nitrogen fixation, nutrient scavenging, forage production, or pollinator services? Species selection and management and their effect on these goals are also discussed in other chapters in this book. The best establishment method will be based on the cover crop species or mixture, the previous and subsequent cash crop, and – most critical – climate and growing season.

2 General considerations

A significant barrier to successful adoption of cover cropping in grain production systems is the length of the growing season, particularly the window after cash crop harvest until establishment of the following crop (Hamilton et al., 2017). The optimal window for cover crop establishment (and termination) often conflicts with cash crop harvest (and planting), respectively, especially in more northern regions. This challenge is compounded by labor availability at these busiest times of the year, consistently reported as a barrier to the use of cover crops (Plastina et al., 2020, 1999; CTIC, 2014; CTIC, 2016; CTIC, 2023; CTIC, 2015). While more southern regions have a long enough growing season that, in theory, allows greater opportunities to use cover crops, these environments pose their own challenges including cover crops potentially replacing cash crops that could be grown, moisture deficits in subsequent cash crops caused by preceding cover crops, poor fertility of highly weathered soils, greater pest pressure and overly rapid cover crop residue decomposition and nutrient release (Balkcom et al., 2007). Other barriers to cover cropping include labor to seed and manage the cover crop, concern about impacts on planting the cash crop, equipment needs, and uncertainty about the effect on economics (CTIC 2014, 2015, 2016, 2023, and others).

Timely seeding is critical to fall-seeded cover crop success. They perform best when seeded while temperature and moisture conditions allow them to establish well before the onset of winter (Brinsfield and Staver, 1991; Balkcom et al., 2023). Nutrient scavenging and total biomass and nutrient accumulation are highly dependent on timely planting and decrease rapidly with delayed establishment (Lawson et al., 2015). Early planting enhances other ecosystem services as well, including erosion control, competition with weeds, rooting depth and compaction amelioration, nitrogen scavenging, and total nitrogen fixation (by legumes). Earlier planting fits well with many diversified vegetable systems. Such systems are also quite flexible with respect to cover crop termination as well and offer windows in summer for cover crops between cash crops.

Cool season cereal and other grass cover crops can be planted later than legumes and other dicot cover crops. Cereal rye (*Secale cereale*, L.) is the most common cover crop in the USA (Appelgate et al., 2017, 1999; CTIC, 2014; CTIC, 2016; CTIC, 2023; CTIC, 2015), at least in part because it performs better than legumes and other grain crops at later planting dates due to its ability to germinate and survive low temperatures (Duiker, 2014). In the most northern areas, cereal rye is often the only viable option, but it still benefits from planting early (Brinsfield and Staver, 1991; Balkcom et al., 2023; Duiker, 2014). Winter annual grass cover crops, including cereal grains and annual ryegrass (*Lolium multiflorum* Lam.), can take up significant amounts of soil residual nutrients and thereby reduce leaching and erosion losses, with better establishment and greater nutrient uptake the earlier they are seeded.

Annual legumes do not usually produce much fall biomass or take up large quantities of nutrients prior to winter if planted after summer cash-crop harvest. Earlier planting results in greater establishment success, but usually does not significantly increase fall nutrient uptake. Coombs et al. (2017), however, reported that fall N was 10 kg N ha^{-1} lower with alfalfa (*Medicago sativa* L.), red and crimson clover (*Trifoliateum pratense* L. and *Trifolium incarnatum* L., resp.) than no cover crop, suggesting these cover crops may mitigate N losses somewhat over the winter. Most legume growth occurs in spring and early summer, however (Clark et al., 1994; Clark et al., 1997; Clark et al., 1995; Dean and Weil, 2009), and that is when they fix most nitrogen and take up more nutrients from the soil.

Brassicas (including mustards) are a diverse genus with varying adaptations to different climates. Fall-planted Brassicas should be planted earlier than grass cover crops and can scavenge large quantities of nutrients (Clark, 2007; Hirsh et al., 2021; Weil and Williams, 2007; Dean and Weil, 2009). Some, especially the forage (or daikon) radish (*Raphanus sativus* L.), have been shown to root to depths of a meter or more, break up compaction layers, and have high nutrient uptake (especially nitrogen) from the subsoil, but this is only effective if planted earlier than other cover crops – e.g. by 1 September (Chen and Weil, 2010; Williams and Weil, 2004; Dean and Weil, 2009).

Early work with cereal rye in Maryland demonstrated drastic reductions in biomass production and nitrogen uptake with delays in planting from mid-September to early November (Brinsfield and Staver, 1991). In Pennsylvania, Duiker (2014) seeded cereal rye, wheat, barley (*Hordeum vulgare*, L.), annual ryegrass, hairy vetch (*Vicia villosa* Roth.), crimson clover, and rape (*Brassica napus* L.) on three dates from mid-September to mid-October over three years and determined above-ground biomass at three dates in spring (approximately May 1 to June 1). Study locations were central (latitude 40° 44′

N) and southeastern Pennsylvania (latitude 40° 05′ N). All cover crop species produced the most biomass with the September planting at both locations. Cereal cover crops ranged from 4.3 Mg ha^{-1} to nearly 10 Mg ha^{-1} biomass by June. Biomass production by the legumes declined dramatically as planting date was delayed, sometimes causing them to not survive the winter. Only cereal rye and wheat produced adequate biomass (defined as 1.5 Mg ha^{-1}) on all planting dates and years, while each of the other cover crops produced less biomass or did not survive the winter in some years and locations unless planted in mid-September. This work underscores the need to plant cover crops – especially legumes – early, or to default to cereal cover crops as fall planting date is delayed in more northern latitudes. In the Pacific Northwest, Lawson et al. (2015) tested hairy vetch, cereal rye, and two mixtures and demonstrated that delaying planting by 2.5 weeks from mid-September to early October reduced average winter ground cover of by 65%, biomass by 50%, and cover crop N accumulation by 40%.

Fall establishment and growth of winter annual cover crops can be determined in part by Growing Degree Days (GDD) accumulated after planting, but there is limited research specifically addressing this. Sedghi and Weil (2022) compared various planting dates for a radish, crimson clover–triticale (*Triticosecale*) cover crop mixture across four site-years (including sandy and silty soils). They also compared early-planted pure cereal rye, radish, and a three-species mixture with no cover. The early planting in 2016 (15 August) resulted in 500 more GDD (base 40F) until 1 December than the early planting (12 September) in 2017. Additional GDD available before 1 December increased cover crop biomass linearly up to ~800 GDD. After accumulating 800 GDD, the rate of biomass increase per GDD declined. University of Delaware Cooperative Extension (2021) seeded cereal rye on 30 September, 13 October, and 30 October. The early-planted rye accumulated 616 GDD (base 40°F) in the first 30 days after planting compared to the late-planted cover crop which accumulated 301 GDD. The early planting biomass ranged from 6 to 8 Mg ha^{-1} compared to 2–4 Mg ha^{-1} for the latest planting, with intermediate values for the middle planting. Thapa et al. (2018) summarize GDD effects on hairy vetch–cereal rye mixtures compared to pure stands in their thorough meta-analysis of 20 published studies. Hairy vetch–cereal rye mixtures consistently produced more biomass than both hairy vetch and cereal rye monocultures across a wide range of GDD (e.g. 600–1600). Similarly, hairy vetch–cereal rye mixtures consistently accumulated more N than cereal rye monocultures across a wide range of GDD, but only at lower GDD (<1000) when compared with hairy vetch monocultures. The lower N content of mixtures relative to hairy vetch monocultures at higher GDD (>1200) was most probably due to lower hairy vetch biomass in the mixture. As accumulated GDD increased, the biomass and N content of hairy vetch–cereal rye mixtures decreased relative to the biomass and N content of hairy vetch monocultures.

It is clear then that after harvest of summer cash crops in the fall, the window of opportunity for establishment of many cover crop species has already passed, especially moving North through corn and soybean growing regions. This has stimulated both researchers and practitioners to search for alternative methods to establish cover crops before harvest of the summer cash crop. These include broadcasting seed during combine harvesting, 'overseeding' a few weeks before harvest (e.g. R2 to R6 in corn (*Zea mays*, L.) or R6 to R8 in soybean (*Glycine max* (L) Merr.), or 'interseeding' cover crops into the cash crop early in the growing season (e.g. V3 to V7 in corn), or frost seeding at Feekes 3.0 in winter wheat or barley (Fig. 1). Research and farmer experience indicate that these alternative establishment methods have variable success rates (Youngerman et al., 2018; Belfry and Van Eerd, 2016). See Klopp (2024) for a recent review and comparison of cover crop seeding method research.

We address these challenges and highlight some of the innovative ways that farmers are making cover crops work for them. We discuss each establishment method, timing, and advantages and drawbacks, starting with establishment after harvest and working back progressively earlier in the cash crop growing season. We focus primarily on winter annual cover crops for grain and oilseed crops, but the concepts are broadly applicable to most cropping systems.

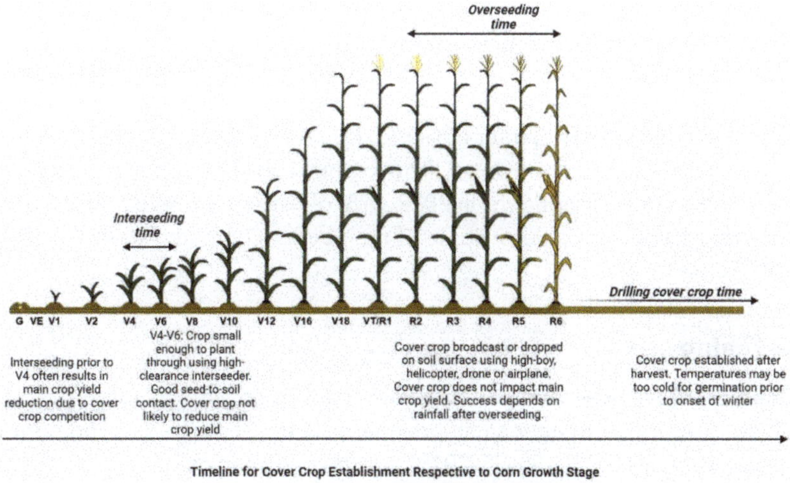

Figure 1 Cover crops can be established in the cash crop growing season or after harvest as illustrated here for corn.

3 Establishment methods: drilling after a cash crop harvest

Drilling cover crops after cash crop harvest results in better establishment than other seeding methods (Moore and Mirsky, 2020; Noland et al., 2018) because it ensures good seed to soil contact, accurate metering and depth placement, and absence of competition with the cash crop (Oliveira et al., 2019; Van Eerd et al., 2023). No-till drills for Conservation Agriculture need to be able to cut through the crop residue of the harvested crop, place cover crop seeds at desired spacing and depth, and close the seed slot (Duiker and Myers, 2006). Conventional drills place the seed in soil that has been prepared to have a firm subsoil and fine, loose surface soil for optimal seed-to-soil contact, depth control, and absence of crop residue at or near the soil surface. Conventional drills are light and do not work well in most no-till situations because of their inability to plant into firm no-till soil with more than 10% crop residue cover.

3.1 Regional considerations

Drilling cover crops in the fall after cash crop harvest is limited in many areas by post-harvest growing season length, as temperatures and growing degree days start to decrease and winter sets in. Most of the South and Southeast USA have a long enough growing season to allow for post-harvest cover crop establishment. Parts of the mid-Atlantic are also suitable, but going further north, e.g. north of latitude 40°N (Duiker, 2014), post-harvest drilling usually gives poor results after crops such as corn grain or soybeans. In the Midwest, post-harvest drilling of cover crops works well in the Southern Corn Belt (some say south of Interstate 80), while it is inconsistent or unsuccessful north of latitude 42°N (Mohammed et al., 2020).

Drilling cover crops after harvest of the cash crop is the most common form of cover crop establishment because it is the most foolproof. Conservation Technology Information Center (CTIC, 2023) reported survey responses from farmers in 49 U.S. states, with Iowa, Illinois, Ohio, and Minnesota accounting for about 30% of total responders. In 2022, 57% of farmers surveyed drilled cover crops after harvest on at least some of their acres, compared to 46% in 2012 (CTIC, 2014).

3.2 Timing

Strategies to plant when temperatures favor cover crop germination and growth include growing them after small grain crops (Fig. 2), early-maturing corn or soybean varieties (USDA-NRCS, 2019), or after early harvested cash crops (e.g. high-moisture corn; corn silage). These tactics broaden the range of cover crop species that can be drilled (Clark, 2007; Wallander, 2021). Wallander (2021)

Figure 2 Farmer no-till drilling cover crop through stubble left after wheat harvest (grain and straw removed). Courtesy Sjoerd W. Duiker.

reported that cover crops were grown more consistently following corn silage (56%) than corn grain (19%) or soybeans (32%). Cotton (*Gossypium hirsutum* L.), which is grown further south with a longer growing season, ranked highest with 69% of the area seeded with cover crops.

3.3 Methodology

Cover crops can be drilled into no-till or conventionally tilled fields. In both cases, residue of the previous crop needs to be uniformly distributed. This is useful for conventional tillage – although the crop residue is incorporated – but essential in Conservation Agriculture systems – where crop residue is left on the soil surface (Prasad and Power, 1991). Specialized drills are essential for no-till drilling of cover crops.

3.4 Species selection

The time of establishment determines which cover crop species can be planted. After the harvest of crops such as corn or soybeans, depending on latitude, there is some flexibility, but species selection may be limited to those that have greatest winter-hardiness such as cereal rye and winter wheat. If harvest of these cash crops can be done earlier than normal (e.g. by harvesting high moisture corn, silage, or early-maturing varieties of soybeans), a window may open for legumes such as hairy vetch, crimson clover, winter peas [*Pisum sativum* subsp. *pisum* var. *arvense* (L.) Poir.], as well as other dicot cover crops. Cocktails or mixtures of grass and legume cover crops need to be planted to accommodate temperature requirements of the legume in the mix (Ranells and Wagger, 1996). After small grain harvest there are many more options: cool season grasses such as oats (*Avena sativa* L.), barley, and annual ryegrass; warm season grasses such as pearl millet, Japanese millet, sudangrass, sorghum-sudangrass, and sorghum; cool season legumes such as hairy vetch and crimson clover; warm season legumes such as cowpeas, mung bean, and sunn hemp; and other forbs such as buckwheat, forage radish, rape, sunflower, and buckwheat. These cover crops can also be used in case of cash crop failure.

3.5 Effects on cash crop yield

Because drilling happens after cash crop harvest, there is no impact on cash crop yield.

3.6 Factors determining success

If cash crop harvest is close to onset of winter, there may not be enough time for a warm-season winter-killed cover crop to generate much growth, or for a winter-hardy species to establish before winter. Large amounts of, or poorly distributed, crop residue at the soil surface can pose a problem for cover crop establishment with no-till practices. Removal of part of the crop residue or some form of shallow tillage that sizes and cuts up residue can address this issue (Siemens and Wilkins, 2006). Cash crop residue needs to be uniformly distributed behind the combine for successful cover crop establishment with a no-till drill (Prasad and Power, 1991). Stalks, stubble, or stems of the cash crop are best left standing and anchored in the soil instead of being mowed, which lays more residue down on the soil surface and often results in windrowing of residue (Siemens and Wilkins, 2006; Prasad and Power, 1991). Seeding depth should be tailored to the cover crop species or mixture; in general, plant seeds at a depth equal to three times their diameter. A common cause of cover crop seeding failure of small-seeded cover crops is that they are planted too deep, and they don't have enough reserves to emerge through a thick soil layer. If planted in mixtures, the depth should be set closer to that required for the smallest seed in the mixture. Finally, adequate soil moisture is needed for germination of the cover crop; if the soil is very dry, it will compromise or delay cover crop emergence. However, if the soil is poorly drained and saturated for extended periods after planting the cover crop, that may also lead to a poor cover crop stand.

4 Establishment methods: broadcasting after a cash crop with or without incorporation

Cover crop drilling is a relatively slow process, so a time-saving alternative is broadcasting. Cover crop seed is broadcast on the soil surface during or after cash crop harvest, with or without shallow incorporation.

4.1 Regional considerations

As described in detail above, rainfall and growing season are the primary factors affecting the timing and success of post-harvest broadcast seeding. In particular, adequate rainfall following seeding is critical for germination and early growth of broadcast cover crops. Shallow incorporation may increase the success rate, but soil moisture is often inadequate following harvest of the cash crop.

4.2 Methodology

The cover crop seeds are broadcast with a spinner or pendulum spreader, from a drop box, or from the air by drones, helicopters, or airplanes. A seed box with a metering device can be mounted on combine heads, and seed is blown through flexible tubes onto the soil underneath the harvesting head (Fig. 3).

Figure 3 Gandy seeder box mounted on a combine, blowing cover crops seed through tubes under the head so that they will be covered by corn stover, improving moisture for germination. Courtesy Sjoerd W. Duiker.

Seed that is broadcast on the soil surface after harvest can also be incorporated lightly into the soil with a shallow tillage pass.

4.3 Species selection

Small-seeded species generally work better for broadcast cover crop seeding without incorporation because a larger proportion of the seed is in contact with the soil to imbibe water for germination (Fisher et al., 2011). If the seed is covered by crop residue after it is broadcast, cover crops with larger seeds are desirable with enough reserves to emerge through the residue, especially if large quantities of crop residue cover the seeds. Cover crop species adapted to broadcasting followed by shallow incorporation need to be able to emerge from variable depths. Eventually, only the most winter-hardy species such as cereal rye can be used as soil temperatures decrease below 4–6°C.

4.4 Effects on cash crop yield

Cover crops are established after cash crop harvest, so there is no impact on cash crop yield.

4.5 Factors determining success

Broadcasting seeds without incorporation results in poor seed-soil contact which reduces the seeds' ability to take up soil moisture necessary for germination (Haramoto, 2019; Fisher et al., 2011). Rainfall or irrigation shortly after broadcasting increases the success of establishment. Germination is often delayed until surface soil moisture conditions favor it (Fisher et al., 2011). Small-seeded species typically germinate better at the soil surface than large-seeded species (Fisher et al., 2011; Berti et al., 2017). If cover crop seed is broadcast on top of a heavy blanket of residue of a harvested crop, establishment may be further compromised. Farmers/engineers have therefore developed air seeders mounted on combines that can blow the cover crop seed on the soil surface between harvesting head and crop residue spreader (Fig. 3). This guarantees that the seed is in contact with the soil and protected from desiccation by the crop residue that falls on top of it. This equipment and the practice were developed by farmers, and there is no published research on it yet. Farmer organizations such as Practical Farmers of Iowa and Sustainable Chesapeake (see https://m.youtube.com/watch?v=qetSGZtxCCQ) are good sources of information.

To improve seed-to-soil contact, shallow tillage can be used to incorporate the seed after broadcasting. Horizontal distribution and depth placement of the seeds vary depending on the tillage tool and affect establishment of cover crop species differently (Brennan and Leap, 2014). Tillage tools that throw soil sideways (e.g. concave disks) lead to greater horizontal non-uniformity, while some tools (e.g. disks, rototillers) can bury seeds too deep or to varying depths. Establishment of large-seeded species that can emerge from depth will be better with this type of incorporation than that of small-seeded species. To compensate for poorer establishment, seeding rate should be increased by 50–100% compared with drilling (Fisher et al., 2011; Brennan and Leap, 2014).

5 Establishment methods: manure slurry seeding

Manure slurry seeding involves mixing cover crop seed with liquid manure and applying it in late summer or fall after cash crop harvest. Although this seeding method is not a fit for all farming operations, it can be an efficient method where manure and slurry application equipment are available.

5.1 Regional considerations

Research is limited on the use of this seeding technique, with most work done in the upper Midwest, the mid-Atlantic, and Northeast USA, with some work in Canada as well.

5.2 Methodology

First studied extensively at Michigan State University (Harrigan et al., 2006), slurry-seeding is done with a commercially available slurry tanker equipped with a rear-mounted rolling-tine aerator and an SSD (sub-surface deposition) slurry distribution system (Fig. 4). The manure tank needs to have bypass flow capacity, so seeds are kept in suspension in the manure. Cover crop seed is mixed with the manure in the slurry tank and passed through a hydraulically driven, rotating chopper/distributor with radially configured outlets to drop tubes onto the loosened soil behind the aerators. Termed micro-site seeding by Harrigan, cover crop seed settles in soil cracks and fissures and germinates with the aid of manure moisture. Germination and plant stands are often reduced compared to non-manure plots, attributed to ammonia (Suarez, 2013; Hilsey and Bench, ND), or salt content of the manure (https://www.midwestcovercrops.org/planting/). Moisture and nutrients in manure promote rapid cover crop growth, which in turn prevents loss of manure N. As with other post-harvest establishment methods, growing-season length can be a limiting factor.

5.3 Species selection

Although manure slurry seeding can be used for a variety of cover crops, it is generally best suited for grasses, which are well adapted for establishing

Figure 4 Manure slurry tanker equipped with aerator and rotating chopper/distributor that can be used to slurry-seed cover crops. Courtesy Sjoerd W. Duiker.

quickly and scavenging manure nutrients in the fall (Pioneer, n.d.). Cereal rye is the most commonly planted cover crop with well-documented ability to scavenge soil residual N, including in manure-based systems. Given the application of nutrient-rich manure in summer or fall with slurry seeding, the inclusion of a grass cover crop is warranted. Both cereal rye and wheat work well when slurry-seeded (described below), while oats were less successful in some work (Harrigan, 2010).

Early work in Michigan (Harrigan, 2006) evaluated oilseed radish and oriental mustard slurry-seeded into wheat stubble. Plant stands were significantly lower compared to the drilled treatment, but total biomass was equal or greater. In various other studies at several Michigan locations over 3 years (Harrigan et al., 2007), biomass yields of ten cover crop species (cereal rye, wheat, annual ryegrass, oil seed radish, forage rape, forage turnips, oriental mustard, and red, ladino, and crimson clover), seeded into wheat or corn silage stubble, and three manure types (swine, dairy and beef) were equal to or greater when slurry-seeded than no-till drill seeding (Harrigan et al., 2007).

In Ontario (Ministry of Agriculture, 2022), a three-species cover crop mixture (14% Nitro radish, 16% crimson clover, 70% oats planted at 33.5 kg/ha) and a ten-species mixture (33% oats, 4% Nitro radish, 2% *Brassica*, 2% sorghum-Sudangrass, 1% phacelia, 2% sunflowers, 4% Sunn hemp, 5% turnips,

25% crimson clover, 23% Austrian peas planted at 45 kg/ha) were compared to an oat cover using various seeding methods following wheat harvest with and without slurry-seeded manure digestate, which is similar in composition to swine manure. Biomass was greatest in all manure treatments, ranging from 17% to 36% greater compared to no manure. In the ten-species mix with only 17% increase in biomass, rapid and aggressive growth by oilseed radish outcompeted other species, possibly leading to lower overall biomass production.

Cover crop germination is usually reduced when established by manure slurry seeding. In Indiana, Suarez (2013) studied the effects of ammonia and salinity on cover crop germination and biomass production. Ten cover crop species (cereal rye, oats, wheat, annual ryegrass, crimson clover, maris kestrel kale [*Brassica oleracea* (L)], daikon radish, pasja hybrid [*Brassica campestris* (L)]) were slurry-seeded with swine manure and compared to an equivalent salinity level using sodium chloride. Manure slurry reduced germination by up to 50%, while germination was not affected by the salinity treatment, suggesting that it is not the salinity of the manure but the ammonia that causes reduced germination. The amount of time that cover crop seed is in the manure slurry may also impact germination. Suarez (2013) and Hilsey and Bench (2019) recommend less than one hour in the slurry before germination may be reduced.

Slurry seeding was also studied for late-season forage (Harrigan, 2010). Forage rape, forage turnip, brown midrib sorghum-Sudangrass (Sudex), and common oats were sown in untilled wheat stubble on August 8, 2005. The oat stand was significantly lower in the slurry-seeded compared to the no-till drilled treatment, while forage rape and forage turnip were not different. The sorghum-Sudangrass did not establish well with either seeding method. Harrigan (2010) also studied the use of slurry seeding to improve pastures. Red clover and orchard grass (*Dactylus glomerata*) were slurry-seeded into an established bromegrass (*Bromus madritensis*) pasture in late summer, and compared to late summer drilled and winter frost-seeded treatments. No-till and slurry seeding were more effective than frost seeding. Because the addition of manure and moisture stimulates growth of existing grasses, Gramoxone was used to burn down existing bromegrass and improved establishment of the orchard grass but not the red clover. In both studies, cattle preferred the non-manured fields in the fall of the seeding year.

5.4 Effect on cash crop yield

Cover crops are not established into standing cash crops, so there is no effect on the current year cash crop yield.

5.5 Factors affecting success

Slurry seeding is best used on relatively level ground, because seed may not spread evenly when going up and down hills (Hilsey and Bench, 2019). Manure should be continuously agitated, cover crop seeding rate should be increased by 30–50%, and seed should not be left in the slurry for more than about 1 hour (Suarez, 2013; Harrigan, 2006; Hilsey and Bench, 2019). For pasture improvement, existing pasture should be grazed or clipped very low, and the use of a burndown herbicide may improve establishment of the new seeding (Harrigan, 2010).

6 Establishment methods: pre-harvest overseeding

Please see Box 1 for key terms used in this and following sections.

Box 1 Overseeding vs. interseeding

Because cover crop establishment during the cash crop growing season is a relatively new strategy, the terminology has evolved over time. In particular, the terms 'overseeding' and 'interseeding' have been used in different ways in the published literature. For the purposes of this chapter, we will use the definitions described by Silva et al. (2023) based on other recent published research (Fig. 1). As such, **interseeding** is defined as planting a cover crop early in the growing season in rows, such as between corn rows when it is between the V3 and V8 vegetative growth stages (Smith et al., 2019, Youngerman et al., 2018). In contrast, cover crop **overseeding** is typically done by broadcast seeding just before or at crop physiological maturity (Kladivko et al., 2014), e.g. growth stages R4, R5, and R6 for corn, and R6, R7, and R8 for soybean. **Frost seeding** is a specific form of overseeding cover crops into winter cereals such as wheat or barley around Feekes growth stage 3.0. See Klopp (2024) for a recent review of cover crop seeding method research.

Overseeding is done late in the season but before cash crop harvest. The rationale behind this strategy is to allow more time for the cover crop to establish before winter sets in. But because the cash crop is still standing, overseeding is limited to broadcasting the seed because planting equipment would do too much crop damage. Overseeding is primarily used in grain crops planted in spring and harvested in the fall. The cash crops are typically corn or soybeans but can also be vegetable crops or other summer crops such as sorghum or cotton. In mechanized crop production, overseeding equipment needs to be able to drive through the cash crop or apply the seed from the air or the edge of the field. We will focus on overseeding in grain production systems in humid, temperate regions.

6.1 Regional considerations

Cover crops are typically overseeded where the growing season is too short to seed them after cash crop harvest. Most of the Southern USA has a long enough growing season to allow for post-harvest cover crop establishment, so overseeding is not typically used there. Overseeding is favored where drilling post-harvest is challenging because of early onset of winter, i.e. north of latitude 40°N in the eastern USA (Duiker, 2014), and north of 42°N latitude in the Midwest (Mohammed et al. 2020).

6.2 Timing

Overseeding is usually done 1–1.5 months before corn or soybean harvest. While earlier overseeding would give the cover crop more GGDs to allow it to establish before the onset of winter, seedlings would experience greater competition for sunlight, nutrients, and water under the canopy of the cash crop. Mohammed et al. (2020) reported that cover crop plant populations were generally lower with the earliest overseeding date (R4 in corn, R6 in soybean), but significant date by species interactions and differences over 2 years did not allow for definitive overseeding date recommendations. The use of early-maturing cash crops allows the canopy to open up early, so overseeded cover crops put on more growth before the onset of winter. Kandel et al. (2021) evaluated three cover crop treatments and two soybean maturity dates in North Dakota. Cereal rye and winter camelina were overseeded (reported as interseeded) at the R6 soybean growth stage, using two different soybean maturity groups (0.5 vs. 0.9) and two row spacings (30.5 vs. 61 cm) and compared with no cover crop. The early-maturing soybean cultivar (0.5 maturity) resulted in increased cover crop biomass and soil cover, with cereal rye outperforming winter camelina. However, the early-maturing soybean yielded 5% less than the late cultivar. Narrow row spacing had higher soybean yield, but row spacing did not affect cover crop growth. Early-maturing cash crop varieties will improve the likelihood of success with overseeding, but cash crop yield may be reduced.

6.3 Methodology

The cover crop seed is broadcast while the cash crop matures and dries down. This is a time when the crop canopy is thinning and water use is declining, providing better light and moisture conditions for cover crop establishment than during the period of active cash crop growth. Cover crop seed can be flown on (e.g. dropped from an airplane, helicopter, or drone), dribbled or spun on using a high-clearance rig, or on a smaller scale, blown on from the sides of the field. As a general rule, seeding rates should be increased 25%-100% above drilling seeding rates to compensate for poor seed-to-soil contact. The uniformity of application of flown or spun-on seed is poorer than that of dribbled seeding using a high-clearance rig. This practical aspect of overseeding on farm scale is missed in many small plot studies where the researchers make sure the seeds are distributed uniformly. There is typically a period of a few weeks when overseeding cover crops can work in corn as it dries down prior to harvest, compared to soybeans, in which there may only be a 1-week window to successfully overseed (after leaves turn yellow but before leaf drop).

6.4 Species selection

Overseeding is especially promising for winter-hardy, small-seeded legume cover crop species, such as crimson clover (Peterson et al., 2021; Youngerman et al., 2018), red clover (Wallace et al., 2017), and *Brassica* species, such as turnips (*Brassica rapa* L.) or camelina (Berti et al., 2017), established before summer crop harvest to increase growth prior to onset of the winter (Brooker et al., 2020b). In Texas, Keeling et al. (1996) tested 13 cover crop species/cultivars overseeded into cotton at boll opening at 3 locations over 6 years. Species included winter wheat, cereal rye, Austrian winter pea, hairy vetch, subterranean clover (*Trifolium subterraneum* L.; 5 cultivars), rose clover (Trif. hirtum All.; 3 cultivars), crimson clover (2 cultivars), red clover, berseem clover (*Trif. alexandrinum* L.), and barrel medic (*Medicago truncatuill* Gaertn.; 2 cultivars). Successful stands of wheat, cereal rye, winter pea, and hairy vetch were obtained in 69% of the plantings, but in only 53% for the other forages. The larger seeded cover crops succeeded when smaller seeded did not, attributed to low rainfall in those location years. However, in other situations smaller seeded cover crops may perform best because of their ability to get down into the cracks and crevices of the soil compared to a large seed like Austrian winter peas. Koehler-Cole et al. (2023) determined that a legume mix of large-seeded hairy vetch and winter pea produced less than 0.2 Mg ha^{-1} biomass in spring in three locations in Nebraska irrespective of establishment method (overseeding in September or drilling post-harvest in corn or soybeans). Mohammed et al. (2020) compared overseeding pennycress (*Thlaspi arvense* L.), camelina (*Camelina sativa* L.), and cereal rye in the U.S. Midwest (MN, ND, IA) and found that cereal rye produced the most biomass at most location/years, but that pennycress provided the most spring ground cover at the ND site. However, even with small-seeded cover crops, overseeding does not always give better results than post-harvest seeding (Berti et al., 2017), mostly dependent on weather conditions. Overseeding can also be used for establishment of grass cover crop species to speed up their establishment before the onset of winter (Koehler-Cole and Elmore, 2020). Species that winter kill (e.g. oats, radish) are not recommended for overseeding because they will not produce significant growth before a killing freeze (Malone et al., 2022).

6.5 Effect on cash crop yield

Establishment of cover crops before cash crop harvest may introduce the risk of competition of the cover crop with the cash crop for nutrients and water.

However, because overseeding is done close to harvest of the cash crop, it is unlikely to affect cash crop yield. Mohammed et al. (2020) tested the feasibility of overseeding winter camelina, field pennycress, and cereal rye at growth stages R4, R5, and R6 in corn, and R6, R7, and R8 for soybean at four locations in Minnesota, North Dakota, and Iowa. Neither cover crop species nor overseeding date affected corn or soybean yield or soil water content at harvest when compared with a control. Blanco-Canqui et al. (2025) did not measure a corn yield reduction on a sandy loam soil in the western Corn Belt due to broadcast overseeded cereal rye 32–67 days prior to harvest. Peterson et al. (2021) overseeded cover crops into double-cropped soybeans in both 38 cm and 76 cm rows in Maryland. Cereal rye, as well as cereal rye mixtures with hairy vetch, crimson clover, red clover, and winter peas were seeded at soybean R5 (late August or early September). Cover crop biomass was usually greater in cereal rye plus legume mixtures than cereal rye alone. Cover crop establishment was unsuccessful in two of five trials. The cover crop did not reduce soybean yield in any of the trials. Soybean yield in the wider soybean row spacing (78 cm) was not different from narrow row spacing in 6 of 8 site-years, which may increase cover crop overseeding success due to less shading. Corn yield the following season was usually greatest following cereal rye plus legume mixtures. Nelson et al. (2022) did not observe soybean yield reductions due to overseeding rye or ryegrass after R6.5. Occasionally, however, one sees a report of yield reduction due to overseeding, such as with rye in Wisconsin (Malone et al., 2022). In general, however, overseeding can be safely used after R4 for corn and R6 for soybeans without concern about cash crop yield reduction.

6.6 Factors determining success

Overseeding can be expected to have more success in short-statured crops that drop their leaves before harvest, allowing greater light penetration while leaving small amounts of crop residue after harvest that could otherwise cover small seedlings, blocking their ability to photosynthesize. Therefore, cover crops are typically more successful when overseeded into soybeans than into corn (Mohammed et al., 2020; Koehler-Cole and Elmore, 2020; Koehler-Cole et al., 2023). Cover crops such as cereal rye overseeded into corn result in etiolated, spindly seedlings with few tillers compared with dark-green, shorter, robust seedlings when broadcast in soybeans (Koehler-Cole and Elmore, 2020). Hairy vetch also shows more compact, vigorous growth when overseeded in soybeans than in corn. Koehler-Cole and Elmore (2020) reported 57% and 48% more biomass in spring for cereal rye and hairy vetch, respectively, when overseeded in soybeans than in corn.

The greatest challenge with overseeding is poor seed-to-soil contact which compromises moisture uptake for germination, while seedlings are exposed to predators, desiccation, and cold temperatures. Koehler-Cole et al. (2023) reported greater cover crop biomass in spring with overseeding than post-harvest drilling in sub-humid parts of eastern Nebraska that had frequent rainfall at the time of overseeding, but the practice was inferior to post-harvest drilling in a semi-arid western location where dry conditions are common when cover crops are typically overseeded. Often only a small percentage of seeds develop into seedlings, depending on rainfall and soil moisture conditions. Koehler-Cole and Elmore (2020) observed that as few as 3–5% of cereal rye and 15% of hairy vetch seeds overseeded into corn and soybeans in September developed into plants in a south central Nebraska location, while 22% of cereal rye and 42% of hairy vetch seeds became plants in the more humid eastern location. Based on their research, these authors estimated that 28 mm precipitation is needed in the week following overseeding for acceptable establishment. Despite poorer seed emergence, cereal rye produced more biomass in spring than hairy vetch due to its ability to tiller, showing that tillering species are better able to compensate for low seedling establishment. The winter hardiness of cereal rye is another factor affecting its performance if germination is delayed in a dry fall. Cost of seed also has to be taken into account. Koehler-Cole and Elmore (2020) reported that the cost of hairy vetch seed was almost $200 per hectare at the lowest overseeding rate, while that of cereal rye was about $60 per hectare at the highest seeding rate (Koehler-Cole and Elmore, 2020). In summary, overseeding can be expected to be more successful with small-seeded, affordable cover crops, in soybeans rather than in corn, and if overseeding is followed by precipitation shortly after seeding. Nonetheless, only a fraction of the broadcast seeds can be expected to develop into seedlings, so seeding rates should be increased 25–100% compared to normal drilling.

7 Establishment methods: frost seeding

Frost seeding is a long-standing overseeding practice to establish leguminous cover crops in fall-seeded small grain crops in the late winter (Singer et al., 2006). It is also used to establish legumes or grasses in pastures or hayland (Castillo et al., 2022; Undersander et al., 2001; George et al., 2000; Gettle et al., 1996). Small-seeded leguminous cover crops such as red clover, sweet clover (*Melilotus officinalis* L.), or crimson clover are broadcast or drilled onto winter cereals (e.g. wheat or barley) in the late winter months (February, March), usually at about Feekes growth stage 3.0. Once grain and straw from the small grain are harvested in June or July, the leguminous cover crop will already be well established. The legume is able to fix significant amounts of atmospheric nitrogen that can benefit a nitrogen-demanding crop such as corn or sorghum planted about 10 months after small grain harvest.

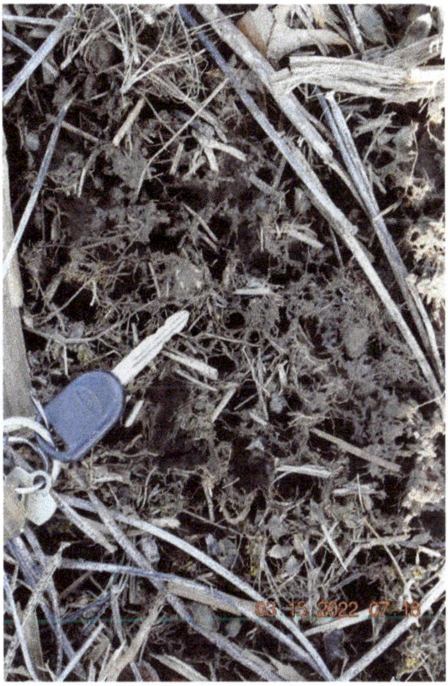

Figure 5 Soil with 'honey comb' surface structure in early morning due to night frost. This is the ideal time to frostseed. Courtesy Sjoerd W. Duiker.

7.1 Timing

Frost seeding is done when the soil goes through freeze-thaw cycles just prior to the onset of spring, when the small grain is in the vegetative stage (typically Feekes stage 3 or 4). At night, the soil develops a special 'honeycomb' surface structure with small openings that allow the broadcast seed to be incorporated into the soil (Fig. 5). When the soil thaws during the day, the soil covers the seeds, improving seed-to-soil contact.

7.2 Methodology

Legume cover crop seed is usually broadcast with a spinner-spreader, a pendulum spreader (Fig. 6), or a drop box spreader. It is also possible to use a conventional or no-till drill to improve seed-to-soil contact (Anderson, 2017). Typically, the cover crop is seeded early in the morning when the soil surface is still frozen and can better support the farm equipment. If the soil is not frozen – e.g. later in the day – the operation can compromise both the small grain stand and cover crop establishment and increase the risk of soil compaction. Seed

Figure 6 A pendulum spreader used to frost seed cover crops into small grain stand. Courtesy Sjoerd W. Duiker.

distribution should be carefully monitored, especially when using spinners or spreaders.

7.3 Species selection

Red and sweet clover are the most common species used for frost seeding, but crimson or white clover (*Trifolium repens* L.) can also be used (Castillo et al., 2022). Red clover is a perennial that often does not establish well after small grain harvest, but if it can be established in February or March of year one, it will have had more than one year to establish itself before termination prior to next crop's seeding. Yang et al. (2024) found that frost-seeding red clover provided more nitrogen to the subsequent corn than red clover planted after wheat harvest. Sweet clover is a biennial with a strong tap root that is vegetative in year one and will produce flowers and seed in year two. However, it should be terminated in year two prior to seed set. These two species are preferred for frost-seeding applications, although some farmers have also experimented with crimson clover (an annual) and white clover (a perennial).

7.4 Effects on cash crop yield

The frost-seeded cover crop should germinate and establish itself, but not become too vigorous and compete excessively with the cash crop. In an organic crop rotation in Ontario, Canada, wheat yield was reduced 10–15% by competition from frost-seeded red clover (Yang et al., 2024). The clover can be suppressed with a herbicide to reduce this competition. Anderson (2017) reported wheat yield reduction when berseem clover was drilled into winter wheat one month after it came out of dormancy due to wheat damage from the no-till drill. Yield loss due to the frost-seeded clover is not usually a major problem, however.

7.5 Factors determining success

Frost seeding in February/March has greater chances of success than broadcast overseeding at other times of the year (e.g. September/October) due to better seed-to-soil contact and typically more favorable soil moisture conditions. The honey-comb soil structure, cool temperatures, and plentiful moisture in spring help achieve greater success with frost seeding than with other forms of overseeding. The greatest challenge to frost-seeding legumes to provide nitrogen for the following crop is non-uniformity of the legume stand. If seed is not distributed uniformly, areas of no cover will then be colonized by weeds. In addition to the potential weed problem in the subsequent cash crop, the areas with no cover crop will lack nitrogen (Gaudin et al., 2013). To improve uniformity of seeding, broadcasting should not be attempted in windy conditions. Also, fluffy seeds will not be spread as far as dense seeds. This can be a problem when spreading cover crop mixtures with a spinner or pendulum spreader because the fluffy seeds are not thrown as far as the dense seeds. Spinner spreaders should rotate at consistent speed to guarantee uniform spreading width, and the gate on the spinner spreader needs to be adjusted so seed is spread evenly to both sides of the spreader. Further, the operator needs to space passes with the spreader at consistent distance from the previous pass to avoid skips.

8 Establishment methods: interseeding

Interseeding establishes a cover crop between the (usually) wide-spaced rows of a cash crop early in the growing season. Because the crop is relatively short and the canopy is not yet closed, it is possible to drive through the crop with a tractor and high clearance seeder that plants the cover crop seeds in the soil (Fig. 7), which improves seed-to-soil contact compared with broadcasting seeds at the surface. However, interseeding can also be done by broadcasting the seeds at the soil surface, either with or without cultivation.

8.1 Regional considerations

Interseeding has been predominantly tested with corn planted in wide rows (e.g. 76 cm row spacing) in the Midwest and the eastern USA. These regions have precipitation throughout the year, but cold winter temperatures pose a challenge to cover crop establishment after harvest of crops such as soybeans or

Figure 7 High-clearance interseeder planting a cover crop at approximately V5 in corn. Three double-disk openers plant the seeds between two corn rows. This interseeder is designed to work in no-till conditions. Courtesy Ron Hoover.

corn, providing the rationale for interseeding. Interseeding seems to be better suited to corn grown in northern locations in soils that have high water holding capacity. The shorter growing season, the more open corn canopies, and the ability to provide water sustain the cover crops until cash crop harvest. Curran et al. (2018) found that location, which ranged from latitudes of 39–43°N, had the greatest impact on cover crop biomass production. Corn was harvested as silage at two northern locations, resulting in greater cover crop biomass, attributed to a longer fall cover crop growth window in which the cover crop is unimpeded by light competition with the cash crop.

8.2 Timing

Interseeding is most common in corn prior to V7 growth stage when the crop is still small enough to drive through it. Interseeding after V4 is unlikely to result in yield reductions of corn due to cover crop competition (Brooker et al., 2020a; Klopp, 2024; Mohammed et al., 2020; Curran et al., 2018). Cover crops are sometimes interseeded into soybeans around V4, but are more commonly overseeded at R4 to R8 (Klopp, 2024). The reason interseeding is more common in corn than in soybeans is that the cover crrop can easily interfere with soybean harvest because the combine head needs to run close to the soil surface. It is also easier to no-till cover crops after soybean harvest than after corn harvest due to lower quantities of crop residue (compared with corn) that can interfere with seed placement in the soil. Nonetheless, cover crops are sometimes seeded at soybean V0 (at soybean planting) to assist with weed control (especially in organic production), with variable success but often a negative effect on soybean yield (Thelen et al., 2004; Brockmueller et al., 2022). Nelson et al. (2011) seeded cereal rye at soybean planting (V0) at seeding rates of 67, 134, and 200 kg/ha. Weed control provided by the rye was compared to a weedy control and weed-free control. Weed-free soybean yield was greater in both years (2450 kg/ha and 3800 kg/ha) than all cereal rye seeding rates (1000–1600 kg/ha and 2600–3100 kg/ha), while the weedy check yield was significantly less than the cover crop treatments. Brockmueller et al. (2022) seeded cereal rye with organic soybeans at V0 in June (later than the standard May planting date) and compared it to the standard May soybean planting date. Under optimal conditions where rye biomass was maximized, interseeding rye adequately suppressed weeds without reducing soybean yields as compared to tilled controls. However, under drier conditions with lower rye production, increased weed pressure and reduced yields emphasize the risks associated with interseeding a cover crop at planting. In short, establishment of cover crops at the same time as the cash crop often incurs a yield penalty, one reason why interseeding is recommended between V4 and V7 in corn (and in soybeans, although overseeding is more common in this crop).

8.3 Methodology

In organic systems interseeding is possible by broadcasting the seeds between the rows and stirring or pressing them into the surface soil at time of last cultivation for weed control with the cultivator shovels or rollers, respectively. Alternatively, cover crops can be drilled between rows in soybeans at V0 (see above). Specialized high-clearance drills have been developed for Conservation Agriculture systems. These are modifications of normal no-till drills that plant two or three rows of cover crop seeds between the rows of the cash crop. Cover crop seeds can also be broadcast on the soil surface with spinner spreaders, airplanes, helicopters, drones, or high-clearance rigs employed with airseeders dropping seeds between crop rows through drop-tubes. Due to improved metering, seed-to-soil contact, and depth control, drilling cover crops between crop rows tends to give better stands than broadcasting with shallow cultivation, while some form of incorporation is again better than broadcasting cover crop seeds at the soil surface. Noland et al. (2018) compared broadcast into the inter-row, broadcast with light incorporation, and a high-clearance drill at two locations in MN (Lamberton and Waseca, MN). The drilled cover crops produced the most biomass in fall, while drilled or incorporated seeds were most productive by spring for all cover crops except for pennycress, which was not affected by planting method. Authors concluded that larger seeds benefitted most from drilling or incorporation, while this was less essential for the small pennycress seeds.

8.4 Species selection

Some cover crops are better adapted to interseeding than others. Cover crop species suited for interseeding have to be shade tolerant and able to survive the hot and sometimes dry conditions of summer. Abdin et al. (1997) seeded 13 different cover crop species 10 or 20 days after corn emergence in Quebec, Canada. For most species, the earliest interseeding date resulted in better cover crop establishment and survival, with the exception of ryegrass/white clover which did better at the later interseeding date. Crimson clover was the best performing cover crop in this interseeding experiment, showing good soil cover and biomass production after corn harvest, with ryegrass/red clover, ryegrass/white clover, and berseem clover also performing well. However, Persian clover (*Trifolium resupinatum* L.), cereal rye, alfalfa, strawberry clover (*Trifolium fragiferum* L.), and hairy vetch were less suited for interseeding in this study. Cereal rye emerged quickly, but then started to die in July. This is a common phenomenon with cool season species that do not tolerate shade or hot/dry conditions.

Moore and Mirsky (2020) tested four cover crops and three establishment methods over three seasons in Maryland. Cover crops were cereal rye, annual ryegrass, a legume mixture including red clover, crimson clover, and hairy vetch, and a grass–legume mixture including annual ryegrass, red clover, crimson clover, and hairy vetch. Establishment methods were drill-interseeding at corn growth stage V4–V6, broadcast-overseeding at corn maturity, and post-harvest drilling. Cereal rye produced the most biomass when drilled post-harvest and was less successful when interseeded compared to annual ryegrass or the legume grass mixture. Other research also suggests that cereal rye is not shade tolerant and therefore may not persist under a dense crop canopy (Abdin et al., 1997). Annual ryegrass biomass was relatively stable across all treatments, so may therefore be a more preferred species for interseeding. Significant cover crop by establishment method by year interactions were partially explained by winter conditions, the timing of precipitation, and the timing of field operations (e.g. late post-harvest drilling due to delayed corn harvest in one year). Isbell et al. (2022) obtained excellent results interseeding annual ryegrass in corn at V5–V6 in corn in central Pennsylvania. Interseeded ryegrass reduced soil nitrate-N concentrations post corn silage harvest (but not after corn grain) in the fall, in contrast to post-harvest drilled cereal rye cover crop. In spring, though, ryegrass produced lower – although still very high (3500—4500 kg ha^{-1}) – amounts of biomass than cereal rye (5000–8000 kg ha^{-1}), but both reduced soil nitrate-N concentrations similarly compared with fallow, which would otherwise be subject to leaching losses.

Curran et al. (2018) evaluated interseeding different cover crops at V5 in corn at 16 locations in New York, Pennsylvania, and Maryland. They tested annual ryegrass, a legume mixture of red clover, crimson clover, and hairy vetch, and a mixture of annual ryegrass and the same three legumes. Across locations, the ryegrass-legume mixture produced most fall and spring biomass. Radish out-produced ryegrass and clover (Brooker et al., 2020a).

8.5 Effect on cash crop yield

While overseeding is done close to harvest of the cash crop and therefore unlikely to affect cash crop yield, interseeding can reduce cash crop yields due to competition because the cover crop is established before cash crop canopy closure (Bich et al., 2014). Much depends on when the cover crops are interseeded. Curran et al. (2018) did not observe corn yield reductions when three different cover crops (annual ryegrass, a legume mix, or a grass-legume mix) were interseeded at V5 (approx. 4–6 weeks after planting) in New York, Pennsylvania, and Maryland. Red clover/ryegrass mix reduced corn yields when interseeded at V2 or V3, but not at V4, V5, or V6 in a Pennsylvania trial (Curran et al., 2018). It is therefore recommended to interseed at or after V4 growth stage in corn, which would be approximately 3–4 weeks after corn planting (Fig. 8). Similarly, Isbell et al. (2022) did not observe yield reductions in corn due to interseeded annual ryegrass at V5–6. Klopp (2024) concluded the same in a review of 70 interseeding studies. Cover crop species can also play a role, with more aggressive species that survive well under the corn canopy more likely

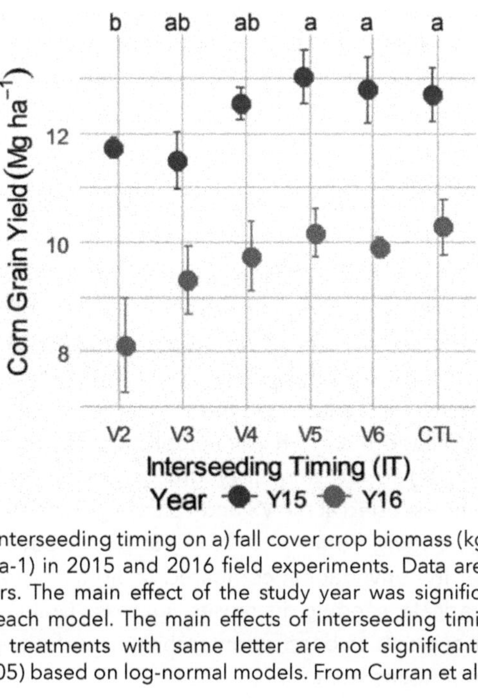

Figure 8 Effect of interseeding timing on a) fall cover crop biomass (kg ha-1) and b) corn grain yields (Mg ha-1) in 2015 and 2016 field experiments. Data are treatment means and standard errors. The main effect of the study year was significant (using Tukey's test; $P < 0.05$) in each model. The main effects of interseeding timing are presented. Interseeder-timing treatments with same letter are not significantly different (using Tukey's test; $P < 0.05$) based on log-normal models. From Curran et al. (2018).

to reduce yields. Abdin et al. (1997) observed occasional corn yield reductions with crimson and sweet clover interseeded 10–20 days after corn emergence in Quebec, but not with 11 other cover crop species. Brooker et al. (2020a) did not observe yield reductions from broadcast interseeding ryegrass, crimson clover, or radish at V3–V6 stage in corn in Michigan. If rainfall or soil moisture is limited, the interseeded cover crop can compete with the cash crop, resulting in cash crop yield reductions (Brooker et al., 2020a). Weed pressure should be low (e.g. through prior cultivation or non-residual herbicide application) at the time of interseeding to favor successful cover crop establishment. In summary, interseeding after V4–V5 leaf stage in corn is unlikely to reduce corn yield.

8.6 Factors determining success

Success with interseeding increases if the cover crops are drilled with a specialized drill that uniformly spaces seeds and plants them at consistent depth between cash crop rows, obtaining good seed-to-soil contact. Second best is broadcast application with shallow incorporation, while the most risky is broadcasting, as shown by Noland et al. (2018). In a Minnesota study, cereal rye (var 'Rymin'), red clover ('Medium'), hairy vetch, field pennycress ('MN-106'), and a mixture of oat, pea, and tillage radish were interseeded in corn at the V7 growth stage using broadcasting, broadcasting+shallow incorporation, or drilling. Drilling resulted in best establishment in the fall, but in spring, broadcasting+shallow incorporation had similar cover crop biomass as drilling for most of the species, with both methods outproducing broadcasting. Cover crop biomass was still relatively low in this grain corn study in the Corn Belt (e.g. approximately 1000 kg ha^{-1} spring biomass). Wallace et al. (2021) interseeded a mix of annual ryegrass, orchardgrass (*Dactylis glomerata*, L.), and forage radish with a drill interseeder or broadcast with a spinner-spreader just after last cultivation of corn on organic farms in Pennsylvania. Fall cover crop biomass and radish proportion were higher with drill-interseeding, but in spring, biomass was similar. If fall nitrate-N uptake is the goal, drill interseeding is therefore preferred over broadcasting (Isbell et al., 2022), but if spring cover crop growth is the goal, broadcast after cultivation can be an economical alternative. Broadcast applications are typically more successful after conventional tillage than in no-till established cash crops. In Michigan, Brooker et al. (2020a) tested annual ryegrass, crimson clover, and oilseed radish broadcast interseeded at the V3 and V6 corn growth stages on nine farm fields (at latitudes ranging from 41° 51′N to 45° 03′N). Cover crops generally established better in conventional compared to no-till systems. Precipitation shortly after broadcast seeding increases success. Drill-interseeded cover crops are less susceptible to dessication at the soil surface, so results can be expected to be more consistent.

If done after V4 in corn (or soybeans), the cover crops are unlikely to reduce cash crop yield. The main challenge, however, has proven to be the intense competition of the cash crop with the cover crops from the time of interseeding to harvest, often causing stands to be thin and variable, especially after corn grain harvest. If cash crop harvest is early, such as when corn is harvested for silage, then the cover crops can be expected to put more growth on in the fall before the onset of winter. If the cash crop canopy is relatively open, as in sweet corn, organic corn, or seed corn (Belfry and Van Eerd, 2016; Isbell et al., 2022), the interseeded cover crops can be expected to do better. While not yet well researched, new short corn varieties may have the same beneficial effect as an open canopy. Another possibility is interseeding into corn planted at wider row spacings, most commonly at 60 inches (150 cm). Some studies show corn yield reductions (up to 39%) due to wider row spacing (Bibby, 2022; Nelson, 2014), but other studies show comparable or better corn yields compared with normal row spacing (Lyon et al., 2009), while cover crop establishment is much improved (Goettl et al., 2024). Corn hybrids adapted to wide-row management should be used in this system, while new programs of weed control need to be developed since crop canopy cover is open during much of the season.

9 Herbicide impacts on cover crop establishment

Herbicides are used by most commercial farmers around the world. They are used by farmers using no-tillage or tillage. Herbicides are used on Genetically Modified (GM) crops resistant to particular herbicides (most well-known being glyphosate-resistant crops), but they are also used for weed control in non-GM crops. Because they are so widely used, their impact on cover crop establishment must be considered because they can facilitate but also seriously compromise cover crop establishment. There are different types of herbicides. Non-residual herbicides do not have soil activity and therefore do not affect establishment of cover crops planted after their application. Examples of these are glyphosate or paraquat. Other herbicides have residual soil activity after they are applied, for example, dicamba, 2-4D, glufosinate, and atrazine. Soil residual herbicides may not affect the cash crop but provide weed control for some period until they dissipate and are degraded by soil microbes or chemically. These herbicides can affect cover crop establishment for some amount of time after application. How long these residual herbicides remain active in the soil is affected by herbicide type, soil texture and organic matter (due to adherence to clay particles or humus, breakdown is inhibited), pH (at pH>6.8 atrazine or sulfonylurea won't chemically break down), rainfall (dry soil inhibiting breakdown), and temperature (low temperature slows breakdown). Residual activity of herbicides is ranked by half-life, a property determined in standardized laboratory procedures measuring the amount of time that passes until half of the herbicide concentration remains.

Herbicides can be applied before planting, pre-emergence, or post-emergence. They control weeds which can help improve cover crop establishment. Abdin et al. (1997) observed that high weed pressure posed a problem for successful cover crop establishment and growth, something that can be addressed with the use of non-selective herbicides without residual activity. However, depending on herbicide half-life, cover crop injury or complete death may occur if the herbicides are still active when the cover crops are planted. Koehler-Cole and Elmore (2020), for example, reported a poor stand of hairy vetch that was broadcast in soybeans in September due to use of the broadleaf herbicide fluvioxazin (trade name Valor) in the soybeans. The effect of herbicides can typically be generalized by their effect on large-seeded cereal cover crops (such as cereal rye, oats, wheat), small-seeded grass cover crops (e.g. annual ryegrass), small-seeded broadleaves (e.g. Brassicas such as radish), or small-seeded legumes such as red or crimson clover. Interestingly, large-seeded winter pea is rated most sensitive but hairy vetch least sensitive to common pre- and post-emergence corn and soybean herbicides (Fig. 9). In general, residual herbicides that have activity on grass weeds can interfere with the establishment of grass cover crop species, especially the smaller seeded ones (such as ryegrass).

The danger of cover crop damage due to residual herbicides dissipates the longer the period between herbicide application and cover crop establishment is. When cover crops are established after crop harvest, the danger of injury from residual herbicides decreases, but doesn't completely disappear. Whalen

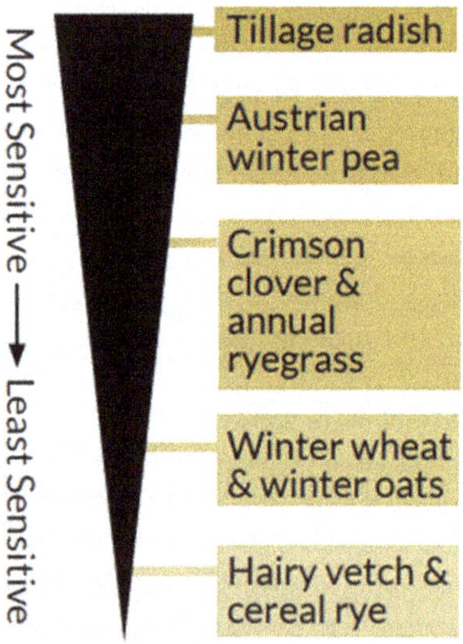

Figure 9 Cover crop species sensitivity rating to common pre- and post-emergence herbicides used in corn and soybean production (Hilsey and Bench, ND).

et al. (2019), for example, evaluated the sensitivity of ten different cover crops to residual herbicides used in soybeans and concluded that certain active ingredients such as fomesafen posed a risk to cover crops, but that in general the risk of injury to cover crops established after soybean harvest was low. Wallace et al. (2017) concluded in their research in U.S. Mid-Atlantic silt-loam soils that annual ryegrass and red clover can be safely interseeded in corn that has received short-lived residual herbicides. In conclusion, management decisions need to account for the potential effect of residual herbicides on the intended cover crop species, with particular attention to the amount of time between herbicide application and cover crop establishment. Before establishing cover crops, a producer should evaluate the herbicide program to determine if it could harm the cover crops.

10 Case study: interseeding cover crops in corn

Authors: Noah Wendt, Huxley, IA; Emma Link, Practical Farmers of Iowa; with Andy Clark.

Noah Wendt farms 2,500 acres (1012 hectares) in Polk and Story counties, Iowa (IA), USA. About 1,000 acres (405 hectares) are in no-till corn/soybean rotation, while the remaining 1,500 acres are in a certified organic corn/soybean rotation. He tries to plant cover crops on all of his land.

He and Caleb Akin of A & W Farms conducted the interseeding study described below on his organic fields in 2023 as part of the Practical Farmers of Iowa Cooperators' Program for farmer-led research. The year before the study, cereal rye was grown for cover crop seed. A diverse legume mix was frostseeded into the rye to provide nitrogen for the subsequent corn crop, the rye grain was then harvested and the straw was baled. The rye stubble plus legume mix were disked and then cultivated two times in the fall.

Corn was planted in spring of 2023, then tine-weeded, rotary hoed three times and row crop cultivated three times. Cover crops were interseeded in late June with the row crop cultivator at the last cultivation. Mr Wendt tested a warm season and a cool season mix and collected data on biomass production and subsequent corn yield. See below for the full project report.

Mr. Wendt also overseeds cover crops into both his organic and conventional fields, usually in August. He broadcasts a mix of cereal rye (*Secale cereal L.*) and rape (*Brassica napus*) with a spinner-spreader before soybeans, or oat (*Avena sativa*) and red clover (*Trifolium praetense*) before corn using a high-clearance rig. Overseeding is about twice as fast as interseeding with cultivation, so he sometimes opts for overseeding in August rather than interseeding due to time and labor constraints or in very wet years when he runs out of time to interseed.

Table 1 Planting and management details at Noah Wendt's farm in 2023

Cash crop	Corn 30 inch rows, 34,500 seeds/ac
Previous crop	Cereal rye
Strip dimensions	900' x 30'
Cool season mix	Annual ryegrass (6 lbs/ac), rapeseed (2 lbs/ac), kale (2 lbs/ac), lentils (3 lbs/ac), purple top turnip (3 lbs/ac)
Warm season mix	Japanese millet (10 lbs/ac), sorghum-Sudangrass (10 lbs/ac)
Interseeding date	Last cultivation (June 30)
Weed control	May 20-June 30: Tine weeded 2 times, rotary hoed 3 times, row crop cultivated 3 times
Fertilizer	March 15, 4 ton/ac chicken dry litter
Corn harvest	Oct 20, 2023
Cover crop biomass cutting	Oct 30, 2023

In his organic system, he aims for at least 3 tons dry matter/ac (6.7 Mg/ha), but sometimes ends up with 5 tons/ac (11.2 Mg/ha). More biomass is better for weed control and organic matter additions. In the study described below, the cool season mix produced 5.2 tons/ac (11.7 Mg/ha).

Mr. Wendt established interseeding treatments at last cultivation of corn in randomized, paired strips with four replications. Planting details and management information are shown in Table 1. Mr Wendt recorded grain yield from each strip using a yield monitor. He also collected aboveground cover crop biomass 10 days after harvest from nine square-feet in one of each strip, dried, and weighed them for cover crop biomass calculations.

Cover crop mix treatment did not affect corn yield (average 154 bu/A), but there were drastic differences in biomass produced in the two treatments. The diverse cool season cover crop mix produced an estimated 5.2 tons/ac of biomass while the warm season sorghum-millet mix only produced 1.1 tons/ac (Fig. 10). Precipitation in July was much higher than average, which may have helped the cool season cover crop do well (Fig. 11). The warm season mix might be more shade-sensitive causing it to produce little biomass. The diverse, cool season seed cost $30/ac while the sorghum/millet warm season seed cost $25/ac, but increased forage production by the cool season mix more than paid for this difference in seed cost. Mr. Wendt reported that these results will likely influence his farming practices and he will likely stick with the cool season mix.

Mr. Wendt was very happy with the results of this study and with the collaboration with PFI. At the time he was also running livestock so was interested in the forage value of the warm season vs. the cool season mix.

(a)

(b)

Figure 10 Corn with interseeded cool season crop mix (a) and warm season cover crop mix (b).

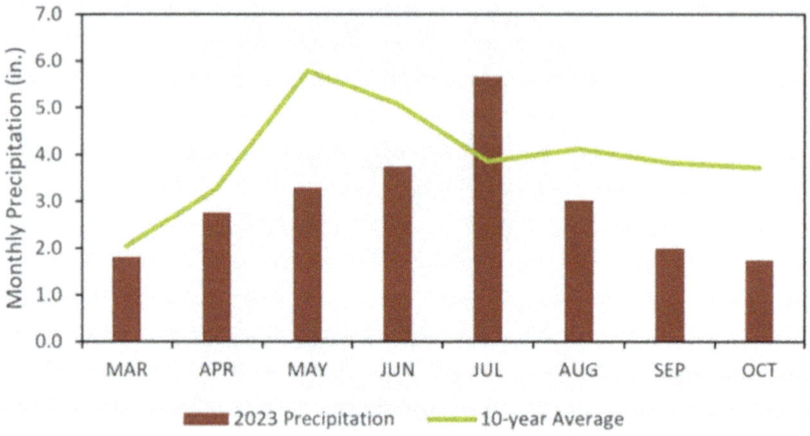

Figure 11 Monthly average precipitation and 2023 precipitation in case study location.

Although he no longer has livestock, the cover crop biomass and associated benefits are still important factors in his operation.

With current (2024-2025) farm economics, he relies on cover crop cost-share payments to make the practice economically viable. With recent reductions in the allowed acreage and the payment amount, he has not planted cover crops on all of his acres. He has also gravitated away from interseeding in favor of overseeding on more acres.

"Overseeding only works really well about 50% of the time," he said. "It is weather dependent. When you are dropping seed on top of the soil, you really need rainfall for it to be successful."

As described elsewhere in this chapter, adequate rainfall is essential for germination and early establishment when broadcasting cover crop seed, whether interseeded, overseeded or broadcast following cash crop harvest. Cost share payments for cover crops are also an important incentive for many farmers (CTIC, 2023), while current levels of investment may not be enough for farmers to continue using cover crops (Sun et al., 2025).

11 Conclusion

Timely establishment is critical to maximize the many benefits of cover crops, while seed-to-soil contact is needed for germination. Where the growing season is long enough, the best methods are: drilling after harvest > broadcasting followed by incorporation > broadcasting without incorporation. If insufficient growing season remains after cash crop harvest, interseeding and overseeding methods are options, but they face the added challenge of competition with the cash crop and the potential for failure. Cover crops are more commonly added to cotton

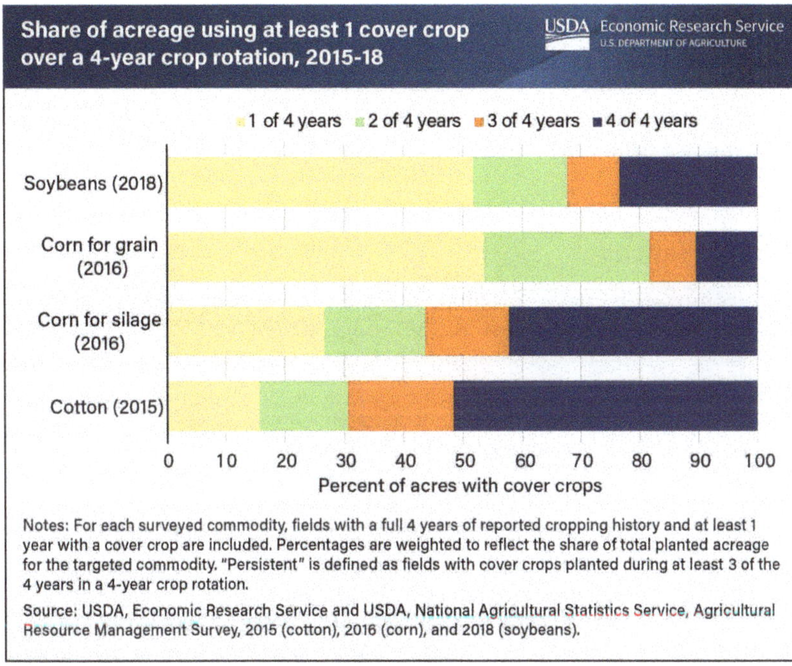

Figure 12 Share of acreage using at least one cover crop over a 4-year crop rotation, 2015-18.

or corn silage systems than corn grain or soybean systems (Fig.12). While non-residual herbicides can assist with cover crop establishment, residual herbicides can compromise it, especially the shorter the period between residual herbicide application and cover crop establishment. Research on cover crops worldwide is addressing many of the challenges described here, but more is still needed on the best species, the optimum timing for alternative seeding methods, cash crop variety selection (tall/short, early or late maturing, etc.), and row geometry (narrow vs wide row spacing), all as influenced by site-specific conditions.

12 Where to look for further information

- Clark, A. 2007. Managing Cover Crops Profitably, 3rd Edition. Sustainable Agriculture Network. Describes cover crop species, their attributes, and suggested seeding and rotation strategies (https://www.sare.org/resources/managing-cover-crops-profitably-3rd-edition/).
- Klopp, H.W. 2024. Effects of Interseeding Cover Crops into Corn and Soybean on Biomass Production, Grain Yields and Ecosystem Services: A Review. Agronomy 14, 1229. https://doi.org/10.3390/ agronomy14061229.

- Wallace, J.M. et al. 2024. 2024. Best Practices for Drill-Interseeding in Mid-Atlantic Crop Systems. Pennsylvania State University Extension. UBR AGR 24-246. https://extension.psu.edu/interseeding-cover-crops-in-corn-production. A thorough management guide to interseeding, applicable to many locations and cropping systems.

13 References

1999. A conservation catalog: practices for the conservation of Pennsylvania's natural resources. https://co.lancaster.pa.us/DocumentCenter/View/146/A-Conservation-Catalog. Practices-for-Conserving-Pennsylvanias-Natural-Resources. USDA Natural Resources Conservation Service.

Abdin, O. A., et al. 1997. Establishment, development and yield of forage legumes and grasses as cover crops in grain corn in Eastern Canada. *Journal of Agronomy and Crop Science*, 179, 19–27.

Anderson, R. L. 2017. Interseeding berseem clover in winter wheat. *Renewable Agriculture and Food Systems*, 32, 573–575.

Appelgate, S. R., et al. 2017. Cover crop options and mixes for upper midwest corn–soybean systems. *Agronomy Journal,* 109, 968–984.

Balkcom, K. S., et al. 2007. Managing cover crops in conservation tillage systems. *In:* Clark, A. (ed.) *Managing Cover Crops Profitably.* Washington, DC: Sustainable Agriculture Network.

Balkcom, K. S., et al. 2023. Rye planting date impacts biomass production more than seeding rate and nitrogen fertilizer. *Agronomy Journal*, 115, 2351–2368.

Belfry, K. D. & Van Eerd, L. L. 2016. Establishment and Impact of Cover Crops Intersown into Corn. *Crop Science*, 56, 1245–1256.

Berti, M., et al. 2017. Integrating winter camelina into maize and soybean cropping systems. *Industrial Crops and Products*, 107, 595–601.

Bibby, S. 2022. *Corn-Alfalfa Intercropping with Different Row Spacings.* M.S. thesis, North Dakota State University.

Bich, A. D., et al. 2014. Corn yield is not reduced by mid-season establishment of cover crops in northern great plains environments. *Crop Management*, 13, CM-2014-0009-RS.

Blanco-Canqui, H., et al. 2025. Interseeding cover crop into an irrigated sandy loam for 6 years: soil, crop, and economic response. *Agronomy Journal*, 117, e70013.

Brennan, E. B. & Leap, J. E. 2014. A comparison of drill and broadcast methods for establishing cover crops on beds. *HortScience Horts*, 49, 441–447.

Brinsfield, R. B. & Staver, K. W. 1991. Use of cereal grain cover crops for reducing groundwater nitrate contamination in the Chesapeake Bay region. *Cover Crops for Clean Water*, 79–82. https://www.swcs.org/resources/publications/cover-crops-for-clean-water.

Brockmueller, B., et al. 2022. Spring-seeded winter rye living mulches enhance crop biodiversity and promote reduced tillage organic soybeans. *Frontiers in Sustainable Food Systems*, 6, 2022.

Brooker, A. P., et al. 2020a. Interseeding cover crops in corn: Establishment, biomass, and competitiveness in on-farm trials. *Agronomy Journal*, 112, 3733–3743.

Brooker, A. P., et al. 2020b. Interseeded annual ryegrass, oilseed radish, and crimson clover tolerance to residual herbicides commonly used in corn. *Weed Technology*, 34, 35–41.

Castillo, M. S., et al. 2022. Clover frost-seeding rate effects on productivity and nutritive value of tall fescue pastures during the year of establishment. *Crop Science*, *62*, 1394–1403.

Chen, G. & Weil, R. R. 2010. Penetration of cover crop roots through compacted soils. *Plant and Soil*, 331, 31–43.

Clark, A. J. 2007. *Managing Cover Crops Profitably*, 3rd ed. Beltsville: Sustainable Agriculture Network.

Clark, A. J., et al. 1994. Seeding rate and kill date effects on hairy vetch-cereal rye cover crop mixtures for corn production. *Agronomy Journal*, 86, 1065–1070.

Clark, A. J., et al. 1995. Hairy vetch kill date effects on soil water and corn production. *Agronomy Journal*, 87, 579–585.

Clark, A. J., et al. 1997. Kill date of vetch, rye, and a vetch-rye mixture: I. Cover crop and corn nitrogen. *Agronomy Journal*, 89, 427–434.

Coombs, C., et al. 2017. Legume cover crop management on nitrogen dynamics and yield in grain corn systems. *Field Crops Research*, 201, 75–85.

CTIC 2014. 2013–2014 cover crop survey report. West Lafayette, IN.

CTIC 2015. 2014–2015 annual report cover crop survey. West Lafayette, IN.

CTIC 2016. Annual report 2015–2016 cover crop survey. West Lafayette, IN.

CTIC 2023. National cover crop survey report 2022–2023. West Lafayette, IN.

Curran, W. S., et al. 2018. Evaluation of cover crops drill interseeded into corn across the mid-Atlantic region. *Agronomy Journal*, 110, 435–443.

Dean, J. E. & Weil, R. R. 2009. Brassica cover crops for nitrogen retention in the mid-Atlantic coastal plain. *Journal of Environmental Quality*, 38, 520–528.

Duiker, S. W. 2014. Establishment and termination dates affect fall-established cover crops. *Agronomy Journal*, 106, 670–678.

Duiker, S. W. & Myers, J. C. 2006. *Steps Toward a Successful Transition to No-Till*. University Park, PA: The Pennsylvania State University.

Fisher, K. A., et al. 2011. Is broadcasting seed an effective winter cover crop planting method? *Agronomy Journal*, 103, 472–478.

Gaudin, A. C. M., et al. 2013. Improving resilience of northern field crop systems using inter-seeded red clover: a review. *Agronomy*, 3, 148–180.

George, J. R., et al. 2000. Growing legumes in mixtures with warm-season grasses. *In:* Moore, K. J. & Anderson, B. E. (eds.) *Native Warm-Season Grasses: Research Trends and Issues*. Madison, WI: Crop Science Society of America and American Society of Agronomy.

Gettle, R. M., et al. 1996. Frost-seeding legumes into established switchgrass: establishment, density, persistence, and sward composition. *Agronomy Journal*, 88, 98–103.

Goettl, B., et al. 2024. Interseeded cover crops in wide-row corn: An opportunity for northern cropping systems. *Crop, Forage & Turfgrass Management*, 10, e20268.

Hamilton, A. V., et al. 2017. The state of the cover crop nation and how to set realistic future goals for the popular conservation practice. *Journal of Soil and Water Conservation*, 72, 111A–115A.

Haramoto, E. R. 2019. Species, seeding rate, and planting method influence cover crop services prior to soybean. *Agronomy Journal*, 111, 1068–1078.

Harrigan, T. M. 2010. Improving pasture and hayground with manure slurry enriched seeding. *Bulletin e3102*.
Harrigan, T. M., et al. 2006. Manure slurry-enriched micro-site seeding of biosuppressive covers. *Applied Engineering in Agriculture*, 22, 827–834.
Harrigan, T. M., et al. 2007. Manure slurry-enriched seeding of cover crops. *Resource*, February 2007.
Hilsey, B. & Bench, C. 2019. *A Practical Guide to No-Till and Cover Crops in the Mid-Atlantic*. Washington, DC: Northeast SARE.
Hirsh, S. M., et al. 2021. Scavenging and recycling deep soil nitrogen using cover crops on mid-Atlantic, USA farms. *Agriculture, Ecosystems & Environment*, 309, 107274.
Isbell, S. A., et al. 2022. Nitrogen services provided by interseeded cover crops in organic corn systems. *Agronomy Journal*, 114, 2458–2472.
Kandel, H. J., et al. 2021. Soybean relative maturity, not row spacing, affected interseeded cover crops biomass. *Agriculture (Basel)*, 11, 441.
Keeling, J. W., et al. 1996. Comparison of interseeded legumes and small grains for cover crop establishment in cotton. *Agronomy Journal*, 88, 219–222.
Kladivko, E. J., et al. 2014. Cover crops in the upper midwestern United States: potential adoption and reduction of nitrate leaching in the Mississippi River Basin. *Journal of Soil and Water Conservation*, 69, 279–291.
Klopp, H. W. 2024. Effects of interseeding cover crops into corn and soybean on biomass production, grain yields and ecosystem services: a review. *Agronomy (Basel)*, 14, 1229.
Koehler-Cole, K. & Elmore, R. W. 2020. Seeding rates and productivity of broadcast interseeded cover crops. *Agronomy (Basel)*, 10, 1723.
Koehler-Cole, K., et al. 2023. Cover crop planting practices determine their performance in the U.S. Corn Belt. *Agronomy Journal*, 115, 526–543.
Lawson, A., et al. 2015. Influence of seeding ratio, planting date, and termination date on rye-hairy vetch cover crop mixture performance under organic management. *PLoS One*, 10, e0129597–e0129597.
Lyon, D. J., et al. 2009. Skip-row planting patterns stabilize corn grain yields in the central great plains. *Crop Management*, 8, 1–8.
Malone, L. C., et al. 2022. Cover crops in a Wisconsin annual cropping system: feasibility and yield effects. *Agronomy Journal*, 114, 1052–1067.
Ministry of Agriculture. 2022. *Slurry Seeding Cover Crops Following Wheat Harvest*. Ontario: Ministry of Agriculture, Food and Agribusiness. https://www.ontario.ca/page/slurry-seeding-cover-crops-following-wheat-harvest.
Mohammed, Y. A., et al. 2020. Establishing winter annual cover crops by interseeding into Maize and Soybean. *Agronomy Journal*, 112, 719–732.
Moore, V. M. & Mirsky, S. B. 2020. Cover crop biomass production across establishment methods in mid-Atlantic corn. *Agronomy Journal*, 112, 4765–4774.
Nelson, K. A. 2014. Corn yield response to the solar corridor in Upstate Missouri. *Agronomy Journal*, 106, 1847–1852.
Nelson, K. A., et al. 2022. Overseed timing of ryegrass and cereal rye in soybean affects rotational crops in upstate Missouri. *Crop, Forage & Turfgrass Management*, 8, e20184.
Nelson, K. A., et al. 2011. Spring-interseeded winter rye seeding rates influence weed control and organic soybean yield. *International Journal of Agronomy*, 2011, 571973.

Noland, R. L., et al. 2018. Establishment and function of cover crops interseeded into corn. *Crop Science*, 58, 863–873.

Oliveira, M. C., et al. 2019. Assessment of cover crop management strategies in Nebraska, US. *Agriculture (Basel)*, 9, 124.

Peterson, C. M., et al. 2021. Evaluation of interseeding cover crop mixtures in Mid-Atlantic double-crop soybean. *Agronomy Journal*, 113, 3935–3951.

Pioneer. n.d. *Managing Winter Cover Crops in Corn and Soybean Cropping Systems*. https://www.pioneer.com/us/agronomy/managing_winter_cover_crops.html.

Plastina, A., et al. 2020. Cover crops use in Midwestern US agriculture: perceived benefits and net returns. *Renewable Agriculture and Food Systems*, 35, 38–48.

Prasad, R. & Power, J. F. 1991. Crop residue management. *Advances in Soil Science*, 15, 205–251.

Ranells, N. N. & Wagger, M. G. 1996. Nitrogen release from grass and legume cover crop monocultures and bicultures. *Agronomy Journal*, 88, 777–882.

Sedghi, N. & Weil, R. 2022. Fall cover crop nitrogen uptake drives reductions in winter-spring leaching. *Journal of Environmental Quality*, 51, 337–351.

Siemens, M. C. & Wilkins, D. E. 2006. Effect of residue management methods on no-till drill performance. *Applied Engineering in Agriculture*, 22, 51–60.

Silva, T. S., et al. 2023. Evaluating cover crop tolerance to corn residual herbicides using field-treated soil in greenhouse bioassay. *Weed Technology*, 37, 500–511.

Singer, J. W., et al. 2006. Wheat effect on frost-seeded red clover cultivar establishment and yield. *Agronomy Journal*, 98, 265–269.

Smith, H. D., et al. 2019. *Interseeding Cover Crops in Row-Cultivated Corn*. Madison, WI: University of Wisconsin.

Suarez, E. A. 2013. *Slurry Seeding Cover Crops*. West Lafayette, IN: Purdue University.

Sun, H., et al. 2025. The flux of agricultural conservation: Understanding changes in Iowa farmer's adoption of cover crops and no-till over time. *Society and Natural Resources*, 1–20.

Thapa, R., et al. 2018. Cover crops reduce nitrate leaching in agroecosystems: a global meta-analysis. *Journal of Environmental Quality*, 47, 1400–1411.

Thelen, K. D., et al. 2004. Utility of interseeded winter cereal rye in organic soybean production systems. *Agronomy Journal*, 96, 281–284.

Undersander, D. J., et al. 2001. Frost seeding into aging alfalfa stands. *Agronomy Journal*, 93, 609–619.

University-Of-Delaware-Cooperative-Extension. 2021. *Plant Cover Crops Early to Increase Chances of Successful Establishment*. https://sites.udel.edu/weeklycropupdate/?p=19502.

USDA-NRCS. 2019. *Guide Sheet: Using a Cover Crop Before Corn (in an Iowa Corn-Soybean Rotation)*. Ames, IA: USDA-NRCS.

Van Eerd, L. L., et al. 2023. Influence of cover crops at the four spheres: A review of ecosystem services, potential barriers, and future directions for North America. *The Science of the total environment*, 858, 159990–159990.

Wallace, J. M., et al. 2017. Tolerance of interseeded annual ryegrass and red clover cover crops to residual herbicides in Mid-Atlantic corn cropping systems. *Weed Technology*, 31, 641–650.

Wallace, J. M., et al. 2021. Drill and broadcast establishment methods influence interseeded cover crop performance in organic corn. *Renewable Agriculture and Food Systems*, 36, 77–85.

Wallander, S. 2021. *Cover Crop Use Is More Persistent on Cotton and Corn Silage Fields*. Washington, DC. https://www.ers.usda.gov/data-products/charts-of-note/chart-detail?chartId=100974.

Weil, R. R. & Williams, S. 2007. *Brassica Cover Crops to Alleviate Soil Compaction*. College Park, MD: University of Maryland.

Whalen, D. M., et al. 2019. Evaluation of cover crop sensitivity to residual herbicides applied in the previous soybean [Glycine max (L.) Merr] crop. *Weed Technology*, 33, 312–320.

Williams, S. M. & Weil, R. R. 2004. Crop cover root channels may alleviate soil compaction effects on soybean crop. *Soil Science Society of America Journal*, 68, 1403–1409.

Yang, X., et al. 2024. Legume cover crop as a primary nitrogen source in an organic crop rotation in Ontario, Canada: impacts on corn, soybean and winter wheat yields. *Organic Agriculture*, 14, 19–31.

Youngerman, C. Z., et al. 2018. Corn density effect on interseeded cover crops, weeds, and grain yield. *Agronomy Journal*, 110, 2478–2487.

Chapter 10

Key considerations in terminating cover crops

Alyssa Essman and Ramawatar Yadav, The Ohio State University, USA; and Travis Legleiter, University of Kentucky, USA

1 Introduction

Successful termination is a critical step in the use and management of cover crops in cropping systems. The method and timing of cover crop termination are two factors that will largely determine the outcome. Selection of an appropriate method and timing will depend on the cover crop species and the goals for using the cover crop (Fig. 1). The benefits of cover crops are correlated with the amount of growth achieved, specifically the biomass and ground covered provided by cover crop vegetation. Among all cover crop management factors, termination timing has the greatest effect on biomass production (Lawson et al., 2015; Mirsky et al., 2011; Mirsky et al., 2013). These decisions often require a tradeoff between the potential benefits of maximizing cover crop biomass while minimizing the risk of potential crop yield reductions associated with high biomass levels. Growing the cover crop for as long as possible can maximize

http://dx.doi.org/10.19103/AS.2025.0162.22

Figure 1 Cover crop termination.

the amount of biomass production and ground cover, but timely management is also necessary because failure to terminate by a certain growth stage can lead to control failures and have a negative effect on the following cash crop.

The most appropriate termination method and timing will maximize benefits from the cover crop while avoiding excessive competition with the cash crop for water, nutrients, and light. The three main methods of cover crop termination are natural, chemical, and mechanical. The selection and success of the different methods depends on cover crop species and growth stage. Termination of cover crops can take place before or after planting of the cash crop. There are benefits and drawbacks of both timings, and the optimal termination strategy will depend on the species being terminated, crop rotation, environmental conditions, and farmer preference.

2 Termination methods: natural termination

Natural termination refers to the use of fall-planted cover crop species that 'winter-kill' or die naturally over the winter due to low temperatures. In a survey of 550 growers, 16% reported *usually* using winter-kill species as a termination strategy, and 36% reported *sometimes* using winter-kill species (SARE, 2023). Species that winter-kill at low temperatures include oats (*Avena sativa* L.), tillage and oilseed radish [*Raphanus sativus* (L.)], Austrian winter pea [*Pisum sativum* (L.) ssp. *sativum* var. *arvense* (may winter-kill if planted early)], and sorghum sudangrass [*Sorghum*

Figure 2 Alfalfa coming up in April through an oat cover crop that was killed by frost.

bicolor (L.) Moench x *S. sudanense* (Piper) Stapf] (Fig. 2). Barley (*Hordeum vulgare* L.) is less winter-hardy than other common grass cover crop species and can winter-kill in some conditions or locations, as can annual ryegrass (*Lolium perenne* L. ssp. *multiflorum* (Lam.) Husnot also referred to as Italian ryegrass), some clover species (*Trifolium* spp.), and canola (*Brassica napus* L.). Winter-killed species often complete their growth at the first frost after planting, but the specific temperature required for plant death is species dependent (Table 1).

The likelihood of a frost-sensitive species being terminated naturally by sufficiently low temperatures can be influenced by cover crop planting date. Cover crop planting date is often dependent on the harvest date of the previous cash crop, and potential growth stages may differ at the time of frost, influencing the termination rate. If planted late, or if the winter is mild, some species may survive the winter and require an alternative termination strategy ahead of the following growing season.

At early vegetative stages, cover crop species that would otherwise winter-kill may be less sensitive to frost and survive the winter, requiring a termination activity in the spring (Fowler et al., 2014). Natural termination via the use of winter-killed species can simplify management in the spring and reduce the number of field activities needed prior to planting the cash crop, but, in general, winter-killed species provide fewer benefits than overwintering species. The short growing period compared to overwintering species, followed by death of small plants, provides a shorter period of soil cover, especially when these

Table 1 Species that can winter-kill in temperate climates and temperature thresholds for winter-kill

Plant family and species	Scientific name	Temperature threshold for winter-kill*
Brassicas		
Canola	*Brassica napus* L.	0 to −15°C[1]
Radish	*Raphanus sativus* (L.) spp.	−9 to −4°C[2,3]
Grasses		
Oat	*Avena sativa* L.	−6°C[4]
Spring barley	*Hordeum vulgare* L.	−8°C[5]
Sudangrass, Sorghum-sudangrass	*Sorghum bicolor* (L.),*Sorghum bicolor* (L.) Moench x *S. sudanense* (Piper) Stapf	0°C[6,7]
Legumes		
Field peas	*Pisum sativum* subsp. *arvense*	−10°C uncovered, −30°C with snow cover[8]
Berseem clover	*Trifolium alexandrinum* L.	−3 to −15°C[9]
Cowpea	*Vigna unguiculata* (L.) Walp.	0°C[10]
Crimson clover	*Trifolium incarnatum* L.	−12°C[11]
Hairy vetch	*Vicia villosa* Roth	−28°C[12]
Sunn hemp	*Crotalaria juncea* L.	0°C[13]
Miscellaneous		
Buckwheat	*Fagopyrum esculentum* Moench	0°C[14]

*References: [1]variety dependent, Secchi et al., 2021; [2]Licht et al., 2024a; [3]Lawley et al., 2012; [4]Eagles et al., 1993; [5]Jacobs, 2016; [6]Licht et al., 2024b; [7]Licht et al., 2024c; [8]Pavek, 2012; [9]variety dependent, Pettit and Haller, 2021; [10]Licht et al., 2024d; [11]Licht et al., 2024e; [12]Licht et al., 2024f; [13]Licht et al., 2024g; [14]Licht et al., 2024h.

species are planted following a late harvest. There is also less root growth and less biomass left on the soil in the spring compared to overwintering species. Growers will sometimes opt to plant winter-killed species in mixtures with overwintering species, in order to capitalize on the benefits of both (Baraibar et al., 2018). Winter hardiness is influenced by variety or cultivar and should be considered when planning cover crop species selection (Vann et al., 2021). Using varieties adapted for local conditions can reduce the likelihood of unintended winter-kill, and early planting date can improve the potential for survival (Duiker, 2014).

3 Chemical termination

Winter-hardy cover crop species usually survive the below-freezing temperatures of winter and must be terminated in order to effectively establish a cash crop in the spring. Often referred to as chemical termination, killing cover crops with herbicides is the most common and effective method of termination for winter-hardy cover crop species. A survey of 550 cover crop users reported that 56% *usually* use herbicides as the termination method, and 10% *sometimes* use chemical termination (SARE, 2023).

Important considerations in planning chemical termination strategies include cover crop species, growth stage, and environmental conditions before and after application. See herbicide labels for specific rate, adjuvant, and tank-mix recommendations. Non-selective herbicides tend to be more effective at terminating cover crops than selective herbicides (Kumar et al., 2023). Non-selective herbicides can include active ingredients that are systemic, such as glyphosate, or contact, such as paraquat and glufosinate. As described in detail below, one of these non-selective herbicides often serves as the 'base' for a termination application, with other active ingredients added to improve efficacy for certain species or weed control at the time of application (added to improve 'burndown') or in the weeks following (added to provide residual control).

3.1 Chemical termination of grass cover crop species

Many of the most common and effective overwintering cover crop species are members of the grass family (Poaceae). In the Eastern Corn Belt and the Mid-Atlantic regions of the United States, cereal rye (*Secale cereale* L.) is the most popular cover crop species in terms of use frequency (SARE, 2023). Other common grass cover crop species include wheat (*Triticum aestivum* L.), barley (*Hordeum vulgare* L.), and annual ryegrass. Annual ryegrass varieties can be problematic because they exhibit numerous weedy characteristics, are difficult to terminate, and the wild biotypes have developed herbicide resistance, including multiple-herbicide resistance (MHR), and thus, can easily escape control and become a weed in cropping systems (Bajwa et al., 2021; Bararpour et al., 2017; Tehranchian et al., 2019) (see Section 5.2).

Grass cover crop species are most easily controlled with applications of glyphosate, or mixes that include glyphosate, applied at the appropriate growth stage. Cereal rye, wheat, and annual ryegrass are most consistently controlled with glyphosate-based herbicide programs (Cornelius and Bradley, 2017; Palhano et al., 2018; Pittman et al., 2020). The addition of clethodim or saflufenacil to glyphosate programs can provide effective termination of grass cover crop species (Whalen et al., 2020a). Other chemical options for the termination of grass cover crop species include paraquat and glufosinate; however, these products are generally less effective, and there is a greater risk for regrowth following application. Paraquat and glufosinate have shown limited efficacy on cereal rye and wheat, but the addition of metribuzin or atrazine can improve the efficacy of paraquat applications (Cornelius and Bradley, 2017; Palhano et al., 2018). In grass species, optimal control occurs when herbicides are applied in vegetative growth stages (Cornelius and Bradley, 2017). Applications made during Feekes 1.0-9.0 (Zadoks 10-41) will be most successful, and efficacy of applications will decline beyond the boot stage (Feekes 10; Zadoks 45).

3.1.1 Cereal rye

Relative to other grass cover crop species, cereal rye is usually well controlled or terminated with timely applications of an appropriate herbicide. Glyphosate alone can result in 87–98% control, with greater control occurring when applied to smaller plants and/or earlier in the season. Reductions in control can occur when group 5 herbicides such as atrazine or metribuzin are included in the termination applications (Cornelius and Bradley, 2017).

3.1.2 Winter wheat

The control of winter wheat with herbicides can be more complicated than that of cereal rye and will be most effective on smaller plants. Glyphosate-based programs are the most effective, and higher rates and/or applications on smaller plants will improve consistency of control. Tank-mixing glyphosate with group 4 herbicides (2,4-D or dicamba) can reduce control of winter wheat when applied at later growth stages, and the inclusion of glufosinate in a tank-mix decreases control relative to glyphosate alone (Cornelius and Bradley, 2017; Palhano et al., 2018). Applications of paraquat or paraquat + metribuzin can also be effective for wheat cover crop termination (Kumar et al., 2023).

3.1.3 Barley

Glyphosate or mixes that contain glyphosate applied to winter barley at the jointing to boot stage can provide 96–98% control of barley as a cover crop, and glyphosate alone can provide a height reduction of 71%, validating efficacy data (Pittman et al., 2020). Glufosinate- and paraquat-based programs provide less control and height reductions than glyphosate-based programs.

3.1.4 Annual ryegrass

Annual ryegrass is a fast-growing species that can be difficult to adequately kill, and incomplete or poor control can lead to escape issues and, therefore, a weed problem in the following years (Van Eerd, 2023). When terminating

Figure 3 Corn no-tilled into a rye cover crop desiccated with a burn-down herbicide.

annual ryegrass, glyphosate-based programs have proven to be more effective than paraquat-based programs (Fig. 3). The most consistently effective treatments for termination of annual ryegrass include glyphosate alone or glyphosate + clethodim or saflufenacil (Cornelius and Bradley, 2017; Whalen et al., 2020a). When using paraquat as the base, the inclusion of atrazine or metribuzin can improve control relative to paraquat alone (Cornelius and Bradley, 2017). Higher rates of glyphosate can lead to improved outcomes. A glyphosate rate of 1.4 kg ae ha^{-1} had greater control relative to 0.84 kg ae ha^{-1} when plants were 15 cm tall, but neither rate provided greater than 87% control when plants were more than 35 cm tall. The addition of clethodim at 0.136 kg ai ha^{-1} improved control by 7–17% (Cornelius and Bradley, 2017).

3.2 Chemical termination of broadleaf cover crop species

Common broadleaf cover crop species in the United States capable of overwintering include hairy vetch (*Vicia villosa* Roth), winter pea species (*Pisum* spp.), clover species (*Trifolium* spp.), and canola. For chemical termination of these species, glyphosate, glufosinate, 2,4-D, dicamba, and paraquat are some of the more common options, and product efficacy will depend on the cover crop species and environmental conditions. For the termination of broadleaf cover crops, including 2,4-D or dicamba can increase the efficacy of control compared to glyphosate, paraquat, or glufosinate applied alone (Cornelius and Bradley, 2017; Palhano et al., 2018; Whalen et al., 2020a).

3.2.1 Clover spp.

Common clover species used as cover crops that can overwinter include crimson clover (*Trifolium incarnatum* L.), red clover (*Trifolium pratense* L.), and subterranean clover (*Trifolium subterraneum* L.). For termination of crimson clover, glufosinate is one of the most effective options (Pittman et al., 2020). Mixtures of glyphosate + 2,4-D, dicamba, or saflufenacil, and paraquat + atrazine are among the most effective treatments at an early termination timing (Cornelius and Bradley, 2017; Pittman et al., 2020).

3.2.2 Canola

Sometimes referred to as rapeseed, canola can be a difficult cover crop species to terminate. A study performed in Virginia found no herbicide treatment resulted in greater than 90% control of canola, and the greatest control came from paraquat and 2,4-D + glufosinate at 84% and 82% 14 DAT, respectively (Kumar et al., 2023). Another study in Virginia revealed that the maximum control provided by single active ingredient applications was 58% and increased to 62% in mixtures (Pittman et al., 2020). In Arkansas, treatments of 2,4-D, dicamba, glufosinate, glyphosate, paraquat, and various tank mixtures were tested for efficacy on canola. No treatments provided adequate control, or above the 71% provided by the mixture of paraquat + metribuzin (Palhano et al., 2018).

3.2.3 Hairy vetch

A regional study across five states found that paraquat and glufosinate alone provide more effective termination of hairy vetch than glyphosate alone (Whalen et al., 2020a). Including 2,4-D or dicamba in applications with any herbicide 'base' program can improve consistency of hairy vetch control (Cornelius and

Bradley, 2017; Whalen et al., 2020a). In Virginia, termination with glyphosate mixtures had improved control relative to paraquat-based mixtures (Pittman et al., 2020).

3.2.4 Winter pea

Overwintering peas are some of the more easily controlled cover crop species. Applied alone, glyphosate, glufosinate, and paraquat can provide 92%, 87%, and 85% control 4 weeks after application, respectively (Pittman et al. 2020). Applications of glyphosate + dicamba or saflufenacil and paraquat + 2,4-D or atrazine provide acceptable control regardless of termination timing (Cornelius and Bradley, 2017). As with hairy vetch, glyphosate alone is not as effective a treatment as applications that include 2,4-D or dicamba for winter pea termination (Whalen et al., 2020a).

3.3 Issues in using herbicides

Chemical programs for the termination of cover crops rely on the use of either glyphosate or paraquat as the foundation for control. These herbicides are non-selective, meaning they will kill any plant species with which they come into contact, except for those that have developed herbicide resistance or have been genetically modified with traits that confer resistance. However, these products have key differences in how they kill plants, which leads to different recommendations and best practices for each. Glyphosate is a systemic herbicide and works by targeting a specific enzyme, enolpyruval shikimate phosphate synthase (EPSPS), produced by plants (Duke, 2018). This means that the rate of kill will be slower, larger spray droplets can be used to reduce risk of drift, and carrier volume can be lower relative to contact herbicides. Paraquat is a contact herbicide, meaning the effects are localized to where spray droplets come into contact with leaf tissue, and killing the plant via cell membrane disruption (Bromilow, 2003). This means that the rate of kill will be much more rapid, droplets should be smaller in size to maximize leaf coverage, and carrier volume should be greater. Always refer to the product label for specific application requirements and tank-mix considerations.

One consideration for cover crop termination with herbicides is whether a residual should be included as part of the tank-mix. There are two primary concerns related to the influence of tank-mix partners: potential antagonistic effects (lower control efficacy than applied alone) and the ability of the residual herbicide to reach the soil surface (tie up in cover crop biomass or residue). Some tank-mixes are known to cause antagonism, which could reduce the efficacy of termination efforts. Products with known antagonistic influence on glyphosate include the Weed Science Society of America Group

5 photosystem II inhibitor herbicides, atrazine and metribuzin (Selleck and Baird, 1981).

There is also a concern about the amount of residual herbicide reaching the soil surface when applications are made to cover crops with high amounts of biomass and extensive ground cover. Residual herbicides can improve weed control when included in termination applications, both preemergence (PRE) and postemergence (POST). However, cover crop biomass can intercept herbicides at the time of application and reduce the concentration of these herbicides in the soil (Nunes et al., 2023a). The amount of herbicide that reaches the soil is affected by cover crop biomass levels and herbicide properties, and the use of residual herbicides at PRE or POST timing had mixed results (Hamberg et al., 2025; Whalen et al., 2020b). Reduced concentrations of residual herbicides may not affect control relative to full concentrations in the absence of a cover crop, but the application of reduced concentrations over time may have implications for herbicide-resistance development in weed populations (Nunes et al., 2023b; Manalil et al., 2011).

4 Mechanical termination

A survey conducted in 2023 found that out of the 550 growers surveyed, mowing was the most common mechanical method (16%/19% usually/sometimes), followed by tillage (16%/16% usually/sometimes), and rolling/crimping (5%/13% usually/sometimes) (SARE, 2023). Performing termination procedures at the correct growth stage, which is species dependent, is especially critical for effective control with mechanical methods.

4.1 Mowing

One option for mechanical termination of a cover crop is mowing. Use of a flail-mower spreads residue across the soil surface more uniformly than other mechanical methods or mower types (Creamer and Dabney, 2002; Kornecki and Kichler, 2002). Mowing can be effective when performed at an appropriate growth stage, which depends on the species used. When mowed too early, regrowth of the cover crop is a risk, especially for grass cover crop species (Creamer and Dabney, 2002). Use of a mower for cover crop termination also leads to more rapid residue breakdown relative to roller-crimping or herbicide applications that leave residue intact (Eivazi et al., 2024). This rapid breakdown can negate some of the benefits associated with cover cropping, such as late-season soil cover and weed suppression.

Cereal rye, barley, and common vetch (*V. sativa* L. ssp. *nigra* L.) can all be terminated with a flail-mower if mowed at later growth stages. Mowing can kill flowering hairy vetch more effectively than when it is in a vegetative growth stage (Teasdale and Rosecrance, 2003). However, vigorous hairy vetch growth including elongated stems and high amounts of biomass can cause incomplete termination with a flail-mower, when mowed at 80% flowering or greater (80% flowering = 4/5 nodes flowering; Mischler et al., 2010; Wayman et al., 2014). Wheat and cereal rye mowed at later growth stages lead to less regrowth (Wilkins and Bellinder, 1996). Mowing cereal rye near anthesis (Feekes 10.51, Zadoks 60) can limit regrowth potential (De Bruin et al., 2005). For brassica cover crop species such as canola, flail-mowing is often followed by tillage for incorporation (Bangarwa and Norsworthy, 2014; Baraibar et al., 2018; Gieske et al., 2016; Haramoto and Gallandt, 2005).

4.2 Tillage

Tillage can terminate cover crops by cutting the root systems and burying plant residue. The efficacy of tillage for cover crop termination is affected by tillage implement or type, timing, and cover crop species. There are tradeoffs associated with the use of tillage as a cover crop termination practice. Incorporation with a disk or plow can increase nutrient breakdown and cycling, due to the reduced amount of time that it takes for the cover crop residue to break down (Lupwayi et al., 2004). This can be beneficial when the goal is to more rapidly recover nutrients from the cover crop, to reduce the amount of biomass in the field at planting, or to reduce competition with young cash crop plants. However, tillage also reduces the amount of ground cover provided by the cover crop residue and can therefore negate many of the benefits associated with cover cropping, including soil preservation and weed suppression. Multiple passes may be required to achieve complete control of cover crops with tillage. Termination via tillage may require follow-up herbicide applications for adequate termination and weed control (Alonso-Ayuso et al., 2020).

4.3 Roller-crimping

Roller-crimping is an attractive option for growers interested in leaving large amounts of residue on the soil surface, and those using organic farming practices for which chemical termination is not an option (Mirsky et al., 2013). Like mowing, the efficacy of roller-crimping depends on the cover crop species and growth stage (Fig. 4). For grass species such as cereal rye, rolling after pollen shed (anthesis; Feekes 10.51, Zadoks 61) leads to the most consistent kill (Mirsky et al., 2009). For legume species such as hairy vetch, rolling in full bloom to early pod set leads to the greatest control (Keene et al., 2017; Mischler et al., 2010). Where multi-species mixes are used, roller-crimping should be conducted at the growth stage that is sufficient for termination of the

Figure 4 Roller crimping to terminate a cover crop.

species with the latest maturity requirement, to avoid variable or poor control and potential seed production of the cover crop (Keene et al., 2017).

4.4 Undercutting

Undercutting cover crops is an option for mechanical termination. It requires the use of an implement with a coulter and blade that severs roots underground, killing the cover crop and leaving the residue on the surface. This method has proven more effective for termination of young cover crops than roller-crimping (Hooks, 2024). Compared to disked cover crops, undercutting can increase soil nitrate levels, soil moisture content, and yield of corn and soybean (Wortman et al., 2012). In one study evaluating cover crop species and termination method, undercutting reduced early season grass biomass and increased corn and soybean yield 23% and 13%, respectively (Wortman et al., 2013).

5 Termination timing

One of the most important considerations when planning cover crop termination practices is the timing of termination activities. The timing of termination influences control efficacy and cover crop outcomes, including biomass production and weed suppression (Wallace et al., 2017; Osipitan et al., 2019). Earlier termination can increase cover crop control efficacy with herbicides and potentially reduce soil moisture losses from the cropping system at the time of cash crop planting where soil moisture is limited (Cornelius and Bradley, 2017; Balkcom et al., 2015; Groff, 2015). Later termination can increase cover crop biomass and thus weed suppressive potential; in wet springs when termination is delayed, later termination may reduce interference with equipment at the time of planting, if performed when the green cover crop is standing vs. planting into a large mat of plants on the soil surface (Kaspar and Licht, 2019). The risk of later termination timing is excessive competition with the emerging cash crop and potential implications for yield (Reed et al., 2019). Also, delayed termination can lead to more advanced growth stages at the time of termination, which is beneficial for some mechanical termination procedures but could potentially lead to seed production by the cover crops (Keene et al., 2017).

5.1 Termination prior to cash crop planting

Terminating cover crops 10 to 14 days before planting the cash crop is the most common recommendation (MCCC, 2025). The benefits of early termination include less risk of termination failure with herbicides, lower amounts of biomass to interfere with cash crop establishment and growth, and less concern about damage from pests such as voles (Essman et al., 2023) because of the lower biomass levels. Potential allelopathic effects of cover crops on weeds and crops are uncertain, and likely not detrimental to yield in field conditions (Dhima et al., 2006; Silva and Bagavathiannan, 2023). Early termination can reduce the risk of nutrient competition, especially nitrogen, if cover crops like cereal rye with high C:N ratios are grown prior to a cash crop with high N needs, such as corn (Quinn et al., 2023).

5.2 Termination after cash crop planting

Also referred to as 'planting green,' delaying termination until after cash crop planting is a viable option, in part because increasingly unpredictable spring weather patterns may otherwise delay termination efforts and cash crop planting

(SARE, 2023). Planting green allows for more timely planting operations than waiting to get into the field to plant after termination activities, especially in years with excess spring precipitation (Groff, 2015). Planting into living residue can potentially improve planting conditions and equipment operation relative to planting into a mulch of decaying cover crop residue where the risk of 'hairpinning' (instead of cutting cover crop and placing seed in the soil, the cover crop residue is pushed into the seedslot and the crop seed does not have good seed to soil contact) is greater, especially in wet springs (Balkcom et al., 2015; Reed et al., 2019). The greater biomass levels that are possible with later termination also increase the weed suppressive potential of a cover crop and can increase the amount of organic matter that goes back into the system (Bloszies et al., 2022; Eivazi et al., 2024; Mirsky et al., 2013; Rosario-Lebron et al., 2019). Larger amounts of biomass left on the soil surface can also lead to increased soil moisture conservation later in the season, when many areas are at greater risk for low precipitation and moisture stress (Kornecki and Kichler, 2023).

6 Risks of incomplete termination

Termination is a critical management factor for cover cropping, and decisions must balance the risks vs. such goals as soil health and weed suppression. Effective termination requires using the right method and the right timing, which will depend on the cover crop species used, the growth stage at the time of termination, the goals of cover cropping, and environmental factors, including weather in the days and weeks before and after termination. Ineffective or incomplete termination of cover crops has two main risks:

1. competition with the current cash crop;
2. possible seed production, and the cover crop becoming a weed problem and competing with the following cash crops.

The benefits of planting green listed above make it an attractive option, but the high biomass levels and advanced maturity of the cover crop can complicate termination, especially with chemical control efforts. As mentioned earlier, annual ryegrass is one species that has a high risk of termination failure. Despite popularity in some regions and situations (i.e., broadcast seeding into standing corn), it is less desirable as a cover crop due to its aggressive nature and tendency to escape chemical control efforts.

Weed suppression, or a reduction in the density and size of weeds at the time of control efforts, is a well-studied benefit of cover cropping (Osipitan et al., 2018, 2019). However, certain species and ineffective termination efforts can lead to the production of hard seed and increase weed issues in the future. Hard seed refers to plant seeds that are impermeable to water. When species that produce hard seed are allowed to reach maturity and produce seed, a certain percentage of that seed may stay in the soil seedbank ungerminated, as opposed to timely, uniform emergence that would be expected with a cash crop (Mirsky et al., 2015). There is also an inherent risk of planting some clover, medic, and vetch species with a percentage of hard seed that does not germinate when planted and could remain in the soil and be a weed concern in the following years, or if they are allowed to set seed in the cover crop season. Hairy vetch is one species in particular that poses a risk of hard seed production. Hard seed (dormant seed that can persist for several years in the soil seedbank) can shatter early in the season and present control issues in the future; approximately 6% of hairy vetch seed can remain dormant and viable 6 months after burial at 3 cm, and 4% at 12 months after burial (Mirsky et al., 2015; Van Eerd et al., 2023).

7 Case study 1: the influence of cover crop termination timing on weed and crop dynamics (Ohio, USA)

Historically, cover crops were terminated 7 to 14 days prior to cash crop planting, depending on the following cash crop. Due in part to increasingly unpredictable spring weather patterns and increased demands on labor, cover crops are sometimes terminated at or after cash crop planting, referred to as 'planting green.' This has potentially significant implications for weed management, as delayed termination often leads to increases in biomass production, which can influence weed suppressive potential and cash crop outcomes.

A 3-year study (2018–2021) at The Ohio State University Western Agricultural Research Station in South Charleston, OH, compared planting green to other common systems for soybeans (*Glycine max* L.), the most commonly used cash crop in delayed termination systems (Essman et al., 2023). Variables studied included cereal rye seeding rate (0, 50, and 100 kg ha^{-1}), management program (consisting of different herbicide programs and timings: preplant, postplant, and delayed), and soybean residual herbicide (flumioxazin + chlorimuron ethyl vs. no herbicide).

The management programs consisted of termination timing (with glyphosate) and herbicide application parameters. In the preplant program, the cereal rye was terminated 7 days before soybean planting (DBP), and the herbicide program included a POST application. The postplant program consisted of cereal rye termination 7 days after soybean planting (DAP) and included a POST application. The delayed system was designed based on the practices of a nearby grower who had been experimenting with ultra delayed cover crop termination and various herbicide inputs. For the delayed system, saflufenacil was applied in early spring, cereal rye was terminated 21 days after soybean planting, and the herbicide program did not include a postemergence application.

The goal for this program was to evaluate whether ultra delayed termination and the expected increase in biomass production would lead to weed suppression that could reduce herbicide inputs, in this case the postemergence application. Data were collected on cover crop biomass, weed density and control, and soybean plant population and yield. The density of giant foxtail (*Setaria faberi* Herrm.) was lower in treatments that included cereal rye compared to treatments without cereal rye, but there was no difference between the two rye seeding rates. The density of giant ragweed (*Ambrosia trifida* L.) was not affected by cereal rye seeding rate. The lowest weed densities were often in the delayed program but were not always significantly different than the postplant program. The residual herbicides used in this trial reduced the density of giant foxtail but did not affect giant ragweed density. The effect of termination timing on soybean yield was variable and dependent on environmental conditions and pest pressure. In the first year of the study, terminating cereal rye after planting increased soybean yield. In the second year, the yield was lower where rye was terminated after soybean planting. Yield was not measured in the third year of the study, due to vole (*Microtus* spp.) damage in the plots that had been terminated late and had high biomass levels. Overall, this study illustrated that changes in cereal rye termination timing can influence weed pressure, and that comprehensive herbicide programs are still needed for acceptable weed control.

8 Case study 2: terminating annual ryegrass cover crops (Kentucky, USA)

Annual ryegrass is a popular choice as a cover crop due to its quick establishment, aggressive growth habit, belowground biomass accumulation, and overall adaptability across environments. Unfortunately, these characteristics of annual ryegrass are also what contribute to this same species, also known as Italian ryegrass, being an extremely problematic weed in the southern United States. Whether present as a weed or cover crop in a field, ryegrass can be difficult to terminate or burndown in the spring prior to cash crop planting and may require multiple applications. When choosing to use annual ryegrass as a cover crop, a producer must take precautions and implement a plan for successful termination to ensure this cover crop does not become a weed in the field. Those producers who have not used annual ryegrass as a cover crop before should start with a small acreage of ryegrass to assure adequate time and effort can be dedicated to terminating the cover crop in the spring. Annual ryegrass should only be used by producers who are experienced in using cover crops and who have the logistical capabilities to dedicate intensive efforts to termination in the spring. The following guidelines for the successful termination of ryegrass in the spring are based on research conducted at Purdue University and the University of Kentucky.

8.1 Plant quality seed

The first step in assuring a ryegrass cover crop does not become a weed is to choose a quality seed source. When purchasing ryegrass seed, verify the seed lot contains a single variety and does not contain perennial ryegrass varieties or weedy ryegrass varieties. Lastly, be aware that ryegrass seed is often grown in regions with known weedy Italian ryegrass with herbicide resistance, and thus there is an inherent risk, although relatively low, of herbicide-resistant seed being present in cover crop ryegrass seed lots.

8.2. Herbicide termination timing

Timing is critical to ensure a herbicide application is successful in terminating a ryegrass cover crop. The following three parameters have been shown to maximize herbicide terminations of ryegrass:

- Ryegrass height is less than 6 to 8 inches (15 to 20 cm) in height.
- Overnight temperatures remain above 7°C for 2 to 3 days prior to and after the application.
- Field soil moisture conditions allow for sprayer traffic to occur.

Unfortunately, these three parameters all occurring at the same time during the spring months can be infrequent. Thus, the suggestion that producers only dedicate a small acreage to ryegrass cover crops as the termination window can be extremely brief. Experienced ryegrass cover crop producers have been known to invest in smaller sprayer equipment to allow for applications when soil conditions may not allow for larger sprayers to cross the field and increase this window on their acres.

While ryegrass termination can be successful outside this window, there is increased risk of escapes and need for follow-up herbicide application.

8.3 Key research findings for successful annual ryegrass termination

- Use at least 1.5 lb ae/a glyphosate (40 fl oz Roundup PowerMax® 3).
  1. This is one of the single biggest factors ensuring successful termination of annual ryegrass in both Purdue University and the University of Kentucky field research evaluating annual ryegrass spring termination.
  2. Lower rates of glyphosate (from farm field observations) often resulted in failed ryegrass termination, whether as a cover crop or weed.
- The addition of 1% v/v methylated seed oil (MSO) to 1.5 lb ae glyphosate results in the consistently greatest ryegrass control.

   1 Initial research at Purdue University and follow-up research at the University of Kentucky found the tank mixes of glyphosate plus saflufenacil (Sharpen) and tiafenacil (Reviton) resulted in the most consistent ryegrass termination. These results were notable, as grass species were not within the known weed control spectrum of saflufenacil and tiafenacil. Further research in 2024 and 2025 found that the inclusion of MSO as an adjuvant in these tank mixes was the leading contributor to the increased consistency in ryegrass control, rather than the inclusion of saflufenacil or tiafenacil.
   2 The inclusion of saflufenacil or tiafenacil is still encouraged in those fields with infestations of difficult-to-control broadleaf species such as horseweed (*Erigeron canadensis* L.).
   3 Ammonium sulfate should also be included in all glyphosate termination treatments to counteract any cations in the spray carrier water.
- Avoid tank mixing atrazine or metribuzin with glyphosate as these products will antagonize glyphosate activity on ryegrass.
- The best non-glyphosate mixture is paraquat plus atrazine or metribuzin plus 2,4-D or dicamba.
   1 Paraquat (Gramoxone) and atrazine or metribuzin are synergistic and increase control as compared to each of the components applied alone. The addition of 2,4-D or dicamba is optional for those fields where troublesome broadleaves like horseweed exist.
   2 These tank mixtures work best on small ryegrass and under warm sunny conditions. A follow-up application to capture any annual ryegrass regrowth should be planned.
- Avoid the use of clethodim or other Group 1 (ACCase-inhibiting) herbicides.

The group 1 herbicides (clethodim, quizalofop, sethoxydim, etc.) work very slowly in comparison to other systemic herbicides. When these products are applied in the spring when temperatures are cool, especially overnight, this only exacerbates the slow activity, and ryegrass almost always escapes application of the group 1 herbicides.

9 Special considerations in using cover crops

9.1 Insurance requirements

In the United States, there are often termination requirements that must be met in order to fulfill the requirements of insurance providers, to be covered in the event of unforeseen circumstances such as flood, drought, and other natural disasters. For this, United States Department of Agriculture Natural Resources

Conservation Service (USDA-NRCS), USDA Risk Management Agency (RMA), and USDA Farm Service Agency (FSA) created a policy with respect to cover crop termination dates, planting dates, and location noted as 'good farming practice' (GFP) in compliance with crop insurance requirements (USDA NRCS, 2019; USDA RMA, 2019). Where the goal of cover cropping is reaping desired benefits and also fulfilling these guidelines, it is recommended to work with a crop insurance agent to ensure requirements are being met for compliance based on current policies.

9.2 Cost-share programs

In some states and regions of the United States, various entities provide cost-share options to incentivize the use of cover crops. The objective is often to provide some sort of environmental benefit, such as soil-preservation or reduced nutrient losses and runoff. These are particularly common in areas near valuable water resources, such as vulnerable lakes and watersheds. One example of this is the H2Ohio program, managed through the Ohio Department of Agriculture (ODA) and local Soil and Water Conservation Districts (SWCD). H2Ohio seeks to reduce nutrient runoff. This was initially introduced in the 24 counties near the Western Lake Erie Basin, which has experienced harmful algal blooms in recent years, and is now offered across 75 counties (H2Ohio, 2025; Ho and Michalak, 2015; Watson et al., 2016). Termination requirements may also be included as part of these programs, and growers should check participation guidelines prior to planning for the implementation and termination of cover crops.

10 Conclusion and future trends

Despite several benefits of using cover crops in cropping systems, less than 5% of the United States cropland was planted with cover crops (USDA, 2022 Census of Agriculture). One of the most commonly cited barriers to cover crop adoption for growers is the additional financial cost (Duke et al., 2022). Research on cover crops and effective termination methods can lead to improved ecosystem services provided by the cover crop and more optimal results for the grower in terms of added benefits and cash crop yield. Current work on different methods and timings of cover crop termination holds promise for a better understanding of the influence of these factors, and how to optimize them in order to maximize cover benefits and minimize competition and yield reductions to the cash crop. Optimizing termination strategies focused on nutrient loss reduction and maximum weed suppression could potentially reduce the external inputs required to produce high-yielding crops and improve the sustainability of cropping systems. By optimizing cover crop termination, increased use of cover crops over a larger area holds promise for long-term sustainability of agricultural and food production systems. One potential outcome of these improved practices is an increase in the amount of production per unit of land. As the most limiting resource for global production, this serves to be a tremendous benefit for food security in the coming years and decades.

The trends in current cover crop and termination research are largely focused on maximizing kill efficacy in varying conditions. Delayed termination of cover crops is desirable for improved weed suppression, but cover crops are hard to kill at later growth stages and make other management practices such as planting cash crops more complex. There are many environmental and agronomic factors that are currently being researched that will improve practices and understanding of cover crop management and termination. Researchers are evaluating novel methods to terminate cover crops, including electrocution treatments (Becerra-Alvarez and Moretti, 2025). Potential barriers to adoption in these areas include fitting the cover crop into existing crop rotations, establishment, weather patterns, and termination (Van Eerd et al., 2023). Species-specific research in these areas includes the different varieties and cultivars that exist, and how they differ in management and termination practices. There are also various barriers to adoption beyond these factors, including financial and the general changes needed within an operation to manage cover crops (Duke et al., 2022). These areas also need research in order to better understand the influence of these barriers and cover crop adoptions on farm sustainability from a profitability perspective. More research is needed to better understand how long-term use of cover crops provides benefits and returns to the system, both agronomically and financially.

11 Where to look for further information

Good introductions to the subject for non-specialists:

- Take action cover crop fact sheet series: https://iwilltakeaction.com/wp-content/legacy/uploads/files/20210623-factsheet-cover-crop-termination-usdadraft.pdf.
- Mid-Atlantic cover crop termination fact sheet series: https://growiwm.org/cover-crop-termination/.

Seminal articles or books:

- Sustainable Agriculture Research and Education (SARE) Managing Cover Crops Profitably, 3rd Edition: https://www.sare.org/resources/managing-cover-crops-profitably-3rd-edition/.

- Sustainable Agriculture Research and Education (SARE) Manage Weeds on Your Farm: https://www.sare.org/resources/manage-weeds-on-your-farm/.

Key societies, professional organizations, or other websites worth visiting:

- Weed Science Society of America (WSSA): https://wssa.net.
- The American Society of Agronomy, Crop Science Society of America, and Soil Science Society of America (ASA/CSSA/SSSA): https://www.sciencesocieties.org.
- American Society for Horticultural Science: https://ashs.org/.
- Crop Protection Network (CPN): https://cropprotectionnetwork.org.
- Getting Rid of Weeds (GROW): https://growiwm.org.

Key journals or conferences:

- Weed Science Society of America (WSSA) Annual Meeting: https://wssa.net/publications/journals/, https://wssa.net/meeting/.
- North Central Weed Science Society (NCWSS) Annual Meeting: https://ncwss.org/annual-meeting/.
- The American Society of Agronomy, Crop Science Society of America, and Soil Science Society of America (ASA/CSSA/SSSA): https://www.sciencesocieties.org, https://www.acsmeetings.org.

International research projects to keep an eye on:

- AgMission (by Foundation for Food & Agriculture Research (FFAR) and World Farmers' Organisation): https://agmission.org/.

The top research centers readers can investigate, e.g. for possible collaboration as well as to keep up with research trends:

- United Soybean Board (USB).
- Take Action Herbicide Resistance Management.
- Getting Rid of Weeds (GROW).

12 References

Bajwa AA, et al. (2021) The remarkable journey of a weed: biology and management of annual ryegrass (*Lolium rigidum*) in conservation cropping systems of Australia. *Plants* 10:1505.

Balkcom KS, et al. (2015) Timing of cover crop termination: management considerations for the Southeast. *Crop Forage Turf Man* 1:1–7.

Bangarwa SK and Norsworthy JK (2014) Brassicaceae cover-crop effects on weed management in plasticulture tomato. *J Crop Improv* 28(2):145–158.

Baraibar B, et al. (2018) Weed suppression in cover crop monocultures and mixtures. *Weed Sci* 66:121–133.

Bararpour MT, et al. (2017) Identification and biological characteristics of ryegrass (*Lolium* spp.) accessions in Arkansas. *Weed Sci* 65(3):350–360.

Becerra-Alvarez A and Moretti M (2025) Exploring electric weed control as a cover crop termination tool for no-till organic systems. https://smallgrains.wsu.edu/weeders-of-the-west/2025/08/14/ewc/.

Bloszies SA, et al. (2022) Legume cover crop type and termination method effects on labile soil carbon and nitrogen and aggregation. *Agron J* 114:1817–1832.

Bromilow RH (2003) Paraquat and sustainable agriculture. *Pest Manag Sci* 60:340–349.

Cornelius CD and Bradley KW (2017) Herbicide programs for the termination of various cover crop species. *Weed Tech* 31:514–522.

Creamer NG and Dabney SM (2002) Killing cover crops mechanically: Review of recent literature and assessment of new research results. *Am J Alternative Agr* 17(1):32–40.

De Bruin JL, et al. (2005) Use of a rye cover crop following corn in rotation with soybean in the upper midwest. *Agron J.* 97:587–598.

Dhima KV, et al. (2006) Allelopathic potential of winter cereals and their cover crop mulch effect on grass weed suppression and corn development. *Crop Sci* 46:345–352.

Duiker SW (2014) Establishment and termination dates affect fall-established cover crops. *Agron J* 106: 670–678.

Duke SO (2018) The history and current status of glyphosate. *Pest Manag Sci* 74:1027–1034.

Duke JM, et al. (2022) Barriers to cover crop adoption: evidence from parallel surveys in Maryland and Ohio. *J Soil Water Conserv* 77(2):198–211.

Eagles CF, et al. (1993) Recovery after freezing in *Avena sativa* L., *Lolium perenne* L. and *L. multiflorum* Lam. *New Phytol* 123:477–483.

Eivazi F, et al. (2024) Comparison of cover crop termination methods for small-scale organic vegetable production: effect on soil fertility and health. *J Plant Nutr* 47(9):1378–1389.

Essman AI, et al. (2023) The effects of cereal rye cover crop seeding rate, termination timing, and herbicide inputs on weed control and soybean yield. *Weed Sci* 71:387–394.

Fowler DB, et al. (2014) Overwinter low-temperature responses of cereals: analyses and simulation. *Crop Sci* 54: 2395–2405.

Gieske MF, et al. (2016) Brassica cover crop effects on nitrogen availability and oat and corn yield. *Agron J* 108:151–161.

Groff S (2015) The past, present, and future of the cover crop industry. *J Soil Water Conserv* 70(6):130A–133A.

Hamberg RC, et al. (2025) Responses of waterhemp and horseweed to cereal rye cover crop termination timing in soybean. *Agrosys Geosci Environ* 8:e70034. https://doi.org/10.1002/agg2.70034.

Haramoto ER and Gallandt ER (2005) Brassica cover cropping: I. Effects on weed and cropestablishment. *Weed Sci* 53:695–701.

H2Ohio (2025) https://h2.ohio.gov/water-quality-projects/managing-nutrients/farming-practices-smart-nutrient-management/farming-practices-smart-nutrient-management.

Ho JC and Michalak AM (2015) Challenges in tracking harmful algal blooms: a synthesis of evidence from Lake Erie. *J Great Lakes Res* 41(2):317–325.

Hooks CR (2024) *Offing Cover Crops for Weed Suppression: Featuring the Roller Crimper and Other Mechanical Contraptions*. University of Maryland Extension. https://extension.umd.edu/resource/offing-cover-crops-weed-suppression-featuring-roller-crimper-and-other-mechanical-contraptions/. Accessed: November 4, 2025.

Jacobs AA (2016) Plant guide for common barley (*Hordeum vulgare* L.). U.S. Department of Agriculture-Natural Resources Conservation Service. https://plants.usda.gov/DocumentLibrary/plantguide/pdf/pg_hovu.pdf. Accessed: May 8, 2025.

Kaspar T and Licht M (2019) Iowa Cover Crop Recipe – Post Corn, Going to Soybean: Use Cereal Rye. Midwest Cover Crops Council (MCCC-103) and Iowa State University Extension (CROP 3158). https://www.midwestcovercrops.org/2019/01/28/iowa-cover-crop-recipe-post-corn-going-soybean-use-cereal-rye/. Accessed: November 4, 2025.

Keene CL, et al. (2017) Cover crop termination timing is critical in organic rotational no-till systems. *Agron J* 109:272–282.

Kornecki TS and Kichler CM (2002) Effectiveness of cover crop termination methods on no-till cantaloupe. *Agriculture* 12(1):66.

Kornecki TS and Kichler CM (2023) Recurring rolling/crimping effects on termination effectiveness of iron clay pea and pearl millet warm-season cover crops. *Agriculture* 13:1949.

Kumar V, et al. (2023) Cover crop termination options and application of remote sensing for evaluating termination efficiency. *PLoS One* 18(4): e0284529.

Lawley YE, et al. (2012) The mechanism for weed suppression by a forage radish cover crop. *Agron J* 104:205–214.

Lawson A, et al. (2015) Influence of seeding ratio, planting date, and termination date on rye-hairy vetch cover crop mixture performance under organic management. *PLoS One* 10(6):e0129597.

Licht M, et al. (2024a) *Radish Cover Crop Fact Sheet for Iowa*. U.S. Department of Agriculture–Natural Resources Conservation Service, Iowa State University. https://www.nrcs.usda.gov/sites/default/files/2024-04/Radish%20Cover%20Crop%20Fact%20Sheet_0.pdf Accessed: May 13, 2025.

Licht M, et al. (2024b) *Sudangrass Cover Crop Fact Sheet for Iowa*. U.S. Department of Agriculture–Natural Resources Conservation Service, Iowa State University. https://www.nrcs.usda.gov/sites/default/files/2024-08/Sudangrass%20Crop%20Fact%20Sheet.pdf Accessed: May 8, 2025.

Licht M, et al. (2024c) *Sorghum-Sudangrass Cover Crop Fact Sheet for Iowa*. U.S. Department of Agriculture–Natural Resources Conservation Service, Iowa State University. https://www.nrcs.usda.gov/sites/default/files/2024-05/Sorghum%20Sudangrass%20Cover%20Crop%20Fact%20Sheet.pdf Accessed: May 8, 2025.

Licht M, et al. (2024d) *Cowpea Cover Crop Fact Sheet for Iowa*. U.S. Department of Agriculture–Natural Resources Conservation Service, Iowa State University. https://www.nrcs.usda.gov/sites/default/files/2024-04/Cowpea%20Crop%20Fact%20Sheet.pdf Accessed May 13, 2025.

Licht M, et al. (2024e) *Crimson Clover Cover Crop Fact Sheet for Iowa*. U.S. Department of Agriculture–Natural Resources Conservation Service, Iowa State University. https://www.nrcs.usda.gov/sites/default/files/2024-04/Crimson%20Clover%20Crop%20Fact%20Sheet.pdf Accessed May 13, 2025.

Licht M, et al. (2024f) *Hairy Vetch Cover Crop Fact Sheet for Iowa*. U.S. Department of Agriculture–Natural Resources Conservation Service, Iowa State University. https://www.nrcs.usda.gov/sites/default/files/2024-04/Hairy%20Vetch%20Crop%20Fact%20Sheet.pdf Accessed May 13, 2025.

Licht M, et al. (2024g) *Sunn Hemp Cover Crop Fact Sheet for Iowa*. U.S. Department of Agriculture–Natural Resources Conservation Service, Iowa State University https://www.nrcs.usda.gov/sites/default/files/2024-07/Sunn%20Hemp%20Cover%20Crop%20Fact%20Sheet.pdf Accessed May 13, 2025.

Licht M, et al. (2024h) *Buckwheat Cover Crop Fact Sheet for Iowa*. U.S. Department of Agriculture–Natural Resources Conservation Service, Iowa State University https://www.nrcs.usda.gov/sites/default/files/2024-05/Buckwheat%20Cover%20Crop%20Fact%20Sheet.pdf Accessed May 13, 2025.

Lupwayi NZ, et al. (2004) Decomposition of crop residues under conventional and zero tillage. *Can J Soil Sci* 84(4):403–410.

Manalil S, et al. (2011) Rapid evolution of herbicide resistance by low herbicide dosages. *Weed Sci* 59:210–217.

Midwest Cover Crop Council [MCCC] (2025) Termination. https://www.midwestcovercrops.org/termination/ Accessed: May 20, 2025.

Mirsky SB, et al. (2009) Control of Cereal Rye with a Roller/Crimperas Infl uenced by Cover Crop Phenology. *Agron J* 101:1589–1596.

Mirsky SB, et al. (2011) Timing of cover-crop management effects on weed suppression in no-till planted soybean using a roller-crimper. *Weed Sci* 59:380–389.

Mirsky SB, et al. (2013) Overcoming weed management challenges in cover crop–based organic rotational no-till soybean production in the Eastern United States. *Weed Tech* 27:193–203.

Mirsky SB, et al. (2015) Hairy vetch seedbank persistence and implications for cover crop management. *Agron J* 107:2391–2400.

Mischler R, et al. (2010) Hairy vetch management for no-till organic corn production. *Agron J* 102:355–362.

Nunes JJ, et al. (2023a) Planting into a living cover crop alters preemergence herbicide dynamics and can reduce soybean yield. *Weed Tech* 37:226–235.

Nunes J, et al. (2023b) Impact of cereal rye cover crop on the fate of preemergence herbicides flumioxazin and pyroxasulfone and control of Amaranthus spp. in soybean. *Weed Sci* 71:493–505.

Osipitan OA, et al. (2018) Cover crop for early season weed suppression in crops: systematic review and meta-analysis. *Agron J* 110(6):2211–2221.

Osipitan OA, et al. (2019) Impact of cover crop management on level of weed suppression: a meta-analysis. *Crop Sci* 59:833–842.

Palhano MG, et al. (2018) Evaluation of chemical termination options for cover crops. *Weed Tech* 32:227–235.

Pavek PLS (2012) *Plant Fact Sheet for Pea* (*Pisum sativum* L.). USDA-Natural Resources Conservation Service. https://plants.sc.egov.usda.gov/DocumentLibrary/factsheet/pdf/fs_pisa6.pdf. Accessed May 13, 2025.

Pettit A and Haller S (2021) *Plant Guide for Berseem Clover* (*Trifolium alexandrinum*). USDA-Natural Resources Conservation Service.

Pittman KB, et al. (2020) Herbicide selection to terminate grass, legume, and brassica cover crop species. *Weed Tech* 34: 48–54.

Quinn DJ, et al. (2023) Corn optimum nitrogen fertilizer rate and application timing when following a rye cover crop. *Field Crops Res* 291:108794.

Reed HK, et al. (2019) Planting green effects on corn and soybean production. *Agron J* 111:2314–2325.

Rosario-Lebron A, et al. (2019) Can winter cover crop termination practices impact weed suppression, soil moisture, and yield in no-till soybean [Glycine max (L.) Merr.]? *Crop Prot* 116:132–141.

Secchi MA, et al. (2021) Winter survival response of canola to meteorological variables and adaptative areas for current canola germplasm in the United States. *Agr Forest Meteorol* 297:108267.

Selleck GW and Baird DD (1981) Antagonism with glyphosate and residual herbicide combinations. *Weed Sci* 29(2):185–190.

Silva GC and Bagavathiannan M (2023) Mechanisms of weed suppression by cereal rye cover crop: a review. *Agron J* 115:1571–1585.

Sustainable Agriculture Research & Education [SARE] (2023) National Cover Crop Survey Report 2022–2023. https://www.sare.org/wp-content/uploads/2022-2023-National-Cover-Crop-Survey-Report.pdf. Accessed: October 11, 2024.

Teasdale JR and Rosecrance RC (2003) Mechanical versus herbicidal strategies for killing a hairy vetch cover crop and controlling weeds in minimum-tillage corn production. *Am J Alternative Agr* 18(2):95–102.

Tehranchian P, et al. (2019) Multiple herbicide resistance in California Italian ryegrass (*Lolium perenne* ssp. multiflorum): characterization of ALS-inhibiting herbicide resistance. *Weed Sci* 67:273–280.

USDA (2022) Census of agriculture: cover crop use continues to be most common in eastern United States. https://www.ers.usda.gov/data-products/charts-of-note/chart-detail?chartId=108950.

USDA NRCS (2019) https://www.nrcs.usda.gov/sites/default/files/2022-09/Termination_Guidelines_Designed_6.28_10.24am_%28002%29.pdf.

USDA RMA (2019) https://www.rma.usda.gov/sites/default/files/2024-02/Cover-Crops-and-Crop-Insurance-Fact-Sheet.pdf.

Van Eerd LL, et al. (2023) Influence of cover crops at the four spheres: a review of ecosystem services, potential barriers, and future directions for North America. *Sci Total Environ* 858(3):159990.

Vann RA, et al. (2021) Differences among eighteen winter pea genotypes for forage and cover crop use in the southeastern United States. *Crop Sci* 61:947–965.

Wallace JM, et al. (2017) Cover crop-based, organic rotational no-till corn and soybean production systems in the Mid-Atlantic United States. *Agriculture-London* 7(34):1–21.

Watson SB, et al. (2016) The re-eutrophication of Lake Erie: harmful algal blooms and hypoxia. *Harmful Algae* 56:44–66.

Wayman S, et al. (2014) The influence of cover crop variety, termination timing and termination methodon mulch, weed cover and soil nitrate inreduced-tillage organic systems. *Renew Agr and Food Syst* 30(5): 450–460.

Whalen DM, et al. (2020a) Herbicide programs for the termination of grass and broadleaf cover crop species. *Weed Tech* 34: 1–10.

Whalen DM, et al. (2020b) Integration of residual herbicides with cover crop termination in soybean. *Weed Tech* 34:11–18.

Wilkins ED and Bellinder RR (1996) Mow-kill regulation of winter cereals for spring no-till crop production. *Weed Tech* 10:247–252.
Wortman SE, et al. (2013) Mechanical termination of diverse cover crop mixtures for improved weed suppression in organic cropping systems. *Weed Sci* 61(1):162–170.
Wortman SE, et al. (2012) Optimizing cover crop benefits with diverse mixtures and an alternative termination method. *Agron J* 104:1425–1435.

Part 3

Cover crop species

Chapter 11

Cool-season leguminous cover crops

Artie Siller, Neda Nikpour-Rashidabad and Masoud Hashemi, University of Massachusetts Amherst, USA

1 Introduction

The agroecological benefits of cover crops, particularly legumes, are well-documented. The environmental and economic costs of intensive synthetic nitrogen fertilizer use highlight the need for more sustainable crop management practices. Legumes have been used as a primary source of nitrogen in farming for thousands of years. For example, Columella, a Roman agricultural writer of the first century, stated that 'Where no kind of manure is to be had, I think the cultivation of lupines [*Lupinus* spp.] will be found the readiest and best substitute. If they are sown about the middle of September in poor soil, and then plowed in, they will answer as well as the best manure.'

In modern agriculture, legumes are widely used not only as a cash crop but also as cover crops or dual-purpose crops (Etemadi et al., 2018a). Legume cover crops offer important benefits, most notably their ability to fix atmospheric nitrogen, but they also pose some challenges for nitrogen management in the

http://dx.doi.org/10.19103/AS.2025.0162.15

environment. Through biological nitrogen fixation (BNF), legumes can greatly reduce the need for nitrogen fertilization in cropping systems. For example, fall-planted faba bean (*Vicia faba* L.) can contribute 133 kg N ha^{-1} (root and aerial biomass) in soils with no history of faba bean cultivation (Ghorbi et al., 2023) and up to 192 kg ha^{-1} (aerial biomass only) in soils where faba bean was planted the previous year (Etemadi et al., 2018b) (Figure 1). Additionally, the nitrogen from legume cover crops is usually less expensive than nitrogen derived from organic fertilizers, making legumes an attractive option for sustainable soil fertility management.

Determining the total amount of nitrogen in cover crop dry matter provides a good starting point for estimating their capacity to provide nitrogen to the succeeding crop. Several factors, including the choice of legume species, time of planting, and the rhizobia species (*Rhizobium* spp.), can greatly influence the efficiency of BNF and legume biomass production. However, it is important to remember that not all nitrogen in legume cover crops is used by following crops. For example, studies found that corn used up to 75% of nitrogen from spring-terminated hairy vetch (Blevins et al., 1990) but only 25% of nitrogen from winterkilled faba bean (Etemadi et al., 2018b).

Strategies leading to production of higher biomass should be identified to maximize nitrogen contribution from legumes. Generally, early planting, selecting the right species, seeding rate and other strategies have been used to increase biomass, thus the potential N contribution. However, in cold regions,

Figure 1 Fall-planted faba bean cover crop grown in Massachusetts, USA. Source: Photo by Masoud Hashemi.

such as Northeast United States, implementation of these strategies is often limited by the relatively short growing season. For example, earlier planting in cold areas may not be feasible if the harvest of the preceding cash crop is delayed. Increasing the seeding rate may compensate to some degree for the shorter growing period. For example, the optimal seeding rate for maximum biomass production of hairy vetch (*Vicia villosa* Roth) in the northern part of the Northeastern U.S., including Massachusetts (MA), New York (NY), and Pennsylvania (PA) was 15–20 kg ha^{-1}. By contrast, the optimal seeding rate in southern part of the region, such as Maryland and North Carolina, was 5–10 kg ha^{-1} (Mirsky et al., 2017) (Fig. 2). Reduced biomass due to late planting was compensated by increased seeding rate only in Northern states such as MA, PA, and NY which have less flexible time of planting. Additionally, in the biological world, there is always a limit to any agroecological benefits. For example, Ghorbi et al. (2023) reported that BNF by faba bean increased with increased population density up to 40 plants m^{-2} with no further increase in the BNF when the density doubled to 80 plants m^{-2}.

Leguminous cover crops often improve soil health and provide a yield boost that goes beyond their ability to fix atmospheric nitrogen. The contributions of cover crops to soil health and, therefore, agricultural system sustainability are often related to their biomass accumulation. Cool-season legumes grow slowly during the fall, which can limit their agroecological benefits, including BNF, soil erosion control, and the support of soil biology. In contrast, grass cover crops

Figure 2 Hairy vetch cover crop grown in Massachusetts, USA. Source: Photo by Masoud Hashemi.

are known for their ability to produce high biomass and are often used when enhancing soil organic carbon is the primary goal. While legume cover crops offer many agroecological benefits, they are typically valued for their BNF capacity (Koudahe et al., 2022; Kaye and Quemada, 2017).

These differing characteristics of cover crops have led to the widespread adoption of mixed-species cover crops to maximize overall benefits. In colder climates, cereal rye (*Secale cereale* L.) is often included in these mixtures due to its quick establishment and winter hardiness. However, the performance of cover crop mixtures is affected by environmental factors and soil conditions. For example, the initial soil inorganic nitrogen level affects the relative composition of legumes and grasses in mixtures (Boniface et al., 2024). As soil inorganic nitrogen stocks increase, the competitiveness and overall shoot biomass of hairy vetch decrease compared to cereal rye or forage colza (rapeseed) (*Brassica rapa* L.) (Boniface et al., 2024). Overall, sites with milder winters allow greater flexibility in species selection, enabling multifunctional mixtures at lower seeding rates, which can significantly reduce seed costs. Understanding the link between climate, soil conditions, and management practices is essential for optimizing cover crop performance and adjusting cover crop seeding rates to align with farmer objectives.

2 Species selection

Cool-season legume cover crops primarily come from four major phylogenetic groups, peavines, clovers, vetches, and medics/sweet clovers, along with a few representatives from other legume genera. Scientific trials around the world provide an important record of average seeding rates and the suitability of various species for fall and spring planting windows (Table 1). Species characteristics determine their utility in different agricultural systems by key characteristics: (1) biomass production, (2) nitrogen content, and (3) cold tolerance.

High biomass production is essential for BNF, physically protecting the soil, and adding carbon to the soil. However, the concentration of nitrogen in legume cover crops can also depend on the growth stage and variation among species (Büchi et al., 2015, Cook et al., 2010). Cold tolerance also plays a significant role, as cover crop productivity is often limited by cold temperatures where cash crops are grown during the warmer periods. Species selection should be based on their expected biomass accumulation, nitrogen content, and cold hardiness.

Table 1 documents the performance of legume species grown as cover crop monocultures. While some species have only been examined in a few trials, some species have been studied extensively. Averaging cover crop performance by species necessarily obscures differences among varieties and

Table 1 Agronomic characteristics of cool-season legume cover crops

Common name	Latin name	Planting time	Mean seeding rate	Mean dry biomass accum.	Mean nitrogen accum.	Lowest average monthly temp.	References
			kg ha^{-1}	Mg DM ha^{-1}	kg N ha^{-1}	°C	
Peavines							
Field pea	*Lathyrus oleraceus* Lam.	Fall, spring	84	2.7	114	−10	Blesh et al., 2019; Boulet et al., 2021; Büchi et al., 2015; Bugg et al., 1996; Burgess et al., 2014; Frye et al., 1988; Heichel, 1987; Holderbaum et al., 1990; Holman et al., 2018; Parr et al., 2011; Power and Koerner, 1994; Power and Zachariassen, 1993; Rijal et al., 2021; Schomberg et al., 2006; Stein et al., 2023; Vanelli et al., 2018; Wortman et al., 2012; Wunsch et al., 2017
Grass pea	*Lathyrus sativus* L.	Fall, spring	93	1.8	84	6	Ahmadnia et al., 2020; Blesh et al., 2019; Büchi et al., 2015; Ordóñez-Fernández et al., 2018; Rao and Northup, 2011; Rijal et al., 2021; Thiessen Martens et al., 2005; Wortman et al., 2012
Tinga pea	*Lathyrus tingitanus* L.	Spring	–	1.8	–	–	Power and Koerner, 1994
Medics and sweetclovers							
Black medic	*Medicago lupulina* L.	Fall	18	0.7	–	−14	Abdin et al., 2000; Brandsæter and Netland, 1999; Brandsæter et al., 2000; May et al., 2022
Alfalfa	*Medicago sativa* L.	Fall	15	2.1	94	11	Abdin et al., 2000; Büchi et al., 2015; Heichel, 1987; Holderbaum et al., 1990
Barrel medic	*Medicago truncatula* Gaertn.	Fall	20	5.8	–	−1	Brandsæter and Netland, 1999; Bugg et al., 1996; Soriano et al., 2023
White sweetclover	*Melilotus albus* Medik.	Fall	19	4.4	105		Büchi et al., 2015; Bugg et al., 1996; Ogunleye et al., 2023; Power and Zachariassen, 1993

(*Continued*)

Table 2 *Continued*

Common name	Latin name	Planting time	Mean seeding rate	Mean dry biomass accum.	Mean nitrogen accum.	Lowest average monthly temp.	References
			kg ha^{-1}	Mg DM ha^{-1}	kg N ha^{-1}	°C	
Yellow sweetclover	*Melilotus officinalis* (L.) Lam.	Fall	15	2.8	103	−4	Abdin et al., 2000; Bair et al., 2008; Brandsæter et al., 2000; Frye et al., 1988; Holderbaum et al., 1990
Clovers							
Berseem clover	*Trifolium alexandrinum* L.	Fall, spring	20	2.8	63	11	Abdin et al., 2000; Büchi et al., 2015; Den Hollander et al., 2007; Frye et al., 1988; Malone et al., 2022; Parr et al., 2011; Read et al., 2011; Rijal et al., 2021; Tadiello et al., 2022; Toom et al., 2019
Rose clover	*Trifolium hirtum* All.	Fall	21	6	–	–	Bugg et al., 1996; Holderbaum et al., 1990
Crimson clover	*Trifolium incarnatum* L.	Fall, spring	21	4.2	112	−5	Abdin et al., 2000; Adeux et al., 2021; Blesh et al., 2018; Boulet et al., 2021; Brandsæter et al., 2000; Büchi et al., 2015; Bugg et al., 1996; Den Hollander et al., 2007; Frye et al., 1988; Gselman and Kramberger, 2008; Kramberger et al., 2009; May et al., 2022; Parr et al., 2011; Perdigão et al., 2021; Power and Zachariassen, 1993; Saini et al., 2005; Schomberg et al., 2006; Tyler, 2020; Wortman et al., 2012
Balansa clover	*Trifolium michelianum* Savi	Fall	9	2.8	73	−1	Parr et al., 2011; Parr et al., 2011; Schomberg et al., 2006
Red clover	*Trifolium pratense* L.	Fall, spring	16	2.8	88	1	Bugg et al., 1996; Den Hollander et al., 2007; Gselman and Kramberger, 2008; Heichel, 1987; Holderbaum et al., 1990; Malone et al., 2022; May et al., 2022
White clover	*Trifolium repens* L.	Fall, spring	15	2.9	110	4	Brandsæter and Netland, 1999; Büchi et al., 2015; Den Hollander et al., 2007; Heichel, 1987; May et al., 2022; Power and Zachariassen, 1993
Persian clover	*Trifolium resupinatum* L.	Fall, spring	15	3.3	110	11	Abdin et al., 2000; Boulet et al., 2021; Büchi et al., 2015; Den Hollander et al., 2007; Von Boberfeld et al., 2005

Squarrose clover	*Trifolium squarrosum* L.	Fall	40	5.4	125	3	Abou Chehade et al., 2023; Adeux et al., 2021
Subterrarean clover	*Trifolium subterraneum* L.	Fall, spring	22	3.6	68	−4	Abdin et al., 2000; Adeux et al., 2021; Büchi et al., 2015; Bugg et al., 1996; Den Hollander et al., 2007; Gselman and Kramberger, 2008; Holderbaum et al., 1990; Kramberger et al., 2009; May et al., 2022; Parr et al., 2011; Scavo et al., 2020
Arrowleaf clover	*Trifolium vesiculosum* Savi	Fall	10	7.3	125	2	Boulet et al., 2021; Holderbaum et al., 1990; Ogunleye et al., 2023; Perdigão et al., 2021
Vetches							
Purple vetch	*Vicia benghalensis* L.	Fall	36	5.8	70	–	Bugg et al., 1996; Tadiello et al., 2022
Ervil	*Vicia ervilia* (L.) Willd.	Fall	97	2.3	–	−5	Abd El Moneim, 1993; Ordóñez-Fernández et al., 2018
Fava bean	*Vicia faba* L.	Fall, spring	144	4.2	152	7	Braiser et al., 2023; Büchi et al., 2015; Bugg et al., 1996; Cai et al., 2018; Etemadi et al., 2018a; Etemadi et al., 2018b; Heichel, 1987; Power and Zachariassen, 1993; Sincik et al., 2008; Stein et al., 2023; Wunsch et al., 2017
Bigflower vetch	*Vicia grandiflora* Scop.	Fall	56	4.6	210	–	Holderbaum et al., 1990
Tiny vetch	*Vicia hirsuta* (L.) Gray	Fall	–	3.5	99	–	Cho et al., 2011a; Lim et al., 2012
Lentil	*Vicia lens* (L.) Coss. & Germ.	Fall, spring	37	2.6	69	6	Blesh et al., 2019; Büchi et al., 2015; Holman et al., 2018; Ogunleye et al., 2023; Rao and Northup, 2011; Rijal et al., 2021; Smith et al., 1987; Thiessen Martens et al., 2018

(*Continued*)

Table 2 *Continued*

Common name	Latin name	Planting time	Mean seeding rate	Mean dry biomass accum.	Mean nitrogen accum.	Lowest average monthly temp.	References
			kg ha^{-1}	Mg DM ha^{-1}	kg N ha^{-1}	°C	
Hungarian vetch	*Vicia pannonica* Crantz	Fall	150	3.8	135	0	Büchi et al., 2015; Stein et al., 2023; Tursun et al., 2018
Common vetch	*Vicia sativa* L.	Fall	69	3.3	138	−5	Abd El Moneim, 1993; Büchi et al., 2015; Bugg et al., 1996; Lacet et al., 2020; Lim et al., 2012; Ordóñez-Fernández et al., 2018; Parr et al., 2011; Sincik et al., 2008; Von Boberfeld et al., 2005; Wunsch et al., 2017
Smooth vetch	*Vicia tetrasperma* (L.) Schreb.	Fall	30	2.5	60	–	Cho et al., 2011a; Cho et al., 2011b; Lim et al., 2012
Hairy vetch	*Vicia villosa* Roth	Fall	62	3.6	143	−9	Abd El Moneim, 1993; Abdin et al., 2000; Adeux et al., 2021; Bair et al., 2008; Brady and Weil, 2002; Büchi et al., 2015; Bugg et al., 1996; Frye et al., 1988; Gselman and Kramberger, 2008; ; Holman et al., 2018; Kandel et al., 2018; May et al., 2022; Ordóñez-Fernández et al., 2018; Parr et al., 2011; Power and Koerner, 1994; Power and Zachariassen, 1993; Schomberg et al., 2006; Toom et al., 2019; Tursun et al., 2018; Wortman et al., 2012
Other legumes							
Chinese milkvetch	*Astragalus sinicus* L.	Fall	45	1.3	98	4	Cai et al., 2018; Ding et al., 2024; Hyun et al., 2005; KBDP et al., 2005; Liu et al., 2022; Nie et al., 2019; Yang et al., 2013; Yang et al., 2019; Yang et al., 2021

Chickpea	*Cicer arietinum* L.	Fall	–	2	53	–	Heichel, 1987; Rose et al., 2016
White lupin	*Lupinus albus* L.	Fall	60	2.9	87	11	Abd El Moneim, 1993; Bhardwaj et al., 2010; Brady and Weil, 2002; Büchi et al., 2015; Carranca et al., 2009; Rosolem et al., 2004; Schomberg et al., 2006; Vanelli et al., 2018
Narrowleaf lupin	*Lupinus angustifolius* L.	Fall	110	5.1	119	2	Bugg et al., 1996; Francis et al., 1998; Kemper et al., 2022; Natera et al., 2023; Parr et al., 2011; Pease, 2020
Serradella	*Ornithopus sativus* Brot.	Fall	20	–	98	–	Perdigão et al., 2021

production methods, but it provides a useful starting point for researchers and growers. There are substantial variations in agronomic performance, even among closely related species and cultivars, making it difficult to extrapolate the characteristics of a species from its better-studied relatives. Although there are cases where a minor cover crop has been grown as a forage or grain crop, data from such uses are not included here.

Key metrics such as planting time, seeding rate, biomass accumulation, and nitrogen accumulation were compiled from reported experimental methods and results. Few experiments directly examined cold tolerance; monthly average temperatures were compiled from reported weather data where cover crops survived. Thus, this method provides a conservative, upper-range estimate of cold climate tolerance. The true cold tolerance of legume cover crops is affected by environmental factors, such as snow cover and plant growth stage.

3 Peavines and medics and sweetclovers

3.1 Peavines

Field pea (*Lathyrus oleraceus* Lam.) is a widely grown annual cover crop that can be planted in either fall or spring. It can accumulate up to 2.7 Mg DM ha^{-1} with over 4% N, equivalent to as much as 114 kg N ha^{-1}. Winter field peas are sturdy semi-upright vines and are well-suited for cold climates. Peas need to be planted slightly later in the fall than many other legumes cover crops to prevent flowering in warmer weather without vernalization. Spring field peas, in contrast, have a vining habit and flower quickly when planted in the spring or early fall. They generally produce less biomass than winter field peas. Other species, such as Tinga pea (*Lathyrus tingitanus* L.) and grass pea (*Lathyrus sativus* L.) have also been grown as cover crops. However, they generally have lower biomass and nitrogen accumulation than field peas.

3.2 Medics and sweetclovers

Both annual and perennial medics have been used as cover crops. While the slow initial growth of perennial medics limits their biomass production during annual cover crop windows, alfalfa (*Medicago sativa* L.) can still produce up to 2.1 Mg DM ha^{-1} and 94 kg N ha^{-1} in such conditions. Black medic (*Medicago lupulina* L.) is highly cold tolerant but less productive than alfalfa. Annual barrel medic (*Medicago truncatula* Gaertn.), on the other hand, is more productive in short periods but has only been minimally explored as a cover crop, while its prostrate growth habit may pose challenges in high weed pressure settings.

White (*Melilotus albus* Medik.) and yellow (*Melilotus officinalis* (L.) Lam.) sweetclovers are winter annuals or biennials that require vernalization to flower in late spring or early summer. They are generally more productive than alfalfa as short-term cover crops with higher biomass up to 4.4 Mg DM ha^{-1} and similar nitrogen accumulation. Both sweetclover species demonstrate comparable agronomic behaviors, cold tolerant to average monthly temperatures of at least −4°C, and have deep taproots. These characteristics make them particularly valuable for biotillage in no-till systems.

4 Clovers, vetches and other legumes

4.1 Clovers

Most clover cover crops are cool-season annuals, though perennial red (*Trifolium pratense* L.) and white (*Trifolium repens* L.) clovers can also be used as annual cover crops. Despite their perennial nature, red and white clover are often grown due to their relatively low seed cost and broader availability than annual clover seed. While they produce less biomass than most annual clovers, they perform well as nitrogen sources, accumulating an average of 110 kg N ha^{-1} compared to 95 kg N ha^{-1} for annual clovers.

There is, however, significant variation among annual clover species. Crimson (*Trifolium incarnatum* L.), squarrose (*Trifolium squarrosum* L.), and arrowleaf (*Trifolium vesiculosum* Savi) clovers accumulate more biomass and nitrogen than other clovers (Fig. 2). In contrast, species such as berseem (*Trifolium alexandrinum* L.), balansa (*Trifolium michelianum* Savi), Persian (*Trifolium resupinatum* L.), and subterranean (*Trifolium subterraneum* L.) clovers produce similar or slightly higher biomass but accumulate less nitrogen than red or white clover.

In general, clovers have smaller seeds than peavines or vetches, with seed size comparable to those of sweet clovers and medics. They are commonly planted in the spring, which has made frost-seeding a common practice. Frost-seeding involves broadcasting clover seeds in early spring to establish clovers without tillage, a particularly effective strategy for wetter soils. Except for Persian and berseem clovers, most clover cover crops are cold tolerant. Species such as crimson (Fig. 3), balansa, and subterranean clovers perform well as cover crops where monthly average temperatures in winter are between −1°C and −6°C.

4.2 Vetches

Vetches are among the most efficient nitrogen-fixing cool-season legumes, with some species accumulating more than 135 kg N ha^{-1} on average (Table 1). Among widely researched species, hairy vetch and common vetch (*Vicia sativa* L.) are recognized for their high nitrogen contributions. Hungarian vetch (*Vicia pannonica* Crantz) and bigflower vetch (*Vicia grandiflora* Scop.) also produce

Figure 3 Crimson clover cover crop in Massachusetts, USA. Source: Photo by Masoud Hashemi.

relatively high amounts of biomass and nitrogen, but these species have been little explored as cover crops.

Faba bean accumulated more biomass and nitrogen than all other vetches besides bigflower vetch but it has lower cold tolerance (surviving monthly average temperatures of only 7°C) than other highly productive vetches like common and hairy vetch. In contrast, hairy vetch is among the most cold-tolerant legume cover crops, surviving average monthly temperatures of −9°C. For these reasons, hairy and common vetch are well suited as overwintering legumes in colder climates. While faba bean is preferred as a spring-planted cover crop, the amount of nitrogen produced by fall-planted faba beans is high enough that there can still be substantial nitrogen credits to summer crops grown after winterkilled faba bean (Etemadi et al., 2018b).

4.3 Other legumes

While the above four groups encompass the majority of cool-season legume cover crops, several other legume species are also productive options. For instance, despite low biomass in published studies, Chinese milkvetch (*Astragalus sinicus* L.) accumulated an average of 98 kg N ha^{-1}. Fall-planted white (*Lupinus albus* L.) and narrowleaf (*Lupinus angustifolius* L.) lupins, as well as serradella (*Ornithopus sativus* Brot.), accumulated 87–119 kg N ha^{-1} with narrowleaf lupin producing 5.1 Mg DM ha^{-1}. Both Chinese milkvetch and narrowleaf lupin displayed moderate cold hardiness, tolerating average monthly temperatures as low as 4°C).

5 Cover crop management

The capacity of winter legume cover crops to provide agronomic and ecological benefits such as BNF, soil conservation and nutrient cycling is highly dependent on their management (Ashford and Reeves, 2003; Islam, 2021). Achieving these objectives necessitates a systematic approach, beginning with integration of cover crops into the overall farm nutrient management plan.

Critical management practices include selecting species tailored to specific agronomic goals and local environmental conditions, identifying optimal planting time, optimizing the seeding rate, and implementing appropriate timing and techniques for cover crop termination. Additional considerations include inoculating leguminous cover crops with the right rhizobia, deciding the best time interval between cover crop termination and cash crop establishment, using cover crop mixtures, and aligning cover crop practices with the cultural requirements of the subsequent cash crop (Shiferaw et al., 2024).

6 Rhizobia inoculation

Legumes form a symbiotic relationship with nitrogen-fixing rhizobia bacteria in the soil, enabling the conversion of atmospheric nitrogen into bioavailable forms, but they perform best if inoculated with the appropriate rhizobia strain. Inoculation promotes effective colonization and symbiotic nitrogen fixation (Albayrak et al., 2006; Vessey, 2004). This practice is particularly advantageous in soils with low rhizobia populations or low soil nitrogen (Vessey, 2004). By boosting nodulation efficiency and nitrogenase activity, inoculated legumes can achieve nitrogen fixation rates exceeding 110 kg N ha^{-1} under optimal conditions (Perrone et al., 2020; Ghorbi et al., 2023).

Cool-season legumes vary considerably in their capacity to fix atmospheric N. The efficiency of BNF depends on many other factors, including legume species, environmental conditions, and management practices (Blesh, 2018). It is therefore difficult to provide reliable estimates of potential nitrogen fixation, mainly because BNF varies with not only species but also between varieties within species. In addition, soil characteristics such as fertility level, pH, salinity level, and abiotic stress influence the capacity of legumes to fix N. While the symbiosis between rhizobia and legume plant is relatively resilient to low or moderate stress, reduction in the BNF could be related to decreased legume productivity under such stress.

Suboptimal temperatures (outside of 15–20°C for many cool-season cover crops) can impose significant limitations on the nitrogen-fixing process by delaying nodulation, reducing the formation and functionality of nodules, and impairing nitrogenase activity (Thurston et al., 2022; Alexandre and Oliveira, 2013). Additionally, low temperatures disrupt the biochemical signaling pathways necessary for effective rhizobia-legume symbiosis, further reducing nitrogen fixation potential (Perrone et al., 2020). To overcome these challenges, inoculating legumes with rhizobia strains that are adapted to cold conditions is crucial. For example, Janczarek et al. (2024) concluded that low temperature can be a stress factor affecting the survival of rhizobia as free-living microorganisms in soil and during various steps of symbiotic interactions. They reported that inoculation of clover with the cold-adapted strain of rhizobia enhanced nodulation and nitrogenase function under low temperature conditions.

Certain cold-tolerant legume species, such as hairy vetch, field peas, faba bean, and white clover, demonstrate superior BNF performance in temperate climates, making them well-suited for such agroecosystems (Tiemann et al., 2015). For instance, Macmillan et al. (2022) evaluated the BNF potential of winter field pea, hairy vetch, crimson clover, and lentil (*Vicia lens* (L.) Coss. and Germ.) in Texas. Field pea consistently outperformed other legumes in both growth and nitrogen fixation, especially in nitrogen-deficient organic wheat (*Tritium aestivum* L.) systems (Fig. 4). Hairy vetch and lentil exhibited moderate performance, whereas crimson clover struggled to achieve substantial growth or nitrogen fixation. These results underscore the necessity of selecting legume species that are adapted to the local climate.

Figure 4 Field pea cover crop grown in Massachusetts, USA. Source: Photo by Masoud Hashemi.

7 Seeding rate and planting time

7.1 Seeding rate

Seeding rate affects the performance and functionality of cool-season legume cover crops, in particular, biomass accumulation. Higher seeding rates typically result in increased biomass production. However, economic factors such as seed costs may discourage growers from adopting higher seeding rates. Douglas et al. (2023) examined the effects of seeding rates on biomass production and weed suppression in cool-season legume monocultures, including berseem clover, common vetch, crimson clover, hairy vetch, and field pea, across the Southeastern, Southern Plains, and Mid-Atlantic US. Lowering legume seeding rates (25–50%) produced statistically similar biomass compared to higher seeding rates (100–150%), therefore providing the highest return on seed investment. Balancing seeding rate and seed cost can maximize the benefits of cover crop systems and ensure cost efficiency.

7.2 Planting time

The timing of sowing winter legume cover crops greatly affects their performance and associated benefits, particularly BNF and biomass production. Early sowing allows legumes to establish well and reach their full potential and is ideally done shortly after the harvest of the preceding crop and with time for establishment before the onset of cold temperatures. For example, early sowing (6 weeks before first frost) of species like subterranean clover, crimson clover, red clover, and hairy vetch can yield up to 4.4 Mg DM ha^{-1} of biomass and fix as much as 179 kg N ha^{-1} of symbiotically fixed nitrogen by spring. In contrast, late sowing significantly reduces these benefits, with late-planted legumes fixing as little as 11.7 kg N ha^{-1} compared to 87.2–114.5 kg N ha^{-1} achieved with early planting (Gselman and Kramberger, 2008).

The timing of the preceding cash crop harvest is often the primary determinant of the cover crop seeding date. Early harvested crops, such as barley (*Hordeum vulgare* L.) or wheat create opportunities for late-summer cover crop sowing, ensuring sufficient time for establishment and growth. In contrast, later-harvested crops, such as grain corn (*Zea mays* L.), shorten the available window for sowing, limiting the benefits of cover crops at some latitudes. Understanding these constraints and selecting appropriate cover crop species for delayed planting can help optimize performance across diverse farming systems.

8 Mixes versus monocultures

Monoculture and mixed-culture systems represent two different strategies for managing cool-season legume cover crops, each offering distinct agronomic and ecological benefits and disadvantages. Monocultures of cool-season legumes, such as hairy vetch, crimson clover, and faba bean, excel in atmospheric nitrogen fixation (MacMillan et al., 2022). They are particularly beneficial when the primary goal is to optimize nitrogen availability or address-specific agronomic needs. Conversely, to achieve multiple objectives besides BNF (such as weed suppression or nitrogen scavenging) mixes of legumes and other species are preferable (Perrone et al., 2020).

Mixed cropping systems, including bi-cultures and multi-species systems, provide a broader range of ecosystem services by utilizing the complementary traits of different plant species. These systems improve nutrient cycling, increase biomass production, suppress weeds, and enhance system stability. For instance, in a cereal rye-hairy vetch bi-culture, rye scavenges nutrients, suppresses weeds, and provides structural support for the growth of hairy vetch. Concurrently, hairy vetch specializes in BNF. This functional complementarity allows mixed cultures to address multiple agronomic objectives, including optimizing nitrogen cycling, increasing soil organic matter, and effectively managing weed pressure (Perrone et al., 2020).

Studies show that mixed cropping systems often outperform monocultures in delivering long-term ecosystem services (Thapa et al., 2018; Wolters, 2020; Feng et al., 2020). Grass–legume mixtures generally produce more biomass and increase the nitrogen content of cover crop residues compared to monocultures. The integration of grasses in cool-season legume cover crop systems helps moderate the rapid nitrogen mineralization of legume residues, leading to a slower and more sustained nitrogen release because grasses, with higher carbon-to-nitrogen (C:N) ratios, can have slower decomposition and nitrogen release rates (Perrone et al., 2020; Furey et al., 2021).

Mixed cultures also contribute significantly to other ecosystem functions, including improved soil health, enhanced biodiversity, and increased resilience to environmental stressors. The diversity inherent in these systems promotes niche differentiation and resource use efficiency, enabling plants to optimize water, nutrients, and light utilization. Additionally, the stability conferred by species diversity acts as an ecological insurance mechanism, ensuring consistent system performance under variable environmental conditions (Lavergne et al., 2021). Despite these advantages, mixed cultures are not without challenges. Interspecies competition can limit legume growth and BNF efficiency, particularly when grasses dominate the mixture. Therefore, careful management of species proportions is essential to optimizing the benefits of mixed systems (Perrone et al., 2020). Since legumes are generally weaker competitors, maintaining their seeding rates close to monoculture levels helps ensure successful establishment (White and Barbercheck, 2017). In contrast, grass seeding rates can be safely reduced to between one-half and one-quarter of their monoculture rates to promote a balanced stand with legumes (Bariabar et al., 2018).

9 Termination time and method

Cover crops are grown primarily for their agroecological benefits rather than for direct harvest, making termination timing critical for maximizing these benefits (Lacey et al., 2020; Sunuwar et al., 2023). Termination timing of legumes influences total biomass (both aboveground and belowground), decomposition rate, and nitrogen contribution to the succeeding cash crop.

Cool-season legumes, such as faba beans, can potentially fix nearly 200 kg N ha^{-1} before being winter-killed (Etemadi et al., 2018a; Ghorbi et al., 2023). However, a significant portion of this nitrogen accumulated in their roots and shoots may be released before the succeeding crop can use it in the spring. For example, Etemadi et al. (2018b) concluded that only 50 kg N ha^{-1} of the 200 kg N ha^{-1} produced by faba beans was ultimately available to the subsequent sweet corn crop.

Not all cool-season legumes are winter-killed. Legumes such as hairy vetch and many clover species survive winter and require deliberate termination to ensure timely planting of the cash crop. Additionally, natural termination of some cool-season legumes varies in different regions. For example, while faba bean or crimson clover may be killed by freezing conditions in Northeast U.S., they may survive in areas with milder winter conditions. If the cool-season legumes survive the winter or are planted in spring, termination timing becomes a crucial management practice. If terminated too early, their nitrogen-fixing potential is diminished. Additionally, their low C:N ratio accelerates decomposition, which, if not synchronized with the succeeding cash crop establishment, may result in a considerable amount of the released nitrogen lost through leaching or N_2O emissions.

In contrast, winter annual grasses like cereal rye typically have a higher C:N ratio, making their timing of termination critical for nitrogen cycling (Clark et al., 1994). However, for cool-season legumes with generally C:N ratio below the desired threshold of 25:1, termination timing is less restrictive. Delayed termination, if it does not interfere with the optimal planting date of the succeeding cash crop, can maximize BNF and nitrogen benefit. Decomposition rate of cover crop residues also significantly changes depending on the tillage system. Residues left on the soil surface result in slower decomposition (Etemadi et al., 2018b; Tonitto et al., 2006) and often reduce nitrogen losses by aligning nitrogen release with subsequent crop nitrogen needs.

Cover crop termination methods include chemical and mechanical approaches, each with distinct advantages and limitations. Chemical termination, particularly herbicides, prevents legume regrowth and is suited to no-till cropping systems. While many reports indicated that the use of herbicide such as glyphosate has no large negative impact on soil biology (Busse et al., 2001; Nguyen et al., 2016), other research demonstrated that use of glyphosate could negatively influence soil biology, for example, by disrupting earthworm ecology (Rose et al., 2016). Chemical termination of cover crops is not allowed in organic farming systems.

Mechanical methods, such as disking, physically incorporate cover crop residues into the soil, accelerating decomposition, disturbing soil structure, and severely impacting soil biology. This method of cover crop termination is incompatible with no-till systems. Incorporating legume residues into the soil accelerates decomposition, which leads to faster nitrogen release that can potentially result in higher nitrogen losses (Etemadi et al., 2018b; Tonitto et al., 2006).

Flail mowing, another mechanical strategy, terminates legume while evenly distributing residues without soil disturbance. Similarly, roller crimpers, another mechanical device, also terminate cover crops without disturbing the soil. However, roller crimpers may not thoroughly kill legume cover crops unless termination is delayed (Mischler et al., 2010).

10 Harvesting legume cover crops

Traditionally, cover crops are grown to improve soil fertility and for other agroecological benefits but not for harvest. However, harvesting legume cover crops as forage or partially harvesting as vegetables is gaining popularity as a strategy to enhance farm profitability. In this system, the cover crop roots and stubble remain in the field, enriching the soil with nutrients, particularly carbon and nitrogen, and promoting overall soil health.

The use of cool-season legumes as dual-purpose cover crops is not a novel concept. Like grass cover crops, legumes such as clovers and alfalfa have been cultivated for combined purposes of forage production and winter cover crops (Möller et al., 2008). In this system, the aerial parts are harvested as forage at specific times while the remaining stubble and roots provide nitrogen and carbon benefits to the soil and the subsequent cash crop. Although significant amounts of nutrients are removed from the field by harvesting cover crop aboveground biomass, the overall farm economics may justify such an approach. Additionally, in sandy soil, harvesting aerial parts of the cover crop as supplemental forage may also reduce the nitrate leaching risk.

Pod legume cover crops like faba beans and peas can also be grown for combined use (Etemadi et al., 2018a; Brasier et al., 2023). Some of the lower pods of the cover crops that are produced first can be harvested as fresh vegetables, while the remaining residues, including remaining upper pods, aerial biomass, and roots, are left in the field to contribute to the nitrogen needs of subsequent crops. Depending on the extent of pod removal, the overall nitrogen contribution of the cover crop is diminished since up to 57–72% of nitrogen can be in the seeds of peas and faba beans (Etemadi et al., 2018c;

Jensen et al., 2010). Despite this limitation, the approach offers increased grower income while benefiting general soil health.

11 Conclusion

Leguminous cover crops are first and foremost chosen for their ability to fix atmospheric nitrogen for the following crops. They are less effective for weed control, nutrient scavenging, or carbon sequestration. Management decisions interact with inherent cover crop traits to affect the utility of legumes. Key decision factors are (1) species or variety selection, (2) seeding rate, (3) planting time, and (4) termination time.

Selecting the species or variety of cool-season legume cover crop that is suited for a particular climate is the first step. Appropriate seeding rates are also critical, and seed size must be considered. Large-seeded legumes (e.g. field peas and faba bean) require higher seeding rates to produce the same populations as smaller seeded varieties of the same species or small-seeded species (e.g. balansa clover and yellow sweet clover). Most legume cover crops can be successfully planted at a range of seeding rates. When legumes are part of mixed cover crops, their seeding rate should be similar to the monoculture rate or only be slightly reduced.

Planting time varies by location, but generally cool-season legumes should be planted in late summer, early fall, or early spring. Late summer or early fall planting is most applicable for legume species that survive the winter, while early spring planting is best used for cover crops that would not survive the winter or where cash crops were grown late into the fall growing season. Generally, fall-planted, winter-hardy legumes will flower earlier and produce more biomass than if planted in spring. However, legume cover crops that winterkill may lose much of their nitrogen before cash crop planting. While legumes can also be planted in late fall, this is only suitable for mild climates where active growth is possible during the winter. Some legumes (e.g. sweetclovers) require vernalization to flower and can only be fall planted. Similarly, species with distinct spring and winter varieties (e.g. field pea) must be planted in their appropriate season to be most productive.

Allowing legumes sufficient growing time before termination increases their nitrogen contribution. Among species that are also used for grain or forage, maturity rates can vary widely, making variety selection crucial. When legumes are grown in a mixture, the best termination time may shift to balance the legume's nitrogen contribution with the benefits from companion cover crops.

Overall, the management guidelines presented here are valuable for adapting cool-season legume cover crops in agricultural systems. However, good agronomic practices can only fine tune the growth of different species.

Seasonally and regionally appropriate species and variety selection remains critical to maximize the agronomic, environmental, and economic benefits of cool-season legume cover crops. Continued research into cool-season cover crop species selection and management is needed to increase cover crop adoption as one important component of sustainable agricultural systems.

12 Where to look for further information

- Herridge, D. F., et al. (2008). Global inputs of biological nitrogen fixation in agricultural systems. *Plant and Soil*, 311, 1–18.
- Peng, Y., et al. (2024). Global synthesis of cover crop impacts on main crop yield. *Field Crops Research*, 310, 109343.
- Zhao, J., et al. (2022). Global systematic review with meta-analysis reveals yield advantage of legume-based rotations and its drivers. *Nature Communications*, 13(1), 4926.

13 References

Abd El Moneim, A. M. (1993). Agronomic potential of three vetches (*Vicia* spp.) under rainfed conditions. *Journal of Agronomy and Crop Science*, 170(2), 113–120.

Abdin, O. A., et al. (2000). Cover crops and interrow tillage for weed control in short season maize (Zea mays). *European Journal of Agronomy*, 12(2), 93–102.

Abou Chehade, L., et al. (2023). Legume cover crop alleviates the negative impact of no-till on tomato productivity in a Mediterranean organic cropping system. *Agronomy*, 13(8), 2027.

Adeux, G., et al. (2021). Cover crops promote crop productivity but do not enhance weed management in tillage-based cropping systems. *European Journal of Agronomy*, 123, 126221.

Ahmadnia, F., et al. (2020). Evaluation of short-term effect of oat (*Avena sativa* L.) and chickling pea (*Lathyrus sativus* L.) cover crops on improving some soil properties. *Applied Soil Research*, 9(1), 72–87.

Albayrak, S., et al. (2006). Effects of inoculation with rhizobium on seed yield and yield components of common vetch (*Vicia sativa L.*). *Turkish journal of Agriculture and Forestry*, 30(1), 31–37.

Alexandre, A. and Oliveira, S. (2013). Response to temperature stress in rhizobia. *Critical Reviews in Microbiology*, 39(3), 219–228.

Ashford, D. L. and Reeves, D. W. (2003). Use of a mechanical roller-crimper as an alternative kill method for cover crops. *American Journal of Alternative Agriculture*, 18(1), 37–45.

Bair, K. E., et al. (2008). Release of available nitrogen after incorporation of a legume cover crop in concord grape. *HortScience*, 43(3), 875–880.

Bariabar, B., et al. (2018). Growing degree days and cover crop type explain weed biomass in winter cover crops. *Agronomy for Sustainable Development*, 38, 1–9.

Bhardwaj, H. L., et al. (2010). Preliminary evaluation of white lupin (*Lupinus albus* L.) as a forage crop in the mid-atlantic region of the United States of America. *Journal of Agricultural Science*, 2(4), 13.

Blesh, J. (2018). Functional traits in cover crop mixtures: biological nitrogen fixation and multifunctionality. *Journal of Applied Ecology*, 55(1), 38–48.

Blesh, J., et al. (2019). Managing ecosystem services with cover crop mixtures on organic farms. *Agronomy Journal*, 111(2), 826–840.

Blevins, R. L., et al. (1990). Legume cover crops as a nitrogen source for no-till corn and grain sorghum. *Agronomy Journal*, 82(4), 769–772.

Boniface, H. S., et al. (2024). Environmental conditions outweigh seeding rates for cover crop mixture performance across the Northeast US. *Field Crops Research*, 318, 109564.

Boulet, A. K., et al. (2021). Introduction of legume cover crops practice in intensive grain corn crop system to mitigate soil threats in the Mediterranean region (No. EGU21-6199). *Copernicus Meetings*.

Brady, N. C. and Weil, R. R. (2002) *The Nature and Properties of Soils*, 13th ed. Upper Saddle River, NJ: Prentice Hall, p. 960.

Brandsæter, L. O. and Netland, J. (1999). Winter annual legumes for use as cover crops in row crops in northern regions: I. Field experiments. *Crop Science*, 39(5), 1369–1379.

Brandsæter, L. O., et al. (2000). Winter annual legumes for use as cover crops in row crops in northern regions: II. Frost resistance study. *Crop Science*, 40(1), 175–181.

Brasier, K., et al. (2023). Potential of Faba Bean (*Vicia faba* L.) for dual-purpose vegetable production and cover cropping. *HortScience*, 58(1), 23–31.

Büchi, L., et al. (2015). Accumulation of biologically fixed nitrogen by legumes cultivated as cover crops in Switzerland. *Plant and Soil*, 393, 163–175.

Bugg, R. L., et al. (1996). Comparison of 32 cover crops in an organic vineyard on the north coast of California. *Biological Agriculture & Horticulture*, 13(1), 63–81.

Burgess, M., et al. (2014). Tillage of cover crops affects soil water, nitrogen, and wheat yield components. *Agronomy Journal*, 106(4), 1497–1508.

Busse, M. D., et al. (2001). Glyphosate toxicity and the effects of long-term vegetation control on soil microbial communities. *Soil Biology and Biochemistry*, 33(12–13), 1777–1789.

Cai, S., et al. (2018). Winter legume-rice rotations can reduce nitrogen pollution and carbon footprint while maintaining net ecosystem economic benefits. *Journal of Cleaner Production*, 195, 289–300.

Carranca, C., et al. (2009). White lupine as a beneficial crop in Southern Europe: I. Potential for N mineralization in lupine amended soil and yield and N2 fixation by white lupine. *European Journal of Agronomy*, 31(4), 183–189.

Cho, J. L., et al. (2011a). Growth of Vicia tetrasperma and V. hirsuta as affected by seeding condition and estimated N production. *Korean Journal of Weed Science*, 31(1), 84–88.

Cho, J. L., et al. (2011b). Red pepper cultivation using a native legume cover crop in Korea. *Korean Journal of Organic Agriculture*, 19(spc), 247–250.

Clark, A. J., et al. (1994). Seeding rate and kill date effects on hairy vetch-cereal rye cover crop mixtures for corn production. *Agronomy Journal*, 86(6), 1065–1070.

Cook, J. C., et al. (2010). Optimizing vetch nitrogen production and corn nitrogen accumulation under no-till management. *Agronomy Journal*, 102(5), 1491–1499.

Den Hollander, N. G., et al. (2007). Clover as a cover crop for weed suppression in an intercropping design: I. Characteristics of several clover species. *European Journal of Agronomy*, 26(2), 92–103.

Ding, K., et al. (2024). Long-term cover crop monocultures and mixtures improve soil multifunctionality and shift microbial communities and functions. *European Journal of Agronomy*, 157, 127188.

Douglas, J., et al. (2023). *Seeding Rates of Annual Cool Season Legume Cover Crops in the Southern and Eastern US*. Fort Worth, TX: USDA-Natural Resources Conservation Service.

Etemadi, F., et al. (2018a). Phenology, yield and growth pattern of faba bean varieties. *International Journal of Plant Production*, 12, 243–250.

Etemadi, F., et al. (2018b). Nitrogen contribution from winter-killed faba bean cover crop to spring-sown sweet corn in conventional and no-till systems. *Agronomy Journal*, 110(2), 455–462.

Etemadi, F., et al. (2018c). Nutrient accumulation in faba bean varieties. *Communications in Soil Science and Plant Analysis*, 49(16), 2064–2073.

Feng, H., et al. (2020). Soil quality indicators as influenced by 5-year diversified and monoculture cropping systems. *The Journal of Agricultural Science*, 158(7), 594–605.

Francis, G. S., et al. (1998). The effect of winter cover crop management on nitrate leaching losses and crop growth. *The Journal of Agricultural Science*, 131(3), 299–308.

Frye, W. W., et al. (1988). Role of annual legume cover crops in efficient use of water and nitrogen. In Hargrove, W. L. (Ed.), *Cropping Strategies for Efficient Use of Water and Nitrogen*. American Society of Agronomy: Madison, WI, Spec. Publ. No. 51, pp. 129–154.

Furey, G.N., et al. 2021. Substituting vetch and chicory for rye in a cover crop mixture enhanced nutrient release. *Canadian Journal of Soil Science*, 101(2), 339–343.

Ghorbi, S., et al. (2023). The use of faba bean cover crop to enhance the sustainability and resiliency of no-till corn silage production and soil characteristics. *Agronomy*, 13(8), 2082.

Gselman, A. and Kramberger, B. (2008). Benefits of winter legume cover crops require early sowing. *Australian Journal of Agricultural Research*, 59(12), 1156–1163.

Heichel, G. H. (1987) Legume nitrogen: Symbiotic fixation and recovery by subsequent crops. In Helsel, Z. R. (Ed.), *Energy in Plant Nutrition and Pest Control*. Amsterdam: Elsevier Scientific Publishers, pp. 63–80.

Holderbaum, J. F., et al. (1990). Fall-seeded legume cover crops for no-tillage corn in the humid east. *Agronomy Journal*, 82(1), 117–124.

Holman, J. D., et al. (2018). Can cover or forage crops replace fallow in the semiarid central Great Plains?. *Crop Science*, 58(2), 932–944.

Hyun, L. J., et al. (2005). Overwintering capacity affected by seeding time and method of chinese milk vetch, *Astragalus sinicus* L., in Upland field. *Korean Journal of Crop Science*, 50(2), 67–72.

Islam, R. ed. (2021). *Cover Crops and Sustainable Agriculture*. Boca Raton, FL: CRC Press.

Janczarek, M., et al. (2024). Adaptation of Rhizobium leguminosarum sv. trifolii strains to low temperature stress in both free-living stage and during symbiosis with clover. *Science of the Total Environment*, 951, 175554.

Jensen, E. S., et al. (2010). Faba bean in cropping systems. *Field Crops Research*, 115(3), 203–216.

Kandel, T. P., et al. (2018). Nitrous oxide emissions as influenced by legume cover crops and nitrogen fertilization. *Nutrient Cycling in Agroecosystems*, 112, 119–131.

Kaye, J. P. and Quemada, M. (2017). Using cover crops to mitigate and adapt to climate change. A review. *Agronomy for Sustainable Development*, 37(1), 4.

KBDP, S., et al. (2005). Effect of Chinese milk vetch (*Astragalus sinicus* L.) as a cover crop on weed control, growth and yield of wheat under different tillage systems. *Plant Production Science*, 8(1), 79–85.

Kemper, R., et al. (2022). Root traits in cover crop mixtures of blue lupin and winter rye. *Plant and Soil*, 475(1), 309–328.

Koudahe, K., et al. (2022). Critical review of the impact of cover crops on soil properties. *International Soil and Water Conservation Research*, 10(3), 343–354.

Kramberger, B., et al. (2009). Effects of cover crops on soil mineral nitrogen and on the yield and nitrogen content of maize. *European Journal of Agronomy*, 31(2), 103–109.

Lacey, C., et al. (2020). Carbon and nitrogen release from cover crop residues and implications for cropping systems management. *Journal of Soil and Water Conservation*, 75(4), 505–514.

Lavergne, S., et al. (2021). Using fall-seeded cover crop mixtures to enhance agroecosystem services: a review. *Agrosystems, Geosciences & Environment*, 4(2), e20161.

Lim, K. H., et al. (2012). Evaluation of potential nutrient contribution of overwintering cover crops in organic orchards. *Korean Journal of Organic Agriculture*, 20(2), 173–183.

Liu, S., et al. (2022). *Astragalus sinicus* incorporated as green manure for weed control in corn. *Frontiers in Plant Science*, 13, 829421.

MacMillan, J., et al. (2022). Biological nitrogen fixation of cool-season legumes in agronomic systems of the Southern Great Plains. *Agrosystems, Geosciences & Environment*, 5(1), e20244.

Malone, L. C., et al. (2022). Cover crops in a Wisconsin annual cropping system: feasibility and yield effects. *Agronomy Journal*, 114(2), 1052–1067.

May, W. E., et al. (2022). Self-regenerating black medic cover crop provides agronomic benefits at low nitrogen. *Agronomy Journal*, 114(5), 2743–2761.

Mirsky, S. B., et al. (2017). Hairy vetch biomass across the Eastern United States: effects of latitude, seeding rate and date, and termination timing. *Agronomy Journal*, 109(4), 1510–1519.

Mischler, R., et al. (2010). Hairy vetch management for no-till organic corn production. *Agronomy Journal*, 102(1), 355–362.

Möller, K., et al. (2008). Growth, composition, biological N 2 fixation and nutrient uptake of a leguminous cover crop mixture and the effect of their removal on field nitrogen balances and nitrate leaching risk. *Nutrient Cycling in Agroecosystems, 82*, 233–249.

Natera, J. F. Z., et al. (2023). Assessing the biomass yield and nitrogen fixation of *Lupinus angustifolius* varieties as green manure in Jalisco, Mexico. *Eurasian Journal of Soil Science*, 12(4), 320–327.

Nguyen, D. B., et al. (2016). Impact of glyphosate on soil microbial biomass and respiration: a meta-analysis. *Soil Biology and Biochemistry*, 92, 50–57.

Nie, J., et al. (2019). Leguminous cover crop Astragalus sinicus enhances grain yields and nitrogen use efficiency through increased tillering in an intensive double-cropping rice system in southern China. *Agronomy*, 9(9), 554.

Ogunleye, A. M., et al. (2023). Cool-season cover crop effects on forage productivity and short-term soil health in a semi-arid environment. *Renewable Agriculture and Food Systems*, 38, e26.

Ordóñez-Fernández, R., et al. (2018). Legumes used as cover crops to reduce fertilisation problems improving soil nitrate in an organic orchard. *European Journal of Agronomy*, 95, 1–13.

Parr, M., et al. (2011). Nitrogen delivery from legume cover crops in no-till organic corn production. *Agronomy Journal*, 103(6), 1578–1590.

Perrone, S., et al. (2020). Nitrogen fixation and productivity of winter annual legume cover crops in Upper Midwest organic cropping systems. *Nutrient Cycling in Agroecosystems*, 117, 61–76.

Pease, S. M. (2020). Evaluating Lupin's agricultural potential as a cover crop in Vermont. Student research conference, University of Vermont, USA. https://scholarworks.uvm.edu/src/2020/agriculture/8/

Perdigão, A., et al. (2021). A 3-year field study to assess winter cover crops as nitrogen sources for an organic maize crop in Mediterranean Portugal. *European Journal of Agronomy*, 128, 126302.

Power, J. F. and Koerner, P. T. (1994). Cover crop production for several planting and harvest dates in eastern Nebraska. *Agronomy Journal*, 86(6), 1092–1097.

Power, J. F. and Zachariassen, J. A. (1993). Relative nitrogen utilization by legume cover crop species at three soil temperatures. *Agronomy Journal*, 85(1), 134–140.

Rao, S. C. and Northup, B. K. (2011). Grass pea (*Lathyrus sativus* L.) as a pre-plant nitrogen source for continuous conventionally tilled winter wheat. *Crop science*, 51(3), 1325–1333.

Read, J. J., et al. (2011). Nitrogen and winter cover crop effects on spring and summer nutrient uptake. *Grass and Forage Science*, 66(3), 381–390.

Rijal, B., et al. (2021). Effect of Leguminous winter cover crops on soil fertility and yield of summer maize. *Agronomy Journal of Nepal*, 5(01), 186–192.

Rose, M. T., et al. (2016). Impact of herbicides on soil biology and function. *Advances in Agronomy*, 136, 133–220.

Rosolem, C. A., et al. (2004). Nitrogen management in maize cover crop rotations. *Plant and Soil*, 264, 261–271.

Saini, M., et al. (2005). Winter weed suppression by winter cover crops in a conservation-tillage corn and cotton rotation. In *Proceeding of 27th Southern Conservation-Tillage Conference*, pp. 124–128.

Scavo, A., et al. (2020). Improving soil health, weed management and nitrogen dynamics by Trifolium subterraneum cover cropping. *Agronomy for Sustainable Development*, 40, 1–12.

Schomberg, H. H., et al. (2006). Conservation tillage and cover crop influences on cotton production on a southeastern US coastal plain soil. *Agronomy Journal*, 98(5), 1247–1256.

Shiferaw, A., et al. (2024). Optimizing cover crop management in Eastern Nebraska: insights from crop simulation modeling. *Agronomy*, 14(7), 1561.

Sincik, M., et al. (2008). Responses of potato (*Solanum tuberosum* L.) to green manure cover crops and nitrogen fertilization rates. *American Journal of Potato Research*, 85, 150–158.

Smith, M. S., et al. (1987) Legume winter cover crops. *Advances in Soil Science*, 7, 95–139.

Soriano, M. A., et al. (2023). Field evaluation of selected autochthonous herbaceous species for cover crops in Mediterranean woody crops. *European Journal of Agronomy*, 143, 126723.

Stein, S., et al. (2023). Plant and soil N of different winter cover crops as green manure for subsequent organic white cabbage. *Nutrient Cycling in Agroecosystems*, 127(2), 285–298.

Sunuwar, S., et al. (2023). Cover crop termination method and N fertilization effects on sweet corn yield, quality, N uptake, and weed pressure. *Nitrogen*, 4(1), 37–57.

Tadiello, T., et al. (2022). Growth, weed control, and nitrogen uptake of winter-killed cover crops, and their effects on maize in conservation agriculture. *Agronomy for Sustainable Development*, 42(2), 18.

Thapa, R., et al. (2018). Cover crops reduce nitrate leaching in agroecosystems: a global meta-analysis. *Journal of environmental quality*, 47(6), 1400–1411.

Thiessen Martens, J. R., et al. (2005). Legume cover crops with winter cereals in southern Manitoba: fertilizer replacement values for oat. *Canadian Journal of Plant Science*, 85(3), 645–648.

Thurston, C. L., et al. (2022). Cold stress reduces nodulation and symbiotic nitrogen fixation in winter annual legume cover crops. *Plant and Soil*, 481(1), 661–676.

Tiemann, L. K., et al. (2015). Crop rotational diversity enhances belowground communities and functions in an agroecosystem. *Ecology Letters*, 18(8), 761–771.

Tonitto, C., et al. (2006). Replacing bare fallows with cover crops in fertilizer-intensive cropping systems: a meta-analysis of crop yield and N dynamics. *Agriculture, Ecosystems & Environment*, 112(1), 58–72.

Toom, M., et al. (2019). Selecting winter cover crop species for northern climatic conditions. *Biological Agriculture & Horticulture*, 35(4), 263–274.

Tursun, N., et al. (2018). Use of living, mowed, and soil-incorporated cover crops for weed control in apricot orchards. *Agronomy*, 8(8), 150.

Tyler, H. L. (2020). Winter cover crops and no till management enhance enzyme activities in soybean field soils. *Pedobiologia*, 81, 150666.

Vanelli, J., et al. (2018). Microbiological assessment of soil planted with cover crops, and Soybean and Maize in succession. *American Journal of Plant Sciences*, 9(4), 659–666.

Vessey, J. K. (2004). Benefits of inoculating legume crops with rhizobia in the northern Great Plains. *Crop Management*, 3(1), 1–8.

Von Boberfeld, W. O., et al. (2005). Nitrogen transfers from *Vicia sativa* L. and *Trifolium resupinatum* L. to the companion grass and the following crop. *Plant, Soil and Environment*, 51(6), 267–275.

White, C. and Barbercheck, M. (2017). *Making the Most of Mixtures: Considerations for Winter Cover Crops*. University Park, PA: Penn State Extension. Accessed online (February 28, 2025). https://extension.psu.edu/making-the-most-of-mixtures-considerations-for-winter-cover-crops.

Wolters, B. R. (2020). *To Mix or Not to Mix: Performance and Management of Diverse Cover Crop Mixtures*. Blacksburg, VA: Virginia Polytechnic Institute and State University. Accessed online (February 28, 2025). https://vtechworks.lib.vt.edu/items/7858ccae-5a2b-40b2-a349-03c8a804524a.

Wortman, S. E., et al. (2012). Cover crop mixtures for the western corn belt: opportunities for increased productivity and stability. *Agronomy Journal*, 104(3), 699–705.

Wunsch, E. M., et al. (2017). Can legumes provide greater benefits than millet as a spring cover crop in southern Queensland farming systems?. *Crop and Pasture Science*, 68(8), 746–759.

Yang, B.-J., et al. (2013). Effects of different multiple cropping systems on paddy field weed community under long term paddy-upland rotation. *Yingyong Shengtai Xuebao*, 24(9), 2533–2538.

Yang, L., et al. (2019). Co-incorporation of rice straw and green manure benefits rice yield and nutrient uptake. *Crop Science*, 59(2), 749–759.

Yang, L., et al. (2021). Biological nitrogen fixation of Chinese Milk Vetch (*Astragalus sinicus* L.) as affected by exogenous carbon and nitrogen input. *Symbiosis*, 85(1), 69–77.

Chapter 12

Warm season leguminous cover crops

Gourav Chahal, Auburn University, USA; and Andrew Price, USDA-ARS National Soil Dynamics Laboratory, USA

1 Introduction

Sustainable intensification will help to meet food, fiber, and fuel needs of a global population projected to reach 12.3 billion by the end of this century (Gerland et al., 2014; Garnett et al., 2013). Nitrogen is a critical plant nutrient, but the environmental externalities of synthetic N production and use are a concern (Godfray et al., 2010). According to Menegat et al. (2022), synthetic N fertilizer production accounts for 10.6% of agricultural greenhouse gas (GHG) emissions and 2.1% of total global GHG emissions. In addition to carbon dioxide emissions in their production, the use of N fertilizer can also increase nitrous oxide emissions, a potent greenhouse gas (Viney et al., 2009). Further, due to low N-use efficiency, losses of N are relatively large and can negatively impact ground and surface water quality (Gaudio et al., 2015; Goolsby et al., 2001; Ran et al., 2019). Warm season leguminous cover crops (CCs) can fix large quantities of N. They differ from cool season legume

http://dx.doi.org/10.19103/AS.2025.0162.16
Published by Burleigh Dodds Science Publishing Limited, 2026.

CCs in their climatic requirements. Fall temperatures slow down metabolic processes and photosynthesis in cool season legumes, affecting overall vegetative growth and nitrogen fixation (Koehler et al., 2023; Kaye et al., 2019; Price and Norsworthy, 2013). Growth and N fixation tend to be more aggressive in warm season legume CCs compared to winter legume CCs (Rao and Northup, 2009a). By providing N to the following crop, N fertilizer use and associated carbon dioxide emissions from N-fertilizer production can be reduced. Nitrous oxide emissions often spike after fertilizer application and may also be reduced if N is supplied through legumes (Mahama et al., 2020). Although legumes are typically not very effective in reducing N-leaching, growing them in mixtures with non-legume CCs can increase their benefits by further reducing N leaching and adding organic carbon (Vogeler et al., 2019; Kakraliya et al., 2018). The combination of legume and non-legume CC has the potential to reduce nitrate (NO_2–) leaching by up to 56% (Thapa et al., 2018). Warm season legumes, characterized by their vigorous growth and low input requirements, present a promising option for intercropping with major cash crops. For example, Manasa et al. (2018) found that maize intercropped with various summer legumes consistently resulted in a combined Land Equivalent Ratio (LER) greater than 1.5, indicating a significant advantage in land-use efficiency, due to maize's wide spacing, slow early growth, and paired row planting that supported better legume development. Also Bybee-Finley et al. (2016) found that a mixture of pearl millet, sorghum sudangrass, and SH exhibited the highest evenness in biomass contribution compared to other intercropping methods. Warm season legume CCs also have the potential to improve overall weed control through competition for resources, allelopathic activity, and acting as physical barrier after termination in conservation tillage systems (Zannopoulos et al., 2024; Adler and Chase, 2007). These legumes can also add large quantities of biomass that may improve soil organic carbon content and soil health, and their taproots may alleviate soil compaction (Blanco et al., 2012; Lal, 2015). Summer legumes also diversify cropping systems, temporally if grown in rotation with main crops and spatially when intercropped in main crops (Bybee-Finley et al., 2016). Finally, they have the potential to generate economic returns if used as forage, if they replace N-fertilizer use for main crops, or if they boost following crop yields through a rotational effect. However, intercropping also presents challenges, such as competition with the cash crop for water, nutrients, and light, as well as complications associated with normal field operations (planting, weed control, harvesting). In this chapter, we will review research showing how warm season legume CCs can play an important role in creating sustainable solutions that decrease our dependence on synthetic fertilizers and alleviate the environmental impacts of crop production.

2 Warm season legume species

Summer legume cover crops are herbaceous plants grown in the summer, either between or intercropped with main crops in a non-competitive way, to improve soil health, add nitrogen, provide forage, and suppress weeds (Wang and Nolte, 2010; Scholberg et al., 2010; Lu et al., 2000).

The selection of an appropriate warm season legume cover crop species is crucial for maximizing its benefits within a specific agricultural system (Snapp et al., 2005). Key considerations include the species' temperature requirements, biomass production potential, and suitability for various cropping contexts. The following table summarizes several warm season legume cover crop species, such as cowpea, mung bean, pigeon pea, and velvet bean, which are native to tropical climates. These species are well-suited to survive in higher temperatures (Saikia et al., 2020). The table also includes some non-tropical legumes that can be grown in the summer, such as soybean, alfalfa, and lespedeza species. Note that this list is not exhaustive; it does not cover all legume options, but it does include information on their optimal temperature ranges, typical biomass yields, and relevant references for the included legumes (Table 1).

Table 1 Key characteristics of several warm season legume cover crops, including optimal temperatures, biomass, and nitrogen fixation yields. This list is not exhaustive

Cover crop	Temperature (°C)Min; Max; Optimal	Biomass (lb/ac)	N fixation (lbs/ac)	References
Cowpea (*Vigna unguiculata)*	15; 35; 25–35	2000–5000	100–150	Power and Koerner, 1994; Ayerdi and Marraccini, 2022; Wang and Nolte, 2010
Sunn Hemp (*Crotalaria juncea*)	20; 35; -	3000–7500	100–130	Wang et al., 2022a; Hariniharisma et al., 2025; Mansoer et al., 1997
Velvetbean (*Mucuna pruriens*)	5; -; 19–27	5000–10 000	100–140	Weaver, 2016; Ayerdi and Marraccini, 2022; Clark, 2007; Harrison et al., 2004
Soybeans (*Glycine max*)	18; 36; 26–30	3000–4500	80–110	Hatfield et al., 2011; Rao and Northup, 2009a
Mung Bean (*Vigna radiata*)	7.5; 40; 30	2000–5500	50–70	Chauhan and Williams, 2018; Rao and Northup, 2009a
Pigeon Pea (*Cajanus cajan* [L.])	10; 45; 26–36	3000–7000	60–70	Carberry et al., 2001; Rao and Northup, 2009a
Guar (*Cyamopsis tetragonobloba* [L.])	20; 35–40; 25–32	2500–6500	60–80	Rao and Northup, 2009b, 2013; Gresta et al., 2018
Sesbania (*Sesbania* spp.)	10–15; 40; 25–35	4000–6500	60–150	Azim et al., 2024; Chanda et al., 2019; Wang and Nolte, 2010
Lespedeza (*Lespedeza* spp.)	10; 35; 10–30	1000–4000	90–150	Kalburtji et al., 2007; Mosjidis, 1990; Garten et al., 2008; Shin et al., 2018

Published by Burleigh Dodds Science Publishing Limited, 2026.

3 Nitrogen dynamics

3.1 N fixation and fertilizer value

Legume CCs can use, or fix, atmospheric N through their symbiotic relationship with Rhizobium bacteria (Udvardi and Poole, 2013). Upon decomposition, the fixed N is then released, becoming available to the following non-legume crop. Legume CCs maintain lower carbon-to-nitrogen (C:N) ratios as they mature compared to non-legumes and therefore tend to decompose quickly (Gomes et al., 2009). This rapid decomposition provides N to a subsequent cash crop (Blanco-Canqui et al., 2013; Balkcom and Reeves, 2005a). Herridge et al. (2008) estimated that biological N fixation by legumes has the potential to contribute approximately 40 million tonnes of nitrogen to global agricultural systems annually. Recent global meta-analyses of 462 studies across 53 countries found that a main crop following legumes had on average a 20% yield advantage, which was predominantly a result of its N-fertilizer value (Zhao et al., 2022).

Specific examples demonstrate this N contribution, such as SH CC grown in late summer of the previous year supplied significant N to corn on loamy sand in Alabama (Balkcom and Reeves, 2005b). SH produced 7.6 Mg ha^{-1} of biomass with an N content of 144 kg ha^{-1}, producing the equivalent corn yield as 112 kg ha^{-1} fertilizer N (Fig. 1).

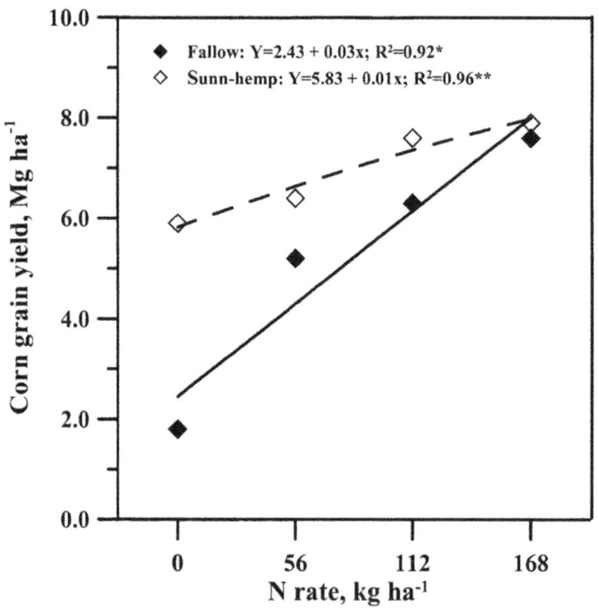

Figure 1 Corn grain yields in 1991 following sunn hemp and fallow plots measured across four N rates at the E.V. Smith Research and Extension Center in Shorter, AL (Balkcom and Reeves, 2005a).

Published by Burleigh Dodds Science Publishing Limited, 2026.

Similarly, in a no-till wheat-sorghum rotation, Blanco-Canqui et al. (2012) found that sorghum yields after SH or LMS CCs were comparable to those achieved with 66–100 kg ha^{-1} of fertilizer N in control plots. While these CCs did not supply N to wheat planted immediately after sorghum, they significantly increased soil N content in the spring following the final sorghum year of a 7-year trial (Fig. 2). This suggests that the N benefit from warm season legume CCs can persist for more than one year; for instance, SH cultivation led to N accumulation ranging from 190 kg ha^{-1} to 319 kg ha^{-1} in a subsequent processing tomato (*Solanum lycopersicum* L.) crop, significantly boosting yield over two consecutive years (Wang et al., 2009). Long-term use of annual legumes can also enhance soil humus, thereby increasing the soil's N-supplying capacity (Blanco-Canqui et al., 2012). When legumes are grown in mixtures, they can meet their N needs through atmospheric N fixation, while a grass companion can efficiently utilize soil N, particularly at low soil N concentrations (Blanco-Canqui al., 2015).

However, various biotic and abiotic factors can limit the potential N fixation in legumes, such as shortage of molybdenum, iron, potassium, sulfur, and zinc needed for optimal functioning by acting as cofactors for the enzyme nitrogenase (Schrer et al., 1996; Islam et al., 2018a; Bursakov, 2023; Khalili, 2024a). Furthermore, according to Linn et al. (1984), N fixation is maximized at field capacity and at a pH range of 6 to 8. In contrast to legumes, grass CCs such as rye, ryegrass (*Lolium perenne* L.), or barley (*Hordeum vulgare* L.) are winter annuals usually planted before the main crop, but they can immobilize available N. This immobilization is greater in drier weather, as found in early summers in the southeastern United States (Nevins et al., 2020). Warm season

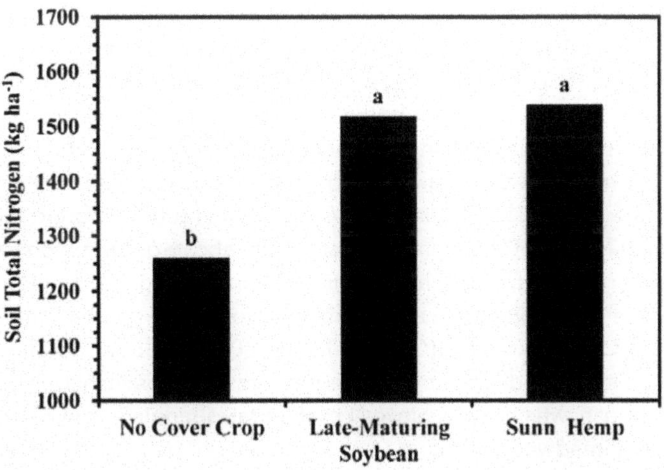

Figure 2 Mean soil total N content for each cover crop treatment in the 0- to 7.5-cm soil depth. Bars accompanied by different lowercase letters are significantly different (Blanco et al., 2012).

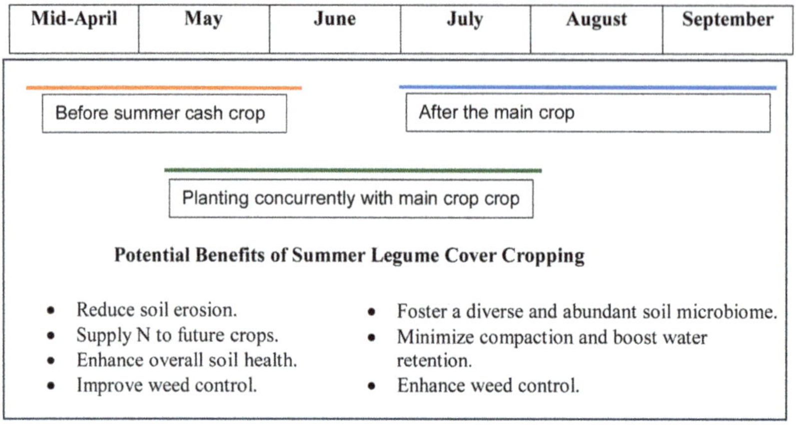

Figure 3 The seasonal availability for summer legumes to integrate into the existing cropping system of corn or soybean in the southeastern United States, highlighting potential benefits noted at the bottom.

legumes, like cowpea (CP) or soybean, fix N that upon decomposition is likely to be more available to the subsequent crop (Dabney et al., 2001; Schomberg et al., 2007) (Fig. 3).

3.2 N synchronization

For maximizing the benefit of N provided by legumes, synchronization of this N release with crop need is necessary. Crops grown after legumes can take up at least 30% to 60% of the N that the legume produces (Clark, 2007; Murungu et al., 2010; Yang et al., 2023). Conversely, aggressive tilling can lead to rapid decomposition that results in excessive availability of soil N over very short periods (Mansoer et al., 1997; Evans et al., 2001a; Liu et al., 2019a). Under optimal conditions for N fixation, legumes play a significant role in nutrient management by retaining N in their roots (approximately 15% to 30% of total plant N), thereby preventing gaseous loss (Gautrat et al., 2020; Oelmann et al., 2005; Silva et al., 2021). For example, Aguiar et al. (2001) reported that in CP, approximately 90% of the plant's N is stored in above-ground biomass, while the remaining 10% is found in roots. If grown alone during the summer in North Carolina, the shoot biomass can accumulate up to 75 kg N ha^{-1} in about 70 days, with a C:N ratio of 15:1 (O'Connell et al. 2015). In South Florida, summer provides an ideal window for tropical legume cover crops such as SH, cowpea, and velvet bean, which can achieve N fixation rates ranging from 63 kg ha^{-1} to 278 kg ha^{-1} (Treadwell, 2006).

Despite their low C:N ratio, N release from legume cover crops may not perfectly synchronize with cash crop demand, potentially necessitating additional fertilizer applications (Campiglia et al., 2011). For instance, crops like cabbage exhibit rapid nitrogen (N) uptake soon after transplanting (Fink and Feller, 1998; Duarte et al., 2019), whereas peak nitrogen release from cover crops typically occurs around five weeks post-termination (O'Connell et al., 2015). This asynchrony can be particularly challenging in organic systems with limited N options.

4 Rotations for summer legume cover crops

Integrating summer legumes effectively into existing crop rotations is a viable option for increasing productivity while efficiently and sustainably utilizing resources (Willey, 1979; Saikia et al., 2020). Their aggressive growth habit and warm season requirements allow for flexible placement within various cropping systems, particularly in regions with long growing seasons like the Southeastern United States. The timing of planting and termination of these cover crops is critical to avoid competition with cash crops and to synchronize nutrient release with crop demand (Wiley and Rao, 1980).

4.1 Before or after cash crop

Summer legumes can be strategically planted during fallow periods following the harvest of early cash crops, such as small grains like wheat in the Midwest (Blanco et al., 2012) or either before or after main summer crops like corn (Balkcom and Reeves, 2005b; Mansoer et al., 1997) in the Southeastern United States where growing season of over 200 frost-free days allows this rotational flexibility. This allows the cover crop to grow and accumulate biomass and nitrogen during the summer, which might benefit subsequent crops (Holman et al., 2018; Clark, 2007).

For example, a 15-year study in Kansas reported by Blanco-Canqui et al. (2012) showed that summer legume cover crops such as SH and LMS improved wheat and sorghum yields, particularly under low nitrogen input conditions. However, O'Dea (2011) in his thesis reported that integrating early-terminated, no-till legume green manures during summer fallow in wheat rotations can improve long-term soil quality, though short-term yield reductions may hinder adoption.

In some systems, it's more appropriate to grow warm season legumes before the main cash crop. This approach allows the cover crop to contribute nitrogen and improve soil health prior to the planting of the subsequent crop (Creamer and Baldwin, 2000; O'Connell et al., 2015). The timing of termination is key to ensure the cover crop residue is managed effectively and does not interfere with cash crop planting or early growth (Kornecki and Price, 2019). In northern climates (USDA plant hardiness zones 1–4), summer cover crops must be integrated into tight rotation windows between spring and fall vegetable production. Although summer cover crops are typically grown for more than 2 to 3 months to achieve high biomass yields (Boyhan et al., 2016; Stute and Shekinah, 2019), shorter durations may be necessary to align with the constraints of cool season vegetable systems. For example, Wauters et al. (2021) found that 30- to 50-day summer cover crops like CP–sudex mixtures and SH in northern Minnesota and Wisconsin produced high biomass, but weed pressure and delayed broccoli planting reduced overall system productivity. However, in the upper Midwest, specifically Iowa, integrating cowpea into rotation followed by lettuce (Kruse and Nair, 2016) and mungbean, SH, and CP followed by cabbage (Bilenky et al., 2022) has been shown to increase the yield of cash crops.

4.2 Intercropping with main crops

Intercropping involves growing two or more crop species together in the same field for a substantial portion of their growth cycles, promoting resource efficiency and ecological intensification (Willey, 1979; Niggli et al., 2009; Bedoussac et al., 2015). Including summer legumes in this strategy can enhance land-use efficiency and provide benefits such as weed suppression and additional nitrogen for the main crop (Agegnehu et al., 2008; Hauggaard-Nielsen et al., 2009). For instance, maize intercropped with various summer legumes has shown a significant advantage in land-use efficiency (Manasa et al., 2018). Intercropping requires careful species selection and management to minimize competition for resources like water, nutrients, and light, especially in drier environments (Bedoussac et al.,2015). In a review by Bedoussac et al. (2015) reported that cereal–legume intercropping increased total grain yield in 91% of cases, with an average relative yield of 1.30, especially benefiting low-yield or low-nitrogen systems. Similarly, the purpose of intercropping corn with warm season forages for silage production has been investigated, showing that intercropping systems generally do not reduce total silage productivity (La Guardia Nave and Corbin, 2018). Nave and Corbin (2018) showed that intercropping corn with warm season legumes like SH and CP sustained total silage yield, highlighting benefits for crop-livestock integration. Intercropping design significantly influences yield outcomes, highlighting the importance of spatial arrangement for optimizing crop performance. Dalal (1974) demonstrated that maize yields decreased when grown with pigeon pea in the same row, suggesting competition for resources. In contrast, Mohta and De (1980) showed that wider row arrangements (2×2) in maize/soybean and sorghum/soybean mixtures maintained grain yields of cereals while enhancing soybean productivity. These findings underscore that careful configuration of intercrops can reduce competition and enhance overall system efficiency. However, intercropping adoption in the highly mechanized agricultural system, like in the United States, faces equipment-related challenges due to machinery designed primarily for monoculture, making diverse planting and harvesting inefficient and costly (Bourke et al., 2021; Eure et al., 2015). Precision tools for varying crop needs are limited and often unaffordable for many producers (Bybee-Finley and Ryan, 2018). Harvesting multiple crops with different maturity stages adds further complexity (Pierre et al., 2022). A lack of technical training on intercropping equipment also restricts effective adoption (Jiang et al., 2023).

4.3 Overseeding into main crops

Overseeding involves broadcasting warm season legume cover crops into the main crop late in the season, often prior to harvest, for example, via aerial seeding using drones (Blanco et al., 2017). This allows the cover crop to establish under the main crop canopy and begin growth as the main crop gets harvested (Nelson et al., 2022; Noland et al., 2018). This approach utilizes degree days and photosynthetically active radiation to enhance plant growth, extend nitrogen fixation in tropical legumes, and suppress late-season weeds, ultimately diminishing the weed seed bank (Baker and Griffis, 2009; Scursoni et al., 2007; Recker et al., 2015). Wilson and Allan (2014) reported that, prompted by incentives, aerial seeding of cover crops has significantly increased. For instance, in Maryland, it grew from 9,850 ha in 2007 to 167,900 ha in 2012 due to expanded cost-share programs (MDA, 2010, 2014).

Blanco-Canqui et al. (2017) reported that aerially seeded cereal rye (*Secale cereale* L.) in corn achieved good growth while maintaining corn yield. In contrast, Smeltekop et al. (2002) found that inter-seeded annual medic (*Medicago scutellata* Mill. 'Sava') cover crops in South Dakota reduced corn growth and yields due to competition for nitrogen and water, despite increasing water infiltration. The strategy emphasizes the crucial role of precise timing in aerial seeding, particularly in southeastern states such as Alabama. Specifically, it gives an opportunity to integrate summer legumes, such as SH, which could be aerially seeded into April-planted corn after the corn has passed its vulnerable growth stages related to water stress. The nitrogen fixed by the summer legume would subsequently become available to a winter grass cover crop, enhancing its growth. This approach offers several benefits, including improved ground cover and better nutrient cycling, which are critical for sustainable agriculture in the region.

The specific rotation chosen will depend on the cash crop, the desired ecosystem services, and the agroecological environment. For instance, in the Southeastern United States, the long-growing season often allows for more flexibility in integrating summer legumes after early-season crops. In contrast, in the Great Plains, water availability is a more significant limiting factor, requiring careful consideration of cover crop water use (Nielsen et al., 2016a).

5 Impacts on soil health

5.1 Soil erosion protection

Warm season legumes are fast-growing and therefore able to cover soil more quickly than cool season legumes. Bare soil is vulnerable to erosion and runoff caused by water and wind (Blanco and Lal, 2008). In humid temperate or subtropical regions, warm season CCs can effectively cover the soil during the fallow period after early harvested crops such as small grains or vegetables (Holman et al., 2018; Clark, 2007). During this period, the soil is susceptible to moisture loss due to evaporation and soil loss due to erosion (Layek et al., 2018; Meena et al., 2020). Warm season CCs can act as a shield against the destructive forces of rain and wind, their robust roots anchoring the soil to prevent soil erosion, while above-ground biomass provides soil protection from raindrop impact and slows down runoff (Blanco-Canqui et al., 2019; Gałązka et al., 2020; Cárceles Rodríguez et al., 2021). In addition, Mansoer et al. (1997) reported that winter-killed SH biomass, resulting from August planting and averaging 5.9 Mg ha^{-1} in December, in a no-till situation could persist on the ground for over 16 weeks. More than 60% of this initial residue remained by early April, demonstrating a significant amount of residual biomass (Fig. 4). This provided soil coverage that varied between 55% and 78% by April, a level generally considered sufficient to substantially reduce soil erosion (Kaspar and Singer, 2011). Thus, winter-killed CC residue not only enhances soil protection during the off-season but also provides soil-erosion protection until spring.

5.2 Increasing soil organic matter

Warm season legume CCs can accumulate large quantities of biomass in a short period of time, which may help to build soil organic matter. Yadvinder et al. (1992) reported that tropical legumes, like SH, CP, and MB, produced 2.9–8.9 Mg dry matter ha^{-1} by 50 to 60 days after planting, compared with winter legumes that produced 1.7 to 6.7 Mg dry matter ha^{-1} in approximately 6 months (Nov–May). Blanco et al. (2011) noted that SH and LMS CCs planted after wheat in Nebraska could produce more than 9.47 Mg ha^{-1} of biomass. Legume CC residue is well-known to decompose quickly, and some studies suggest that due to their rapid decomposition, they may not contribute significantly to the stable organic matter fraction (Silva et al., 2021; Blesh, 2019). The residue decomposition process can be slowed down by using no-till practices (Fig. 4; Balkcom et al., 2007; Thapa et al., 2022; Kichler et al., 2023). Long-term legume cover cropping has been shown to increase soil organic matter content (Blanco et al., 2012, 2019; Wander et al., 1994; Clark,

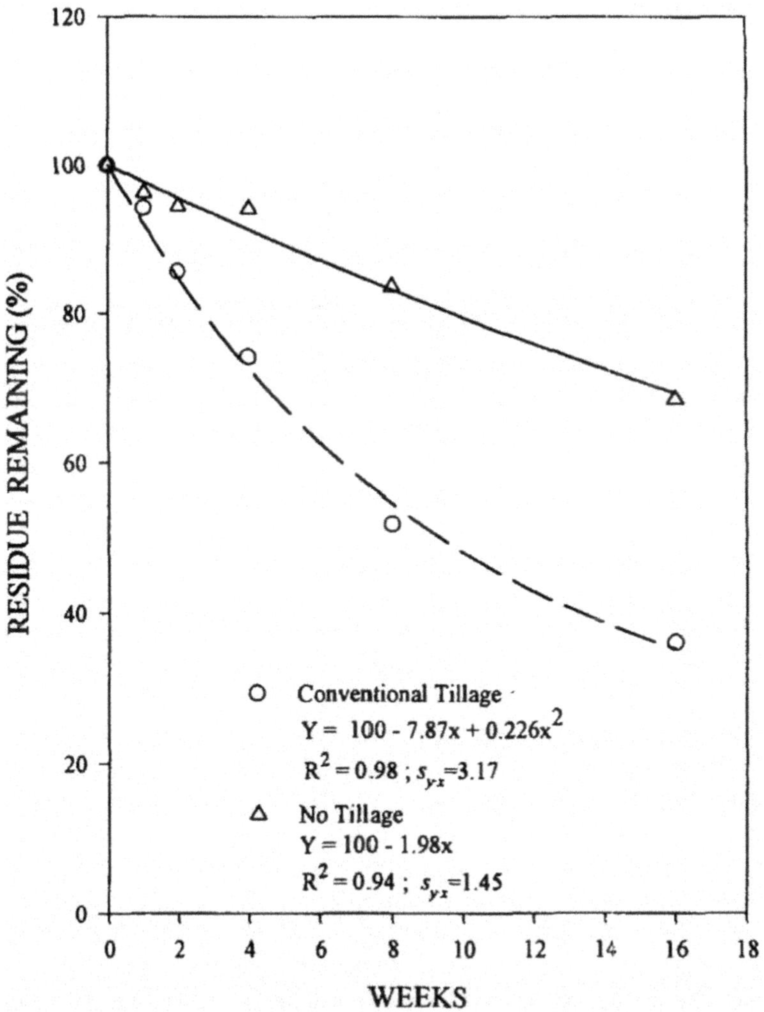

Figure 4 Decomposition of overwintered sunn hemp residue from 17 April to 7 August 1992 (normal corn growing season) at E.V. Smith Center Shorer, AL. (Mansoer et al., 1997).

2007; Blanco and Jasa, 2019). After 10 years of cover cropping SH and LMS after wheat in a wheat-sorghum rotation in Nebraska, the average soil organic carbon concentration was 1.25 times higher to a depth of 0 to 7.5 cm in CC plot compared to control (Blanco-Canqui et al., 2011). Attia et al. (2024) reported from a four-year modeling study in southwest Germany that legume CCs significantly increased soil organic carbon while non-legumes reduced N leaching.

5.3 Microbial activity

The large quantities of easily decomposable residue of warm season legume CCs serve as a food source for microbes, enhancing their activity in soil. Both the above- and below-ground CC biomass contribute to improved soil health. However, the relationship between annual legumes and soil microbial communities is complex and multifaceted (Bowles et al., 2016). Incorporating legumes into a cropping system can enhance microbial diversity and activity, thereby improving nutrient cycling and soil structure. Warm season legume CCs release exudates from their roots while growing, which act as catalysts that promote decomposition (Eisenhauer et al., 2017), enhance aggregation and soil structure (Ma et al., 2023), elevate organic carbon level in deeper layers (Veloso et al., 2018), and improve nutrient cycling (Fortier et al., 2013), ultimately leading to better overall soil health (Lin et al., 2011). However, root exudates from legume CCs and competition for water can also reduce main crop yield, as noted by Rose et al. (2024b). Legumes can enhance the abundance of plant-growth-promoting rhizobacteria (PGPR), which improve plant health by enhancing nutrient uptake and providing protection against pathogens (Bellabarba et al., 2019).

Adding legume residues can provide substrates for microbial mineralization, which may lead to spikes in emissions of greenhouse gases such as CO_2 and N_2O due to increased microbial activity (Bosco et al., 2019). Legume CCs contribute to both the diversity and abundance of mycorrhizal fungi. Their roots tend to support large populations of these beneficial organisms (Kim et al., 2020), and they have also been shown to increase the abundance of mycorrhizal spores in soil (Njeru et al., 2014; Moura et al., 2022). Somenahally et al. (2018) demonstrated that winter wheat cash crops had 30–70% higher levels of arbuscular mycorrhizal fungi (AMF) colonization when planted following CP, compared to summer fallow on the southern Great Plains of the United States. Mycorrhizae produce the glycoprotein glomalin that promotes soil aggregation, another way that warm season legume CCs can enhance water infiltration and improve the storage of water and soil nutrients, even when there is no detectable increase in soil organic matter, although this was questioned by Irving et al. (2021). Runion et al. (2004) reported that under no-till conditions, the presence of legumes during the summer led to a younger and more viable microbial population compared to conventional tillage. Dehydrogenase data suggested that the low and stable microbial activity under conventional tillage might reflect a mature microbial population in a stationary phase of growth, while the increase under no-till would reflect the younger, more viable growing population. They also reported an increase in gram-negative bacteria, which can enhance macronutrient cycling, particularly the solubilization of phosphorus (Gannett et al., 2024). However, the rapid nutrient

Published by Burleigh Dodds Science Publishing Limited, 2026.

release associated with legume decomposition can alter the composition of microbial communities, favoring fast-growing species that may not contribute to the long-term health of the soil (Singh et al., 2023). Kokalis-Burelle et al. (2017) found that SH CC reduced gram-positive spore-forming bacteria and fluorescent pseudomonads compared to check treatment. Kim et al. (2020) and Rose et al. (2022a) observed that legume CCs increased microbial biomass C and N by 30% and 28% compared to fallow control plots and also increased respiration and enzyme activity, specifically β-glucosidase and phosphatase, by up to 35%. Understanding the potential for both positive and negative impacts of warm season legumes in agricultural systems is crucial for developing management practices that maximize the benefits while minimizing potential drawbacks.

5.4 Improvement of soil physical properties

Warm season legume CCs can improve soil physical properties. Long-term use of these CCs increased soil organic matter (Blanco et al., 2012, 2019). Organic matter can enhance soil structure, reduce compaction, and improve water retention (Lal, 2015). By promoting the formation of well-aggregated soil, SH increased soil macroporosity to 0.20 m depth, improving soil aeration compared to fallow (Calonego et al., 2017). Well-aggregated soils are less prone to compaction (Kumar et al., 2018). In Australia's northern cropping zone, warm season legume CCs have resulted in improved water infiltration following intense rainfall events, suggesting that they can facilitate water entry into soils, particularly in non-wetting sands (Whish et al., 2009a). However, while the introduction of CCs can enhance water retention, their water use must also be considered, as that can reduce yields of subsequent cash crops (Hunt et al., 2013; Nielsen et al., 2016b). In summary, warm season legume CCs have the potential to enhance soil health and optimize water management, but further research is needed to fully understand the balance between these factors and their implications for crop yield, especially in water-limited environments (Rose et al., 2022b).

6 Weed control

Warm season legume CCs (e.g. CP, SH, VB, LMS) and mixtures of summer grasses and legumes have shown potential to reduce weed growth as compared to fallow (Creamer and Baldwin, 2000; Price et al., 2012a; Zannopoulos et al., 2024). Cool season legumes typically grow slowly and don't compete well with weeds (Gamble and Price, 2021). Warm season legumes like SH, however, are aggressive growers and can produce biomass of up to 7.6 Mg ha^{-1} in only two months, and therefore, compete vigorously with weeds (Balkcom and Reeves, 2005a; Osipitan et al., 2018) while also possessing allelopathic potential (Adler and Chase, 2007). Morris et al. (2015) and Zannopoulos et al. (2024) therefore propose this warm season legume CC as a viable choice for integrated weed management (IWM), potentially reducing the need for herbicides or tillage for weed control (Price et al., 2012a; Heap, 2014; Blanco-Canqui et al., 2020).

6.1 Warm season legumes can be used for weed control in several ways within cropping systems

6.1.2 Off-season weed suppression

When grown after the main crop or during fallow periods, warm season legumes effectively compete with and smother summer weeds, reducing their seed bank by competing with them while growing and acting as a physical barrier that blocks light. This is particularly relevant in climates where sufficient Growing Degree Days (GDDs) remain after the timely harvest of crops like corn or soybeans (Balkcom and Reeves, 2005b). Leaving the field fallow, in contrast, increases the risk of weed growth (Abouziena et al., 2014), contributing to a larger weed seed bank and negative impacts on farm economics (Cechin et al., 2020). Late-season weed seed production, from weeds that germinated after the critical period for weed control, might not affect crop yield in that season, but can still contribute to the seed bank and affect future crops (Price et al., 2018; Kumari et al., 2023).

6.1.3 Intercropping and overseeding for in-season weed control

Alternatively, aerial seeding via drones allows for the broadcasting of warm season legume CCs prior to corn harvest (Blanco et al., 2017). This has the potential to enhance weed control by enabling cover crops to compete with weeds that evade herbicide applications (Scursoni et al., 2007) and late-season flushes (Recker et al., 2015). If weeds are controlled solely by herbicides, the likelihood of herbicide-resistant weeds increases, thereby complicating future weed management strategies (Bagavathiannan and Norsworthy, 2012).

6.1.4 Shading and allelopathy

Brainard et al. (2005) found that in shady environments or low Red to Far-Red light (R:FR) ratios, weeds like Powell amaranth (*Amaranthus powellii*) produce seeds with a 40–50% reduction in seed germinability compared to normal light conditions. Comparable findings were reported by Jha et al. (2010) for Palmer amaranth (*Amaranthus palmeri*), a small-seeded broadleaf weed resistant to multiple herbicides, which can cause up to 80% yield loss if left uncontrolled (Norsworthy et al., 2016; Basinger et al., 2019; Moore et al., 2021). However,

Residue source	Germination (%)^z Week 1	Week 2	Week 3	Plant ht (%)^z	Dry wt (%)^z
			Goosegrass		
Cowpea	75 a^y	65 a	68 a	106 a	123 a
Sunn hemp	62 a	62 a	65 a	94 a	81 b
Velvetbean	71 a	78 a	80 a	92 a	85 b
			Smooth amaranth		
Cowpea	54 a	71 a	78 ab	86 a	114 a
Sunn hemp	15 b	39 b	61 b	52 b	20 b
Velvetbean	71 a	95 a	87 a	75 a	79 a

[z]Percentage of control.
[y]Means in columns followed by the same lowercase letter are not significantly different as determined by Fisher's protected least significant difference test ($P \leq 0.05$).

Figure 5 Effect of cover crop residue source on germination, plant height, and dry weight of goosegrass and smooth amaranth (adapted from Rugare et al. 2021).

weeds that germinate even in dark conditions, such as ivyleaf morning glory (*Ipomoea hederacea*), may not be controlled by this strategy (Singh, 2012). Apart from providing competition to weeds, warm season legumes can also control weeds through non-competitive measures such as phytotoxicity. Adler and Chase (2007) evaluated the phytotoxicity of aqueous foliar extracts and ground dried residues from SH, CP, and VB on the germination and growth of various crops and weeds (Fig. 5). CP extracts at 10% concentration resulted in the lowest germination rates for goosegrass, dropping to 75%, while SH showed the highest germination rates. Germination of livid amaranth (*Amaranthus blitum*) decreased to 50% with CP and SH extracts, and further to 22% with VB. Ground-dried residues significantly inhibited growth; for smooth amaranth (*Amaranthus hybridus*), SH reduced germination and biomass, while CP decreased tomato dry weight to 69% of control.

Overall, warm season legumes exhibited greater phytotoxic effects on weeds than on crops, suggesting their potential in integrated weed management strategies while necessitating careful consideration of their impacts on the cash crop. With the increasing prevalence of herbicide-resistant weeds, more research is needed on the potential for CCs to impact weed composition and growth dynamics compared to the use of herbicides or other weed management practices (Norsworthy et al., 2012). Finally, when water is limiting, warm season CCs can compete with weeds but also reduce main crop yield if they reduce its water availability (Rose et al., 2022a, 2024a).

7 Forage use

Beyond their role as soil improvers, warm season legumes can also be utilized for forage, either as monocultures or in mixtures with warm season grasses. This dual-purpose use provides an additional economic return, particularly valuable in regions like the Southern Great Plains, where forage demand often increases during the late grazing season (Horn et al., 2021; Baath et al., 2018).

Intercropping annual grasses with legumes has shown LERs greater than one, indicating more efficient use of land and resources for forage production (Bybee-Finley et al., 2023). Warm season legumes such as CP, SH, and lablab have demonstrated strong forage yields and nutritive value under both dry and normal growing conditions, making them suitable for drought-prone grazing systems due to their nitrogen-fixing ability and drought tolerance (Butler et al., 2013). For example, Nave and Corbin (2018) reported that SH produced the highest biomass under drought conditions, while cowpea had the highest crude protein concentration due to robust nitrogen fixation. Because legumes are typically rich in protein, they can also enhance the nutritional quality of grasses in legume–grass forage mixtures. Specifically, CP, SH, and lablab have been shown to increase crude protein levels, improve protein digestibility, and enhance overall forage quality when combined with grass silage (Oskey et al., 2023; Salama and Zied, 2016; La Guardia Nave and Corbin, 2018).

In the southern Great Plains of the United States, where forage supply is often constrained by summer drought, legumes such as CP, tepary bean (*Phaseolus acutifolius* [A.] Gray), and moth bean (*Vigna aconitifolia* [Jacq.] Marechal) show promise due to their superior drought tolerance and forage quality compared to soybean, the more commonly used legume forage crop (Horn et al., 2021; Baath et al., 2018). La Guardia Nave and Corbin (2018) compared the intercropping of crabgrass, CP, and SH with corn to evaluate impacts on silage production. Their findings showed that while intercropping with SH resulted in greater forage biomass, it negatively affected corn growth, whereas CP maintained higher forage quality with less impact on the corn component.

8 Conclusion

The integration of warm season leguminous cover crops into agricultural systems offers a suite of potential ecological and agronomic benefits. As delineated in this chapter, their capacity for biological nitrogen fixation presents a significant opportunity to reduce reliance on synthetic nitrogen fertilizers, thereby mitigating associated environmental impacts such as greenhouse gas (GHG) emissions. Beyond nitrogen contribution, warm season legumes can enhance soil health through multiple pathways: reducing erosion, improving soil organic matter over the long term, stimulating beneficial microbial communities, including arbuscular mycorrhizal fungi, and improving soil structure and water infiltration. Furthermore, certain species exhibit considerable potential for weed suppression via competition and allelopathic mechanisms, contributing valuable options for integrated weed management strategies, particularly in the face of increasing herbicide resistance. Moreover, they can provide high-quality forage, offering a crucial resource, particularly during periods of traditional forage scarcity in regions like the Southern Great Plains.

9 Challenges

Despite their many benefits, integrating warm season leguminous cover crops presents several notable challenges. One major hurdle is competition with cash crops for vital resources like water, nutrients, and light. The vigorous growth of warm season legumes, while good for biomass, can reduce main crop yields, especially in dry conditions (Bedoussac et al., 2015; Hunt et al., 2013; Nielsen et al., 2016a; Rose et al., 2022a, 2024a). Also, the nitrogen release from decomposing legume residues often doesn't perfectly match the nutrient needs of fast-growing cash crops, potentially requiring extra fertilization (Campiglia et al., 2011; Fink and Feller, 1998; Duarte et al., 2019; O'Connell et al., 2015). Rapid decomposition, particularly with tillage, can even lead to nitrogen surpluses and losses (Mansoer et al., 1997; Evans et al., 2001b; Liu et al., 2019b). While no-till can slow this down and improve residue retention (Balkcom et al., 2007; Kichler et al., 2023), more research is needed on how these legumes affect soil microbes and nitrogen dynamics.

Equipment limitations also hinder widespread adoption of intercropping warm season legumes in highly mechanized systems, like those in the United States (Bourke et al., 2021; Eure et al., 2015). Machinery designed for monocultures makes diverse planting and harvesting costly and inefficient, and specialized tools remain limited and expensive (Bybee-Finley and Ryan, 2018). Technical training gaps and the complexities of harvesting crops with different maturity stages also restrict broader implementation (Jiang et al., 2023; Pierre et al., 2022).

Finally, their performance is heavily influenced by environmental conditions, including soil pH, nutrient availability (e.g. Mo, Fe, K), and moisture levels (Schrer et al., 1996; Islam et al., 2018b; Bursakov, 2023; Khalili, 2024b; Linn et al., 1984). Though they improve soil health and water infiltration (Whish et al., 2009b), understanding their water use and potential yield trade-offs in dry environments remains a key challenge (Rose et al., 2022b).

10 Future research

Continued research is vital to refine management practices and fully realize the benefits of warm season leguminous cover crops for sustainable agriculture. Key areas include:

- **Understanding Drawbacks:** Further investigating competition with main crops for moisture and the varied effects on soil microbial activity and nitrogen release dynamics.
- **Optimizing Management:** Developing precise guidelines for species selection, planting dates, and termination methods tailored to specific crops and environments.

- **Exploring Innovation:** Researching innovative approaches and technologies, such as drone seeding (*Blanco et al., 2017*), to overcome logistical challenges in complex cropping systems.
- **Assessing Long-Term Impact:** Evaluating the long-term economic viability and societal benefits, including carbon sequestration and overall farm sustainability.

By addressing these research areas, warm season legumes can significantly contribute to reducing reliance on synthetic fertilizers and mitigating agriculture's environmental impact.

11 Where to look for further information

- Blanco-Canqui, H., et al. (2012), 'Summer cover crops fix nitrogen, increase crop yield, and improve soil–crop relationships', *Agronomy Journal*, 104(1), 137–147.
- Zannopoulos, S., et al. (2024), 'The potential of three summer legume cover crops to suppress weeds and provide ecosystem services–A review', *Agronomy*, 14(6), 1192.
- Blanco-Canqui, H. and Jasa, P.J. (2019), 'Do grass and legume cover crops improve soil properties in the long term?', *Soil Science Society of America Journal*, 83(4), 1181–1187.
- Clark, A. (Ed.) (2007), *Managing Cover Crops Profitably*, 3rd ed., Sustainable Agriculture Research and Education (SARE), Washington, DC, USA.
- Saikia, P., et al. (2020), 'Tropical legumes: Status, distribution, biology and importance', In: M. Hasanuzzaman (Ed.), *The plant family fabaceae: biology and physiological responses to environmental stresses*, Springer, Singapore, pp. 27–41.

12 References

Abouziena, H. F., et al. (2014), 'Water loss by weeds: a review', *International Journal of ChemTech Research*, 7(01), 323–336.

Adler, M. J. and Chase, C. A. (2007), 'Comparison of the allelopathic potential of leguminous summer cover crops: Cowpea, sunn hemp, and velvetbean', *HortScience*, 42(2), 289–293.

Agegnehu, G., et al. (2008), 'Yield potential and land-use efficiency of wheat and faba bean mixed intercropping', *Agronomy for Sustainable Development*, 28, 257–263.

Aguiar, J. L., et al. (2001), 'Factor for estimating nitrogen contribution of cowpea as a cover crop', *Journal of Agronomy and Crop Science*, 186, 145–149.

Attia, A., et al. (2024), 'Do rotations with cover crops increase yield and soil organic carbon? A modeling study in southwest Germany', *Agriculture, Ecosystems & Environment*, 375, 109–167.

Ayerdi Gotor, A. and Marraccini, E. (2022), 'Innovative pulses for western european temperate regions: a review', *Agronomy*, 12(1), 170.

Azim, A., et al. (2024), 'Evaluation of growth and biomass production of three Sesbania species for their potential as green manure in Bangladesh', *Journal of Agricultural Sciences*, 69(1), 45–56.

Baath, G. S., et al. (2018), 'Forage potential of summer annual grain legumes in the southern great plains', *Agronomy Journal*, 110(6), 2198–2210.

Bagavathiannan, M. V. and Norsworthy, J. K. (2012), 'Late-season seed production in arable weed communities: management implications', *Weed Science*, 60(3), 325–334.

Baker, J. M. and Griffis, T. J. (2009), 'Evaluating the potential of winter cover crops in corn-soybean systems for sustainable co-production of food and fuel', *Agric. and Forest Meterorol,* 149(12): 2120–3132.

Balkcom, K. S. and Reeves, D. W. (2005a), 'Sunn hemp as a nitrogen source for corn production in the southeastern United States', *Agronomy Journal*, 97(6), 1600–1606.

Balkcom, K. S. and Reeves, D. W. (2005b), 'Sunn-hemp utilized as a legume cover crop for corn production', *Agronomy Journal*, 97(1), 26–31.

Balkcom, K. et al. (2007), 'Managing Cover Crops in Conservation Tillage Systems'. In Clark, A. (Ed.), Managing Cover Crops Profitably Third Edition. Sustainable Agriculture Research and Education (SARE) Program, University of Maryland (Maryland), USA.

Basinger, N. T., et al. (2019), 'Large crabgrass (*Digitaria sanguinalis*) and Palmer amaranth (*Amaranthus palmeri*) intraspecific and interspecific interference in soybean', *Weed Science*, 67(6), 649–656.

Bedoussac, L., et al. (2015), 'Ecological principles underlying the increase of productivity achieved by cereal-grain legume intercrops in organic farming. A review', *Agronomy for Sustainable Development*, 35(3), 911–935.

Bellabarba, A., et al. (2019), 'Deciphering the symbiotic plant microbiome: translating the most recent discoveries on rhizobia for the improvement of agricultural practices in metal-contaminated and high saline lands', *Agronomy*, 9(9), 529.

Bilenky, M. T., et al. (2022), 'Effect of summer cover crops on cabbage yield, weed suppression, and N mineralization in a low input cropping system', *Frontiers in Sustainable Food Systems,* 6, 1021639.

Blanco, H. and Lal, R. (2008), *Principles of soil conservation and management*, Springer, New York, NY.

Blanco-Canqui, H. and Jasa, P. J. (2019), 'Do grass and legume cover crops improve soil properties in the long term?', *Soil Science Society of America Journal*, 83(4), 1181–1187.

Blanco-Canqui, H. and Wortmann, C.S. (2020), 'Does occasional tillage undo the ecosystem services gained with no-till? A review', *Soil and Tillage Research*, 198, 104534.

Blanco-Canqui, H., et al. (2012), 'Summer cover crops fix nitrogen, increase crop yield, and improve soil–crop relationships', *Agronomy Journal*, 104(1), 137–147.

Blanco-Canqui, H., et al. (2013), 'Replacing fallow with cover crops in a semiarid soil: effects on soil properties', *Soil Science Society of America Journal*, 77(3), 1026–1034.

Blanco-Canqui, H., et al. (2011), 'Addition of cover crops enhances no-till potential for improving soil physical properties', *Soil Science Society of America Journal*, 75(4), 1471–1482.

Blanco-Canqui, H., et al. (2017), 'Aerial seeding of cover crops: efficacy and corn yield response', *Agronomy Journal,* 109(6), 2634–2642.

Blanco-Canqui, H., et al. (2015), 'Cover crops and ecosystem services: insights from studies in temperate soils', *Agronomy Journal*, 107(6), 2449–2474.

Blesh, J. (2019), 'Feedbacks between nitrogen fixation and soil organic matter increase ecosystem functions in diversified agroecosystems'. *Ecol. Appl.* 29(8): e01986.

Bosco, S., Volpi, I.,et al. (2019), 'Greenhouse gas emissions from soil cultivated with vegetables in crop rotation under integrated, organic, and organic conservation management in a Mediterranean environment', *Agronomy*, 9(8), 446.

Bourke, P. M., et al. (2021), 'Breeding beyond monoculture: putting the "intercrop" into crops', *Frontiers in Plant Science*, 12.

Bowles, T. M., et al. (2016), 'Ecological intensification and arbuscular mycorrhizas: a meta-analysis of tillage and cover crop effects', *Journal of Applied Ecology*, 54(6), 1785–1793.

Boyhan, G. E., et al (2016), 'Evaluation of cool-season vegetable rotations in organic production', *HortTechnology*, 26, 637–646.

Brainard, D. C., et al. (2005), 'Effects of canopy shade on the morphology, phenology, and seed characteristics of Powell amaranth (*Amaranthus powellii*)', *Weed Science*, 53(2), 175–186.

Bursakov, S. A. (2023), 'Molybdenum in nitrogenase and beyond', *Coordination Chemistry Reviews*, 481, 214953.

Bursakov, S. A., et al. (2023), 'Tracing the element: the molecular bases of molybdenum homeostasis in legumes', *Agronomy*, 13(9), 2300.

Butler, T. W. et al. (2013), 'Switchgrass biomass and nitrogen yield with over-seeded cool-season forages in the Southern Great Plains'. *BioEnergy Res.* 6(1): 44–52.

Bybee-Finley, K. A. and Ryan, M. R. (2018), 'Advancing intercropping research and practices in industrialized agricultural landscapes', *Agriculture*, 8(6), 80.

Bybee-Finley, K. A., et al. (2016), 'Functional diversity in summer annual grass and legume intercrops in the northeastern United States', *Crop Science*, 56(5), 2775–2790.

Calonego, J. C., et al. (2017), 'Soil compaction management and soybean yields with cover crops under no-till and occasional chiseling', *European Journal of Agronomy*, 85, 31–37.

Campiglia, E., et al. (2011), 'Nitrogen mineralization from hairy vetch (Vicia villosa Roth) in Mediterranean organic cropping systems', *European Journal of Agronomy*, 35(5–6), 333–342.

Carberry, P. S., et al. (2001), 'Response of sorghum and pigeonpea to environmental and management factors in Queensland and New South Wales, Australia', *Australian Journal of Agricultural Research*, 52(6), 661–670.

Cárceles Rodríguez, B. et al. (2021), 'Soil Erosion and the Efficiency of the Conservation Measures in Mediterranean Hillslope Farming (SE Spain)'. Eurasian Soil Science 54(5): 792–806.

Cechin, J., et al. (2020), 'Weed seed viability affected by pre-harvest herbicide', *Communications*, 10, 46–51.

Chanda, S. C., et al. (2019), 'Phenology and reproductive biology of three Sesbania species', *International Journal of Minor Fruits, Medicinal and Aromatic Plants,* 5(1), 29–37.

Chauhan, Y. S. and Williams, R. (2018), 'Physiological and agronomic strategies to increase mungbean yield in climatically variable environments of Northern Australia', *Agronomy*, 8(6), 83.

Cherr, C. M., et al. (2006), 'Green manure as nitrogen source for sweet corn in a warm–temperate environment', *Agronomy Journal*, 98(5), 1173–1180.

Clark, A. (Ed.) (2007), *Managing cover crops profitably*, 3rd ed., Sustainable Agriculture Research and Education (SARE), Washington, DC.

Creamer, N. G. and Baldwin, K. R. (2000), 'An evaluation of summer cover crops for use in vegetable production systems in North Carolina', *HortScience*, 35, 600–603.

Dabney, S. M., et al. (2001), 'Using winter cover crops to improve soil and water quality', *Communications in Soil Science and Plant Analysis*, 32(7–8), 1221–1250.

Dalal, R. C. (1974), 'Effects of intercropping maize with pigeon peas on grain yield and nutrient uptake', *Experimental Agriculture*, 10, 219–224.

Duarte, J., et al. (2019), 'Nitrogen fertilization in cabbage crops: effects on yield, quality, and nitrogen use efficiency', *Horticulturae*, 5(2), 33.

Eisenhauer, N., et al. (2017), 'Root biomass and exudates link plant diversity with soil bacterial and fungal biomass', *Scientific Reports*, 7(1).

Eure, P. M., et al. (2015), 'Weed control, crop response, and profitability when intercropping cantaloupe and cotton', *Weed Technology*, 29(2), 217–225.

Evans, J., et al. (2001a), 'Net nitrogen balances for cool-season grain legume crops and contributions to wheat nitrogen uptake: a review', *Australian Journal of Experimental Agriculture*, 41(3), 347–359.

Evans, S. D., et al. (2001b), 'Nitrogen release from hairy vetch and rye cover crop residues', *Agronomy Journal*, 93(2), 404–411.

Fink, M. and Feller, C. (1998), 'Nitrogen uptake and accumulation in cabbage (Brassica oleracea L. convar. capitata L.)', *Journal of Plant Nutrition and Soil Science*, 161(2), 177–182.

Fortier, J., et al (2013), 'Root biomass and soil carbon distribution in hybrid poplar riparian buffers, herbaceous riparian buffers and natural riparian woodlots on farmland', *Springer Plus*, 2, 1–19.

Gałązka, A., et al. (2020), 'Evaluation of changes in glomalin-related soil proteins (GRSP) content, microbial diversity and physical properties depending on the type of soil as the important biotic determinants of soil quality', *Agronomy*, 10(9), 1279.

Gamble, A. and Price, A. (2021), 'The intersection of integrated pest management and soil quality in the resistant weed era', *Italian Journal of Agronomy*, 16(4), 1875.

Gannett, M. et al. (2024), 'Microbial nitrogen immobilization as a tool to manage weeds in agroecosystems'. *Agric., Ecosyst. and Environ.* 366: article 108904.

Garnett, T., et al. (2013), 'Sustainable intensification in agriculture: premises and policies', *Science*, 341(6141), 33–34.

Garten Jr, C. T., et al. (2008), 'Role of N2-fixation in constructed old-field communities under different regimes of [CO2], temperature, and water availability', *Ecosystems*, 11(1), 125–137.

Gaudio, N., et al. (2015), 'Combined effect of atmospheric nitrogen deposition and climate change on temperate forest soil biogeochemistry: a modeling approach', *Ecological Modelling*, 306, 24–34.

Gautrat, P., et al. (2020), 'Compact root architecture 2 promotes root competence for nodulation through the mir2111 systemic effector', *Current Biology*, 30(7), 1339–1345.e3.

Gerland, P., et al. (2014), 'World population stabilization unlikely this century', *Science*, 346(6206), 234–237.

Godfray, H. C. J., et al. (2010), 'Food security: the challenge of feeding 9 billion people', *Science*, 327(5967), 812–818.

Gomes, J., et al. (2009), 'Soil nitrous oxide emissions in long-term cover crops-based rotations under subtropical climate', *Soil and Tillage Research*, 106(1), 36–44.

Goolsby, D. A., et al. (2001), 'Nitrogen input to the Gulf of Mexico', *Journal of Environmental Quality*, 30(2), 329–336.

Gresta, F., et al. (2018), 'Germination of guar (Cyamopsis tetragonoloba (L.) Taub.) genotypes with reduced temperature requirements', *Australian Journal of Crop Science*, 12(06), 954–960.

Hariniharisma, V.P., et al. (2025), 'Sunn hemp – an under-exploited and versatile crop for soil nitrogen transformation and fibre production'. *Plant Science Today* 12(3), 1–8.

Harrison, H. F., et al. (2004), 'Broccoli production in cowpea, soybean, and velvetbean cover crop mulches', *HortTechnology*, 14(4), 484.

Hatfield, J. L., et al. (2011), 'Climate impacts on agriculture: implications for crop production', *Agronomy Journal*, 103(2), 351–370.

Hauggaard-Nielsen, H., et al. (2009), 'Pea-barley intercropping for efficient symbiotic N2-fixation, soil N acquisition and use of other nutrients in European organic cropping systems', *Field Crops Research*, 113, 64–71.

Heap, I. (2014), 'Global perspective of herbicide-resistant weeds', *Pest Management Science*, 70(9), 1306–1315.

Holman, J., et al. (2018), 'Can cover or forage crops replace fallow in the semiarid central Great Plains?', *Crop Science*, 58(2), 932–944.

Horn, K. M., et al. (2021), 'Evaluating cover crops forage nutritive value in Oklahoma winter wheat systems', *Agronomy Journal*, 113(4), 3361–3371.

Hunt, L. J., et al. (2013), 'Impact of winter cover crops on soil water content in semi-arid and sub-humid environments: a review', *Agricultural Water Management*, 116, 21–30.

Irving, T. B., et al. (2021), 'A critical review of 25 years of glomalin research: a better mechanical understanding and robust quantification techniques are required', *New Phytologist*, 232(4), 1572–1581.

Islam, M. M., et al. (2018a), 'Effect of molybdenum and boron on growth, yield and nutrient uptake by mungbean (*Vigna radiata* L.)', *Journal of Applied Biology & Biotechnology*, 6(1), 22–29.

Islam, M., et al. (2018b), 'Impacts of trace element addition on lentil (*Lens culinaris* L.) agronomy', *Agronomy*, 8(7), 100.

Jha, P., et al. (2010), 'Annual changes in temperature and light requirements for germination of Palmer amaranth (*Amaranthus palmeri*) seeds retrieved from soil', *Weed Science*, 58(4), 426–432.

Jiang, M., et al. (2023), 'Material and carbon footprints of machinery capital', *Environmental Science & Technology*, 57(50), 21124–21135.

Kakraliya, S. K., et al. (2018), 'Nitrogen and legumes: a meta-analysis', In: R. W. Meena, et al. (Eds.), *Legumes for soil health and sustainable management*, Springer, Singapore, pp. 277–314.

Kalburtji, K., et al. (2007), 'Effects of day-night temperature combinations under constant day length on emergence and early growth of sericea lespedeza genotypes', *Canadian Journal of Plant Science*, 87(1), 77–81.

Kaspar, T. C. and Singer, J. W. (2011), 'The Use of Cover Crops to Manage Soil', In Hatfield, J. L. and Sauer, T. J. (eds), Soil Management: A Stable Base for Agriculture. American Society of Agronomy and Soil Science Society of America, Madison (Wisconsin), USA.

Kaye, J. et al. (2019). 'Managing nitrogen through cover crop species selection in the U.S. mid-Atlantic'. *PLOS ONE,* 14(4): e0215448.

Kennedy, A. C. and Papendick, R. I. (1995), 'Microbial characteristics of soil quality', *Journal of Soil and Water Conservation*, 50(4), 243–248.

Khalili, A. (2024a), 'Temporal synchronization of nitrogen and sulfur fertilization: impacts on nutrient uptake, use efficiency, productivity, and relationships with other micronutrients in soybean', *Agronomy*, 14(3), 570.

Khalili, M. (2024b), 'Effect of sulfur on nitrogenase activity and nitrogen fixation in legumes: a review', *Plant and Soil*, 494(1), 1–15.

Kichler, J. M., et al. (2023), 'Cover crop termination timing effects on cotton growth and yield under conservation tillage', *Agronomy Journal*, 115(2), 850–862.

Kim, N., et al. (2020), 'Do cover crops benefit soil microbiome? A meta-analysis of current research', *Soil Biology and Biochemistry*, 142, 107701.

Kocira, A., et al. (2020), 'Legume cover crops as one of the elements of strategic weed management and soil quality improvement: a review', *Agriculture*, 10(9), 394.

Koehler-Cole, K., et al. (2023), 'Cover crop planting practices determine their performance in the US Corn Belt', *Agronomy Journal*, 115(2), 526–543.

Kokalis-Burelle, N. et al. (2017), 'Rhizosphere microorganisms affected by soil solarization and cover cropping in *Capsicum annuum* and *Phaseolus lunatus* agroecosystems'. *Applied Soil Ecology,* 119: 64-71.

Kornecki, T. S. and Price, A. (2019), 'Management of high-residue cover crops in a conservation tillage organic vegetable on-farm setting', *Agronomy,* 9(10), 640.

Kruse, R. and Nair, A. (2016), 'Summer cover crops and lettuce planting time influence weed population, soil nitrogen concentration, and lettuce yields', *HortTechnology*, 26(4), 409–416.

Kumari, A., et al. (2023), 'Effect of crimson clover on the critical period of weed control in conservation tillage corn', *Frontiers in Agronomy*, 4, 1068365.

La Guardia Nave, R. and Corbin, M. D. (2018), 'Forage warm-season legumes and grasses intercropped with corn as an alternative for corn silage production', *Agronomy*, 8(10), 199.

Lal, R. (2015), 'Restoring soil quality to mitigate soil degradation', *Sustainability*, 7(5), 5875–5895.

Lal, R. (2020), 'Soil organic matter and water retention', *Agronomy Journal*, 112(5), 3265–3277.

Layek, J., et al. (2018), 'Cereal+ legume intercropping: an option for improving productivity and sustaining soil health', In R. S. Meena, et al. (Eds.), *Legumes for soil health and sustainable management*, Springer, Cham, Switzerland, pp. 347–386.

Lin, Y., et al. (2011), 'Combining legume groundcovers and low nitrogen fertilization in a short-rotation fraser fir (*Abies fraseri*) cropping system: effect on productivity, soil fertility, and nutrient leaching', *HortScience*, 46(3), 481–486.

Linn, D. M. and Doran, J. W. (1984), 'Effect of water-filled pore space on carbon dioxide and nitrous oxide production in tilled and nontilled soil', *Soil Science Society of America Journal*, 48, 1267–1272.

Linn, D. M., et al. (1984), 'Effect of soil moisture tension and pH on nitrogenase activity in clover-ryegrass associations', *Soil Science Society of America Journal*, 48(4), 819–823.

Liu, L., et al. (2019a), 'A side-by-side comparison of biological nitrogen fixation and yield of four legume crops', *Plant and Soil*, 442, 169–182.

Liu, X., et al. (2019b), 'Effects of tillage on carbon and nitrogen mineralization in agricultural soils: a meta-analysis', *Soil and Tillage Research*, 186, 175–184.

Lu, Y.-C., et al. (2000), 'Cover crops in sustainable food production', *Food Reviews International*, 16, 121–157.

Ma, Q., et al. (2023), 'Agricultural soil aggregation is affected by the crop root biomass rather than morphological characteristics', *Journal of Plant Nutrition and Soil Science*, 186(3), 339–350.

Mahama, G. Y., et al. (2020), 'Reduction of nitrogen fertilizer requirements and nitrous oxide emissions using legume cover crops in a no-tillage sorghum production system', *Sustainability*, 12(11), 4403.

Manasa, P., et al. (2018), 'Effect of summer maize-legume intercropping system on growth, productivity and competitive ability of crops', *International Journal of Management, Technology and Engineering*, 8(12), 2871–2875.

Mansoer, Z., et al. (1997), 'Suitability of sunn hemp as an alternative late-summer legume cover crop', *Soil Science Society of America Journal*, 61(1), 246–253.

Maryland Department of Agriculture (MDA). (2014), *MACS 2013 annual report: powering the bay cleanup*, Report No: MDA 15.03.14, Annapolis, MD.

Maryland Department of Agriculture (MDA). (2010), *MACS 2009 annual report*, Report No: MDA 15.07.10, Annapolis, MD.

Meena, R., et al. (2020), 'Effect of rice residue retention and foliar application of K on water productivity and profitability of wheat in north west India', *Agronomy*, 10(3), 434.

Menegat, S., et al. (2022), 'Greenhouse gas emissions from global production and use of nitrogen synthetic fertilisers in agriculture', *Scientific Reports*, 12(1), 1–13.

Mohta, N. K. and De, R. (1980), 'Intercropping maize and sorghum with soya beans', *Journal of Agricultural Science*, 95, 117–122.

Moore, L., et al. (2021), 'Safety and efficacy of linuron with or without an adjuvant or s-metolachlor for post control of palmer amaranth (*Amaranthus palmeri*) in sweetpotato', *Weed Technology*, 35(3), 471–475.

Morris, J., et al. (2015), 'Effect of sunn hemp (*Crotalaria juncea*) cutting date and planting density on weed suppression in georgia, usa', *Journal of Environmental Science and Health Part B*, 50(8), 614–621.

Mosjidis, J. A. (1990), 'Daylength and temperature effects on emergence and early growth of sericea lespedeza', *Agronomy Journal*, 82(5), 923–926.

Moura, J. B., et al. (2022), 'Arbuscular mycorrhizal fungi in the soil using cover crops with and without nitrogen addition', *Frontiers in Plant Science*, 13.

Murphy, B. W. (2015), 'Impact of soil organic matter on soil properties–a review with emphasis on Australian soils', *Soil Research*, 53(6), 605–635.

Murungu, F. S., et al. (2010), 'Decomposition, nitrogen and phosphorus mineralization from winter-grown cover crop residues and suitability for a smallholder farming system in south africa', *Nutrient Cycling in Agroecosystems*, 89(1), 115–123.

Nelson, K. A., et al. (2022), 'Overseed timing of ryegrass and cereal rye in soybean affects rotational crops in upstate missouri', *Crop, Forage & Turfgrass Management*, 8(2).

Nevins, C. J., et al. (2020), 'The synchrony of cover crop decomposition, enzyme activity, and nitrogen availability in a corn agroecosystem in the Midwest United States', *Soil and Tillage Research*, 197, 104518.

Nielsen, D. C., et al. (2016a), 'Cover crop effect on subsequent wheat yield in the central Great Plains', *Agronomy Journal*, 108(1), 243–256.

Nielsen, K. L., et al. (2016b), 'Cover crop effects on soil water and subsequent crop yields in rainfed systems: a review', *Agronomy Journal*, 108(4), 1600–1610.

Niggli, U., et al. (2009), *Low greenhouse gas agriculture: mitigation and adaptation potential of sustainable farming systems*, Food and Agriculture Organization of the United Nations, Rome, Italy.

Njeru, E. M., et al. (2014), 'First evidence for a major cover crop effect on arbuscular mycorrhizal fungi and organic maize growth', *Agronomy for sustainable Development*, 34, 841–848.

Noland, R., et al. (2018), 'Establishment and function of cover crops interseeded into corn', *Crop Science*, 58(2), 863–873.

Norsworthy, J. K., et al. (2016), 'Emergence date influences growth and fecundity of Palmer amaranth in cotton', *Journal of Cotton Science*, 20, 263–270.

Norsworthy, J. K., et al. (2012), 'Reducing the risks of herbicide resistance: best management practices and recommendations', *Weed Science*, 60(SP1), 31–62.

O'Connell, S., et al. (2015), 'Short-term nitrogen mineralization from warm-season cover crops in organic farming systems', *Plant and Soil*, 396, 353–367.

O'Dea, J. K. and Miller, P. R. (2011), *Greening summer fallow: agronomic and edaphic implications of legumes in dryland wheat agroecosystems*, M.Sc. Thesis, Montana State University, Bozeman, MT.

Oelmann, Y., et al. (2005), 'Nitrogen-15 in no characterises differently reactive soil organic N pools', *Rapid Communications in Mass Spectrometry*, 19(22), 3177–3181.

Osipitan, O. A., et al. (2018), 'Cover crop for early season weed suppression in crops: systematic review and meta-analysis', *Agronomy Journal*, 110(6), 2211–2221.

Oskey, M. A., et al. (2023), 'Yield, nutritional composition, and digestibility of conventional and brown midrib (bmr) pearl millet as affected by planting and harvesting dates and interseeded cowpea', *Animals*, 13(2), 260.

Pierre, J. F., et al. (2022), 'Maize legume intercropping systems in southern mexico: a review of benefits and challenges', *Ciência Rural*, 52(11), e20210409.

Power, J. F. and Koerner, P. T. (1994), 'Cover crop production for several planting and harvest dates in eastern Nebraska', *Agronomy Journal*, 86(6), 1092–1097.

Price, A. J. and Norsworthy, J. K. (2013), 'Cover crops for weed management in southern reduced-tillage vegetable cropping systems', *Weed Technology*, 27(1), 212–217.

Price, A. J., et al. (2018), 'Influence of a cereal rye cover crop and conservation tillage on the critical period for weed control in cotton', *Weed Technology*, 32(6), 683–690.

Price, A. J., et al. (2012a), 'Herbicide and cover crop residue integration for *Amaranthus* control in conservation agriculture cotton and implications for resistance management', *Weed Technology*, 26(3), 490–498.

Price, A. J., et al. (2012b), 'Utilization of sunn hemp for cover crops and weed control in temperate climates'. https://www.intechopen.com/books/weed-control/utilization-of-sunn-hemp-for-cover-crops-and-weed-control-in-temperate-climates.

Raimbault, B. A. and Vyn, T. J. (1991), 'Crop rotation and tillage effects on corn growth and soil structural stability', *Agronomy Journal*, 83(6), 979–985.

Ran, L., et al. (2019), 'An integrated agriculture, atmosphere, and hydrology modeling system for ecosystem assessments', *Journal of Advances in Modeling Earth Systems*, 11(12), 4645–4668.

Rao, S. C. and Northup, B. K. (2009a), 'Capabilities of four novel warm-season legumes in the southern great plains: grain production and quality', *Crop Science*, 49(3), 1103–1108.

Rao, S. C. and Northup, B. K. (2009b), 'Water use by five warm-season legumes in the Southern Great Plains', *Crop Science*, 49(6), 2317–2324.
Rao, S. C. and Northup, B. K. (2013), 'Biomass production and quality of Indian-origin forage guar in Southern Great Plains', *Agronomy Journal*, 105(4), 945–950.
Rawls, W. J., et al. (2003), 'Effect of soil organic carbon on soil water retention', *Geoderma*, 116(1–2), 61–76.
Recker, R. A., et al. (2015), 'Late-season weed escape survey reveals discontinued atrazine use associated with greater abundance of broadleaf weeds', *Weed Technology*, 29(3), 451–463.
Rose, M. T., et al. (2022a), 'Do cover crops really improve soil water relations in semi-arid and Mediterranean climates? A critical review', *Agriculture, Ecosystems & Environment*, 337, 108044.
Rose, T. J., et al. (2022b), 'Prospects for summer cover crops in southern Australian semi-arid cropping systems', *Agricultural Systems*, 200, 103415.
Rose, M. T., et al. (2024a), 'A critical review of cover crop impacts on soil water in temperate dryland cropping systems', *Advances in Agronomy*, 184, 1–60.
Rose, T. J., et al. (2024b), 'Summer cover crop and temporary legume-cereal intercrop effects on soil microbial indicators, soil water and cash crop yields in a semi-arid environment', *Field Crops Research*, 312, 109384.
Rugare, J. T., et al. (2021), 'Allelopathic potential of green manure cover crops on germination and early seedling development of goose grass [*Eleusine indica* (L.) Gaertn] and blackjack (*Bidens pilosa* L.)', *International Journal of Agronomy*, 2021(1), 6552928.
Runion, G. B., et al. (2004), 'Microbial responses to wheel-traffic in conventional and no-tillage systems', *Communications in soil science and plant analysis*, 35(19–20), 2891–2903.
Saikia, P., et al. (2020), 'Tropical legumes: status, distribution, biology and importance', In: M. Hasanuzzaman (Ed.), *The plant family fabaceae: biology and physiological responses to environmental stresses*, Springer, Singapore, pp. 27–41.
Salama, H. and Zeid, M. (2016), 'Hay quality evaluation of summer grass and legume forage monocultures and mixtures grown under irrigated conditions', *Australian Journal of Crop Science*, 10(11), 1543–1550.
Scholberg, J. M., et al. (2010), 'Cover crops for sustainable agrosystems in the Americas', In: *Genetic engineering, biofertilisation, soil quality and organic farming*, Springer, Dordrecht, pp. 23–58.
Schomberg, H. H., et al. (2007), 'Potential for using sunn hemp as a source of biomass and nitrogen for the Piedmont and Coastal Plain regions of the southeastern USA', *Agronomy Journal*, 99(6), 1448–1457.
Schrer, K., et al. (1996), *Principles of plant nutrition*, Kluwer Academic Publishers, Dordrecht, Netherlands.
Scursoni, J. A., et al. (2007), 'Weed escapes and delayed weed emergence in glyphosate-resistant soybean', *Crop Protection*, 26(3), 212–218.
Shin, C., et al. (2018), 'Evaluating the effect of bacterial inoculation and fertilization on the soil nutrient status of coal mine soil by growing soybean (Glycine max) and shrub lespedeza (Lespedeza bicolor)', *Sustainability*, 10(12), 4793.
Silva, J. P. d., et al. (2021), 'Decomposition and nutrient release from legume and non-legume residues in a tropical soil', *European Journal of Soil Science*, 73(1).

Singh, A. K., t al. (2023), 'Understanding soil carbon and phosphorus dynamics under grass-legume intercropping in a semi-arid region', *Agronomy*, 13(7), 1692.
Singh, M., et al. (2012), 'Factors affecting the germination of tall morningglory (*Ipomoea purpurea*)', *Weed Science*, 60(1), 64–68.
Smeltekop, H. et al. (2002), 'The Impact of Intercropping Annual 'Sava' Snail Medic on Corn Production', *Agron. J.* 94(4): 917–924.
Snapp, S. S., et al. (2005), 'Evaluating cover crops for benefits, costs and performance within cropping system niches', *Agronomy Journal*, 97(1), 322–332.
Somenahally, A., et al. (2018), 'Microbial communities in soil profile are more responsive to legacy effects of wheat-cover crop rotations than tillage systems', *Soil Biology and Biochemistry*, 123, 126–135.
Stute, J. K. and Shekinah, D. E. (2019), 'Planting date and biculture affect sunn hemp productivity in the Midwest', *Sustainable Agriculture Research*, 8, 26–35.
Thapa, R., et al. (2018), 'Cover crops reduce nitrate leaching in agroecosystems: a global meta-analysis', *Journal of Environmental Quality*, 47(6), 1400–1411.
Thapa, R. et al. (2022), 'Cover crop residue decomposition in no-till cropping systems'. *Agric., Ecosyst. and Environ.* 326(3): 107823.
Treadwell, D. D. (2006), 'A review of cover crop research in Florida: contributions to nutrient cycling in vegetable production', *Proceedings of the Florida State Horticultural Society*, 119, 255–257.
Udvardi, M. and Poole, P. S. (2013), 'Transport and metabolism in legume-rhizobia symbioses', *Annual Review of Plant Biology*, 64(1), 781–805.
Veloso, M. G., et al. (2018), 'High carbon storage in a previously degraded subtropical soil under no-tillage with legume cover crops'. *Agric., Ecosyst. and Environ.* 268: 15–23.
Viney, P., et al. (2009), 'Effects of agriculture upon air quality and climate: research, policy, and regulations', *Environmental Science & Technology*, 43(12), 4234–4240.
Vogeler, I., et al. (2019), 'Legumes in catch crop mixtures: effects on nitrogen retention and availability, and leaching losses', *Journal of Environmental Management*, 239, 324–332.
Wander, M. M., et al. (1994), 'Organic and conventional management effects on biologically active soil organic matter pools', *Soil Science Society of America Journal*, 58(4), 1130–1139.
Wang, G. and Nolte, K. (2010), *Summer cover crop use in Arizona vegetable production systems* (University of Arizona Cooperative Extension Publication AZ1519). College of Agriculture and Life Sciences, University of Arizona, Tucson, AZ. https://hdl.handle.net/10150/144794 (Accessed: July 28, 2025).
Wang, Q., et al. (2009), 'Cover crops and organic mulch to improve tomato yields and soil fertility', *Agronomy Journal*, 101, 345–351.
Wang, Q., et al. (2022a), *Sunn Hemp–A promising cover crop in Florida*, University of Florida IFAS Extension, Gainesville, FL.
Wang, T., et al. (2022b), 'Effects of soil acidification on bacterial and fungal communities in the Jiaodong Peninsula, northern China', *Agronomy*, 12(4), 927.
Wauters, V. M., et al. (2021), 'Ecosystem services and cash crop tradeoffs of summer cover crops in northern region organic vegetable rotations', *Frontiers in Sustainable Food Systems*, 5, 635955.
Weaver, D. (2016), Velvet Bean (Mucuna pruriens var. utilis) [Fact Sheet No. 04t], USDA Agricultural Research Service, National Soil Dynamics Laboratory, Washington D.C., USA.

Whish, J. P. M., et al. (2009a), 'The effect of cover crops on soil water dynamics in a semi-arid environment', *Australian Journal of Agricultural Research*, 60(11), 1017–1028.

Whish, J. P. M., et al. (2009b), 'Do spring cover crops rob water and so reduce wheat yields in the northern grain zone of eastern Australia?', *Crop and Pasture Science*, 60(6), 517–525.

Willey, R. W. (1979), 'Intercropping its importance and research needs. I. Competition and yield advantages', *Field Crop Abstracts*, 32, 1–10.

Willey, R. W. and Rao, M. R. (1980), 'A competitive ratio for quantifying competition between intercrops', *Experimental Agriculture*, 16, 105–111.

Wilson, M. L. and Allan, D. (2014), 'Aerially seeding cover crops in the northern US corn belt', *J. Soil and Water Conserv*, 69(3): 62A–72A.

Wolfe, D. (1997), 'Soil compaction: Crop response and remediation', Report No. 63, Cornell University, Department of Fruit and Vegetable Science, Ithaca, N.Y.

Wright, S. F. and Nichols, K. A. (2002), 'Glomalin: hiding place for a third of the world's stored soil carbon', *Agricultural Research*, 50(9), 4.

Yadvinder, S., et al. (1992). 'Nutrient transformations in soil amended with green manures', *Advances in Soil Science*, 20, 237–309.

Yang, X., et al. (2023), 'Impacts of summer-seeded legume cover crops and termination-tillage method on nitrogen availability to corn in rotation', *Frontiers in Soil Science*, 2: article 1082940.

Zannopoulos, S., et al. (2024), 'The potential of three summer legume cover crops to suppress weeds and provide ecosystem services–A review', *Agronomy*, 14(6), 1192.

Zhao, J., et al. (2022), 'Global systematic review with meta-analysis reveals yield advantage of legume-based rotations and its drivers', *Nature Communications*, 13(1), 4926.

Published by Burleigh Dodds Science Publishing Limited, 2026.

Chapter 13

Cool season grass cover crops

H. Blanco-Canqui, University of Nebraska-Lincoln, USA; S. J. Ruis, USDA-ARS National Laboratory for Agriculture and the Environment, USA; H. Klopp, South Dakota State University, USA; and A. Jhala, J. McMechan, C. Proctor and V. de Sousa Ferreira, University of Nebraska-Lincoln, USA

1 Introduction

Cover crops have potential to control soil erosion, improve soil properties, promote nutrient cycling, control pests and diseases, suppress weeds, increase or maintain subsequent cash crop yields, and provide numerous other benefits (Scavo et al., 2022; Camargo et al., 2023). However, the extent of cover crop benefits depends on many factors (Scavo et al., 2022). In particular, the cover crop functional group – e.g., grasses vs. legumes – affects nearly all of these benefits (Marcillo and Miguez, 2017). Grass cover crops differ from legume cover crops in such functions which include N fixation, decomposition rates, nutrient cycling, growing season, biomass production, and weed suppression. Grass cover crops include winter rye, wheat, triticale, barley, oats, black oats, and annual ryegrass. One of the most common cool season grass cover crops in the U.S. Midwest and other cool temperate regions is winter rye (Ruis et al., 2019). The rapid growth and winter hardiness of this species favor its selection,

http://dx.doi.org/10.19103/AS.2025.0162.17

particularly in corn-soybean and continuous corn systems where winter rye is planted in fall, either before or after crop harvest, and terminated in spring of the following year before planting crops.

While grass cover crops provide several services, their benefits vary depending on species (i.e. overwintering vs. non-overwintering species), biomass production, and weather, among others (Table 1), which deserve further discussion based on case studies (McClelland et al., 2021; Scavo et al., 2022; Brzozowski et al., 2023). In some cases, grass cover crops may compete with the main crops for water and nutrients and reduce subsequent crop yields. However, at the same time, cool season grass cover crops can provide other benefits to support soil productivity and overall agricultural sustainability (Marcillo and Miguez, 2017). Grass cover crops can also compete more aggressively with weeds and thus suppress weeds more effectively than legume cover crops (Camargo et al., 2023).

In this chapter, we discuss the potential benefits and limitations of cool season grass cover crops in order to better understand how these cover crops differ in their benefits from other functional groups, which are discussed in the companion chapters. The specific objectives of this chapter are to discuss: (1) how cool season grass cover crops affect soil properties, nutrient cycling, pest and disease control, weed control, and subsequent cash crop yields and (2) key issues or challenges in integrating grass cover crops into crop rotations. Because soil erosion control benefits of cover crops are relatively well known, especially when cover crops produce abundant biomass, this chapter focuses more on other soil and crop parameters.

Table 1 Summary of cool season grass cover crop species, overwintering ability, and conditions each species can tolerate (Midwest Cover Crops Council, 2014)

Grass cover crop species	Overwintering	Tolerance
Annual ryegrass	Sometimes, depending on weather and snow cover	Shade and flooding
Barley (winter or spring)	Sometimes, depending on weather and snow cover	Drought, some shade tolerance
Oats	No	Some flooding
Winter rye	Yes	Drought and shade, some brief flooding
Triticale	Winter triticale overwinters unlike spring triticale	Drought and shade, some brief flooding
Wheat	Winter wheat overwinters unlike spring wheat	Drought and shade, some brief flooding

2 Soil physical properties

Soil should be able to retain water, allow air and water movement through the soil, and allow root penetration (Indoria et al., 2017). Cool season grass cover crops such as oats, wheat, barley, and rye have many fine fibrous roots near the surface, which could improve near-surface soil physical properties (0–10 cm depth; Blanco-Canqui and Ruis, 2020). Cover crop roots can reduce soil bulk density and create pores for water and air movement. The addition of roots and C additions to soil from cover crops can lead to improvements in soil physical properties (Rillig, 2004; Blanco-Canqui and Ruis, 2020; Haruna et al., 2020). Grass cover crops may increase soil water retention, porosity, wet aggregate stability, saturated hydraulic conductivity, and infiltration rates and reduce soil penetration resistance and bulk density (Blanco-Canqui and Ruis, 2020; Haruna et al., 2020).

Grass cover crops improve many soil physical properties, but not in all cases (Blanco Canqui and Ruis, 2020; Haruna et al., 2020). For example, a global review by Blanco-Canqui and Ruis (2020) found that planting grass cover crops can have differing effects on soil compaction indicators, depending on the indicator in question. Cover crops had no effect on bulk density in 21 of 26 studies and reduced bulk density in five studies. Cool season grass cover crops reduced penetration resistance in five of nine studies by 5–22%, but did not affect it in four. Therefore, depending on the compaction indicator used, cool season cover crops can potentially reduce soil compaction. Cover crops also increased wet aggregate stability in 17 of 30 studies but had limited effects on dry aggregate stability (no effect in five of seven studies). Cover crops frequently increased soil infiltration rate (7 of 10 studies) and saturated hydraulic conductivity (increase in four of seven studies) but did not generally affect the amount of plant available water (three of five studies, Blanco-Canqui and Ruis, 2020). During the spring, cover crops can reduce soil temperature due to the increased soil cover (Blanco-Canqui and Ruis, 2020). The studies listed in Table 2 show that planting cool season grass cover crops can improve some soil physical properties, but effects vary depending on site and other factors.

The ability of cool season grass cover crops to positively impact soil physical properties is dictated by many factors, including total biomass production and the number of years of cover crop use. For example, Ruis et al. (2023) found that terminating rye late in spring led to more benefits to soil physical properties compared with early termination by increasing the amount of biomass produced. Higher biomass production leads to greater C input into the soil, which alters soil physical properties. An increase in soil C under cover crops can lead to improvement in soil physical properties (Blanco-Canqui and Ruis, 2020). For example, an increase in soil C under cover crops can be

Table 2 Summary of the impacts of cool season grass cover crops on key metrics of soil physical properties

Location	Duration (yr)	Cover crop species	Crop rotation	Effect on soil property	Reference
Mississippi, USA	5	Cereal rye and winter wheat	Continuous soybean	• No effect on wet aggregate stability • Rye reduced bulk density, but winter wheat had no effect	Dai et al., 2024
Brazil	1	Cereal rye and oat	Corn	• Rye increased saturated hydraulic conductivity, but oat had no effect • No effect on bulk density	Gongora et al., 2022
Japan	19	Cereal rye	Rice and soybean	No effect on bulk density, wet-aggregate stability, and plant available water	Hashimi et al., 2023
Alabama, USA	2 and 4	Cereal rye	Cotton-peanut	No effect on water stable aggregates or penetration resistance	Johnson et al., 2021; Decker et al., 2022
Tennessee, USA	2 and 4	Cereal rye	Cotton-soybean	• No effect on water stable aggregates • Reduced penetration resistance	Johnson et al., 2021; Decker et al., 2022
Nebraska, USA	10	Cereal rye	Continuous corn	• Increased wet aggregate stability and penetration resistance • Reduced bulk density • No effect on water infiltration, hydraulic conductivity, plant available water, and dry aggregate stability	Klopp et al., 2023, 2024
Missouri, USA	3	Cereal rye	Corn and soybean-based rotations	Reduced bulk density	Panday and Nkongolo, 2021
Nebraska, USA	6	Early and late terminated cereal rye	Continuous corn	No effect on bulk density, penetration resistance, wet aggregate stability, initial water infiltration, and plant available water	Ruis et al., 2023

Nebraska, USA	6	Early and late terminated cereal rye	Continuous corn	• Late-terminated cover crop reduced bulk density • Late-terminated cover crop increased wet aggregate stability • No effect on initial water infiltration, penetration resistance, and plant available water	Ruis et al., 2023
South Dakota, USA	9	Cereal rye	Corn and soybean-based	No effect on water stable aggregates, saturated hydraulic conductivity, and bulk density	Saha et al., 2024
Kansas, USA	12	Triticale	Wheat-sorghum-fallow	• Reduced bulk density • Increased dry aggregate stability • No effect on wet aggregate stability	Simon et al., 2022
Nebraska, USA	8	Cereal rye	Corn-soybean	• Increased wet aggregate stability • No effect on bulk density, penetration resistance, water infiltration, and plant available water	Blanco-Canqui et al., 2023
Nebraska, USA	8	Cereal rye	Corn-soybean	• Increased wet aggregate stability • Reduced bulk density and penetration resistance • No effect on water infiltration and plant available water	Blanco-Canqui et al., 2023
Japan	18	Cereal rye	Rice and soybean	• Reduced bulk density • No effect on penetration resistance	Wulanningtyas et al., 2021

related to reductions in soil bulk density (compaction indicator) and increases in soil wet-aggregate stability (Blanco-Canqui and Ruis, 2020). Planting and successfully establishing cover crops consecutively for many years may be needed to observe changes in soil physical properties (Klopp et al., 2024). Other factors include cover crop species, crop rotation, and tillage practice (Blanco-Canqui and Ruis, 2020; Kim et al., 2020). Overall, cool season cover crops can improve some soil physical properties, but the amount of biomass produced, and duration of cover crop are determining factors.

3 Soil chemical and biological properties

3.1 Soil chemical properties

Soil chemical properties include soil pH, electrical conductivity, and concentrations of nutrients like N, P, K, S, Ca, Mg, and soil C concentration. Since soil C is a key driver of many soil properties and processes, we focus on the impacts of cool season grass cover crops on C. Cool season grass cover crops can enhance soil C due to increased C input from their biomass. Cover crops can increase soil C by 0.32–0.88 Mg ha^{-1} yr^{-1} in the upper 30 cm of the soil regardless of cover crop species (Poeplau and Don, 2015; Ruis and Blanco-Canqui, 2017; Joshi et al., 2023, Table 3). While these values suggest cover crops can have large potential to sequester C, the reality is that there are many factors involved in such gains. Often, such gains occur in the surface 10 cm, particularly under no-till. As discussed with soil physical properties, two key factors affecting C gains under cover crops are cover crop biomass production and duration of management.

The number one factor affecting soil C under cool season grass cover crops is the amount of biomass produced. Indeed, in corn-based systems, >3 Mg ha^{-1} yr^{-1} of cover crop biomass was required to increase soil C stocks by about 11.5% (Joshi et al., 2023). A review concluded that at least 2 Mg ha^{-1} yr^{-1} of cover crop biomass was needed to increase soil C in temperate regions (Blanco-Canqui, 2022a). The key with cover crop biomass production impacts on soil C is that it must be relatively consistent from year to year. For example, winter rye cover crop under no-till continuous corn must consistently produce high biomass (minimum of 3 Mg ha^{-1} yr^{-1}) to exert gains in soil C (Ruis et al., 2023). Cover crop effects on soil C take time to develop, so C gains are often not realized until after 5 years of consistent management (Blanco-Canqui, 2022a). Wheat and oat cover crops producing >3 Mg ha^{-1} yr^{-1} of biomass increased soil C in an Argentinian soybean system after 6 years (Duval et al., 2016), but rye and oat cover crops (<2.1 Mg ha^{-1} yr^{-1} biomass) had no effect in an Iowa, USA, corn-soybean system after the same management duration (Kaspar et al., 2006). The key difference between these two studies lies in the amount of biomass produced and less so on the climatic conditions which were relatively similar. Thus, because biomass is key, if a non-overwintering cover crop can produce sufficient biomass, it may contribute to C gains like that of an overwintering cover crop.

3.2 Soil biological properties

Soil biological properties are many, but here, we focus on microbial biomass and earthworm abundance. Grass cover crops can increase soil total microbial biomass by about 18–20%, bacteria by 10%, fungi by 26%, and arbuscular mycorrhizal fungi by 26% compared with no cover crop (Kim et al., 2020; Fohrafellner et al., 2024; Table 2). The increase in arbuscular mycorrhizal fungi is particularly important as they can increase P uptake, stress tolerance, and crop yields (Muhammad et al., 2021). Because the cover crop can enhance the fungal community more than the bacterial community, there is greater potential for increases in soil organic matter, particularly in the more persistent mineral-associated pools (Six et al., 2006; Muhammad et al., 2021). These increases in microbial biomass are attributable to greater duration of a living root in the soil (i.e. greater addition of root exudates), addition of residues, and moderation of soil temperature and water content (Kim et al., 2020; Muhammad et al., 2021; Fohrafellner et al., 2024).

While soil microbial indicators are important, they are not easily observable, unlike earthworms. Grass cover crops increased earthworm abundance by 17–93 individuals m^{-2} across four of five studies (Blanco-Canqui, 2022b; Table 2). Planting a winter rye cover crop (3 Mg ha^{-1} yr^{-1} biomass) following corn silage and soybean for 14 years increased earthworm abundance (42 to 135 individuals m^{-2}) (Korucu et al., 2018; Crespo et al., 2024a). Even though earthworms may prefer lower C:N ratio residues (i.e. legumes), grass cover

Table 3 Summary of the impact of grass cover crops on soil chemical and biological properties

Soil properties	Overall effect of grass cover crops	Reference
Soil C	• Across species, cover crops can increase soil C by 0.32–0.88 Mg ha^{-1} yr^{-1}. • To increase soil C, sufficiently high and consistent cover crop biomass production and time (>5 years) are needed.	Poeplau and Don, 2015; Ruis and Blanco-Canqui, 2017; Blanco-Canqui, 2022a; Joshi et al., 2023
Microbial biomass	Grass cover crops increase soil microbial biomass, particularly soil fungi.	Kim et al., 2020; Muhammad et al., 2021; Fohrafellner et al., 2024
Earthworms	Grass cover crops can increase earthworm abundance.	Korucu et al., 2018; Blanco-Canqui, 2022b

crops can have relatively lower C:N ratios when younger and cover the soil following termination, thereby moderating fluctuations in temperature and soil water content, thus favoring earthworm activity (Blanco-Canqui, 2022b). These increases in earthworm abundance often co-occur with increased soil C, soil aggregation, water infiltration, and reduced soil bulk density (Blanco-Canqui, 2022b).

In summary, cool season grass cover crops can increase soil C stocks, provided consistent and sufficiently high (>2–3 Mg ha^{-1} yr^{-1}) biomass is produced and cover crops are used for at least 5 years. These cool season grass cover crops can enhance soil microbial biomass and earthworm abundance. The potential for increased soil C, microbial biomass, and earthworm abundance highlights the benefits of cool season grass cover crops for soil chemical and biological properties (Table 3).

4 Nutrient cycling

In this section, we focus on cover crops and N cycling, although cover crops can also alter the cycling of other nutrients like P and K (Poudel et al., 2023; Crespo et al., 2024b). Losses of N can be particularly large between crop harvest and crop planting the following spring. Grass cover crops, typically grown during this window, can significantly alter such N losses through immobilization and mineralization.

Grass cover crops are known to develop extensive fibrous root systems that can scavenge residual N following cash crop harvest. One of the primary soil nutrients that grass cover crops capture is NO_3^-. Nitrate is a product of the nitrification process mediated by soil microbes that convert NH_4^+ to NO_3^-. Nitrification depends on aerobic soil conditions, ammonium availability, soil water content, and microbial activity. Nitrate, due to lack of fixation to soil particles (unlike PO_4^{3-}), its negative charge, and high water solubility, is highly susceptible to leaching. Grass cover crops can capture surplus NO_3^-, mitigating losses. A meta-analysis showed non-legume cover crops can reduce NO_3^- losses by about 56% (Thapa et al., 2018). Cover crop biomass production is the key factor dictating reductions in NO_3^- losses. Indeed, 1 Mg ha^{-1} of cover crop biomass production can reduce NO_3^- losses by 44%, 2 Mg ha^{-1} by 55%, 3 Mg ha^{-1} by 64%, and 4 Mg ha^{-1} by 71% (Thapa et al., 2018).

While grass cover crops have the ecological benefit of scavenging NO_3^-, this N may be immobilized and therefore unavailable for plant uptake. As the cover crop residue decomposes, mineralization converts the organic N to inorganic N. The mineralization process is dictated in part by the quality or C:N ratio of the residue. In general, high C:N ratios can lead to too little N released for microbial use, while low C:N ratio residues supply microbes with sufficient N plus leaving additional N for plant uptake.

The C:N ratio varies by plant species and increases as the maturity of cool season grass cover crops increases. For example, the C:N ratio of cereal rye at early vegetative stage (Feekes growth stage 2–5) is about 17:1, at stem elongation stage (Feekes growth stage 6–8) 25:1, and increases to 35:1 at boot stage (Feekes stage 10). Termination at these different growth stages impacts the N release rate and availability for subsequent cash crop. Due to the potential for N immobilization from grass cover crops, some crops may require supplemental N during the early season to avoid N stress until organic N mineralizes later in the growing season. Thus, the timing of grass cover crop termination becomes an important factor for nutrient cycling. Even though some research shows grass cover crops reduced soil N concentration (Qin et al., 2021), resulting in low soil N during early plant growth stages (Clark et al., 2007), the slow release of N from cover crop residue decomposition can be slow and synchronized with the cash crop (Singh et al., 2020).

Another key loss pathway for N that cover crops could affect is N_2O fluxes. Grass cover crops may not only reduce soil NO_3^- concentrations but also reduce soil water content, both of which should aid in reducing N_2O fluxes (Van Eerd et al., 2023). However, literature shows that non-legume cover crops may not always affect N_2O fluxes compared with no cover crop (Ruis et al., 2018; Abdalla et al., 2019). Indeed, 9 of 11 studies showed grass cover crops had no effect on N_2O fluxes relative to no cover crops (Ruis et al., 2018). The inconsistent grass cover crop effects on N_2O fluxes may be partially due to the variable nature of N_2O fluxes, which vary significantly over space and time.

In summary, incorporating cool season grass cover crops into agricultural systems can improve nutrient cycling by scavenging excess N, reducing N leaching, and possibly reducing gaseous losses. Terminating grass cover crops at the right growth stage can ensure adequate N release for the subsequent cash crop, although supplemental N fertilizer application may be needed to offset N immobilized by cover crop decomposition.

5 Weed management

Cover crops have been adopted by growers for various reasons, including weed suppression. The evolution of herbicide-resistant weeds and their widespread occurrence in agronomic crops such as corn and soybean have encouraged growers to explore non-chemical weed management options (Kumar et al., 2025). Cover crops can serve as an effective non-chemical tool for suppressing herbicide-resistant weeds, including Palmer amaranth (*Amaranthus palmeri* S. Watson) and waterhemp (*Amaranthus tuberculatus* (Moq.) Sauer), in corn-soybean production systems (Whalen et al., 2020; Fig. 1).

Figure 1 Palmer amaranth emergence in soybean without cereal rye cover crop (a), and in soybeans planted into cereal rye that was terminated 7 days before soybean planting (b). The rye cover suppressed Palmer amaranth in a soybean field in Nebraska (b) (Photos by A. Jhala).

Cool season grass cover crops are particularly effective for weed suppression. Depending on the crops and weather, most cool season grass cover crops survive through winter and provide early spring growth. They are typically planted in the fall, either before or after crop harvest, and help suppress winter annual weeds in the fall and early spring (Nichols et al., 2020). Fall-planted cover crops can also suppress summer annual weeds or delay their emergence if sufficient residue is present in early summer (Nunes et al., 2024; Stephens et al., 2024).

Cool season grass cover crops can have a high C:N ratio at termination, which means they break down more slowly compared to cool season legume cover crops. This slower decomposition results in longer-lasting residue that effectively suppresses weeds (Kumar et al., 2025). Furthermore, cool season grass cover crops produce more biomass and residue than cool season legume cover crops, enhancing their ability to suppress weeds more effectively.

5.1 Weed suppression ability of cool season grass cover crops

The weed suppression ability of cool season grass cover crops depends on several factors, including the grass species planted, their biomass production, the weed species present in the field (Kumari et al., 2024), planting and termination times (Mirsky et al., 2011), seeding rate (Bish et al., 2021), termination method (Nichols et al., 2020), and the time after crop planting (Curran et al., 1994). Palhano et al. (2017) reported that cereal rye decreased Palmer amaranth density by 90% four weeks after planting cotton compared to a 44% reduction at eight weeks after planting in a study conducted in Arkansas. The effect of cool season grass cover crops on weed suppression is species-specific, as some weed species are more susceptible than others. In Illinois, Crawford et al. (2018) reported that cereal rye and radish (*Raphanus sativus* L.) reduced broadleaf weed density but did not affect grass weed density. In Mississippi, Reddy (2003) reported a 50% reduction in smooth pigweed density with cereal rye residues compared to a conventionally tilled no-cover-crop treatment, with no effect on other broadleaf weeds.

Kumar et al. (2025), in a meta-analysis, reported that cover crop biomass of 5.35 Mg ha^{-1} is required for a 50% biomass reduction of *Amaranthus* species. Nichols et al. (2020) found that 5 Mg ha^{-1} of cover crop biomass can reduce weed biomass by 75%, which is higher than the 50% reduction in *Amaranthus* species biomass. This discrepancy might be because Nichols et al. (2020) included studies from the Midwestern United States, whereas Kumar et al. (2025) included studies from the Southeastern United States. Rapid accumulation of heat units and high annual rainfall in the Southeastern United States make it favorable for weed growth, development, and higher weed biomass accumulation (Reinhardt Piskackova et al., 2021; Weisberger et al., 2023). This results in higher cover crop biomass requirements for weed biomass reduction in the Southeastern United States (Weisberger et al., 2023). Higher biomass of cool season grass cover crops, such as cereal rye, can significantly reduce weed biomass, density, and seed production (Stephens et al., 2024; Nunes et al., 2024) (Figs. 1 and 2).

5.2 Termination of cool season grass cover crops

Cool season grass cover crops should be terminated in a timely fashion to avoid interference with cash crops. Historically, cool season grass cover crops have been terminated before planting cash crops such as corn or soybean (Rosa et al., 2021). However, in recent years, growers have started planting cash crops directly into cool season grass cover crops and terminating the cover crop on the day of planting (Grint et al., 2022) or several days after planting the cash crop (Stephens et al., 2024), a practice known as "planting green." Planting green allows cover crops to grow longer and produce more biomass, which enhances their weed suppression capabilities (Fig. 2) compared to terminating them before crop planting (Blanco-Canqui and Jhala, 2024).

For example, in Nebraska, cereal rye terminated 2 weeks after planting soybean improved Palmer amaranth suppression compared to terminating it two weeks before planting soybean (Stephens et al., 2024). Nunes et al. (2024) found that planting soybean green increased cereal rye biomass production by 33% compared to early termination in a study conducted

Figure 2 Cool-season cover crop (cereal rye) terminated 2 weeks after planting soybean combined with a residual herbicide applied at soybean planting provided excellent Palmer amaranth suppression in a study conducted in Nebraska (Photo by A. Jhala).

across 19 site-years in the United States. However, cereal rye terminated a few days after planting soybean reduced grain yield (Stephens et al., 2024; Nunes et al., 2024). More research is needed to determine whether crop yield loss is due to early-season competition between the cover crop and the cash crop or to signals produced by the cover crop, such as light quality alterations and/or the release of volatile chemicals that may affect crop growth and development (Horvath et al., 2023).

6 Insect and disease management

Grass cover crops can support a diverse range of insect species and plant diseases. The beneficial or detrimental impact of these organisms on the subsequent cash crop depends on several factors, including the taxonomic similarity between the cover and cash crops, the host range of the pest, the timing between cover crop termination and cash crop planting, pesticide applications, and prevailing weather conditions.

6.1 Insects in cool season cover crops

Inveninato Carmona et al. (2021) conducted a systematic review of insects in cover crops and found that 55% of the 37 research studies since 1950 have used only a grass cover crop. Mixed-species cover crops containing a grass species accounted for 37% of the remaining studies. Most of these studies focused on cereal rye, wheat, and triticale, with a few investigating oat, barley, timothy (*Phleum pratense* Hurd), fescue (*Festuca* spp.), wheatgrass (*Thinopyrum intermedium*), and orchardgrass (*Dactylis glomerata* L.) (Inveninato Carmona et al., 2021).

Insect pests such as wheat stem maggot (*Meromyza americana* Fitch) (Inveninato Carmona et al. 2019), true armyworm (*Mythimna unipuncta* Haworth) (Dunbar et al. 2016), and seed corn maggot (*Delia platura* Meigen) (Hammond 1991) may increase in corn and soybeans planted after grass cover crops. The frequency and intensity of these pests vary significantly across locations and years, influenced by factors such as the stage and mobility of the pest at the time of cover crop termination, the cover crop termination method, and environmental conditions.

For wheat stem maggot, infestations in corn can be linked to the termination of the cover crop after corn planting. Cool and wet weather can reduce the efficacy of glyphosate, resulting in an additional herbicide application (Inveninato Carmona et al., 2019). As the cover crop dies, wheat stem maggot larvae can move from the wheat or rye cover crop to corn to complete their development, preferentially feeding on the growing point of corn, resulting in the death of the main stem and excessive tillering (Fig. 3). Some pest outbreaks in cover crop systems could be caused by a combination of management practices and environmental conditions. Erratic and unexpected pest pressure in cover crop systems has the potential to increase insecticide use. However, applying insecticides as a tank mix during cover crop termination or delaying applications in the absence of pest pressure often has little to no economic benefit, can negatively impact beneficial insects, and increase the likelihood of secondary pest outbreaks.

Several studies have reported decreased pest pressure and increased beneficial insect activity with cover crop use. Koch et al. (2015) observed a reduction in soybean aphid (*Aphis glycines* Matsumura) populations in five of six locations where a grass cover crop preceded soybeans. Significant reductions in soybean aphid only occurred when cereal rye was mowed between the rows after soybean emergence (Koch et al., 2015). Predator densities did not vary between treatments, and no yield differences were observed. In corn, Lungren and Fergen (2010) found that wheatgrass reduced both root injury and the abundance of third-instar western corn rootworm (*Diabrotica virgifera* LeConte), a major pest of continuous corn systems. Predator abundance strongly correlated with reduced rootworm populations, suggesting that both predation and wheatgrass played key roles in rootworm suppression. An early cover crop planting date in the fall combined with a late termination date in the spring resulted in the greatest abundance of spiders (Carmona et al., 2022).

Figure 3 Wheat stem maggot resulted in death of the main stem and excessive tillering in corn following a rye cover crop in south central Nebraska. (Photo by J. McMechan).

Since the same cover crop management tactics can support both beneficial and pest insects, regular scouting after planting the cash crop is important (Fig. 6).

6.2 Plant diseases in cool season cover crops

No systematic reviews have been conducted on plant pathogens in grass cover crop systems, but research on different cover crop and cash crop combinations indicate both positive and negative effects on plant disease incidence. Reductions in plant disease incidence from cover crop use could be attributed to decreased pathogen dispersal, microclimatic changes unfavorable to disease development, or biochemical exudates from cover crops that affect pathogens (Ristaino et al., 1997; Mills et al., 2002). Pethybridge et al. (2020) found that roller-crimped rye decreased sclerotia germination and increased the frequency of non-functional sclerotia, reducing white mold incidence in soybean and dry bean. Similarly, Ristaino et al. (1997) reported a 28% reduction in *Phytophthora* incidence in bell peppers when stubble from a fall-planted wheat cover crop was used compared to bare ground. However, the close phylogenetic relationship between corn and rye can facilitate pathogen carryover.

Fusarium spp. (Chongo et al., 2001), *Pythium* spp. (Vestberg, 1990), and *Rhizoctonia* spp. (Neate, 1989) can persist in such systems. Bakker et al. (2016) isolated *Fusarium* and *Pythium* species from cereal rye roots and found that root lengths in corn were reduced by 20–60% compared to control treatments. These pathogens also occur in soybean, but soybean generally compensates for early-season plant losses more effectively than corn (Bakker et al., 2016). Acharya et al. (2020) evaluated the effects of cereal rye as a cover crop prior to corn or soybean. In both study years, corn seedlings exhibited significantly greater *Pythium* clade B density when cereal rye was used as a cover crop. However, no significant differences in *Pythium* incidence were observed in soybean. Green cover crops may also offer an escape for some pathogens from the highly competitive environment of decomposing plant residue (Bakker et al., 2016). As facultative saprotrophs, plant pathogens have the potential to flourish on cover crops prior to their termination if they are displaced by other more highly competitive saprotrophs on residue (Bakker et al., 2016). These findings underscore the importance of considering both the benefits and potential risks of the temporal occurrence of cover crops when integrating them into a cropping system.

7 Crop yield

Grass cover crops can have positive, negative, or neutral effects on cash crop yield (Fig. 4). A recent global meta-analysis showed non-legume cover crops (grass and brassica cover crops) can have neutral effects on crop yield overall (Peng et al., 2024). Another review, across the United States and Canada, showed that grass winter cover crops had no effect on corn grain yield (Marcillo and Miguez, 2017). Implementing cover crops in soils under long-term no-till had neutral effects on crop yield (Peng et al., 2024). However, use of these non-legume cover crops in short-term no-till in fine-textured soils can result in yield reductions (Peng et al., 2024), in part due to the "yield drag" associated with transitioning to no-till (Pittlekow et al., 2015). Research reviewed by Peng et al. (2024) showed the variable nature of cash crop yield responses to cover crops, with positive effects on corn and barley, neutral effects on soybean and rice, and negative effects on wheat yield. In a review specific to semi-arid regions, Blanco-Canqui et al. (2022) showed that cover crops had no effect on subsequent crop yield in 56% of studies but reduced yield in 38%.

The impact of grass cover crops on crop yield is site-specific and influenced by weather and management, which explains the variability in crop yield response. For example, a winter rye cover crop had no effect on continuous corn grain yield in Iowa, USA, except in two years with drought conditions (Martinez-Feria et al., 2016). In these two years, the rye cover crop reduced corn yield by 1.9–2.6 Mg ha^{-1}. In contrast, a seven-site study under corn-soybean rotation with and without an annual ryegrass or cereal rye cover crop in Illinois, USA, showed that these grass cover crops can reduce corn yield by 3.9 ± 3% but have no effect on soybean yield (Qin et al., 2021). Qin et al. (2021)

Peng et al., 2024	• Global meta-analysis reported non-legume cover crops have neutral effects on crop yields.
Marcillo and Miquez, 2017	• Meta-analysis across U.S. and Canada reported winter grass cover crops had no effect on corn yields.
Blanco-Canqui et al., 2022	• In semi-arid regions, cover crops had no effect on crop yields in most cases.
Martinez-Feria et al., 2016	• In Iowa, a winter rye cover crop had no effect on continuous corn grain yield except in dry years.
Qin et al., 2021	• Across 7 sites in Illinois, grass cover crops reduced corn yield, but not soybean yield.
Ruis et al., 2023	• In Nebraska, winter rye cover crops had no negative effect on continuous corn yield at two sites except in dry years, even with late termination and high cover crop biomass production.
Preza-Fontes et al., 2022	• In Indiana, winter rye cover crop reduced corn yield at one site but had no effect at other two sites.
Quinn et al., 2023	• In Kentucky, averaged across 7 site years, winter rye cover crop reduced corn yield.
Malone et al., 2022	• In Wisconsin, oat and winter rye cover crop did not affect soybean yield. However, preharvest planted winter rye cover crop reduced corn yield, but post harvest planted winter rye did not affect corn yield. Oat cover crop did not affect corn yield.
Marcos et al., 2023	• In Iowa, winter rye cover crop reduced corn yield in two of six years. In the other four years, there was no effect of winter rye cover crop on corn yield.
Rosa et al., 2021	• In Nebraska, winter rye and spring triticale cover crop reduced corn yield but oat cover crop did not affect corn yield.
Overmyer et al., 2023	• In Ohio, late terminated winter rye cover crop increased or had no effect on soybean yield in 4 site-years. The early terminated had no effect on soybean yield in 3 site-years and reduced in 1 site-year.
Acharya et al., 2022	• In New Mexico, oat cover crop did not affect wheat yield.
Silva et al., 2024	• Across midwestern United States, termination of winter rye cover crop at planting did not affect soybean yield in 25 of 28 site-years.

Figure 4 Summary of the impacts of grass cover crops on cash crop yields from three meta-analyses and 11 case studies.

suggested that terminating the cereal rye one day before corn planting can reduce corn yield by about 10% compared with 5.1% when terminated about 1 month before corn planting even when cover crop biomass is relatively low (<1 Mg ha^{-1}). While late termination (i.e. at corn planting) can reduce corn yield (Qin et al., 2021), this is not always the case. For instance, a winter rye cover crop terminated at corn planting did not affect corn yield in most years compared with no cover crop, even when cover crop biomass was relatively large (3 Mg ha^{-1} yr^{-1}) across 6 years and two locations in Nebraska, USA (Ruis et al., 2023). Fertilizer N applied at different times and rates to corn following a rye cover crop did not prevent reductions in corn grain yield (Pantoja et al., 2015; Patel et al., 2019; Preza-Fontes et al., 2022; Quinn et al., 2023). Factors besides N may explain the reduction in corn yield following a winter rye cover crop. When the winter rye cover crop reduces corn yield, it can be attributed to delayed cover crop termination, seedling disease, allelopathy, and drought periods at times critical to corn growth. The effects of grass cover crops on the yield of other crops are also variable. A review by Toler et al. (2019) showed neutral effects of grass cover crops on cotton lint and seed yield. Winter rye and a barley and oat mixture did not affect tomato yield in Italy (Lenzi et al., 2009). In Australia, an oat cover crop did not affect potato yield (Little et al., 2004), while oat and triticale cover crops did not affect potato yield in Israel (Eshel et al., 2015). In Massachusetts, potato yield was 6.1 Mg ha^{-1} greater following a rye cover crop compared to no cover crop (Jahanzad et al., 2017). While not reviewed here, grass cover crops might have similar impacts on the yield of other cash crops.

Why do grass cover crops reduce crop yield in some situations and not in others? Common reasons cited for crop yield reductions following cover crops are (1) reductions in soil moisture and (2) increased nutrient immobilization by the cover crop (Peng et al., 2024; Qin et al., 2021). Terminating cover crops early to allow recharge of water from rainfall and decomposition of cover crop to release nutrients may help prevent yield reductions. However, flexibility in termination timing is needed. For example, in dry springs, a producer may terminate early, particularly in semi-arid regions to reduce risk of yield reductions. In a wet spring or humid regions, delaying cover crop termination may be suitable to gain additional cover crop benefits without reductions in crop yields (Blanco-Canqui et al., 2022). Three important goals of using cover crops are: (1) covering the soil, (2) reducing nutrient losses, and (3) improving soil properties. Terminating cover crops early, particularly in corn-based systems in cool, temperate regions, results in small cover crop plants that (a) cover very small portions of the soil surface and (b) are not very effective at reducing nutrient losses and protecting soil. As mentioned elsewhere, sufficient cover crop biomass is needed to achieve these and other goals. A cover crop with sufficient biomass production can act as a mulch, moderating soil water and temperature fluctuations. In addition, as discussed by Ruis et al. (2023) and Blanco-Canqui (2022a), this high cover crop biomass (at least 2–3 Mg ha^{-1}) can improve soil properties, which can thereby lead to soils being more resilient to weather-related stressors. Overall, grass cover crops can have neutral to negative effects on subsequent crop yields depending upon termination timing, weather conditions, and other site-specific factors.

8 Challenges in integrating grass cover crops into crop rotations

Integrating grass cover crops into agricultural systems may provide ecological benefits, such as weed suppression, improved soil properties, and capture of excess N. However, there are some challenges with integrating grass cover crops, including limited time for establishment and growth following crop harvest, interaction with herbicide programs, and equipment limitations. Most cover crop benefits are positively correlated with increased biomass production, but to achieve this, planting as early as possible either before crop harvest with specialized equipment or drilling immediately after harvest and delaying termination as long as possible in spring may be necessary. The following sections explore these critical management considerations in greater detail.

8.1 Planting time

The duration of optimal growing conditions for grass cover crops can impact cover crop establishment and development. In temperate regions, the window for cover crop establishment after crop harvest and before the onset of winter is short. Given this limitation, grass cover crops may not establish or accumulate sufficient fall biomass, and significant growth may only occur the following spring. Interseeding cover crops into standing crops at the vegetative stage, or overseeding at the reproductive stage, are strategies used to accumulate more grass cover crop biomass and thus achieve their ecological benefits (e.g. N scavenging). Klopp (2024) evaluated interseeding or overseeding cover crops at the vegetative or reproductive stages of crops, often in corn and soybean. Depending on equipment availability and growth stage of the crop, interseeded cover crops can be broadcast or drilled (Figs. 5 and 6). Research from the mid-Atlantic to the Midwestern United States showed that interseeding grass cover crops into vegetative corn or soybean can improve cover crop biomass with limited impacts on corn yields (Caswell et al., 2019; Mohammed et al., 2019; Brooker et al., 2020). However, drill interseeding annual ryegrass into corn before the V3 growth stage reduced corn grain yield compared with interseeding between the V3 and V6 growth stages (Curran et al., 2018).

Compared with post-harvest drilled grass cover crops, spring cover crop biomass was higher under R5 overseeded cover crops compared with drilled cover crops in two of three locations (Koehler-Cole et al., 2023). Similarly, an overseeded winter rye cover crop had higher biomass production across 6 years in a no-till corn-soybean rotation when planted at R7 for corn and R5 for soybean (Blanco-Canqui et al., 2025). Both Koehler-Cole et al. (2023) and

Figure 5 Interrow cover crop interseeder.

Figure 6 High clearance tractor for broadcast overseeding grass cover crops.

Blanco-Canqui et al. (2025) showed no yield penalty in the cash crop. One of the caveats with seeding into a standing cash crop is that specialized equipment is needed. This seeding method requires the use of a high-clearance tractor, air seeder, high-clearance interrow drill, airplane, helicopter, or UAV (drone). To interseed grass cover crops in corn between V2-V8 growth stage, a high-clearance interrow drill is needed to avoid knocking down corn plants. This drill has row spacing and clearance that allows the grower to drive through the field drilling the cover crop seed. For late season overseeded cover crops, an airplane, UAV, or high-clearance tractor is needed to broadcast grass cover crops into standing corn. While interseeding or overseeding can improve establishment and increase biomass production, the economics of hiring or buying specialized equipment must be considered.

8.2 Herbicide carryover

Herbicide management programs are devised to ensure effective weed control. Residual herbicides play an important role in weed management, but some herbicides persist in the soil and can negatively impact grass cover crop establishment. Brooker et al. (2020) reported that an annual ryegrass cover crop was reduced by more than 60% following application of Group 15 herbicides at corn planting followed by V3 and V6 interseeding timing. Wallace et al. (2017) reported that annual ryegrass could safely be interseeded in V5 corn following preemergence application of dimethenamid-P or acetochlor. Herbicide carryover may negatively impact fall-seeded grass cover crops but will depend on herbicide chemical characteristics and environmental conditions that influence herbicide persistence in the soil. Ley et al. (2023) tested the impact of four residual preemergence herbicides on fall-seeded cereal rye and found no difference in biomass production at termination time the following spring. The degradation rate of herbicides is influenced by chemical properties, soil texture, soil organic matter, pH, soil moisture, temperature, rainfall, and irrigation rate. In western Nebraska, de Sousa Ferreira et al. (2024) evaluated the potential herbicide carryover impact on early-season interseeded grass cover crops (annual ryegrass, cereal rye, and winter wheat) at V2, V4, and V6 corn growth stage. Three residual herbicides were evaluated: saflufenacil+dimethenamid-P, acetochlor+atrazine, and s-metolachlor+atrazine. Biomass accumulation from the V2 interseeding timing differed across herbicides, highlighting the importance of herbicide selection for early season interseeding in corn. Overall, residual herbicides may not kill the interseeded grass cover crops, but biomass reduction may be observed depending on the herbicide and interseeding timing.

8.3 Termination timing

The likelihood of ecological benefits increases with the amount of grass cover crop biomass production. Termination timing plays a role in biomass accumulation in the spring. Growers may decide to terminate grass cover crops a few weeks or days prior to cash crop planting, at planting, or after cash crop emergence. However, excessive grass cover crop residue may negatively impact cash crop emergence and grain yield. Silva et al. (2024) reported that terminating cereal rye at soybean planting increased biomass accumulation compared with termination 39 days prior to soybean planting across 28 site-years in the Midwestern United States. Termination timing did not negatively impact soybean yield in 25 of 28 site-years. In Wisconsin, Grint et al. (2022) compared cereal rye termination 14 days before planting and at planting and found no decrease in soybean yield. Cover crops may reduce cash crop yield when termination is significantly delayed, such as 10 days after planting compared with prior to or at planting (Almeida et al., 2024). However, when grass cover crops are terminated at corn planting, corn yield may not decrease in most years (66%) compared with cover crops terminated 7–22 days prior to planting (Ruis et al., 2023). Such decreases are often attributed to reduced plant available water, but more critical is the timing of dry periods during the cash crop period.

A concern among farmers is the soil water availability for the cash crop following termination of grass cover crops. In semi-arid regions, reductions in soil water content by the cover crops can reduce cash crop yield (Rosa et al., 2021). However, reductions in soil water due to cover crops do not always translate to reduced crop yields. A review concluded that even though cover crops reduced soil water content in 50% of cases, they had no effect on subsequent cash crop yield in 56% of locations in water-limited regions (Blanco-Canqui et al., 2022).

In summary, the effective adoption of grass cover crops into any agroecosystem is impacted by management decisions, including planting time, termination time, herbicide selection, and equipment availability. Interseeding in the early season during the V2-V6 growth stages can increase fall biomass accumulation without negatively impacting yield. However, herbicide selection is critical for successful establishment. Late season overseeding in the reproductive stages can increase spring biomass with minimal yield loss, but timely rainfall or irrigation is needed for effective establishment. Timing of termination is also an important consideration because water use by the cover crop can impact cash crop yields in dry years. Finally, availability of specialized equipment, such as a high-clearance interrow drill for early interseeding or high-clearance equipment for late-season broadcast interseeding is needed.

9 Conclusion

Grass cover crops, which include cereal rye, barley, wheat, oats, and annual ryegrass, can be a useful tool to manage soil erosion, soil properties, nutrients, weeds, pests, and diseases. Grass cover crops can improve soil physical, chemical, and biological properties, which include water infiltration, soil aggregate stability, soil C, microbial biomass, and earthworm population. Such positive effects are linked with biomass C inputs from the cover crops and having a living plant during a greater portion of the year. High biomass producing grass cover crops can reduce soil erosion and improve nutrient cycling by reducing NO_3 leaching and N_2O emissions. However, grass cover crops may have neutral to negative effects on subsequent crop yields depending upon factors including termination timing and weather conditions. If there is a negative effect, it can sometimes be attributed to drought conditions during a key period of the cash crop growing period. Integrating grass cover crops into agricultural systems is not without challenges, which include a short growing window for winter cover crops. Interseeding cover crops into standing crops and terminating at or after cash crop planting can be options to boost biomass production. Despite the potential benefits, in some regions, cover crop adoption is low, with farmers often citing time and knowledge as barriers to adoption. Without equipment changes or clear management options that work most of the time, adoption is likely to remain low. Overall, grass cover crops have potential to provide soil and weed control benefits to agricultural systems.

10 Where to look for further information

The following sources contain additional information on grass cover crops and their implications for soil and crop management:

- Clark, A. (ed.). (2007). *Managing cover crops profitably*. 3rd ed. Sustainable Agriculture Network.
- Blanco-Canqui, H. & Ruis, S. J. (2020). Cover crop impacts on soil physical properties: a review. *Soil Science Society of America Journal*, *84*(5), 1527–1576.
- Haruna, S.I., et al. (2020). Improving soil physical properties through the use of cover crops: a review. *Agrosystems, Geosciences & Environment*, 3, e20105. https://doi.org/10.1002/agg2.20105.
- Peng, Y., et al. (2024). Global synthesis of cover crop impacts on main crop yields. *Field Crops Research*, 310, 109343.
- Marcillo, G.S. and Miguez, F.E. (2017). Corn yield response to winter cover crops: an updated meta-analysis. *Soil and Water Conservation*, 72, 226–239.

11 References

Abdalla, M., et al. (2019). A critical review of the impacts of cover crops on nitrogen leaching, net greenhouse gas balance and crop productivity. *Global Change Biology*, 25, 2530–2543.

Acharya, J., et al. (2020). Cover crop rotation effects on growth and development, seedling disease, and yield of corn and soybean. *Plant Disease*, 104(3), 677–687. https://doi.org/10.1094/PDIS-09-19-1904-RE.

Acharya, P., et al. (2022). Soil profile carbon, nitrogen, and crop yields affected by cover crops in semiarid regions. *Nutrient Cycling in Agroecosystems*, 122, 191–203.

Almeida, T.F., et al. (2024). Effect of cover crop species and termination timing on corn growth and seedling disease. *Agronomy Journal*, 116, 1792–1803.

Bakker, M.G., et al. (2016). The potential for cereal rye cover crops to host corn seedling pathogens. *Phytopathology*, 106(6), 591–601. https://doi.org/10.1094/PHYTO-09-15-0214-R.

Bish, M., et al. (2021). Effects of cereal rye seeding rate on waterhemp (*Amaranthus tuberculatus*) emergence and soybean growth and yield. *Weed Technology*, 35, 838–844.

Blanco-Canqui, H. and Jhala, A.J. (2024). Planting green and ecosystem services in row crop production systems. *CABI Reviews*, 19, 1. https://doi.org/10.1079/cabireviews.2024.0029.

Blanco-Canqui, H. (2022a). Cover crops and carbon sequestration: lessons from U.S. studies. *Soil Science Society of America Journal*, 86, 501–519.

Blanco-Canqui, H. (2022b). Cover crops and soil ecosystem engineers. *Agronomy Journal*, 114, 3096–3117.

Blanco-Canqui, H. & Ruis, S.J. (2020). Cover crop impacts on soil physical properties: a review. *Soil Science Society of America Journal*, 84(5), 1527–1576.

Blanco-Canqui, H., et al. (2023). Cover crops and soil health in rainfed and irrigated corn: What did we learn after 8 years? *Soil Science Society of America Journal*, 87, 1174–1190.

Blanco-Canqui, H., et al. (2022). Can cover crops improve soil ecosystem services in water-limited environments? A review. *Soil Science Society of America Journal*, 86, 1–18.

Blanco-Canqui, H., et al. (2025). Interseeding cover crop into an irrigated sandy loam for 6 years: soil, crop, and economic response. *Agronomy Journal*, 117, e70013.

Brooker, A.P., et al. (2020). Interseeding cover crops in corn. *Agronomy Journal*, 112(1), 139–147.

Brzozowski, L.J. et al. (2023). Breeding cereal rye (Secale cereale) for quality traits. *Crop Science*, 63, 1964–1987.

Camargo Silva, G. & Bagavathiannan, M. (2023). Mechanisms of weed suppression by cereal rye cover crop: a review. *Agronomy Journal*, 115(4), 1571–1585. https://doi.org/10.1002/agj2.21347.

Carmona, G.I., et al. (2022). Impact of the timing and use of an insecticide on arthropods in cover-crop-corn systems. *Insects*, 13(4), 348. https://doi.org/10.3390/insects13040348.

Caswell, K., et al. (2019). Cover crop species and cultivars for drill-interseeding in Mid-Atlantic corn and soybean. *Agronomy Journal*, 111, 1060–1067.

Chongo, G., et al. (2001). Reaction of seedling roots of 14 crop species to Fusarium graminearum from wheat heads. *Canadian Journal of Plant Pathology*, 23(2), 132–137. https://doi.org/10.1080/07060660109506920.

Clark, A.J., et al. (2007). Effects of a grass-selective herbicide in a vetch–rye cover crop system on nitrogen management. *Agronomy Journal*, 99, 36–42.

Crawford, L.E., et al. (2018). An early–killed rye (*Secale cereale*) cover crop has potential for weed management in edamame (*Glycine max*). *Weed Science*, 66, 502–507.

Crespo, C., et al. (2024a). Thermal time and precipitation dictate cereal rye shoot biomass production. *Field Crops Research*, 315, 109473.

Crespo, C., et al. (2024b). Cover crops affect phosphorus fractions in soybean-based sequences with different phosphorus availability in Mollisols. *Soil and Tillage Research*, 240, 106096.

Curran, W.S., et al. (1994). The influence of a hairy vetch (*Vicia villosa*) cover crop on weed control and corn (*Zea mays*) growth and yield. *Weed Technology*, 8, 777–784.

Curran, W.S., et al. (2018). Evaluation of cover crops drill interseeded into corn across the Mid-Atlantic region. *Agronomy Journal*, 110(2), 435–443.

Dai, W., et al. (2024). Soil aggregate stability and erosion resistance in response to integration of cover crops and poultry litter in a no-till rainfed soybean cropping system. *Soil and Tillage Research*, 244, 106245.

de Sousa Ferreira, V., et al. (2024, July 24). *Early season interseeded cover crops in corn and the impact of residual herbicides*. UNL CropWatch. University of Nebraska–Lincoln. https://cropwatch.unl.edu/2024/early-season-interseeded-cover-crops-corn-and-impact-residual-herbicides/.

Decker, H.L., et al. (2022). Cover crop monocultures and mixtures affect soil health indicators and crop yield in the southeast United States. *Soil Science Society of America Journal*, 86, 1312–1326.

Dunbar, M.W., et al. (2016). Increased risk of insect injury to corn following rye cover crop. *Journal of Economic Entomology*, 109(4), 1691–1697. https://doi.org/10.1093/jee/tow101.

Duval, M.E., et al. (2016). Winter cover crops in soybean monoculture: effects on soil organic carbon and its fractions. *Soil and Tillage Research*, 161, 95–105.

Eshel, G., et al. (2015). Benefits of growing potatoes under cover crops in a Mediterranean climate. *Agriculture, Ecosystems & Environment*, 211, 1–9.

Fohrafellner, J., et al. (2024). Cover crops affect pool specific soil organic carbon in cropland: a meta-analysis. *European Journal of Soil Science*, 75, e13472.

Gongora, V.R.M., et al. (2022). Impact of cover crops on soil physical properties, soil loss and runoff in compacted Oxisol of Southern Brazil. *Geoderma Regional*, 31, e00577. https://doi.org/10.1016/j.geodrs.2022.e00577.

Grint, K.R., et al. (2022). Cereal rye cover crop terminated at crop planting reduces early-season weed density and biomass in Wisconsin corn–soybean production. *Agrosystems, Geosciences & Environment*, 5, e20245. https://doi.org/10.1002/agg2.20245.

Hammond, R.B. (1991). Seedcorn Maggot (Diptera: Anthomyiidae) Populations on Ohio Soybean. *Journal of the Kansas Entomological Society*, 64(2), 216–220.

Haruna, S.I., et al. (2020). Improving soil physical properties through the use of cover crops: a review. *Agrosystems, Geosciences & Environment*, 3, e20105. https://doi.org/10.1002/agg2.20105.

Hashimi, R., et al. (2023). No-tillage and rye cover crop systems improve soil water retention by increasing soil organic carbon in Andosols under humid subtropical climate. *Soil & Tillage Research*, 234, e105861. https://doi.org/10.1016/j.still.2023.105861.

Horvath, D.P., et al. (2023). Weed-induced crop yield loss: a new paradigm and new challenges. *Trends in Plant Science*, 28, 5. https://doi.org/10.1016/j.tplants.2022.12.014.

Indoria, A.K., et al. (2017). Conservation agriculture–a panacea to improve soil physical health. *Current Science*, 112, 52–61.

Inveninato Carmona, G., et al. (2019). Wheat Stem Maggot (Diptera: Chloropidae): an emerging pest of cover crop to corn transition systems. *Plant Health Progress*, 20(3), 147–154. https://doi.org/10.1094/PHP-01-19-0009-S.

Inveninato Carmona, G., et al. (2021). Does cover crop management affect arthropods in the subsequent corn and soybean crops in the United States? A systematic review. *Annals of the Entomological Society of America*, 114(2), 151–162. https://doi.org/10.1093/aesa/saaa049.

Jahanzad, E., et al. (2017). Improving yield and mineral nutrient concentration of potato tubers through cover cropping. *Field Crops Research*, 212, 45–51.

Johnson, A.M., et al. (2021). Influence of cover crop mixtures on soil health in south eastern crop production systems. *Agrosystems, Geosciences & Environment*, 4(3), e20202.

Joshi, D.R., et al. (2023). A global meta-analysis of cover crop response on soil carbon storage within a corn production system. *Agronomy Journal*, 115, 1543–1556.

Kaspar, T.C., et al. (2006). Examining changes in soil organic carbon with oat and rye cover crops using terrain covariates. *Soil Science Society of America Journal*, 70, 1168–1177.

Kim, N., et al. (2020). Do cover crops benefits soil microbiome? A meta-analysis of current research. *Soil Biology and Biochemistry*, 142, 107701.

Klopp, H.W. (2024). Effects of interseeding cover crops into corn and soybean on biomass production, grain yields and ecosystem services: a review. *Agronomy*, 14(6), 1229.

Klopp, H.W., et al. (2024). Comparing impacts of corn residue removal and amelioration practices on soil properties after 3, 6, and 10 years. *Soil Science Society of America Journal*, 88(2), 510–526.

Klopp, H.W., et al. (2023). Did cover crop or animal manure ameliorate corn residue removal effects on soil mechanical properties after 10 years? *Journal of Agriculture and Food Research*, 14, 100802.

Koch, R.L., et al. (2015). On-farm evaluation of a fall-seeded rye cover crop for suppression of soybean aphid (Hemiptera: Aphididae) on soybean. *Agricultural and Forest Entomology*, 17(3), 239–246. https://doi.org/10.1111/afe.12099.

Koehler-Cole, K., et al. (2023). Cover crop planting practices determine their performance in the US Corn Belt. *Agronomy Journal*, 115(2), 526–543.

Korucu, T., et al. (2018). Rye cover crop increases earthworm populations and reduces losses of broadcast, fall-applied, fertilizer in surface runoff. *Soil and Tillage Research*, 180, 99–106.

Kumar, V., et al. (2025). Implications of cover crop management decisions on *Amaranthus* species density and biomass in temperate cropping systems: a meta-analysis. *Weed Science*, 73, e28. https://doi.org/10.1017/wsc.2024.99.

Kumari, A., et al. (2024). Integrating cover crops and herbicides for weed control in soybean. *Weed Technology*, 38, 1–8.

Lenzi, A., et al. (2009). Effect of different cover crops on organic tomato production. *Renewable Agriculture and Food Systems*, 24(2), 92–101.

Ley, E., et al. (2023). Residual herbicides in corn and their effect on fall-planted cover crop species. *Agrosystems, Geosciences & Environment*, 6(4), e20446.

Little, S.A., et al. (2004). A preliminary study of the role of cover crops in improving soil fertility and yield for potato production. *Communications in Soil Science and Plant Analysis*, 35(3–4), 471–494.

Lundgren, J.G. and Fergen, J.K. (2010). The effects of a winter cover crop on diabrotica virgifera (Coleoptera: Chrysomelidae) populations and beneficial arthropod communities in no-till maize. *Environmental Entomology*, 39(6), 1816–1828. https://doi.org/10.1603/EN10041.

Malone, LC., et al. (2023). Linking soil health indicators to management history and soybean yield. *Field Crops Research*, 297, 108951.

Marcillo, G.S. and Miguez, F.E. (2017). Corn yield response to winter cover crops: an updated meta-analysis. *Soil and Water Conservation*, 72, 226–239.

Marcos, F.M., et al. (2023). Cereal rye cover crop seeding method, seeding rate, and termination timing effects corn development and seedling disease. *Agronomy Journal*, 115, 1356–1372.

Martinez-Feria, R.A., et al. (2016). Rye cover crop effects on maize: a system-level analysis. *Field Crops Research*, 196, 145–159.

McClelland, S.C., et al. (2021). Management of cover crops in temperate climates influences soil organic carbon stocks: a meta-analysis. Ecol. Appl., 31, e02278, https://doi.org/10.1002/eap.2278.

Mills, D.J., et al. (2002). Factors associated with foliar disease of staked fresh market tomatoes grown under differing bed strategies. *Plant Disease*, 86(4), 356–361. https://doi.org/10.1094/PDIS.2002.86.4.356.

Mirsky, S.B., et al. (2011). Timing of cover-crop management effects on weed suppression in no-till planted soybean using a roller-crimper. *Weed Science*, 59(3), 380–389.

Mohammed, Y.A., et al. (2019). Establishing winter annual cover crops by interseeding into maize and soybean. *Agronomy Journal*, 111, 1–14.

Muhammad, I., et al. (2021). Cover cropping enhances soil microbial biomass and affects microbial community structure: a meta-analysis. *Geoderma*, 381, 114696.

Neate, S.M. (1989). A comparison of controlled environment and field trials for detection of resistance in cereal cultivars to root rot caused by Rhizoctonia solani. *Plant Pathology*, 38(4), 494–501. https://doi.org/10.1111/j.1365-3059.1989.tb01442.x.

Nichols, V., et al. (2020). Cover crops and weed suppression in the U.S. Midwest: a meta-analysis and modeling study. *Agricultural & Environmental Letters*, 5, e20022.

Nunes, J., et al. (2024). Planting soybean green: how cereal rye biomass and preemergence herbicides impact *Amaranthus* spp. management and soybean yield. *Weed Science*, 72, 615–629.

Overmyer, K., et al., (2023). Influence of rye/oat cover crop in a two-year soybean production system. *Agronomy Journal*, 115, 303–313.

Palhano, M.G., et al. (2017). Cover crops suppression of Palmer amaranth (*Amaranthus palmeri*) in cotton. *Weed Technology*, 32, 60–65.

Panday, D. and Nkongolo, N.V. (2021). No tillage improved soil pore space indices under cover crop and crop rotation. *Soil Systems*, 5, 38. https://doi.org/10.3390/soilsystems5030038.

Pantoja, J.L., et al. (2015). Corn nitrogen fertilization requirement and corn–soybean productivity with a rye cover crop. *Soil Science Society of America Journal*, 79(5), 1482–1495.

Patel, S., et al. (2019). Can management practices enhance corn productivity in a rye cover crop system? *Agronomy Journal*, 111(6), 3161–3171.

Peng, Y., et al. (2024). Global synthesis of cover crop impacts on main crop yields. *Field Crops Research*, 310, 109343.

Pethybridge, S.J., et al. (2020). Rolled–crimped cereal rye residue suppresses white mold in no-till soybean and dry bean. *Renewable Agriculture and Food Systems*, 35(6), 599–607. https://doi.org/10.1017/S174217051900022X.

Pittlekow, C.M., et al. (2015). When does no-till yield more? A global meta-analysis. *Field Crops Research*, 183, 156–168.

Poeplau, C. and Don, A. (2015). Carbon sequestration in agricultural soils via cultivation of cover crops: a meta-analysis. *Agriculture, Ecosystems, and Environment*, 200, 33–41.

Poudel, P., et al. (2023). Cover crop residues decomposition and nutrient releases in a sandy ultisols of US coastal plain: impacts of termination timing. *Communications in Soil Science and Plant Analysis*, 54(17), 2394–2411.

Preza-Fontes, G., et al. (2022). Corn yield response to starter nitrogen rates following a cereal rye cover crop. *Crop, Forage & Turfgrass Management*, 8(2), e20187.

Qin, Z., et al. (2021). Assessing the impacts of cover crops on maize and soybean yield in the US Midwestern agroecosystems. *Field Crops Research*, 273, 108264.

Quinn, D.J., et al. (2023). Corn optimum nitrogen fertilizer rate and application timing when following a rye cover crop. *Field Crops Research*, 291, 108794.

Reddy, K.N. (2003). Impact of rye cover crop and herbicides on weeds, yield, and net return in narrow-row transgenic and conventional soybean (*Glycine max*). *Weed Technology*, 17(1), 28–35.

Reinhardt Piskackova, T.A., et al. (2021). Windows of action for controlling Palmer amaranth (*Amaranthus palmeri*) using emergence and phenology models. *Weed Research*, 61, 188–198.

Rillig, M.C. (2004). Arbuscular mycorrhizae, glomalin, and soil aggregation. *Canadian Journal of Soil Science*, 84(4), 355–363.

Ristaino, J.B., et al. (1997). Suppression of phytophthora blight in bell pepper by a no-till wheat cover crop. *Phytopathology*, 87(3), 242–249. https://doi.org/10.1094/PHYTO.1997.87.3.242.

Rosa, A.T., et al. (2021). Implications of cover crop planting and termination timing on rainfed maize production in semiarid cropping systems. *Field Crops Research*, 271, 108251. https://doi.org/10.1016/j.fcr.2021.108251.

Ruis, S.J. and Blanco-Canqui, H. (2017). Cover crops could offset crop residue removal effects on soil carbon and other properties: a review. *Agronomy Journal*, 109, 1785–1805.

Ruis, S.J., et al. (2023). Increasing rye cover crop biomass production after corn residue removal to balance economics and soil health. *Field Crops Research*, 302, e109076. https://doi.org/10.1016/j.fcr.2023.109076.

Ruis, S.J., et al. (2019). Cover crop biomass production in temperate agroecozones. *Agronomy Journal*, 111, 1535–1551.

Ruis, S.J., et al. (2018). Impacts of early- and late-terminated cover crops on gas fluxes. *Journal of Environmental Quality*, 47, 1426–1435.

Saha, A.K., et al. (2024). Impact of no-till, crop rotation, cover crop, and drainage on soil physical and hydraulic properties. *Soil Science Society of America Journal*, 88, 239–257.

Scavo, A., et al. (2022). The role of cover crops in improving soil fertility and plant nutritional status in temperate climates: a review. Agron. Sustain. Dev., 42(5), 1–25.

Silva, T.S., et al. (2024). Cereal rye cover crop termination at or before soybean planting has minimal effect on soybean yield across the midwestern US. *Field Crops Research*, 312, 109393.

Simon, L.M., et al. (2022). Long-term cover crop management effects on soil properties in dryland cropping systems. *Agriculture, Ecosystems & Environment*, 328, 107852.

Singh, G., et al. (2020). Decomposition and nitrogen release of cover crops in reduced-and no-tillage systems. *Agronomy Journal*, 112(5), 3605–3618.

Six, J., et al. (2006). Bacterial and fungal contributions to carbon sequestration in agroecosystems. *Soil Science Society of America Journal*, 70, 555–569.

Stephens, T., et al. (2024). Integrating fall-planted cereal rye cover crop with herbicides for reducing Palmer amaranth seed production in soybean under planting green conditions. *Agrosystems, Geosciences and Environment*, 7, e20507. https://doi.org/10.1002/agg2.20507.

Thapa, R., et al. (2018). Cover crop reduce nitrate leaching in agroecosystems: a global meta-analysis. *Journal of Environmental Quality*, 47, 1400–1411.

Toler, H.D., et al. (2019). Global meta-analysis of cotton yield and weed suppression from cover crops. *Crop Science*, 59(3), 1248–1261.

Van Eerd, L.I., et al. (2023). Influence of cover crops at the four spheres: a review of ecosystem services, potential barriers, and future directions for North America. *Science of the Total Environment*, 859, 159990.

Vestberg, M. (1990). Occurrence and pathogenicity of Pythium spp. in seedling roots of winter rye. *Agricultural and Food Science*, 62(3), 275–284. https://doi.org/10.23986/afsci.72934.

Wallace, J.M., et al. (2017). Tolerance of interseeded annual ryegrass and red clover cover crops to residual herbicides in mid-Atlantic corn cropping systems. *Weed Technology*, 31(5), 641–650.

Weisberger, D.A., et al. (2023). Do cover crops suppress weeds in the U.S. Southeast? A meta-analysis. *Weed Science*, 71(3), 244–254.

Whalen, D.M., et al. (2020). Integration of residual herbicides with cover crop termination in soybean. *Weed Technology*, 34, 11–18. https://doi.org/101017/wet.2019.111.

Wulanningtyas, H.S., et al. (2021). A cover crop and no-tillage system for enhancing soil health by increasing soil organic matter in soybean cultivation. *Soil and Tillage Research*, 205, 104749.

Chapter 14

Warm season grass cover crops

A. Sophie Westbrook, Kansas State University, USA; Jaya Nepal, The Pennsylvania State University, USA; and Matthew R. Ryan, Cornell University, USA

1 Introduction

Warm season grass cover crops are plants in the *Poaceae* family that are typically grown during the summer months or in arid areas. Several species are also commonly grown as grain and forage crops. In tropical and subtropical regions, warm season grass cover crops are used in grain and vegetable cropping systems. For example, in the Brazilian Cerrado, warm season grass cover crops are planted either at the end of the rainy season after corn (*Zea mays* L.) or soybean [*Glycine max* (L.) Merr.] harvest or at the beginning of the rainy season prior to cash crop planting (de Assis, de Freitas and Mason, 2018). In subtropical regions such as South Texas, these cover crops are incorporated into vegetable systems (Soti and Racelis, 2020). Likewise, in temperate regions, they are used between short-duration vegetable crops (Mohler and Johnson, 2009; Besançon, Wasacz, and Heckman, 2021; Wauters et al., 2021; Schoolman and Arbuckle Jr, 2022) and, to a much lesser extent, in grain cropping systems. Lower acreage of warm season grass cover crops compared with winter cover crops in the Upper Midwest and Northeast United States is because of temporal overlap with summer cash crops (Snapp et al., 2005; Moore et al., 2016; Wallander et al., 2021). Despite their limited use relative to other types of cover crops in cooler production regions, warm season grass cover crops

http://dx.doi.org/10.19103/AS.2025.0162.18

can fill an important agroecological niche and provide valuable benefits when strategically integrated into cropping systems.

In this chapter, we focus on three groups of warm season grass cover crops: (1) sorghums (tribe: *Andropogoneae*), (2) millets (tribe: *Paniceae*), and (3) teff (tribe: *Eragrosteae*). Across the three groups, nine common cover crop species are highlighted: (1) sorghum [*Sorghum bicolor* (L.) Moench], (2) sudangrass [*S. sudanense* (Piper) Stapf.], (3) sorghum sudangrass [*S. bicolor* (L.) Moench × *S. sudanense* (Piper) Stapf.], (4) Japanese millet [*Echinochloa esculenta* (A. Braun) H. Scholz], (5) proso millet (*Panicum miliaceum* L.), (6) pearl millet (*Pennisetum glaucum* Brunken), (7) foxtail millet (*Setaria italica* L.), (8) browntop millet (*Urochloa ramosa* L.), and (9) teff [*Eragrostis tef* (Zucc.) Trotter]. All nine of these warm season grass species are annuals of tropical origins that utilize the C_4 photosynthetic pathway (Table 1; Figs. 1 and 2). Although less commonly used as cover crops, several species of warm season perennial grasses have been studied for the potential to improve soil health (Blanco-Canqui, 2010; Paredes et al., 2018; Maltais-Landry et al., 2024). However, this chapter is focused on the nine annual species listed above.

1.1 Sorghums

Cover crops in the *Sorghum* genus share East African origin and are grown for different purposes across the world (Venkateswaran, Elangovan, and Sivaraj, 2019). Sorghum produces round to oval seeds that are often reddish brown (Table 1; Fig. 2). Plants develop a fibrous root system reaching 2.4 m depth (Dial, 2012), supplemented by adventitious roots from the crown and lower nodes. Mature plants of some forage varieties can exceed 4 m in height with thick, pithy stems and broad leaves (Table 1). Sudangrass is a smaller plant than sorghum with comparatively fine stems (Iowa State University and USDA NRCS, 2024d). Sorghum sudangrass hybrids are intermediate between the parental lines in many of their morphological characteristics. For instance, they are taller than sudangrass with coarser stems and sturdy roots. Sorghum sudangrass hybrids are distinguished by their high biomass production, often 4500–5500 kg ha^{-1} but potentially up to 20 000 kg ha^{-1} (Clark, 2007). This biomass production contributes to their robust performance as cover crops (Clark, 2007; Boren et al., 2022; Griffiths et al., 2022).

Although sorghum sudangrass stands out for producing the highest biomass per hectare at the lowest cost, all sorghums provide multiple potential benefits such as smothering weeds, suppressing nematodes, and alleviating subsoil compaction (Clark, 2007; Dial, 2012; Boren et al., 2022). They are able to resist drought through adaptations such as waxy leaf surfaces and vigorous root systems (Clark, 2007; Dial, 2012). Sorghum cover crops can also be used

Table 1 Characteristic traits of nine warm season grass cover crop species

		Sorghum	Sudangrass	Sorghum sudangrass
Phylogeny	Species	*Sorghum bicolor* (L.) Moench	*Sorghum sudanense* (Piper) Stapf.	*Sorghum bicolor* (L.) Moench × *Sorghum sudanense* (Piper) Stapf.
	Subfamily	Panicoideae	Panicoideae	Panicoideae
	Tribe	Andropogoneae	Andropogoneae	Andropogoneae
	Origin	Likely East Africa[1]	East Africa[7]	See parent species[7]
Morphology	Seed size (l × w)	4.0–4.7 × 2.9–4.0 mm[2]	3.5–6 × 2–2.5 mm[2]	Intermediate[10]
	Seed description	Elliptical, often red to reddish brown[3]	Oblong, pointed at one end, brown-tan[8]	See parent species
	Height	1.5–4.5 m (forage)[4]	2–3 m[9]	1.5–3.7 m[11]
	Leaf description	Linear or lance-shaped, hairless, up to 110 cm long, 10 cm wide[5]	Smaller, often 15–30 cm long, 0.8–1.2 cm wide[9]	Intermediate[12]
	Rooting depth	2.4 m[3]	Otherwise consistent with sorghum,[3] often seeded at a higher rate	Otherwise consistent with sorghum,[3] often seeded at a higher rate
Physiology	Photosynthesis	C4		
	Drought tolerance	High[6]		
	Allelochemical (if applies)	Multiple, notably sorgoleone (high confidence)[3]		
Management	Seeding depth	1.9–5.1 cm[3]		
	Seeding rate	Up to 39–56 kg/ha[3] but often lower		
	Temperature for seeding	Soil 16–21 C[3]		
	pH preference	6.0–6.5[3]		
	Potential concerns	Risk of prussic acid or nitrate toxicity to livestock[3]		
	Benefits or use	Suppress weeds and nematodes; reduce compaction[3]		

(Continued)

Table 1 (*Continued*)

		Japanese millet	Proso millet	Pearl millet
Phylogeny	Species	*Echinochloa esculenta* (A. Braun) H. Scholz	*Panicum miliaceum* L.	*Pennisetum glaucum* Brunken
	Subfamily	Panicoideae	Panicoideae	Panicoideae
	Tribe	Paniceae	Paniceae	Paniceae
	Origin	East Asia[13]	East Asia[17]	African Sahel[21]
Morphology	Seed size (l × w)	2.0–3.0 × 1.0–2.0 mm[14]	2.1–4.8 × 1.0–2.2 mm^2	2.0–3.9 × 1.0–2.1 mm^2
	Seed description	Darker than congener barnyardgrass [*Echinochloa crus-galli* (L.) P. Beauv.], no awns[15]	Shiny, smooth, oval, yellow to brown[18]	Cylindrical, white or yellow, sometimes brown or purple[21]
	Height	0.6–1.2 m[15]	0.3–1.5 m[18]	1.2–2.4 m[21]
	Leaf description	Hairless, thickened margins, 20–50 cm long by 1.2–2.5 cm wide[5]	Linear to lance-shaped, maybe hairy, 15–40 cm long by 0.7–2.5 cm wide[5]	Flat, 30–100 cm long by 2–7 cm wide[22]
	Rooting depth	0.5 m[15]	0.9 m[19]	1–1.8 m[23]
Physiology	Photosynthesis	C4	C4	C4
	Drought tolerance	Low[15]	Low[6]	Medium[6]
	Allelochemical (if applies)	Unknown	Allelopathy has been suggested (compound unknown)[20]	Potentially pallidol along with others (medium confidence)[24]
Management	Seeding depth	0.6–1.3 cm[15]	1.3–1.9 cm[18]	1.3–1.9 cm[25]
	Seeding rate	22–28 kg/ha[15]	34–45 kg/ha[18]	13–45 kg/ha[21]
	Temperature for seeding	Soil 16–21 C[16]	Soil 13–18 C[18]	Above 12 C[21]
	pH preference	4.6–7.4[15]	5.8–6.8[6]	5.5–6.8[6]
	Potential concerns	Can become weedy[15]	Can become weedy[18]	Risk of nitrate toxicity to livestock[21]
	Benefits or use	Grow in flooded soil; suppress weeds; scavenge N; support wild birds[15]	Catch crop; filter strips; support wild birds[18]	Suppress diseases; increase organic matter; reduce compaction; support wild birds[21]

		Foxtail millet	Browntop millet	Teff
Phylogeny	Species	*Setaria italica* L.	*Urochloa ramosa* L.	*Eragrostis tef* (Zucc.) Trotter
	Subfamily	Panicoideae	Panicoideae	Chloridoideae
	Tribe	Paniceae	Paniceae	Eragrosteae
	Origin	East Asia[26]	Africa to South Asia[30]	East Africa[33]
Morphology	Seed size(l × w)	1.6–2.5 × 1.1–1.7 mm^2	3.0 × 1.8 mm[31]	0.9–1.1 × 0.5–0.6 mm^2
	Seed description	Oval, convex, various colors[27]	Ellipsoid, tan[32]	Very small, white to dark brown[34]
	Height	0.6–1.5 m[27]	0.9 m[32]	0.6 m[22]
	Leaf description	Flat, usually rough, 15–45 cm long by 0.6–2 cm wide[22]	Lance-shaped, hairless, up to 25 cm long by 1.4 cm wide[32]	Linear, hairless, up to 30 cm long by 0.4 cm wide[5]
	Rooting depth	0.5 m[28]	0.6 m[32]	0.9–1.2 m[35]
Physiology	Photosynthesis	C4	C4	C4
	Drought tolerance	Low[6]	Low[6]	High[36]
	Allelochemical (if applies)	Allelopathy has been suggested (compound unknown)[29]	No evidence of allelopathy[32]	Allelopathy has been suggested (compound unknown)[37]
Management	Seeding depth	0.6–1.3 cm[27]	1.3–2.5 cm[32]	0.3–0.6 cm[36]
	Seeding rate	17–34 kg/ha[27]	16–34 kg/ha[32]	6–11 kg/ha[38]
	Temperature for seeding	Soil above 18 C[27]	Air 17–26 C[32]	Soil above 18 C[36]
	pH preference	5.5–7.0[27]	5.5–6.9[6]	5.5–6.5[39]
	Potential concerns	Risk of nitrate toxicity to livestock; can become weedy[27]	Risk of nitrate toxicity to livestock; can become weedy[32]	NA
	Benefits or use	Suppress weeds; prevent erosion; support wild birds[27]	Catch crop; suppress diseases; support wild birds[32]	Rapidly suppress weeds; reduce compaction[36]

[1] Venkateswaran, Elangovan and Sivaraj, 2019; [2] Ganhão and Dias, 2019; [3] Dial, 2012; [4] Oregon State University, 2022; [5] World Flora Online, 2025; [6] USDA NRCS, 2025; [7] Boonman, 1993; [8] Purdue University, n.d; [9] Hitchcock, 1971; [10] Heath, Metcalfe and Barnes, 1973; [11] Clark, 2007; [12] Knowles and Ottman, 1997; [13] Yabuno, 1987; [14] Sood et al., 2015; [15] Sheahan, 2014d; [16] Brown, 2017a; [17] Lu et al., 2009; [18] Sheahan, 2014f; [19] Lyon et al., 2008; [20] Ahmed and Hou, 2021; [21] Sheahan, 2014e; [22] Fish et al., 2015; [23] Ausiku et al., 2022; [24] Mantovanelli et al., 2025; [25] Lee et al., 2024; [26] European and Mediterranean Plant Protection Organization, 2025; [27] Sheahan, 2014c; [28] Matsuura et al., 2012; [29] Dong et al., 2020; [30] Royal Botanic Gardens, Kew, 2025; [31] Nagaraju et al., 2020; [32] Sheahan, 2014b; [33] Barretto et al., 2021; [34] Woldeyohannes et al., 2022; [35] Ayele, Blum and Nguyen, 2001; [36] Iowa State University and USDA NRCS, 2024e; [37] Gebrehiwot et al., 2020; [38] Björkman and Shail, 2010; [39] UN Food and Agriculture Organization, 2025.

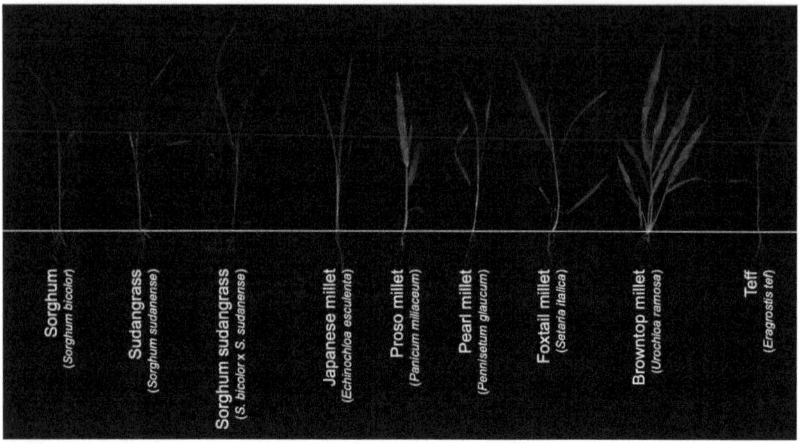

Figure 1 Seedlings of warm season grass cover crops that were grown in a greenhouse (~22°C) in plastic pots in a mix of 50% medium sand (0.5–2 mm) and 50% silt loam soil and scanned 21 days after sowing. Credits: Keith Berns provided seeds and Craig Cramer scanned plants.

as an emergency forage, although there is a risk of prussic acid and nitrate poisoning (Dial, 2012).

1.2 Millets

Millets, including Japanese millet, proso millet, pearl millet, foxtail millet, and browntop millet, comprise a diverse group of small-seeded annual plants spanning multiple genera within the *Poaceae* family. They primarily originate from Asia, except for pearl millet, which is native to Africa (Myers, 2018). While commonly grouped together based on their seed morphology and agronomic uses, these species are phylogenetically distinct (Sheahan, 2014a). Despite their taxonomic diversity, they share convergent phenotypic traits, and most are adapted to arid environments. To reduce potential confusion stemming from the word 'millet' being shared among species, they are grouped together here and discussed within the context of their use as warm season cover crops.

Among the millets, pearl millet is a relatively large plant, reaching up to 2.4 m in height with roots up to 1.8 m deep (Table 1). The deep roots contribute to higher drought tolerance in pearl millet compared with other millets. In general, millets have finer roots than sorghums, which have thicker, more wiry roots (Fig. 3). However, the fibrous roots of pearl millet are dense, and large amounts of root dry matter are produced (Sheahan, 2014a). Pearl millet also produces substantial amounts of aboveground dry matter, over 10 000 kg ha^{-1} in some varieties (Sheahan, 2014a). As a cover crop, pearl

Figure 2 Seeds of nine warm season grass cover crops. Credits: Keith Berns provided seeds and Craig Cramer assisted in photographing seeds.

millet therefore plays a role like sorghum sudangrass, valued for its high biomass production (Myers, 2018).

Among the other millets, plant height does not exceed 1.5 m and rooting depths are shallower as well (Table 1; Sheahan, 2014a). Japanese millet tolerates poorly drained soils and can thrive in cooler climates and flooded conditions (Sheahan, 2014a). Browntop millet is a similar size and can be grown in many of the same regions but prefers drier soils than Japanese millet (Myers, 2018). Proso millet is often found in comparatively dry conditions, such as the western United States (Myers, 2018). This species is more commonly used as a grain crop than a cover crop, partially because it is often the least productive of these species, but it can add diversity and ecological benefits (Myers, 2018). Foxtail

Figure 3 Roots from a mixture of millets (left) and a mixture of sorghums (right). Both mixtures were grown in a greenhouse (~22°C) in pots with 50% medium sand (0.5–2 mm) and 50% silt loam soil and photographed 8 weeks after sowing. Credits: Keith Berns provided seeds and Craig Cramer assisted with photography.

millet works well as a fast-growing cover crop after early-summer vegetable harvests (Sheahan, 2014a; Myers, 2018).

Given the wide range of environmental conditions where different millet species grow, millets are an important group of cover crops that can be utilized to provide ecosystem services across various ecoregions and cropping systems. Compared with the sorghums, they are less likely to provide benefits such as reducing subsoil compaction and suppressing soil pathogens (Griffiths et al., 2022). However, they do still provide many benefits, including protection from erosion and nutrient scavenging (Griffiths et al., 2022).

1.3 Teff

Teff, native to Ethiopia, is increasingly used as a cover crop due to its versatility and tolerance to high temperatures and drought (Iowa State University and USDA NRCS, 2024e; Oregon State University, 2025). The plants have many fine stems tillering from a single crown and narrow leaves (Iowa State University and USDA NRCS, 2024e; Oregon State University, 2025). Despite its relatively small stature (Table 1), teff develops a fine, extensively branched root system that contributes to drought tolerance once established (Brown, 2017b; Iowa State University and USDA NRCS, 2024e; Oregon State University, 2025). Seedlings begin to exhibit drought tolerance after 3 weeks (Björkman and Shail, 2010). Teff is effective at alleviating soil compaction, suppressing weeds, and controlling erosion while requiring minimal maintenance (Björkman and Shail, 2010; Iowa State University and USDA NRCS, 2024e).

2 Management considerations

As with other types of cover crops, the integration of warm season grass cover crops into cropping systems requires careful consideration. The optimal cover crop selection and management regime will depend on environmental context and the goals for using the cover crop.

2.1 Species selection

Selecting a species and variety adapted to local environment is crucial. The warm season grass cover crops described in this review exhibit limited cold tolerance and vary substantially in their tolerance to other stresses (Table 1).

Pest and disease resistance are important considerations in species and variety choice. All three groups of warm season grass cover crops can host pests such as armyworms and grasshoppers. Depending on the environment, these species may also be susceptible to a range of foliar diseases including leaf spots and rusts (Newman et al., 2010; Dial, 2012). Although pest and disease issues usually do not affect the biomass production of these warm season grass cover crops, they can harbor inoculum that can infect subsequent cash crops. Sorghums face additional pest challenges from chinch bug, sorghum midge, corn leaf aphid, corn earworm, greenbugs, sorghum webworm (Clark, 2007), and sugarcane aphid. Key problems in millets include root knot nematodes in Japanese millet; fall armyworm, leaf spot, and rust in pearl millet; leaf spot in foxtail millet, which also hosts the mite causing wheat streak mosaic; and armyworms, grasshoppers, and mung bean yellow mosaic virus in browntop millet (Sheahan, 2014a). Proso millet has fewer pest problems but can be infested by thrips, grasshoppers, and European corn borer (Sheahan, 2014a). Teff generally demonstrates good resistance to pests and diseases, with exceptions such as occasional rust infestations (Cheng et al., 2017).

2.2 Soil fertility

Soils with higher nutrient levels may have greater cover crop establishment and biomass production. Although cover crops are not typically fertilized, the application of manure or mineral fertilizer may be particularly beneficial if high biomass levels are desired. Indeed, many ecosystem services are strongly correlated with biomass production (Schipanski et al., 2014). On the other hand, high biomass production can also present challenges for residue management and may lead to nitrogen immobilization if vegetation has a high carbon-to-nitrogen (C:N) ratio (Wagger, Cabrera, and Ranells, 1998; O'Connell et al., 2015; Besançon, Wasacz, and Heckman, 2021; Allar and Maltais-Landry, 2022).

Sorghums are best adapted to fertile soils, although they can grow under low fertility conditions (Dial, 2012). Standard biomass production usually requires fertilizer applications similar to corn, including approximately 84–112 kg ha^{-1} nitrogen (Clark, 2007). Millets may have more moderate nutrient requirements and may be a better option for nutrient poor soils (Sheahan, 2014a). For teff, 45–67 kg ha^{-1} nitrogen has been recommended for good biomass production (Björkman and Shail, 2010).

2.3 Establishment

Due to their tropical origins, warm season grass cover crops are grown during summer months in temperate climates. For most of these species, planting in the spring should be delayed until soil temperatures reach 15–20 °C to ensure good germination (Table 1). For instance, a recommendation from Utah, United States, is to wait at least 2 weeks after the optimal corn planting date before planting sorghum sudangrass (Boren et al., 2022). Warm season grass cover crops can also be planted later in the summer, which can reduce the need to terminate the crop in temperate regions with freezing temperatures in the fall. If sorghums are planted within 7 weeks prior to frost, they do not need to be mowed because they will winterkill before producing seed and excessive amounts of biomass (Clark, 2007).

Most warm season grass cover crops can be either broadcast or drilled. Broadcasting can require up to twice the seeding rate of drilling, so represents the higher end of seeding rate ranges (Table 1). Seeding rate should also match production goals. Sorghum sudangrass stem diameter decreases with increased plant population, so it is possible to reduce drying time by increasing seeding rate when growing this species as a forage crop (Boren et al., 2022). High seeding rates can hasten canopy closure and increase the competitive ability of the cover crop relative to weeds.

Teff has very small seeds (Fig. 2; Table 1) that require careful planting, e.g. firm seed beds and shallow seeding depth (Björkman and Shail, 2010). Planting techniques may include specialized equipment (e.g. Brillion seeder) or broadcasting followed by rolling to maximize seed-to-soil contact (Björkman and Shail, 2010). In addition, the seeds are often coated with inert materials to increase their size for planting (Brown, 2017b).

2.4 Mowing

Warm season grass cover crops vary in their response to mowing. In sorghums, especially sorghum sudangrass, a key goal is to prevent the plants from producing large amounts of residue that will be difficult to incorporate into soils (Clark, 2007; Dial, 2012) and slow to decompose due to its high lignin content (Stefaniak et al., 2012; Ntonta et al., 2024). The recommendation is to mow or graze when stems are 0.9–1.2 m tall (Clark, 2007). This practice will promote tillering and help reduce accumulation of fibrous structural materials (Clark, 2007). At least 15 cm of stubble should remain after mowing or grazing to ensure regrowth (Clark, 2007; Dial, 2012).

Compared with the sorghums, millets and teff produce less biomass and vary in their ability to regrow after mowing. Proso millet, browntop millet, and foxtail millet exhibit little regrowth after mowing, so are typically cut only once at termination (Sheahan, 2014a, 2014f, 2014b, 2014c). Japanese millet is comparatively resilient to mowing and can recover as long as the mowing is performed at the right stage, approximately 50–90 cm height with 15–25 cm stubble left behind (Sheahan, 2014d; Myers, 2018). Pearl millet, the millet with the greatest biomass production, can also be mowed or grazed down to approximately 15–20 cm (Sheahan, 2014e, 2014a). Teff can recover from a lower mowing height, with recommendations from New York, United States, stating that teff should be mowed to 8–10 cm every 7–8 weeks to prevent seed production (Björkman and Shail, 2010).

2.5 Termination

Options for cover crop termination include tillage, herbicides, rolling-crimping, mowing, and exposure to cold temperatures (Dial, 2012; Sheahan, 2014b, 2014c, 2014d, 2014e, 2014f; Ryan et al., 2024). When terminating sorghums with tillage, the dense root system may require several passes to fully incorporate residue and prepare a fine seedbed (Clark, 2007). The timing and method of termination can also affect nutrient availability. In general, decomposition will be faster and nitrogen immobilization will be less persistent if tillage occurs while plants are still vegetative and mowed prior to incorporation (Dover, Wang, and Mcsorley, 2004; Clark, 2007). With their high biomass production and high C:N ratios (10:1 to 30:1 for sorghum sudangrass; Iowa State University and USDA NRCS, 2024c), sorghums pose significant nitrogen immobilization risks. The optimal termination methods for sorghums depend on which ecosystem services are desired. For instance, suppression of nematodes and soilborne diseases is most effective when biomass is incorporated into the soil while still green (Dover, Wang and Mcsorley, 2004; Clark, 2007), whereas weed suppression may be most effective when residues are left on the soil surface (Stapleton et al., 2010).

Proso millet, with its origins in relatively cool areas of central and east Asia, tolerates cooler temperatures than other millets (Myers, 2018). This species exhibits some herbicide resistance (Iowa State University and USDA NRCS, 2024b). Pearl millet and foxtail millet can produce high biomass that makes good surface mulches; however, in some cases, these mulches cause nitrogen immobilization or become too thick for cash crop planting (Sheahan, 2014a, 2014c, 2014e; Myers, 2018). The millets have high C:N ratios, increasing with maturity and reaching 40:1 in proso millet to 50:1 in pearl millet (Iowa State University and USDA NRCS, 2024a, 2024b). For teff, tillage and mowing are commonly used, but rolling-crimping may not provide effective termination (Brown, 2017b). However, teff will winterkill and is easy to terminate with herbicides (Björkman and Shail, 2010; Iowa State University and USDA NRCS, 2024e).

3 Potential benefits

Benefits from cover cropping are inherently context dependent and influenced by many factors, including climate, soil conditions, pests, and cropping system as well as the type of cover crop and management. This section focuses on the potential for warm season grass cover crops to improve soil health and suppress pests, which are two of the most common reasons for introducing these cover crops into cropping systems.

3.1 Soil physical properties

Warm season grass cover crops can be used to rebuild soil structure and improve physical soil health indicators. These species are distinguished from other cover crops by their rapid biomass production and extensive fibrous root systems in the top layers of the soil profile (Rosolem, Foloni, and Tiritan, 2002; Clark, 2007; Dial, 2012; Magdoff and van Es, 2021). Throughout their life cycles and especially after termination, warm season grass cover crops can contribute large amounts of organic matter to the soil (Clark, 2007; Mohler and Johnson, 2009; Magdoff and van Es, 2021; Nyabami et al., 2024). Thus, these cover crops can enhance soil microorganisms and mesofauna. The relatively high C:N ratios of these cover crops would tend to favor fungi over bacteria, compared with legume cover crops (Muhammad et al., 2021). Plant roots physically stabilize the soil and release exudates that feed soil microorganisms and bind soil particles, which promotes aggregation (Mohler and Johnson, 2009; Magdoff and van Es, 2021). By improving aggregation and porosity, warm season grass cover crops may alleviate compaction resulting from repeated tillage, heavy traffic over the field, or field operations occurring when the soil is wet (Clark, 2007; Wick et al., 2017; Magdoff and van Es, 2021).

Field studies in Brazil testing the impacts of warm season grass cover crops on soil physical properties have reported a mixture of positive and neutral effects. In an upland rice system in Goiás, Brazil, Nascente et al. (2015) compared a pearl millet cover crop with a fallow, no-till control. The cover crop had little effect on aggregation but reduced soil bulk density compared with the control throughout the soil profile and also increased soil organic carbon in aggregates at the soil surface (Nascente, Li, and Crusciol, 2015). In São Paulo, Brazil, spring cover crops including pearl millet and sorghum sudangrass were tested as alternatives to fallow between fall/winter and summer cash crops (Garcia, Li, and Rosolem, 2013). These cover crops increased root dry matter and root length density relative to fallow throughout the soil profile. This root growth resulted in increased porosity, percentage of aggregates above 2 mm, and organic carbon, relative to fallow (Garcia, Li, and Rosolem, 2013). Another study in São Paulo tested cover crops including pearl millet and sorghum within no-till corn and soybean systems (Seben, Corá, and Lal, 2014). For the most part, they reported no differences in soil physical properties among cover crops, although sorghum did achieve a greater rate of carbon sequestration than sunflower (*Helianthus annuus* L.). In a soybean production system in Mato Grosso do Sul, Brazil, pearl millet reduced bulk density and increased both macroporosity and microporosity compared with fallow at 20–30 cm depth (Teixeira et al., 2016).

Reduced compaction has been widely reported as a benefit of sorghums in particular (Clark, 2007; Grubinger, 2008; Magdoff and van Es, 2021), likely because these species are known for having strong, wiry roots that extend deeper into the soil compared with other common grass cover crops (Figs. 3 and 4). The extent to which this benefit occurs may be dependent on management: Root system development and therefore compaction reduction are thought to increase with mowing (Clark, 2007). Mowing sorghum sudangrass whenever stalks reach 0.9–1.2 tall was reported to increase root mass by five to eight times compared with no mowing (Clark, 2007). The mowing recommendation is based partially on research with sudangrass in New York, United States (Wolfe, 1998). Given that soil compaction is a widespread problem (Magdoff and van Es, 2021), we feel that quantifying the possible compaction-related benefits of warm season grass cover crops under different management regimes is an important goal for further research.

3.2 Soil water

Cover crops can affect water availability to the subsequent cash crops in several ways. In wet regions, cover crop transpiration can help remove excess water (Magdoff and van Es, 2021). However, it is important to note that cover crops can also deplete the soil profile of moisture that would otherwise be

Figure 4 Annual ryegrass (*Lolium multiflorum* Lam.) with a fibrous root system (left) and sorghum sudangrass with fewer, but thicker roots (right).

stored for the cash crop, which is a particular concern for cropping systems in arid regions where bare fallows are used to recharge water in the soil profile (Magdoff and van Es, 2021; Rose et al., 2022). In general, cover crop benefits related to water availability include reduced water runoff with surface residue, improved infiltration by increasing porosity, and enhanced water holding capacity with organic matter additions (Magdoff and van Es, 2021). Cover crops may also increase water uptake by the subsequent cash crop, e.g. by increasing mycorrhizal availability to the cash crop or enabling deep cash crop rooting by improving soil physical properties (Magdoff and van Es, 2021).

In a dryland wheat cropping system in Queensland State, Australia, Zhang et al. (2023) compared a sorghum cover crop with fallow. The cover crop was terminated at early (flag leaf emergence), mid (soft dough), or late (hard dough) stages. Benefits of the cover crop, including organic matter additions and increased arbuscular mycorrhizal fungi concentrations, increased with the length of the cover crop growing period. However, water availability at wheat sowing decreased with the length of the cover crop growing period. Early termination of the cover crop increased wheat water use efficiency, yield, and grain protein by 10%, 12%, and 5%, respectively, compared with fallow, whereas later termination had negative effects on the subsequent wheat crop (Zhang et al., 2023).

Other studies from Australia have tested millets as cover crops in arid dryland wheat cropping systems. Wunsch et al. (2017) compared Japanese millet with warm season and cool season legumes for use as spring cover crops. Japanese millet outperformed legumes in biomass production and ground cover. Although Japanese millet extracted more water, the soil accumulated water faster than it did following legume cover crops between cover crop termination and the end of the subsequent fallow period. In New South Wales, Australia, Japanese millet was screened against other cover crops (both legume and non-legume) and confirmed to have high biomass accumulation (Parvin, Condon, and Rose, 2023). It also had a dense, shallow root system that primarily used water at the soil surface, with 3–4 roots per cm^2 in the top 0–20 cm and a sharp decrease below 40 cm (Parvin, Condon, and Rose, 2023). A study conducted across 31 commercial fields over 3 years tested whether proso millet can be used in place of fallow without compromising subsequent wheat yield (Whish, Price and Castor, 2009). When the proso millet was terminated by December (30–90 days after sowing), there was no significant difference in plant available water at the time of subsequent wheat sowing. However, if the cover crop was allowed to grow to maturity, it reduced plant-available water at wheat sowing. Simulations showed that the best approach to maintain wheat yield while still receiving benefits of the cover crop was to sow the cover crop when greater than 25 mm of plant available water was present between September 1 and December 1, terminating the cover crop when it reached 50% ground cover or began flowering. This approach increased wheat yield and increased infiltration (Whish, Price, and Castor, 2009).

In a semi-arid no-till wheat system in Texas, United States, warm season cover crop mixtures including pearl millet, foxtail millet, browntop millet, sorghum sudangrass, and sorghum reduced soil water by 22–26%, measured at the time of cover crop termination (Mubvumba, DeLaune, and Hons, 2021). However, in the final 2 years of the study, soil water recharge between cover crop termination and wheat planting was two to four times greater compared with the no cover crop control, suggesting improved water infiltration and retention (Mubvumba, DeLaune, and Hons, 2021).

In some dryland environments, cover crops reduce soil water loss by creating shade and minimizing evaporation from the soil. In the arid state of New Mexico, United States, four cover crop species (Japanese millet, pearl millet, browntop millet, and sorghum sudangrass) were tested for their ability to conserve water and prevent wind erosion under full or deficit irrigation (Darapuneni et al., 2021). Sorghum sudangrass achieved the greatest ground cover (sometimes equivalent to Japanese millet and pearl millet), height, and biomass at 30, 60, and 80 days after planting. Browntop millet had the least growth and did not cover the ground effectively, resulting in the lowest soil moisture in this treatment by 80 days after planting. The dry, unprotected soil in the browntop millet treatment was also more vulnerable to wind erosion (Darapuneni et al., 2021).

In one of the few long-term studies conducted, Blanco-Canqui and Jasa (2019) reported results from a 12-year field experiment in Nebraska, United States. They showed a 31–45% increase in the proportion of large water-stable soil aggregates (2–8.0 mm) at 0–15 cm depth in plots where cereal rye (*Secale cereale* L.) and sorghum were seeded after corn and winter wheat harvest, respectively, in a corn–wheat rotation. Grass cover crops, unlike legume cover crops, also increased organic matter by 11% at 0–7.5 cm depth. However, there was no effect of cover crops on water infiltration, water retention, or plant available water. The authors suggested that high organic matter levels in the fallow treatment and limited cover crop biomass production might have contributed to these results (Blanco-Canqui and Jasa, 2019). Additional long-term research is needed to explore the impacts of warm season grass cover crops on water dynamics over time under various environmental and management conditions, given that soil-related effects of cover crops may manifest over an extended period (Wood and Bowman, 2021; Bielenberg et al., 2023).

3.3 Nutrient cycling

Cover crops can prevent nutrient loss or increase nutrient availability in several ways. The capacity for nutrient, particularly nitrogen, scavenging is one of the key advantages of grasses compared with legumes (Clark, 2007). By taking up nitrogen and also by using water, they minimize the potential for nitrate leaching (Clark, 2007; Sheahan, 2014a). The rapid growth and extensive root systems of warm season grass cover crops, especially sorghums, may strengthen these benefits compared with other cover crops (O'Connell et al., 2015). As deep-rooted crops (Table 1), the warm season species also have the potential to bring nutrients traveling with water, such as calcium and potassium, up from deep in the soil profile (Clark, 2007). As hosts for mycorrhizal fungi (Soudzilovskaia et al., 2020), they can contribute to phosphorus capture as well. This benefit is often stronger in grass cover crops compared with broadleaf species (Wick et al., 2017). The role of cover crops in preventing erosion is also important to maintaining availability of all nutrients (Darapuneni et al., 2021; Magdoff and van Es, 2021).

Multiple studies have investigated how warm season grass cover crops influence nitrogen availability to subsequent cash crops. On a sandy soil in Florida, United States, Allar and Maltais-Landry (2022) found no difference between sorghum sudangrass and weedy fallow in terms of soil inorganic nitrogen, potential net nitrogen mineralization, nitrogen uptake, or yields of subsequent organic vegetable crops. On a muck soil in Michigan, United States, sorghum sudangrass had similar biomass production and nitrogen, phosphorus, and potassium content to brassica cover crops (Wang, Ngouajio, and Warncke, 2008). Sorghum sudangrass had higher magnesium but lower calcium and sulfur compared with the brassicas. At subsequent onion (*Allium cepa* L.) planting, there were no differences in soil nutrient availability between the sorghum sudangrass, brassica cover crop, and fallow control treatments (Wang, Ngouajio, and Warncke, 2008).

In addition to nitrogen, warm season grass cover crops can alter phosphorus uptake and availability. In a field experiment conducted in Tsukuba, Japan, sorghum sudangrass had higher biomass, phosphorus uptake, phosphorus-solubilizing fungal population, total fungal population, and total bacterial population than buckwheat (*Fagopyrum esculentum* Moench), groundnut (*Arachis hypogaea* L.), crotalaria (*Crotalaria spectabilis* Roth), or the control (Karasawa and Takahashi, 2015). Sorghum sudangrass incorporation increased arbuscular mycorrhizal fungal colonization of succeeding wheat. In the sorghum sudangrass treatment, which did not receive fertilizer, phosphorus uptake and shoot dry weight of wheat were not significantly different from the fertilized control without cover crops (Karasawa and Takahashi, 2015). Similarly, millets and teff have the capacity to form arbuscular mycorrhizal fungal associations increasing phosphorus availability (Arruda et al., 2021; Mythili and Ramalakshmi, 2022; Gebremeskel et al., 2025).

3.4 Insects and disease

Cover crops, including warm season grass cover crops, can contribute to insect pest management by supporting beneficial insects that serve as predators or parasitoids of pest insects (Clark, 2007). A study in Texas, United States, found that beneficial insect abundance did not differ between cover crop and control treatments; however, a treatment with a mix of sudangrass, sunn hemp (*Crotalaria juncea* L.), and cowpea [*Vigna unguiculata* (L.) Walp.] had fewer insect pests than the control (Martinez et al., 2020). Sorghum can be used as a refuge for natural enemies to manage cotton (*Gossypium hirsutum* L.) pests (Prasifka et al., 1999). Sorghums can also serve as trap crops for pests of cotton and other crops (Tillman and Mullinix, 2004; Ratnadass et al., 2012).

An important impact of warm season grass cover crops on pest and disease management is the ability of sorghums to suppress undesirable soil organisms, especially plant-parasitic nematodes (Mojtahedi, Santo, and Ingham, 1993; Viaene and Abawi, 1998; Dover, Wang, and Mcsorley, 2004; Kratochvil et al., 2004; Khanal and Harshman, 2022; but see also MacGuidwin and Layne, 1995). This ability is largely attributable to allelochemicals, mostly dhurrin, and may reflect other factors as well (Curto et al., 2012; Paudel, Waisen, and Wang, 2021). Low- and high-dhurrin sorghums are both able to control nematodes (Dover, Wang, and Mcsorley, 2004; Djian-Caporalino et al., 2019). Crop management factors influence the extent to which nematodes are controlled. The ability of sorghum plants to suppress nematodes via allelopathy declines with increasing age of the plants at incorporation, although older plants might still help regulate nematode populations by enhancing soil biological health (Viaene and Abawi, 1998; Paudel, Waisen, and Wang, 2021). One recommendation is to cultivate sorghum cover crops for a maximum of 1 month, incorporate, and allow 1 month for biofumigation prior to cash crop planting (Djian-Caporalino et al., 2019). Termination needs to occur before frost, and mowed biomass should be incorporated immediately (Clark, 2007).

The ability of sorghums to kill nematodes, or at least not host them, can vary depending on the sorghum and nematode species and varieties (Dover, Wang, and Mcsorley, 2004). Mojtahedi et al. (1993) tested several sudangrass and sorghum sudangrass varieties as potential nematode hosts, reporting that two races of Columbia root-knot nematode could reproduce on certain varieties. However, the northern root-knot nematode could not reproduce on any of the varieties tested. Another study found that some plant-parasitic nematodes were less abundant with a sorghum sudangrass cover crop relative to velvetbean [*Mucuna pruriens* (L.) DC.], but other nematodes were less abundant with velvetbean relative to sorghum sudangrass (Crow et al., 2001). No difference between the two cover crop treatments was observed in the yield of the subsequent potato crop (Crow et al., 2001).

Like sorghums, millet cover crops may be a good choice for fields infested with plant-parasitic nematodes. Pearl millet, browntop millet, and Japanese millet have all been reported to be non-hosts for guava root-knot nematode, much like sorghum and sorghum sudangrass (Khanal and Harshman, 2022). In no-till systems covering over 5 million ha across Brazil, pearl millet has been used to manage nematode infestations in soybean and corn rotations (de Assis, de Freitas, and Mason, 2018). Pearl millet is reported to suppress at least six nematode species in these systems, including southern root-knot nematode, coffee root-knot nematode, Javanese root-knot nematode, ring nematode, reniform nematode, and root lesion nematode (de Assis, de Freitas, and Mason, 2018).

In addition to direct effects such as allelopathy, cover crops may contribute to disease management in several other ways. By reducing the need for pesticide applications (e.g. by smothering weeds), they help minimize damage to crop cuticles and beneficial microorganisms in the phyllosphere that help prevent disease (Clark, 2007). Similarly, increased levels of organic matter may help reduce plant disease incidence by supporting beneficial microorganisms in the soil (Clark, 2007). Warm season grass crops including sorghum and millet species may also be used as barrier crops, reducing pest and pathogen transmission (Ntahimpera et al., 1998; Ratnadass et al., 2012). Pearl millet and sorghum sudangrass cover crops have shown promise as carbon sources for anaerobic soil disinfestation, a technique in which carbon addition, irrigation, and a polyethylene mulch are used to create anaerobic conditions resulting in death of plant pathogens, nematodes, and weeds (Butler et al., 2012; Shrestha, Augé, and Butler, 2016).

Compared with other cover crops, warm season grass cover crops such as sorghum sudangrass carry relatively little risk of increasing disease in subsequent cash crops (Clark, 2007). Understanding mechanisms of disease suppression by these cover crops in different agroecosystems around the world will help support effective cover crop selection by farmers.

3.5 Weeds

Cover crops can contribute to weed management through a variety of mechanisms (Zimdahl, 2018). They can compete with weeds for resources while they are growing, exhibit allelopathic effects, and – when used as mulches after termination – eliminate weed germination cues or physically block emergence (Creamer et al., 1996; Teasdale, 1996; Teasdale and Mohler, 2000; Teasdale et al., 2007). Compared with broadleaf cover crops, grasses are particularly effective at exploiting these mechanisms of weed management (Osipitan et al., 2019).

Sorghums can provide strong weed suppression due to their high biomass production (McKenzie-Gopsill et al., 2022). In addition to competing for resources, sorghums produce allelochemicals. These chemicals include sorgoleone, a benzoquinone, in combination with other compounds such as dhurrin, the cyanogenic glycoside responsible for prussic acid poisoning of livestock (Weston, 1996; Czarnota et al., 2001; Weston, Alsaadawi, and Baerson, 2013; Gruss et al., 2023). Both root exudates and shoot residues can exhibit allelopathic effects against weeds (Wortman, Schmidt, and Lindquist, 2014). Allelochemical concentrations vary by species (e.g. sorghum contains more dhurrin than sudangrass; Undersander, n.d.) and variety (Kremer and Reinbott, 2021). In field studies, it is difficult to disentangle the extent to which competitive vs. allelopathic effects explain observed weed suppression; this question would be a fruitful area for further research.

When managed properly, sorghums often suppress weeds very well. In North Carolina, United States, Creamer and Baldwin (2000) tested a variety of summer cover crops, including sorghum sudangrass, sudangrass, Japanese millet, pearl millet, and foxtail millet, as well as various broadleaf species. After approximately 2 months, sorghum sudangrass had the greatest biomass, followed by pearl millet and then sudangrass. All grass species significantly reduced weed biomass relative to the weedy control, with 54–254 kg ha^{-1} weeds in grass cover crops vs. 2185 kg ha^{-1} weeds in the control (Creamer and Baldwin, 2000). In Hawaii, United States, multi-year mean weed cover was

4% in a sorghum sudangrass treatment, compared with 74% in a bare-ground control (Valenzuela and Smith, 2002). In a similar study, weed cover was 2% (sorghum sudangrass) vs. 90% (control) 6 weeks after planting (Valenzuela and Smith, 2002).

Sorghum suppression of weeds can carry into subsequent growing seasons. In Illinois, United States, mown sudangrass provided effective suppression of Canada thistle [*Cirsium arvense* (L.) Scop.], largely because it produced large amounts of mulch (Bicksler and Masiunas, 2009). Sudangrass produced 3.7 Mt ha^{-1} after one mowing and 4.4–4.9 Mt ha^{-1} after two mowings. The lignified stems of sudangrass contributed to residue persistence, with more than 3 Mt ha^{-1} surface residues remaining 3 weeks after the second mowing. Reductions in thistle shoot mass, as a percentage of initial values, continued into a second growing season in which soybeans were grown (Bicksler and Masiunas, 2009). In another study, sorghum sudangrass reduced weed density, but not weed biomass, during the onion growing season in the year after the cover crop was grown (Wang, Ngouajio, and Warncke, 2008). However, yellow nutsedge (*Cyperus esculentus* L.) density was higher in the sorghum sudangrass treatment than in the no cover crop control, and the reason for this effect was not known (Wang, Ngouajio, and Warncke, 2008). In New Jersey, United States, Besançon et al. (2021) compared sorghum sudangrass with sunn hemp for broadleaf weed suppression. Sorghum sudangrass provided better suppression of common lambsquarters (*Chenopodium album* L.) and smooth pigweed (*Amaranthus hybridus* L.) but similar suppression of hairy galinsoga (*Galinsoga quadriradiata* Ruiz & Pav.) and large crabgrass [*Digitaria sanguinalis* (L.) Scop.]. In the season after cover cropping, sorghum sudangrass reduced broadleaf weed biomass by 74% compared with sunn hemp in June and 56% compared with sunn hemp in July (Besançon, Wasacz, and Heckman, 2021).

Like sorghums, millets and teff provide strong weed suppression through rapid biomass accumulation. They are less strongly allelopathic than sorghums (with potential exceptions; Table 1). A study in Australia found that proso millet and teff reduced initial weed emergence by 76–80% and 84–85%, respectively, relative to the no cover crop control (Shabbir, Hickman, and Walsh, 2022). Weed biomass was reduced by 77–92% for proso millet and 90–95% for teff. In the subsequent wheat crop, teff reduced weed density by 73% at one site but was not significantly different from the control at the other site (Shabbir, Hickman, and Walsh, 2022). In another study, sorghum and Japanese millet significantly suppressed the growth of jungle rice [*Echinochloa colona* (L.) Link], reducing leaf and tiller number per plant by >85% compared with no cover crop (Matloob and Chauhan, 2021). Studies in Ohio, United States, on several teff varieties demonstrated reductions in Canada thistle biomass compared with the no cover crop control in one site-year but no effect in the other site-year (Wedryk and Cardina, 2012). Most teff varieties reduced Canada thistle ground cover spread in at least one site-year. Most teff varieties also reduced annual weed biomass in both site-years, and all varieties reduced annual weed ground cover spread in both site-years compared with the control (Wedryk and Cardina, 2012). Variety-focused studies like this one are relatively rare in the literature on warm season grass cover crops and would be a valuable avenue for exploration in additional species.

4 Challenges and limitations

In temperate climates, a significant limitation of warm season grass cover crops is that germination and growth require warm temperatures and the grasses do not tolerate frost (Clark, 2007; Dial, 2012; Sheahan, 2014a; Scavo et al., 2022; Iowa State University and USDA NRCS, 2024e). These cover crops are therefore grown mostly in the summer, often leading to temporal overlap with spring-planted cash crops. Some possible strategies include planting warm season grass cover crops after harvest of cool season crops such as winter wheat or integrating them into short-season vegetable rotations. In tropical climates, temperatures may be suitable for warm season grass cover crops year-round (Clark, 2007).

In addition to temperature constraints, moisture constraints may hamper cover crop establishment and growth, especially in dry climates such as southern Australia (Rose et al., 2022). In this region, a modeling exercise based on historical weather data showed that summer cover crop sowing opportunities existed in only 16–65% of years, depending on location (Rose et al., 2022). In areas with infrequent cover crop sowing opportunities, alternatives to cover cropping such as retaining cash crop stubble may be a better option. Even in wetter climates, moisture constraints may compromise establishment for less drought-tolerant species such as Japanese millet (Sheahan, 2014d).

Moisture constraints are a concern not only because they may inhibit cover crop development but also because water use by a cover crop can compromise water availability to the subsequent cash crop. This concern is particularly pressing when cover crops replace fallow periods intended to store water for cash crops (Magdoff and van Es, 2021; Rose et al., 2022). In general, an early cover crop termination time is crucial to minimizing water depletion and stress to the cash crop (Whish, Price, and Castor, 2009; Zhang et al., 2023). Other strategies include selecting cover crops with less capacity for water extraction deep in the soil profile, such as Japanese millet (Wunsch, Bell, and Bell, 2017; Parvin, Condon, and Rose, 2023), and planting cover crops only in years with higher soil moisture.

Another potential challenge is that the high C:N ratios of warm season grass cover crops could lead to nitrogen immobilization (Wagger, Cabrera, and Ranells, 1998; Iowa State University and USDA NRCS, 2024a, 2024b,

2024c, 2024d, 2024e). The risk of nitrogen immobilization is higher with plants that also have high concentrations of structural carbohydrates and lignin (Wagger, Cabrera, and Ranells, 1998). Particularly in sorghums, large amounts of high-carbon residue, containing high levels of structural materials, can tie up nitrogen for long time periods (Clark, 2007; Stefaniak et al., 2012; Ntonta et al., 2024). Large amounts of residue can also interfere with subsequent cash crop planting and delay soil warming. Management strategies to address nitrogen immobilization include supplemental fertilizer, although this approach can be cost prohibitive and less feasible for low-input or organic farming systems where synthetic fertilizers are limited. Other strategies include terminating the grass cover crop earlier or adding legume cover crops, either simultaneously as a mixture or sequentially (Fageria, Baligar, and Bailey, 2005; Clark, 2007). To accelerate sorghum decomposition, one can mow or graze stalks during the vegetative period, reduce the residue size as much as possible at termination, then immediately incorporate the residue fully (Clark, 2007). In no-till systems, the incorporation step can be skipped, but the points about mowing and minimizing residue size still apply (Clark, 2007).

Other potential drawbacks of warm season grass cover crops exist. Allelochemicals from sorghum species have the potential to compromise subsequent cash crop yields for crops such as cabbage (*Brassica oleracea* L.) and lettuce (*Lactuca sativa* L.) (Finney et al., 2009; Kruse and Nair, 2016). Regarding pest issues, the introduction of any new management technique may shift the composition of the pest community rather than eliminating pests altogether. For example, cover cropping practices that leave large amounts of residue might be good for suppressing small-seeded annual weeds, but can exacerbate problems with pathogens that thrive in the residue or weed species, including perennials, that can grow through the residue (Teasdale, 1998; Teasdale et al., 2007). Long-term pest management therefore requires continuous adjustment in response to emerging issues. It is often beneficial to diversify the cropping system, e.g. incorporating cover crops of different functional types, rather than relying on warm season grass cover crops alone. Lastly, the costs, time commitment, and uncertainty associated with cover cropping sometimes discourage adoption by farmers who may already be operating on tight margins in a risky business.

5 Case study

Weed control is a major barrier to adoption of organic no-till grain cropping systems. This case study describes an experiment conducted in two fields in New York, United States, in which winter wheat was no-till planted into rolled-crimped cover crops (Ryan et al., 2024). The experimental factors were cover crop species and cover crop planting date. Cover crop species were buckwheat, radish (*Raphanus sativus* L.), sorghum sudangrass, soybean, sunn hemp, and a control without cover crops. Cover crop planting dates were July 12 ('early-planted'), July 30 ('mid-planted'), and August 20 ('late-planted'). On September 29, cover crops in all treatments were terminated with a roller-crimper and the control treatment was chisel plowed and cultimulched. Winter wheat was planted on the same date in all treatments.

Cover crop biomass at termination showed a three-way interaction between field, cover crop species, and cover crop planting date. Early-planted sorghum sudangrass had the greatest biomass (857 g m^{-2} for one field and 669 g m^{-2} for the other field). Cover crop biomass was lowest in the late-planted sunn hemp treatment (25 g m^{-2}) or the late-planted soybean treatment (22 g m^{-2}).

Weed biomass at wheat planting also showed a three-way interaction. Among cover crop treatments, it was greatest in the early-planted treatments (6–245 g m^{-2}). All mid-planted treatments (2–58 g m^{-2}) and late-planted treatments (2–68 g m^{-2}) suppressed weed biomass relative to the control (225–277 g m^{-2}). These findings likely reflect the fact that weeds had more time to grow in the early-planted and control treatments than the mid-planted and late-planted treatments. Sorghum sudangrass had low weed biomass at wheat planting in all six combinations of planting date and field (3–31 g m^{-2}).

Wheat seedling density, measured 31 days after planting, was influenced by a two-way interaction between cover crop species and cover crop planting date. Among cover crops, wheat seedling density was lowest in early-planted sorghum sudangrass (243 seedlings m^{-2}) and highest in early-planted buckwheat (333 seedlings m^{-2}). Wheat seedling density was 236 seedlings m^{-2} in the control, perhaps reflecting poor soil conditions formed when rain followed chisel plowing at wheat planting.

The three-way interaction influenced weed biomass at wheat harvest. Early-planted sorghum sudangrass was the only treatment that reduced weed biomass at wheat harvest relative to the control in both fields. Early-planted buckwheat, mid-planted radish, and early-planted sunn hemp reduced weed biomass at wheat harvest relative to the control in one field. Overall, weed biomass at wheat harvest decreased with increasing cover crop biomass.

Wheat grain yield was affected by the three-way interaction as well. In one field, wheat yield was lowest in the early-planted sorghum sudangrass treatment (286 g m^{-2}), and this treatment was the only one that differed from the control. In the other field, wheat yield was lowest in the late-planted sorghum sudangrass treatment (175 g m^{-2}) and most cover crop treatments, except early-planted soybean and early-planted buckwheat, reduced yield relative to the control. No cover crop treatment significantly increased wheat yield relative to the control.

Findings demonstrate both the potential of warm season grass cover crops for weed suppression and associated challenges. Sorghum sudangrass was distinguished by its high biomass production and provided weed suppression in all planting date treatments. In the early-planted treatment, sorghum sudangrass residue provided effective weed suppression throughout the wheat growing season, outperforming the chisel-plowed control. However, this treatment also had lower wheat seedling density than other cover crop treatments. Reduced nitrogen availability and negative plant-soil feedbacks likely contributed to reduced wheat grain yield after a sorghum sudangrass cover crop (Ryan et al., 2024).

6 Variety development

Thus far, breeding efforts for sorghums, millets, and teff have focused primarily on grain and forage applications of these crops rather than cover crop applications. Nevertheless, previous and ongoing breeding projects provide a solid foundation for improving the key traits of warm season grass cover crops.

Breeding programs for sorghums are well established and have focused on key traits such as abiotic stress resistance, disease and insect pest resistance, productivity, male sterility, grain and forage quality, and eliminating dhurrin (Chen et al., 2019; Hao et al., 2021; Gruss et al., 2023, 2024). Millets vary in the availability of genetic and genomic resources; browntop millet and Japanese millet are examples of species for which breeding programs are newer (Bhavani et al., 2024; Pradhan et al., 2024). In pearl millet, breeding targets include stress tolerance, plant stature, maturity, photoperiod response, male sterility, and apomixis (Serba and Yadav, 2016; Serba et al., 2017). There is also interest in stress tolerance in other millets, such as foxtail millet (Moharil et al., 2019; Panchal, Singh, and Prasad, 2023). Proso millet breeding may alleviate limitations to grain production, which include relatively low yields and problems with lodging and seed shattering (Santra, Khound, and Das, 2019). Teff is a crop with high levels of germplasm diversity, including a wide range of landraces (Woldeyohannes et al., 2022). This diversity may be leveraged to improve drought resistance, pest resistance, lodging resistance, phenology, plant stature, yield-related panicle traits, and seed traits (Woldeyohannes et al., 2022). New teff varieties include drought-tolerant and lodging-resistant, semi-dwarf varieties (Cannarozzi et al., 2018; Numan et al., 2021; Roeber, 2023).

Most of the breeding underway for grain and forage applications will also lead to better cover performance. In particular, given the hot and dry conditions under which warm season grass cover crops are often grown, improved heat and drought tolerance is almost universally helpful if it does not come at the expense of biomass production or worsen soil water depletion. For a few traits, usefulness may depend on cover cropping context. For example, the desirability of dhurrin-free sorghum as a cover crop would depend on whether a farmer is more interested in disease suppression or a potential emergency forage resource.

In addition to stress tolerance and biomass accumulation, traits that would be helpful for cover crop applications might include faster development, increased water use efficiency, reduced subsoil water depletion, enhanced root architecture (Griffiths et al., 2022), either increased or reduced allelopathic activity, and lower C:N ratios or less lignification to minimize periods of nitrogen immobilization. As gene editing technology is becoming available in all three groups of warm season grass cover crops (Assem et al., 2023; Roeber, 2023; Sapara et al., 2024), it may soon become easier to achieve these breeding goals and develop varieties specifically for cover cropping.

7 Conclusion and future directions

Warm season grass cover crops, including sorghums, millets, and teff, are valuable tools for diversifying cropping systems and enhancing agricultural system resilience. These crops can be strategically employed to provide rapid ground cover, improve soil health, and enhance pest management. However, warm season grass cover crops may not be feasible in all environments or cropping systems.

A top priority for researchers studying warm season grass cover crops should be to enhance decision support resources available to farmers. It would be particularly valuable to develop models based on large-scale historical or simulated data sets to understand how factors such as weather, soil conditions, and cropping system management may interact to produce desirable or undesirable cover cropping outcomes. Modeling approaches can also be used to identify regions where a particular cover crop species or group might be more or less suitable (Rose et al., 2022). To further support farmer decision-making, researchers can pursue quantification of ecosystem services (Schipanski et al., 2014; Finney et al., 2017), including services such as carbon sequestration that may be variable or difficult to assess (Poeplau and Don, 2015; Blanco-Canqui, 2022). Economic analyses are also crucial and should be conducted when possible (Bergtold et al., 2019; Van Eerd et al., 2023).

In the field, we recommend capitalizing on the increasing availability of precision agriculture technologies, which are beginning to be used for cover crop as well as cash crop management (Melkonian et al., 2017; Roth and Streit, 2018; Kumar et al., 2023; Vavlas et al., 2024). Because some of the key barriers to implementation of warm season grass cover crops relate to resource availability in the subsequent cash crop, this challenge might be alleviated with a combination of sensing (remote or proximal) and use of variable-rate seeding, fertilization, or irrigation.

Warm season grass cover crops have the potential to help address major challenges, including pesticide resistance and soil degradation, but there is also significant variability in their reported effects. Thus, there is a need for more systems-level field research across diverse environments and multi-year timeframes. Research conducted with farmers and other stakeholders as a community of practice will help increase the utility and adoption of warm season grass cover crops around the world.

8 Where to look for further information

The following books, both freely available online, provide a good general overview.

- Clark, A. (Ed.). (2007). *Managing cover crops profitably* (3rd ed.). Sustainable Agriculture Research and Education (Handbook Series, bk. 9). Available at: https://www.sare.org/wp-content/uploads/Managing-Cover-Crops-Profitably.pdf.
- Magdoff, F. and van Es, H. (2021). *Building soils for better crops: Ecological management for healthy soils* (4th ed.). Sustainable Agriculture Research and Education (Handbook Series, bk. 10). Available at: https://www.sare.org/resources/building-soils-for-better-crops/.

USDA NRCS resources are also available for most major cover crops.

- Dial, H. (2012). *Plant guide for sorghum (Sorghum bicolor L.)*. USDA-Natural Resources Conservation Service, Tucson Plant Materials Center. Available at: https://plants.usda.gov/DocumentLibrary/plantguide/pdf/pg_sobi2.pdf.
- Iowa State University & USDA NRCS. (2024). *Teff cover crop fact sheet* (CROP 3248). Available at: https://www.nrcs.usda.gov/sites/default/files/2024-08/Teff%20Crop%20Fact%20Sheet.pdf.
- Sheahan, C.M. (2014). *Comparison of five millet species for conservation use in the United States* (Plant Materials Technical Note No. 2). U.S. Department of Agriculture, Natural Resources Conservation Service. Available at: https://www.nrcs.usda.gov/plantmaterials/njpmctn12841.pdf.

9 References

Ahmed, A.I. and Hou, F.J. (2021) "Allelopathic effects of proso millet (*Panicum miliaceum* L.) extracts on seed germination and seedling growth of alfalfa and vetch," *Allelopathy Journal*, 53(1), pp. 219–230. Available at: https://doi.org/10.26651/allelo.j/2021-53-2-1339.

Allar, J. and Maltais-Landry, G. (2022) "Limited benefits of summer cover crops on nitrogen cycling in organic vegetable production," *Nutrient Cycling in Agroecosystems*, 122(2), pp. 119–138. Available at: https://doi.org/10.1007/s10705-021-10189-8.

Arruda, B. et al. (2021) "Cover crop species and mycorrhizal colonization on soil phosphorus dynamics," *Rhizosphere*, 19, 100396. Available at: https://doi.org/10.1016/j.rhisph.2021.100396.

Assem, S.K. et al. (2023) "Development of an in vitro regeneration system from immature inflorescences and CRISPR/Cas9-mediated gene editing in sudangrass," *Journal of Genetic Engineering and Biotechnology*, 21(1), 58. Available at: https://doi.org/10.1186/s43141-023-00517-6.

Ausiku, P.A. et al. (2022) "Crop model parameterisation of three important pearl millet varieties for improved water use and yield estimation," *Plants*, 11(6), 806. Available at: https://doi.org/10.3390/plants11060806.

Ayele, M., et al. (2001) "Diversity for osmotic adjustment and root depth in tef [*Eragrostis tef* (Zucc) Trotter]," *Euphytica*, 121(3), pp. 237–249. Available at: https://doi.org/10.1023/A:1012099914738.

Barretto, R. et al. (2021) "Teff (*Eragrostis tef*) processing, utilization and future opportunities: a review," *International Journal of Food Science and Technology*, 56(7), pp. 3125–3137. Available at: https://doi.org/10.1111/ijfs.14872.

Bergtold, J.S. et al. (2019) "A review of economic considerations for cover crops as a conservation practice," *Renewable Agriculture and Food Systems*, 34(1), pp. 62–76. Available at: https://doi.org/10.1017/S1742170517000278.

Besançon, T.E., et al. (2021) "Weed suppression, nitrogen availability, and cabbage production following sunn hemp or sorghum-sudangrass," *HortTechnology*, 31(4), pp. 439–447. Available at: https://doi.org/10.21273/HORTTECH04811-21.

Bhavani, P. et al. (2024) "Brown-top millet: an overview of breeding, genetic, and genomic resources development for crop improvement," *Planta*, 260(1), 10. Available at: https://doi.org/10.1007/s00425-024-04446-7.

Bicksler, A.J. and Masiunas, J.B. (2009) "Canada thistle (*Cirsium arvense*) suppression with buckwheat or sudangrass cover crops and mowing," *Weed Technology*, 23(4), pp. 556–563. Available at: https://doi.org/10.1614/WT-09-050.1.

Bielenberg, H. et al. (2023) "Cover crop composition in long-term no-till soils in semi-arid environments do not influence soil health measurements after one year," *Soil Science Society of America Journal*, 87(3), pp. 528–540. Available at: https://doi.org/10.1002/saj2.20523.

Björkman, T. and Shail, J.W. (2010) "Cornell cover crop guide for teff," *Cornell University*, Ver. 1.100716. Available at: https://covercrop.org/wp-content/uploads/teff.pdf.

Blanco-Canqui, H. (2010) "Energy crops and their implications on soil and environment," *Agronomy Journal*, 102(2), pp. 403–419. Available at: https://doi.org/10.2134/agronj2009.0333.

Blanco-Canqui, H. (2022) "Cover crops and carbon sequestration: lessons from U.S. studies," *Soil Science Society of America Journal*, 86(3), pp. 501–519. Available at: https://doi.org/10.1002/saj2.20378.

Blanco-Canqui, H. and Jasa, P.J. (2019) "Do grass and legume cover crops improve soil properties in the long term?," *Soil Science Society of America Journal*, 83(4), pp. 1181–1187. Available at: https://doi.org/10.2136/sssaj2019.02.0055.

Boonman, J.G. (1993) *East Africa's grasses and fodders: Their ecology and husbandry*. Kluwer Academic Publishers.

Boren, D. et al. (2022) *Sorghum-sudangrass production guide*. Utah State University Extension. Available at: https://digitalcommons.usu.edu/cgi/viewcontent.cgi?article=3314&context=extension_curall.

Brown, R. (2017a) *Using Japanese millet as a summer cover crop*. University of Rhode Island. Available at: https://digitalcommons.uri.edu/riaes_bulletin/22/.

Brown, R. (2017b) *Using teff as a summer cover crop*. University of Rhode Island. Available at: https://digitalcommons.uri.edu/cgi/viewcontent.cgi?article=1023&context=riaes_bulletin.

Butler, D.M. et al. (2012) "Exploring warm-season cover crops as carbon sources for anaerobic soil disinfestation (ASD)," *Plant and Soil*, 355(1–2), pp. 149–165. Available at: https://doi.org/10.1007/s11104-011-1088-0.

Cannarozzi, G. et al. (2018) "Technology generation to dissemination: lessons learned from the tef improvement project," *Euphytica*, 214(2), p. 31. Available at: https://doi.org/10.1007/s10681-018-2115-5.

Chen, J. et al. (2019) "Identification of the first nuclear male sterility gene (*Male-sterile 9*) in sorghum," *The Plant Genome*, 12(3), 190020. Available at: https://doi.org/10.3835/plantgenome2019.03.0020.

Cheng, A. et al. (2017) "Diversifying crops for food and nutrition security – a case of teff," *Biological Reviews*, 92(1), pp. 188–198. Available at: https://doi.org/10.1111/brv.12225.

Clark, A. (ed.) (2007) *Managing cover crops profitably*. 3rd ed. Sustainable Agriculture Research and Education (Handbook series, bk. 9). Available at: https://www.sare.org/wp-content/uploads/Managing-Cover-Crops-Profitably.pdf.

Creamer, N.G. et al. (1996) "Mechanisms of weed suppression in cover crop-based production systems," *HortScience*, 31(3), pp. 410–413. Available at: https://doi.org/10.21273/HORTSCI.31.3.410.

Creamer, N.G. and Baldwin, K.R. (2000) "An evaluation of summer cover crops for use in vegetable production systems in North Carolina," *HortScience*, 35(4), pp. 600–603. Available at: https://doi.org/10.21273/HORTSCI.35.4.600.

Crow, W.T. et al. (2001) "Effect of sorghum-sudangrass and velvetbean cover crops on plant-parasitic nematodes associated with potato production in Florida," *The Journal of Nematology*, 33(4S), pp. 285–288.

Curto, G. et al. (2012) "Evaluation of the activity of dhurrin and sorghum towards *Meloidogyne incognita*," *Nematology*, 14(6), pp. 759–769. Available at: https://doi.org/10.1163/156854112X627291.

Czarnota, M.A. et al. (2001) "Mode of action, localization of production, chemical nature, and activity of sorgoleone: a potent PSII inhibitor in *Sorghum* spp. root exudates," *Weed Technology*, 15(4), pp. 813–825. Available at: https://doi.org/doi:10.1614/0890-037X(2001)015[0813:MOALOP]2.0.CO;2.

Darapuneni, M.K. et al. (2021) "Growth characteristics of summer cover crop grasses and their relation to soil aggregate stability and wind erosion control in arid southwest," *Applied Engineering in Agriculture*, 37(1), pp. 11–23. Available at: https://doi.org/10.13031/aea.13972.

de Assis, R.L., et al. (2018) "Pearl millet production practices in Brazil: a review," *Experimental Agriculture*, 54(5), pp. 699–718. Available at: https://doi.org/10.1017/S0014479717000333.

Dial, H. (2012) "Plant guide for sorghum (*Sorghum bicolor* L.)," *USDA-Natural Resources Conservation Service, Tucson Plant Materials Center, Tucson, AZ*. Available at: https://plants.usda.gov/DocumentLibrary/plantguide/pdf/pg_sobi2.pdf.

Djian-Caporalino, C. et al. (2019) "Evaluating sorghums as green manure against root-knot nematodes," *Crop Protection*, 122, pp. 142–150. Available at: https://doi.org/10.1016/j.cropro.2019.05.002.

Dong, S.-Q. et al. (2020) "Allelopathic effects of water extracts from different parts of foxtail millet straw on three kinds of weeds," *Ying Yong Sheng tai xue bao = The Journal of Applied Ecology*, 31(7), pp. 2243–2250. Available at: https://doi.org/10.13287/j.1001-9332.202007.009.

Dover, K., et al. (2004) *Nematode management using sorghum and its relatives*. University of Florida IFAS Extension Bulletin ENY-716. Available at: http://doi.org/10.32473/edis-in531-2004.

European and Mediterranean Plant Protection Organization (2025) *EPPO global database*. Available at: https://gd.eppo.int.

Fageria, N.K., et al. (2005) "Role of cover crops in improving soil and row crop productivity," *Communications in Soil Science and Plant Analysis*, 36(19–20), pp. 2733–2757. Available at: https://doi.org/10.1080/00103620500303939.

Finney, D.M. et al. (2009) "Sorghum sudangrass as a summer cover and hay crop for organic fall cabbage production," *Renewable Agriculture and Food Systems*, 24(3), pp. 225–233. Available at: https://doi.org/10.1017/S174217050999007X.

Finney, D.M. et al. (2017) "Ecosystem services and disservices are bundled in simple and diverse cover cropping systems," *Agricultural & Environmental Letters*, 2(1), 170033. Available at: https://doi.org/10.2134/ael2017.09.0033.

Fish, L. et al. (2015) *Identification guide to southern African grasses: an identification manual with keys, descriptions and distributions*. South African National Biodiversity Institute.

Ganhão, E. and Dias, L.S. (2019) "Seed volume dataset–an ongoing inventory of seed size expressed by volume," *Data*, 4(2), p. 61. Available at:https://doi.org/10.3390/data4020061.

Garcia, R.A., et al. (2013) "Soil organic matter and physical attributes affected by crop rotation under no-till," *Soil Science Society of America Journal*, 77(5), pp. 1724–1731. Available at: https://doi.org/10.2136/sssaj2012.0310.

Gebrehiwot, H.G. et al. (2020) "Allelopathic potential of teff varieties and effect on weed growth," *Agronomy*, 10(6), p. 854. Available at: https://doi.org/10.3390/agronomy10060854.

Gebremeskel, K. et al. (2025) *Arbuscular mycorrhizal fungi reduce lodging of teff [Eragrostis tef (Zucc.) Trotter] under N fertilizer applications*. Arid Land Research and Management. Available at: https://doi.org/10.1080/15324982.2025.2518417.

Griffiths, M. et al. (2022) "Optimisation of root traits to provide enhanced ecosystem services in agricultural systems: a focus on cover crops," *Plant, Cell & Environment*, 45(3), pp. 751–770. Available at: https://doi.org/10.1111/pce.14247.

Grubinger, V.P. (2008) *Sorghum-sudangrass, a vigorous cover crop*. University of Vermont Extension. Available at: https://scholarworks.uvm.edu/cgi/viewcontent.cgi?article=1027&context=extfac.

Gruss, S.M. et al. (2023) "Dhurrin stability and hydrogen cyanide release in dried sorghum samples," *Field Crops Research*, 291, p. 108764. Available at: https://doi.org/10.1016/j.fcr.2022.108764.

Gruss, S.M. et al. (2024) "Preference of dhurrin-free sorghum by ewes," *Crop, Forage & Turfgrass Management*, 10(1), p. e20259. Available at: https://doi.org/10.1002/cft2.20259.

Hao, H. et al. (2021) "Sorghum breeding in the genomic era: opportunities and challenges," *Theoretical and Applied Genetics*, 134(7), pp. 1899–1924. Available at: https://doi.org/10.1007/s00122-021-03789-z.

Heath, M.E., et al. (1973) *Forages*. 3rd ed. Iowa State University Press.

Hitchcock, A.S. (1971) *Manual of the grasses of the United States*. 2nd ed. Courier Corporation.

Iowa State University and USDA NRCS (2024a) "Pearl millet cover crop fact sheet." Available at: https://www.nrcs.usda.gov/sites/default/files/2024-07/Pearl%20Millet%20Crop%20Fact%20Sheet.pdf.

Iowa State University and USDA NRCS (2024b) "Proso millet cover crop fact sheet." Available at: https://www.nrcs.usda.gov/sites/default/files/2024-07/Proso%20Millet%20Crop%20Fact%20Sheet_0.pdf.

Iowa State University and USDA NRCS (2024c) "Sorghum-sudangrass cover crop fact sheet." Available at: https://www.nrcs.usda.gov/sites/default/files/2024-05/Sorghum%20Sudangrass%20Cover%20Crop%20Fact%20Sheet.pdf.

Iowa State University and USDA NRCS (2024d) "Sudangrass cover crop fact sheet." Available at: https://www.nrcs.usda.gov/sites/default/files/2024-08/Sudangrass%20Crop%20Fact%20Sheet.pdf.

Iowa State University and USDA NRCS (2024e) "Teff cover crop fact sheet." Available at: https://www.nrcs.usda.gov/sites/default/files/2024-08/Teff%20Crop%20Fact%20Sheet.pdf.

Karasawa, T. and Takahashi, S. (2015) "Introduction of various cover crop species to improve soil biological P parameters and P uptake of the following crops," *Nutrient Cycling in Agroecosystems*, 103(1), pp. 15–28. Available at: https://doi.org/10.1007/s10705-015-9715-4.

Khanal, C. and Harshman, D. (2022) "Evaluation of summer cover crops for host suitability of *Meloidogyne enterolobii*," *Crop Protection*, 151, p. 105821. Available at: https://doi.org/10.1016/j.cropro.2021.105821.

Knowles, T.C. and Ottman, M.J. (1997) *Sudangrass hay production in the irrigated deserts of Arizona and California*. University of Arizona Cooperative Extension. Available at: https://repository.arizona.edu/handle/10150/147028?show=full.

Kratochvil, R.J. et al. (2004) "Evaluation of crop rotation and other cultural practices for management of root-knot and lesion nematodes," *Agronomy Journal*, 96(5), pp. 1419–1428. Available at: https://doi.org/10.2134/agronj2004.1419.

Kremer, R.J. and Reinbott, T.M. (2021) "Allelopathy of sorghum residues on weed establishment is affected by hybrid, phenolic acid contents and environment," *Journal of Plant Sciences*, 9(1), 25. Available at: https://doi.org/10.11648/j.jps.20210901.14.

Kruse, R. and Nair, A. (2016) "Summer cover crops and lettuce planting time influence weed population, soil nitrogen concentration, and lettuce yields," *HortTechnology*, 26(4), pp. 409–416. Available at: https://doi.org/10.21273/HORTTECH.26.4.409.

Kumar, V. et al. (2023) "Cover crop termination options and application of remote sensing for evaluating termination efficiency," *PLoS One*, 18(4), p. e0284529. Available at: https://doi.org/10.1371/journal.pone.0284529.

Lee, D. et al. (2024) *Pearl millet for grain.* University of Georgia Bulletin 1216. Available at: https://secure.caes.uga.edu/extension/publications/files/pdf/B%201216_4.PDF.

Lu, H. et al. (2009) "Earliest domestication of common millet (*Panicum miliaceum*) in East Asia extended to 10,000 years ago," *Proceedings of the National Academy of Sciences*, 106(18), pp. 7367–7372. Available at: https://doi.org/10.1073/pnas.0900158106.

Lyon, D.J. et al. (2008) *Producing and marketing proso millet in the Great Plains.* University of Nebraska Lincoln Extension (EC137). Available at: https://www.ars.usda.gov/research/publications/publication/?seqNo115=227495.

MacGuidwin, A.E. and Layne, T.L. (1995) "Response of nematode communities to sudangrass and sorghum-sudangrass hybrids grown as green manure crops," *Journal of Nematology*, 27(4S), pp. 609–616. Available at: https://www.ncbi.nlm.nih.gov/pmc/articles/PMC2619645/.

Magdoff, F. and van Es, H. (2021) *Building soils for better crops: ecological management for healthy soils.* 4th ed. Sustainable Agriculture Research and Education (Handbook series, bk. 10). Available at: https://www.sare.org/resources/building-soils-for-better-crops.

Maltais-Landry, G. et al. (2024) "Long-term integration of bahiagrass into a cover-cropped and strip-tilled peanut–cotton rotation has a limited effect on soil carbon and other soil properties," *Soil Science Society of America Journal*, 88(1), pp. 185–191. Available at: https://doi.org/10.1002/saj2.20602.

Mantovanelli, G.C., et al. (2025) "Pearl millet cover crop extract inhibits the development of the weed *Ipomoea grandifolia* by inducing oxidative stress in primary roots and affecting photosynthesis efficiency," *Plants*, 14(2), p. 222. Available at: https://doi.org/10.3390/plants14020222.

Martinez, L. et al. (2020) "Impact of cover crops on insect community dynamics in organic farming," *Agriculture*, 10(6), p. 209. Available at: https://doi.org/10.3390/agriculture10060209.

Matloob, A. and Chauhan, B.S. (2021) "Utilization of the neighborhood design to evaluate suitable cover crops and their density for *Echinochloa colona* management," *PLoS One*, 16(7), p. e0254584. Available at: https://doi.org/10.1371/journal.pone.0254584.

Matsuura, A. et al. (2012) "Effect of pre- and post-heading water deficit on growth and grain yield of four millets," *Plant Production Science*, 15(4), pp. 323–331. Available at: https://doi.org/10.1626/pps.15.323.

McKenzie-Gopsill, A. et al. (2022) "The importance of species selection in cover crop mixture design," *Weed Science*, 70(4), pp. 436–447. Available at: https://doi.org/10.1017/wsc.2022.28.

Melkonian, J. et al. (2017) "Estimating nitrogen mineralization from cover crop mixtures using the precision nitrogen management model," *Agronomy Journal*, 109(5), pp. 1944–1959. Available at: https://doi.org/10.2134/agronj2016.06.0330.

Moharil, M.P., et al. (2019) "Foxtail millet (Setaria italica L.): potential of smaller millet for future breeding", in Al-Khayri, J.M., et al. (eds.) *Advances in plant breeding strategies: Cereals.* Volume 5. Springer International Publishing, pp. 133–163.

Mohler, C.L. and Johnson, S.E. (eds.) (2009) *Crop rotation on organic farms: a planning manual.* Sustainable Agriculture Research and Education. Available at: https://www.sare.org/wp-content/uploads/Crop-Rotation-on-Organic-Farms.pdf.

Mojtahedi, H., et al. (1993) "Suppression of *Meloidogyne chitwoodi* with sudangrass cultivars as green manure," *The Journal of Nematology*, 25(2), pp. 303–311.

Moore, V.M. et al. (2016) "Cover crop adoption and intensity on Wisconsin's organic vegetable farms," *Agroecology and Sustainable Food Systems*, 40(7), pp. 693–713. Available at: https://doi.org/10.1080/21683565.2016.1181694.

Mubvumba, P., et al. (2021) "Soil water dynamics under a warm-season cover crop mixture in continuous wheat," *Soil and Tillage Research*, 206, p. 104823. Available at: https://doi.org/10.1016/j.still.2020.104823.

Muhammad, I. et al. (2021) "Cover cropping enhances soil microbial biomass and affects microbial community structure: a meta-analysis," *Geoderma*, 381, p. 114696. Available at: https://doi.org/10.1016/j.geoderma.2020.114696.

Myers, R. (2018) *Growing millets for grain, forage or cover crop use*. University of Missouri. Available at: https://extension.missouri.edu/publications/g4164.

Mythili, M. and Ramalakshmi, A. (2022) "Unraveling the distribution of AMF communities and their metabolites associated with soils of minor millets," *Rhizosphere*, 21, p. 100473. Available at: https://doi.org/10.1016/j.rhisph.2022.100473.

Nagaraju, M. et al. (2020) "Physical properties of an underutilized crop: Browntop millet (*Urochloa ramose*)," *International Journal of Chemical Studies*, 8(6), pp. 192–197. Available at: https://doi.org/10.22271/chemi.2020.v8.i6c.10768.

Nascente, A.S., et al. (2015) "Soil aggregation, organic carbon concentration, and soil bulk density as affected by cover crop species in a no-tillage system," *Revista Brasileira de Ciência do Solo*, 39, pp. 871–879. Available at: https://doi.org/10.1590/01000683rbcs20140388.

Newman, Y. et al. (2010) *Pearl millet (Pennisetum glaucum): overview and management*. University of FLorida IFAS Extension. Available at: https://edis.ifas.ufl.edu/publication/ag347.

Ntahimpera, N. et al. (1998) "Effects of a cover crop on splash dispersal of *Colletotrichum acutatum* conidia," *Phytopathology*, 88(6), pp. 536–543. Available at: https://doi.org/10.1094/PHYTO.1998.88.6.536.

Ntonta, S. et al. (2024) "Variability in decomposition rate of sorghum cultivar residues linked to lignin content," *Rhizosphere*, 29, p. 100850. Available at: https://doi.org/10.1016/j.rhisph.2024.100850.

Numan, M. et al. (2021) "From traditional breeding to genome editing for boosting productivity of the ancient grain tef [Eragrostis tef (Zucc.) Trotter]," *Plants*, 10(4), p. 628. Available at: https://doi.org/10.3390/plants10040628.

Nyabami, P. et al. (2024) "Three years of cover crops management increased soil organic matter and labile carbon pools in a subtropical vegetable agroecosystem," *Agrosystems, Geosciences & Environment*, 7(1), p. e20454. Available at: https://doi.org/10.1002/agg2.20454.

O'Connell, S. et al. (2015) "Short-term nitrogen mineralization from warm-season cover crops in organic farming systems," *Plant and Soil*, 396(1–2), pp. 353–367. Available at: https://doi.org/10.1007/s11104-015-2594-2.

Oregon State University (2022) *Sorghum*. Forage Information System. Available at: https://forages.oregonstate.edu/forages/sorghum.

Oregon State University (2025) *Teff*. Forage Information System. Available at: https://forages.oregonstate.edu/forages/teff.

Osipitan, O.A. et al. (2019) "Impact of cover crop management on level of weed suppression: a meta-analysis," *Crop Science*, 59(3), pp. 833–842. Available at: https://doi.org/10.2135/cropsci2018.09.0589.

Panchal, A., et al. (2023) "Recent advancements and future perspectives of foxtail millet genomics," *Plant Growth Regulation*, 99(1), pp. 11–23. Available at: https://doi.org/10.1007/s10725-022-00858-1.

Paredes, S.S. et al. (2018) "Perennial warm-season grass monocultures and mixtures: biomass production and soil improvement in semiarid and shallow soil conditions," *Journal of Arid Environments*, 154, pp. 82–88.

Parvin, S., et al. (2023) "Rooting depth and water use of summer cover crops in a semi-arid cropping environment," *European Journal of Agronomy*, 147, p. 126847. Available at: https://doi.org/10.1016/j.eja.2023.126847.

Paudel, R., et al. (2021) "Exploiting the innate potential of sorghum/sorghum–sudangrass cover crops to improve soil microbial profile that can lead to suppression of plant-parasitic nematodes," *Microorganisms*, 9(9), p. 1831. Available at: https://doi.org/10.3390/microorganisms9091831.

Poeplau, C. and Don, A. (2015) "Carbon sequestration in agricultural soils via cultivation of cover crops: a meta-analysis," *Agriculture, Ecosystems & Environment*, 200, pp. 33–41. Available at: https://doi.org/10.1016/j.agee.2014.10.024.

Pradhan, P.P. et al. (2024) "Genetic improvement of barnyard millet (*Echinochloa esculenta*): prospective and challenges," *Technology in Agronomy*, 4(1), p. e027. Available at: https://doi.org/10.48130/tia-0024-0024.

Prasifka, J.R. et al. (1999) "Predator conservation in cotton: using grain sorghum as a source for insect predators," *Biological Control*, 16(2), pp. 223–229. Available at: https://doi.org/10.1006/bcon.1999.0760.

Purdue University (n.d.) *Purdue forage ID*. Available at: https://ag.purdue.edu/department/agry/agry-extension/forages/forage-id/.

Ratnadass, A. et al. (2012) "Plant species diversity for sustainable management of crop pests and diseases in agroecosystems: a review," *Agronomy for Sustainable Development*, 32(1), pp. 273–303. Available at: https://doi.org/10.1007/s13593-011-0022-4.

Roeber, K. (2023) *USDA clears Danforth Center's genome edited teff*. Danforth Plant Science Center. Available at: https://www.danforthcenter.org/news/usda-clears-danforth-centers-genome-edited-teff/.

Rose, T.J. et al. (2022) "Prospects for summer cover crops in southern Australian semi-arid cropping systems," *Agricultural Systems*, 200, p. 103415. Available at: https://doi.org/10.1016/j.agsy.2022.103415.

Rosolem, C.A., et al. (2002) "Root growth and nutrient accumulation in cover crops as affected by soil compaction," *Soil and Tillage Research*, 65(1), pp. 109–115. Available at: https://doi.org/10.1016/S0167-1987(01)00286-0.

Roth, L. and Streit, B. (2018) "Predicting cover crop biomass by lightweight UAS-based RGB and NIR photography: an applied photogrammetric approach," *Precision Agriculture*, 19(1), pp. 93–114. Available at: https://doi.org/10.1007/s11119-017-9501-1.

Royal Botanic Gardens, Kew (2025) *Plants of the world online*. Available at: https://powo.science.kew.org/.

Ryan, M.R. et al. (2024) "Rolled-crimped cover crops for organic no-till planted winter wheat (*Triticum aestivum*) in the Northeast region of the United States," *Agrosystems, Geosciences & Environment*, 7(4), p. e70015. Available at: https://doi.org/10.1002/agg2.70015.

Santra, D.K., et al. (2019) 'Proso millet (Panicum miliaceum L.) breeding: progress, challenges and opportunities', in Al-Khayri, J.M., et al. (eds.) *Advances in plant breeding strategies: Cereals*. Springer International Publishing, pp. 223–257. Available at: https://doi.org/10.1007/978-3-030-23108-8_6.

Sapara, V. et al. (2024) "Gene editing tool kit in millets: present status and future directions," *The Nucleus*, 67(1), pp. 157–179. Available at: https://doi.org/10.1007/s13237-024-00485-3.

Scavo, A. et al. (2022) "The role of cover crops in improving soil fertility and plant nutritional status in temperate climates. A review," *Agronomy for Sustainable Development*, 42(5), p. 93. Available at: https://doi.org/10.1007/s13593-022-00825-0.

Schipanski, M.E. et al. (2014) "A framework for evaluating ecosystem services provided by cover crops in agroecosystems," *Agricultural Systems*, 125, pp. 12–22. Available at: https://doi.org/10.1016/j.agsy.2013.11.004.

Schoolman, E. and Arbuckle Jr, J. (2022) "Cover crops and specialty crop agriculture: exploring cover crop use among vegetable and fruit growers in Michigan and Ohio," *Journal of Soil and Water Conservation*, 77(4), pp. 403–417. Available at: https://doi.org/10.2489/jswc.2022.00006.

Seben, G.F., et al. (2014) "Effect of cropping systems in no-till farming on the quality of a Brazilian Oxisol," *Revista Brasileira de Ciência do Solo*, 38, pp. 1268–1280. Available at: https://doi.org/10.1590/S0100-06832014000400023.

Serba, D.D. et al. (2017) "Status of global pearl millet breeding programs and the way forward," *Crop Science*, 57(6), pp. 2891–2905. Available at: https://doi.org/10.2135/cropsci2016.11.0936.

Serba, D.D. and Yadav, R.S. (2016) "Genomic tools in pearl millet breeding for drought tolerance: status and prospects," *Frontiers in Plant Science*, 7, p. 1724. Available at: https://doi.org/10.3389/fpls.2016.01724.

Shabbir, A., et al. (2022) "The weed-suppressive ability of summer cover crops in the Northern Grains Region of Australia," *Agronomy*, 12(8), p. 1831. Available at: https://doi.org/10.3390/agronomy12081831.

Sheahan, C.M. (2014a) *Comparison of five millet species for conservation use in the United States*. U.S. Department of Agriculture Natural Resources Conservation Service. Available at: https://www.nrcs.usda.gov/plantmaterials/njpmctn12841.pdf.

Sheahan, C.M. (2014b) *Plant guide for browntop millet (Urochloa ramosa)*. U.S. Department of Agriculture Natural Resources Conservation Service. Available at: https://plants.sc.egov.usda.gov/DocumentLibrary/plantguide/pdf/pg_urra.pdf.

Sheahan, C.M. (2014c) *Plant guide for foxtail millet (Setaria italica)*. U.S. Department of Agriculture Natural Resources Conservation Service. Available at: https://plants.sc.egov.usda.gov/DocumentLibrary/plantguide/pdf/pg_seit.pdf.

Sheahan, C.M. (2014d) *Plant guide for Japanese millet (Echinochloa esculenta)*. U.S. Department of Agriculture Natural Resources Conservation Service. Available at: https://plants.usda.gov/DocumentLibrary/plantguide/pdf/pg_eces.pdf.

Sheahan, C.M. (2014e) *Plant guide for pearl millet (Pennisetum glaucum)*. U.S. Department of Agriculture Natural Resources Conservation Service. Available at: https://plants.usda.gov/DocumentLibrary/plantguide/pdf/pg_pegl2.pdf.

Sheahan, C.M. (2014f) *Plant guide for proso millet (Panicum miliaceum).* U.S. Department of Agriculture Natural Resources Conservation Service. Available at: https://www.nrcs.usda.gov/Internet/FSE_PLANTMA.

Shrestha, U., et al. (2016) "A meta-analysis of the impact of anaerobic soil disinfestation on pest suppression and yield of horticultural crops," *Frontiers in Plant Science*, 7, 1254. Available at: https://doi.org/10.3389/fpls.2016.01254.

Snapp, S.S. et al. (2005) "Evaluating cover crops for benefits, costs and performance within cropping system niches," *Agronomy Journal*, 97(1), pp. 322–332. Available at: https://doi.org/10.2134/agronj2005.0322a.

Sood, S. et al. (2015) "Barnyard millet – A potential food and feed crop of future," *Plant Breeding*, 134(2), pp. 135–147. Available at: https://doi.org/10.1111/pbr.12243.

Soti, P. and Racelis, A. (2020) "Cover crops for weed suppression in organic vegetable systems in semiarid subtropical Texas," *Organic Agriculture*, 10(4), pp. 429–436. Available at: https://doi.org/10.1007/s13165-020-00285-4.

Soudzilovskaia, N.A. et al. (2020) "FungalRoot: Global online database of plant mycorrhizal associations," *New Phytologist*, 227(3), pp. 955–966. Available at: https://doi.org/10.1111/nph.16569.

Stapleton, J.J. et al. (2010) "Deleterious activity of cultivated grasses (Poaceae) and residues on soilborne fungal, nematode and weed pests," *Phytoparasitica*, 38(1), pp. 61–69. Available at: https://doi.org/10.1007/s12600-009-0070-3.

Stefaniak, T.R. et al. (2012) "Variation in biomass composition components among forage, biomass, sorghum-sudangrass, and sweet sorghum types," *Crop Science*, 52(4), pp. 1949–1954. Available at: https://doi.org/10.2135/cropsci2011.10.0534.

Teasdale, J.R. (1996) "Contribution of cover crops to weed management in sustainable agricultural systems," *Journal of Production Agriculture*, 9(4), pp. 475–479. Available at: https://doi.org/10.2134/jpa1996.0475.

Teasdale, J.R. (1998) 'Cover crops, smother plants, and weed management', in Hatfield, J.L., et al. (eds.) *Integrated weed and soil management.* Ann Arbor Press, pp. 247–270.

Teasdale, J.R. et al. (2007) 'Cover crops and weed management', in Upadhyaya, M.K. and Blackshaw, R.E. (eds.) *Non-chemical weed management: principles, concepts and technology.* CABI, pp. 49–64.

Teasdale, J.R. and Mohler, C.L. (2000) "The quantitative relationship between weed emergence and the physical properties of mulches," *Weed Science*, 48(3), pp. 385–392. Available at: https://doi.org/10.1614/0043-1745(2000)048[0385:TQRBWE]2.0.CO;2.

Teixeira, R.B. et al. (2016) "Tillage systems and cover crops on soil physical properties after soybean cultivation," *Revista Brasileira de Engenharia Agrícola e Ambiental*, 20, pp. 1057–1061. Available at: https://doi.org/10.1590/1807-1929/agriambi.v20n12p1057-1061.

Tillman, P.G. and Mullinix, B.G. (2004) "Grain sorghum as a trap crop for corn earworm (Lepidoptera: Noctuidae) in cotton," *Environmental Entomology*, 33(5), pp. 1371–1380. Available at: https://dx.doi.org/10.1603/0046-225X-33.5.1371 (Accessed: September 11, 2025).

UN Food and Agriculture Organization (2025) *GAEZ v4 data portal.* Available at: https://gaez.fao.org/.

Undersander, D. (n.d.) *Sorghums, sudangrass, and sorghum-sudan hybrids.* University of Wisconsin-Madison Extension. Available at: https://cropsandsoils.extension.wisc.edu/articles/sorghums-sudangrass-and-sorghum-sudan-hybrids/.

USDA NRCS (2025) *USDA plants database*. Available at: https://plants.usda.gov/.

Valenzuela, H. and Smith, J. (2002) *Sorghum-sudangrass hybrids*. University of Hawai'i. Available at: https://scholarspace.manoa.hawaii.edu/server/api/core/bitstreams/f7d2ecd8-4f21-4301-afdf-74ef4f86cb80/content.

Van Eerd, L.L. et al. (2023) "Influence of cover crops at the four spheres: a review of ecosystem services, potential barriers, and future directions for North America," *Science of The Total Environment*, 858, p. 159990. Available at: https://doi.org/10.1016/j.scitotenv.2022.159990.

Vavlas, N.-C. et al. (2024) "Cover crop impacts on soil organic matter dynamics and its quantification using UAV and proximal sensing," *Smart Agricultural Technology*, 9, p. 100621. Available at: https://doi.org/10.1016/j.atech.2024.100621.

Venkateswaran, K., et al. (2019) "Origin, domestication and diffusion of *Sorghum bicolor*," in Aruna, C., et al. (eds.) *Breeding sorghum for diverse end uses*. Elsevier, pp. 15–31.

Viaene, N.M. and Abawi, G.S. (1998) "Management of *Meloidogyne hapla* on lettuce in organic soil with sudangrass as a cover crop," *Plant Disease*, 82(8), pp. 945–952. Available at: https://doi.org/10.1094/PDIS.1998.82.8.945.

Wagger, M.G., et al. (1998) "Nitrogen and carbon cycling in relation to cover crop residue quality," *Journal of Soil and Water Conservation*, 53(3), pp. 214–218. Available at: https://doi.org/10.1080/00224561.1998.12457221.

Wallander, S. et al. (2021) *Cover crop trends, programs, and practices in the United States*. U.S. Department of Agriculture Economic Research Service Bulletin Number 222. Available at: https://ageconsearch.umn.edu/record/309562/?v=pdf.

Wang, G., et al. (2008) "Nutrient cycling, weed suppression, and onion yield following Brassica and sorghum-sudangrass cover crops," *HortTechnology*, 18(1), pp. 68–74. Available at: https://doi.org/10.21273/HORTTECH.18.1.68.

Wauters, V.M. et al. (2021) "Ecosystem services and cash crop tradeoffs of summer cover crops in northern region organic vegetable rotations," *Frontiers in Sustainable Food Systems*, 5, p. 635955. Available at: https://doi.org/10.3389/fsufs.2021.635955.

Wedryk, S. and Cardina, J. (2012) "Evaluation of tef as a smother crop during transition to organic management," *Weed Technology*, 26(1), pp. 102–109. Available at: https://doi.org/10.1614/WT-D-11-00042.1.

Weston, L.A. (1996) "Utilization of allelopathy for weed management in agroecosystems," *Agronomy Journal*, 88(6), pp. 860–866. Available at: https://doi.org/10.2134/agronj1996.00021962003600060004x.

Weston, L.A., et al. (2013) "Sorghum allelopathy–from ecosystem to molecule," *Journal of Chemical Ecology*, 39(2), pp. 142–153. Available at: https://doi.org/10.1007/s10886-013-0245-8.

Whish, J.P.M., et al. (2009) "Do spring cover crops rob water and so reduce wheat yields in the northern grain zone of eastern Australia?" *Crop and Pasture Science*, 60(6), p. 517. Available at: https://doi.org/10.1071/CP08397.

Wick, A. et al. (2017) "Integration of annual and perennial cover crops for improving soil health," in Al-Kaisi, M.M. and Lowery, B. (eds.) *Soil health and intensification of agroecosytems*. Elsevier, pp. 127–150.

Woldeyohannes, A.B. et al. (2022) "Value of teff (*Eragrostis tef*) genetic resources to support breeding for conventional and smallholder farming: A review," *CABI Agriculture and Bioscience*, 3(1), p. 27. Available at: https://doi.org/10.1186/s43170-022-00076-9.

Wolfe, D.W. (1998) *Final report: management strategies for improved soil quality with emphasis on soil compaction. LNE94-044*. Sustainable Agriculture Researech and

Education Grant Management System. Available at: https://projects.sare.org/project-reports/lne94-044/.
Wood, S.A. and Bowman, M. (2021) "Large-scale farmer-led experiment demonstrates positive impact of cover crops on multiple soil health indicators," *Nature Food*, 2(2), pp. 97–103. Available at: https://doi.org/10.1038/s43016-021-00222-y.
World Flora Online (2025) *World Flora Online*. Available at: https://www.worldfloraonline.org/.
Wortman, S.E., et al. (2014) "Weed suppressive potential of sudangrass is driven by interactions of root exudates and decomposing shoot residue," *Crop Management*, 13(1), CM-2013-0037-RS. Available at: https://doi.org/10.2134/CM-2013-0037-RS.
Wunsch, E.M., et al. (2017) "Can legumes provide greater benefits than millet as a spring cover crop in southern Queensland farming systems?" *Crop and Pasture Science*, 68(8), p. 746. Available at: https://doi.org/10.1071/CP17223.
Yabuno, T. (1987) "Japanese barnyard millet (*Echinochloa utilis*, Poaceae) in Japan," *Economic Botany*, 41(4), pp. 484–493. Available at: https://doi.org/10.1007/BF02908141.
Zhang, H. et al. (2023) "Cover cropping impacts on soil water and carbon in dryland cropping system," *PLoS One*, 18(6), p. e0286748. Available at: https://doi.org/10.1371/journal.pone.0286748.
Zimdahl, R.L. (2018) *Fundamentals of weed science*. 5th ed. Academic Press.

Chapter 15

Non-leguminous broadleaf cover crops

Christina Curell and Madelyn Celovsky, Michigan State University, USA

1 Introduction

Non-leguminous broadleaf cover crops (sometimes called forbs) are often species in the Brassicaceae (mustard) family. Common non-leguminous broadleaf cover crop species include mustard (*Brassica juncea and Sinapis alba*), rapeseed/canola (*Brassica napus*), and turnip (*Brassica rapa*). However, other broadleaf species outside these groups may also be suitable depending on a farmer's specific agronomic goals (MCCC, 2014). Other common non-leguminous broadleaf cover crop species include: buckwheat (*Fagopyrum esculentum*), oilseed/forage radish (*Raphanus sativus*), and winter camelina (*Camelina sativa*). This group may also contain such species as flax (*Linum usitatissimum*), phacelia (*Phacelia tanacetifolia*), safflower (*Carthamus tinctorius*), and sunflower (*Helianthus annuus*). Adding non-leguminous broadleaf species to a cover crop rotation introduces new plant families that can boost overall system diversity and resilience.

As discussed elsewhere in this book, there are various criteria for selecting a particular cover crop or cover crop mix. One set of criteria relates to the

http://dx.doi.org/10.19103/AS.2025.0162.19

potential soil health and other benefits they can provide (Snapp et al., 2023) such as the following:

- soil physical health, e.g. preventing erosion, reducing compaction, promoting water infiltration;
- soil chemical health, e.g. acting as nutrient scavengers in capturing and retaining nutrients such as nitrogen (N) in the soil (making them available to a following crop and reducing leaching); improving soil organic matter and carbon sequestration; improving mineralization to promote nutrient availability;
- soil biological health, e.g. promoting beneficial soil microbial and fungal activity;
- playing a role in insect pest, disease, and weed control (e.g. through allelopathic effects or disruption of pest/disease life cycles, as in the case of nematode control); and
- promoting biodiversity (e.g. by providing a food source and/or habitat for beneficial species

Although research on non-legumes is more limited than for legumes and grasses, they have been found to improve soil physical structure (Hao et al., 2023). Because of deeper root systems, non-leguminous broadleaf cover crop species help reduce soil compaction and improve soil structure, with tap roots in some species that create voids that improve soil aeration and water infiltration (White et al., 2016).

In terms of soil chemistry, compared to grass cover crops, non-leguminous broadleaf cover crops tend to have a lower carbon:nitrogen ratio, which means biomass tends to decompose more quickly. While this promotes more rapid release of nutrients such as nitrogen (N), nutrient immobilization may decrease as the biomass decomposes, though some research suggests non-leguminous species still play an important role in preventing N leaching (Finney et al., 2016; Abdalla et al., 2019; Koudahe et al., 2022). These crops have also been associated with increased phosphorus availability (White et al., 2016) as well as with improving nutrient release from manure application when the two practices are combined (Cottney et al., 2020).

Some non-leguminous broadleaf species can promote beneficial soil microbial activity (Finney at al., 2017). In contrast, non-leguminous broadleaf *Brassica* species have also been associated with improved control of soil pests, e.g. by releasing glucosinolates which have a disease-suppressing effect in suppressing pests such as plant parasitic nematodes (Kruger et al., 2013; Sassenrath et al., 2019; Chekanai et al., 2024). Species such as buckwheat are also a potential source of food for insect pollinators such as bees (Mallinger et al., 2019).

Another set of criteria relates to the suitability of a cover crop to particular abiotic and biotic conditions, e.g. to different soil types and regional climates (in terms of their ability to grow within particular temperature ranges, patterns of rainfall, etc.) and to cultivation at particular times of the year, as well as any potential vulnerability to local pests and diseases. The degree of vulnerability to lower winter temperatures can be a major factor in ensuring the cover crop is easily terminated before a following spring cash crop. Non-leguminous broadleaf cover crops are relatively adaptable to a range of environmental conditions (SARE, 2007). Brassicas, for example, typically thrive in cool weather conditions with a relatively broad temperature range (around 5°C to around 30°C), allowing growth to occur, although optimum temperatures are typically between 15–21°C (Borges et al., 2023).

A further set of criteria relates to their growth characteristics, both positive and negative. The rapid germination and vigorous growth of many non-leguminous broadleaf cover crops makes them a flexible addition to a rotation. Including these species supports diversified cropping systems and thus can be a valuable strategy for long-term sustainable land management. However, these growth and other characteristics can also create potential problems, such as:

- a cover crop providing a potential habitat for some pests;
- management conflicts with cash crops (e.g. the potential to become a weed if not properly managed and easily terminated); and
- allelopathic or other effects that potentially inhibit subsequent cash crop growth

Recent evidence suggests that, in some circumstances, broadleaved and other cover crop species may negatively affect cash crop yields at least in the short term, though the right combination of leguminous and broad-leaved species can boost yields (Won et al., 2024).

A livestock nutritionist should be consulted when planning on grazing cover crops. Grazing brassicas as a monocrop presents a bloating risk for livestock, so it is recommended to plant brassicas in a mix with grasses when using them for forage.

Table 1 summarizes some of the main pros (and some cons) of common non-leguminous broadleaf cover crops, but this chapter does not cover all available non-leguminous broadleaf options. Depending on the region, soil type, and production goals, other species may be better suited to provide desired ecosystem services.

2 Buckwheat

Buckwheat (Fig. 1) belongs to the Polygonaceae family. Buckwheat is a vigorous, warm, or cool-season cover crop often planted between cash crops

Table 1 Non-leguminous broadleaf cover crop characteristics

	Buckwheat	Mustard	Radish	Rapeseed/ canola	Turnip	Winter camelina	Phacelia	Sunflowers
Usually overwinters				X		X		
Quick growth	X	X	X	X	X	X		X
Weed suppression*	X	X	X		X		X	
Breaks shallow compaction	X	X	X		X		X	X
Breaks deep compaction			X		X			X
Summer/fall nutrient scavenger	X	X	X		X		X	X
Spring nutrient scavenger				X		X		
Good forage value for livestock			X	X	X			
Good for pollinators	X	X					X	X
Pathogenic nematode suppression*	X	X						
Potential weed problem	X	X		X	X			

*Effectiveness of suppression may vary by management, species, and local conditions.

during summer fallow periods (SARE, 2007; Björkman et al., 2008). Buckwheat establishes quickly (reaching maturity in 70–90 days) and performs well in cool, moist but well-drained soils (e.g. loams) but not in heavy, wet soils. Buckwheat thrives in low-fertility soils, making it good for rejuvenating poor or overworked soils. However, it is sensitive to drought and frost and needs to be planted to avoid extreme heat but while it can still mature before the first frost.

Buckwheat grows quickly with larger leaves developing within 2 weeks to create a dense canopy to protect soil, thus promoting water infiltration. It develops a fine, dense, fibrous root system that helps alleviate surface compaction. This root structure, together with the competition provided by the dense canopy, makes it very effective in suppressing weeds (Bulan et al., 2015; Vieites-Álvarez et al., 2024; Worthing et al., 2025). Additionally, buckwheat can mine soil phosphorus by solubilizing P from soil minerals by releasing acid phosphatase (Amann and Amberger, 1989). When the plant starts to die down, absorbed phosphorus is released and can be used by the next crop. For best

Figure 1 Buckwheat in bloom. Photo credit: Madelyn Celovsky, Michigan State University Extension.

results, terminate buckwheat before seed set to prevent volunteer growth (and the risk of the crop becoming a weed threat) and to maximize nutrient release (McKenzie-Gopsill, 2025).

In addition to its agronomic benefits, buckwheat is an excellent pollinator crop. Its small white flowers attract a wide range of beneficial insects, supporting biodiversity and improving pollination in nearby crops, as well as attracting pest predator species such as parasitic wasps.

3 Mustard

Mustards are fast-growing, warm-season brassicas commonly used as cover crops between summer annuals or in early spring (Brown et al., 2022). They are sensitive to cold and will not survive temperatures below −6°C, so proper timing is essential for successful establishment and growth (SARE, 2007).

Two main types of mustard are used in cover cropping:

- Indian and brown mustard (*Brassica juncea*); and
- white and yellow mustard (*Sinapis alba*).

Figure 2 shows an example of a white mustard. Mustards have deep-rooted systems with taproots that can reach up to 120 cm, helping to reduce soil compaction, improve water infiltration, and moisture retention. Mustards produce a large amount of biomass and, given their deep roots, are good nutrient scavengers of nitrogen, phosphorus, and potassium, which can then become available for the following crop (Björkman et al., 2015).

Given their rapid growth and dense canopy cover, mustards can outcompete and suppress weeds (Haramoto and Gallandt, 2004; Johnson et al., 2020). Mustards are also rich in glucosinolates – sulfur-containing compounds naturally found in plants. Due to their high glucosinolate content, mustards are effective biofumigants and can be used to suppress soil-borne pests such as nematodes and disease-causing fungi as well as some weeds (Jones et al., 2020; Harper et al., 2022).

To maximize their biofumigant benefits and prevent mustard from becoming a weed, it should be terminated before going to seed (e.g. via mowing). The cover crop residue needs to be chopped up and immediately incorporated into the soil for best results. This guarantees that the glucosinolates are released

Figure 2 Flower buds form on white mustard. Photo credit: Madelyn Celovsky, Michigan State University Extension.

and subsequently hydrolyzed into isothiocyanates that have biocidal activity (dos Santos et al., 2021).

4 Oilseed and forage radish

Oilseed radish (*Raphanus sativus* var. *oleiformis*) and forage radish (*R. sativus* var. *niger*) are similar, fast-growing winter annuals. Oilseed radish has a somewhat stubbier root and is marginally more winter hardy than forage radish (Fig. 3). These radishes are used as cover crops across the United States, with many cultivars originally developed in Europe (Ngouajio et al., 2004; Weil et al., 2006; Sundermeier, 2008). Commonly planted varieties include *Adagio*, *Arena*, *Colonel*, *Common*, *Daikon* (Fig. 4), *Remonta*, *Revena*, *Rimbo*, and *Ultimo*. Each cultivar offers unique traits such as root structure, glucosinolate production, and aboveground biomass production that make them well-suited to specific cover crop goals (as described below).

Figure 3 Oilseed and forage radish (Weil et al., ND).

Figure 4 Daikon forage radish in late fall. Photo credit: Madelyn Celovsky.

Radishes produce a deep taproot, which reduces soil compaction and improves soil structure (Blaanco-Canqui and Ruis, 2020). Root channels in the subsoil created by radishes have been found to increase deep rooting of the following summer crops such as corn and soybeans (Chen and Weil, 2011). Deep roots allow plants to mine nutrients from deeper in the soil profile, while high above- and belowground biomass makes radishes good scavengers of phosphorus and nitrogen, thus reducing nitrate leaching (White et al., 2016; Cavadini and Kladivko, 2025). Nitrate taken up from the subsoil is converted into protein in roots and leaves, and nitrogen is subsequently released upon decomposition in the topsoil where the next crop can utilize it as long as the decomposition does not occur too long before main crop planting. Rapid growth and good canopy cover also make radishes effective weed suppressants (Lawley et al., 2012). Partly due to its glucosinolate production, radishes have been associated with suppression of plant-parasitic nematodes, although their vulnerability to some root diseases means they can also pass disease on to a following crop (Taning et al., 2023; Wallenhammar et al., 2024). Radishes are also suitable as a forage crop.

Radishes are seeded in late summer or early fall to establish before winter. They germinate quickly in cool conditions. Radish plants typically winterkill at temperatures below −6°C. However, in northern regions if the ground has not frozen before snowfall, they may overwinter. When this occurs, plants often resume growth in the early spring. Instead of putting energy into root and tuber development, overwintered or spring-planted radishes tend to bolt rapidly, producing flowers and seeds, which then make them a potential weed. In the fall, they typically winterkill before flowering.

5 Rapeseed/canola

Brassica napus is a hybrid species derived from *Brassica rapa* and *Brassica oleracea* to produce rapeseed (Fig. 5). One of the most popular varieties for cover cropping is Dwarf Essex rapeseed, known for its ability to overwinter reliably. Due to a greater tolerance to low temperatures, rapeseed is a particularly versatile cover crop, which can be used as an autumn/fall-seeded winter cover crop. Canola is a specialized cultivar of rapeseed bred for

Figure 5 Rapeseed. Photo credit: Madelyn Celovsky, Michigan State University Extension.

reduced glucosinolate and erucic acid content, making it suitable for both livestock feed and as a cover crop (Barry, 2013; Begna et al., 2017; Kirkegaard et al., 2008).

Rapeseed establishes quickly in the autumn/fall and, when it overwinters, it is highly effective at scavenging residual nitrogen from the soil in the spring. This helps prevent nitrate leaching and thus protects water quality and enhances nutrient availability for the following cash crop as the biomass decomposes (Nouri et al., 2022; see also: https://ahdb.org.uk/knowledge-library/cover-crops-reduce-nutrient-losses-and-leaching). Given its glucosinolate content, rapeseed varieties (excluding canola) have also shown potential for suppressing some plant-parasitic nematode species as well as suppressing weeds (Asaduzzamin et al., 2011; Eugui et al., 2022). As with mustards and other Brassicas, residue needs to be chopped up and incorporated into the soil for best results.

6 Turnip

Turnip is a fast-growing, cool-season Brassica species commonly used in cover cropping systems (Fig. 6). While it establishes quickly, it lacks winter hardiness and typically succumbs to temperatures below −4°C. When planted in late summer or early fall, turnips can effectively extend the grazing season by providing added forage (Björkman and Shall, 2010; Mäkelä et al., 2011).

Figure 6 Purple top turnip. Photo credit: Liz Schultheis, Michigan State University.

The species produces large foliage that suppresses weed growth through shading (Petersen et al., 2001). Its taproots and fibrous secondary roots contribute to soil structure improvement by alleviating compaction and minimizing wind and water erosion. Additionally, turnip roots are efficient nutrient scavengers, capturing residual soil nutrients and storing them in the biomass, which are released back into the soil upon decomposition. Variety selection is important when turnips are used for forage to ensure best performance and palatability.

7 Winter camelina

Winter camelina is a cold-hardy, overwintering oilseed species in the Brassica family that offers multiple potential agronomic and environmental benefits when used as a cover crop (Leavitt et al., 2025) (Fig. 7). When seeded in the early autumn/fall, it establishes quickly, germinates, and survives freezing temperatures, resuming growth in early spring (https://forevergreen.umn.edu/crops/winter-camelina).

Its early spring development provides ground cover that suppresses weeds and reduces soil erosion by wind and water (Johnson et al., 2021; Ghidoli et al., 2023). Winter camelina also efficiently uses residual nutrients left in the soil after the primary cash crop harvest, thereby minimizing nutrient leaching and runoff into nearby water bodies. If allowed to reach the flowering stage, it serves as a valuable forage resource for pollinators, contributing to biodiversity and other ecosystem services (Pessereau et al., 2025).

Figure 7 Winter camelina. Photo credit: Marisol Berti, North Dakota State University.

8 Phacelia

Phacelia (Fig. 8) is a cover crop originating in the southwestern United States (Kirk, 2005). It is now used widely as a cover crop in Europe. It is a cool-season annual, tolerant of temperatures as low as −8°C (18°F) (Kilian, 2016). It flowers for up to 8 weeks and provides abundant nectar and pollen for pollinator species. It can be a useful rotation crop because it is in a different plant family (Hydrophyllaceae) and therefore doesn't share pests and diseases with commonly grown commercial crops. Phacelia is drought-resistant and has a dense root system that improves soil tilth. In a high tunnel experiment in the spring in Poland, it produced more than 6 Mg ha^{-1} (3 t/ac) dry matter in 3 months (Lalewicz et al., 2024). It absorbed high quantities of N, Ca, P, S, B, and Cu, and reduced bulk density while it improved formation of large (>1 mm) water-stable aggregates compared with no-cover control. Phacelia can effectively reduce nitrate leaching (Herrera et al., 2010). Phacelia germinates and grows more slowly than common brassica cover crops (Brust et al., 2014). However, once established, it can effectively outcompete other vegetation, such as volunteer small grain plants (Brant et al., 2009).

Figure 8 Honeybee on phacelia flower (photo courtesy USDA-NRCS). https://www.nrcs.usda.gov/plantmaterials/mtpmctn12938.pdf.

9 Sunflower

Sunflowers are a summer annual forb in the Asteraceae family (Fig. 9). Black oilseed varieties are typically used for cover crop sunflower use, because they have small seeds, and seed costs are relatively low, resulting in more plants per unit area and many smaller flowers. Sunflowers germinate at soil temperatures of 16°C/60°F, prefer well-drained soil, and have a deep taproot that can help combat soil compaction and improve pore space and infiltration. They die at first frost. Sunflower roots can penetrate as deep as 2.70 m by mid-summer (Jaafar et al., 1993). Sunflowers can withstand high heat, moderate drought, and grow in low-fertility soil (Alapati et al., 2025). They extract moisture from greater depths than small grain crops (Halvorson et al., 1999). Sunflowers are effective at capturing nitrogen from the soil, thus reducing nitrate leaching losses (Herrera et al., 2010). Biomass production is modest, typically amounting to 1–5 Mg ha^{-1} (0.5–2.5 t/ac). Sunflowers have thick stems that decompose slowly and thus protect soil from erosion while also reducing water evaporation losses. Leaves decompose rapidly, however. After they die, soil cover is therefore rapidly reduced, with only the stems that don't cover the soil extensively. If left to flower, sunflowers provide pollinator habitat and aesthetic value.

Figure 9 Sunflowers. Courtesy Sjoerd W. Duiker.

10 Conclusion

Non-leguminous broadleaf cover crops serve as important components for diversifying agricultural production systems while contributing a broad array of agronomic functions. Species within the Brassicaceae family, along with other broadleaf options such as buckwheat, phacelia, and sunflower, can be strategically integrated into cover crop mixtures to achieve specific management objectives, including weed suppression, nutrient scavenging, and pest mitigation. As with all cultivated plants, these cover crop species may require appropriate fertilization to ensure successful germination and stand establishment. When incorporated into rotations with grasses and legumes, broadleaf cover crops enhance functional diversity, disrupt pest and disease cycles, and strengthen the overall resilience and sustainability of cropping systems.

11 Where to look for further information

- Managing Cover Crops Profitably, Sustainable Agriculture Research and Education 2007: https://www.sare.org/resources/managing-cover-crops-profitably-3rd-edition/Contact your regional cover crops council (United States and Canada).
- Midwest Cover Crops Council: https://www.midwestcovercrops.org/.
- Northeast Cover Crops Council: https://northeastcovercrops.com/.
- Southern Cover Crops Council: https://southerncovercrops.org/.
- Western Cover Crops Council: https://westerncovercrops.org.

12 References

Abdalla M. et al. 2019. A critical review of the impacts of cover crops on nitrogen leaching, net greenhouse gas balance and crop productivity. *Glob. Chang Biol.* 25(8): 2530–2543.

Alapati, D. K. et al. 2025. Sunflower and sunn hemp potential as summer cover crops in Southern Texas. *Agronomy (Basel)* 15: 986.

Amann, C. and Amberger, A. 1989. Phosphorus efficiency of Buckwheat (Fagopyrum esculentum). *Zeitschrift für Pflanzenernährung und Bodenkunde* 152: 181–189.

Asaduzzamin, M. et al. 2011. *Allelopathy in Canola: Potential for Weed Management.* Australian Research Association NSW DPI, New South Wales, Australia.

Barry, T. N. 2013. The feeding value of forage brassica plants for grazing ruminant livestock. *Anim. Feed Sci. Tech.* 181: 15–25.

Begna, S. et al. 2017. Winter Canola: a potential dual-purpose crop for the United States Southern great plains. *Agron. J.* 109(6): 2508–2520.

Blanqui-Campo, H. and Ruis, S. J. 2020. Cover crop impacts on soil physical properties: a review. *Soil Sci. Soc. Am. J.* 84: 1527–1576.

Björkman, T. et al. 2008. *Buckwheat Cover Crop Handbook.* Cornell University, Ithaca, New York.

Björkman, T. and Shall, J. W. 2010. *Cornell University Extension Fact Sheet 6: Cover Crop Guide for Forage Turnip and Rape*. Cornell University, Ithaca, New York.

Björkman, T. et al. 2015. Mustard cover crops for biomass production and weed suppression in the Great Lakes Region. *Agron. J.* 107(4): 1235–1249.

Borges, C. A. et al. 2023. Forecasting Brassica napus production under climate change with a mechanistic species distribution model. *Sci. Rep.* 13: Article 12656.

Brant, V. et al. 2009. Competition of some summer catch crops and volunteer cereals in the areas with limited precipitation. *Plant Soil Environ.* 55: 17–24.

Brust, J. et al. 2014. Growth and weed suppression ability of common and new cover crops in Germany. *Crop Prot.* 63: 1–8.

Bulan, M. T. S. et al. 2015. Buckwheat species as summer cover crops for weed suppression in no-tillage vegetable cropping systems. *Weed Sci.* 63(3): 690–702.

Brown, M. et al. 2022. Mustard cover crops: benefits and implementation in sustainable agriculture. *Agron. J.* 114(1): 305–317.

Cavadini, J. S. and Kladivko, E. J. 2025. Oilseed radish/cereal cover crop bicultures and soil phosphorus distribution. *J. Soil Water Conserv.* 80(1): 76–90.

Chekanai, V. et al. 2024. In vitro nematicidal efficacy of brassica-derived isothiocyanates against the root lesion nematode, *Pratylenchus penetrans. Nematology* 26(8): 899–908.

Chen, G. and Weil, R. 2011. Root growth and yield of maize as affected by soil compaction and cover crops. *Soil Till. Res.* 117: 17–27.

Clark, A. 2012. Managing Cover Crops Profitably. In: *Sustainable Agriculture Research and Education (SARE) Handbook Series Book* 9, Sustainable Agriculture Research and Education (SARE) Program, 1–232.

Cottney, R. et al. 2020. The correct cover crop species integrated with slurry can increase biomass quality and nitrogen cycling to positively affect yields in a subsequent spring barley rotation. *Agronomy* 10: Article 1760.

Dos Santos, C. A. et al. 2021. Biofumigation with species of the Brassicaceae family: a review. *Ciência Rural.* 51 4–6.

Eugui, D. et al. 2022. Glucosinolates as an effective tool in plant-parasitic nematodes control: exploiting natural plant defenses. *Appl. Soil Ecol.* 176: Article 104497.

Finney, D. M. et al. 2016. Biomass production and carbon nitrogen ratio influence ecosystem services from cover crop mixtures. *Agon. J.* 108(1): 39–52.

Finney, D. M. et al. 2017. Living cover crops have immediate impacts on soil microbial community structure and function. *J. Water Soil Conserv.* 72: 361–373.

Ghidoli, M. et al. 2023. Camelina sativa (L.) crantz as a promising cover crop species with allelopathic potential. *Agron. J.* 13(8); article 2187.

Halvorson, A. D. et al. 1999. Sunflower response to tillage and nitrogen fertilization under intensive cropping in a wheat rotation. *Agron. J.* 91: 637–642.

Hao, X. et al. 2023. Are there universal soil responses to cover cropping? A systematic review. *Sci. Total Environ.* 861: Article 160600.

Haramoto, E. R. and Gallandt, E. R. 2004. Brassica cover cropping for weed management: a review. *Renew. Agric. Food Syst.* 19(4): 187–198.

Harper, E. et al. 2022. Mustard as a biopesticide. *J. Sust. Agric.* 46(2): 148–162.

Herrera, J. M. et al. 2010. Root growth and nitrate-nitrogen leaching of catch crops following spring wheat. *J. Environ. Qual.* 39: 845–854.

Jaafar, M. N. et al. 1993. Rooting depth and dry matter development of sunflower. *Agron. J.* 85: 281–286.

Johnson, E. et al. 2020. White mustard cover crop effects on weed suppression and soil health. *Agron. J.* 112(3): 1700–1711.

Johnson, K. L. et al. 2021. Interseeding camelina and rye in soybean provides soil cover without affecting soybean yield. *Agronomy* 11: Article 353.

Jones, B. et al. 2020. Glucosinolates in mustard: biochemical pathways and allelopathic effects. *Phytochem. Rev.* 19(2): 275–290.

Kilian, R. 2016. *Lacy Phacelia: A Native Annual Forb for Conservation Use in Montana and Wyoming*. Plant Material Technical Note No. MT-113, Bridger, MT.

Kirk, W. D. J. 2005. Phacelia. *Bee World* 86: 14–16.

Kirkegaard, J. A. et al. 2008. Dual-purpose canola: a new opportunity in mixed farming systems. *Aust. J. Agric. Res.* 59(4): 291–302.

Koudahe, K. et al. 2022. Critical review of the impact of cover crops on soil properties. *Int. Soil Water Conserv. Res.* 10(3): 343–354.

Kruger, D. H. M. et al. 2013. Cover crops with biofumigation properties for suppression of plant-parasitic nematodes: a review. *S. African J. Enol. Vitic.* 34(2): 1–3.

Lalewicz, P. et al. 2024. Phacelia and buckwheat cover crops' effects on soil quality in organic vegetable production in a high tunnel system. *Agronomy* 14: 1614.

Lawley, Y. E. et al. 2012. The mechanism for weed suppression by a forage radish crop. *Agron J.* 104: 204–214.

Leavitt. M. et al. 2025. *Reflections on the 2024–2025 Piloting of Winter Camelina*. MBOLD Coalition/University of Minnesota Green Initiative. University of Minnesota, St Paul, MN.

Mäkelä, P. S. A. et al. 2011. Revitalizing the winter turnip rape crop in the northern latitudes. *Acta Agric. Scandinavica Section B Soil Plant Sci.* 61(3): 195–201.

Mallinger, R. E. et al. 2019. Annual cover crops for managed and wild bees: optimal plant mixtures depend on pollinator enhancement goals. *Agric. Ecosys. & Environ.* 273: 107–116.

MCCC 2014. *Midwest Cover Crops Field Guide Second Edition*. Midwest Cover Crops Council (MCCC), West Lafayette, IN.

McKenzie-Gopsill, A. 2025. The Critical Period of Cover Crop Management: A Framework for Maximizing Biomass Potential and Minimizing Volunteers with Buckwheat. Agriculture and Agri-Food Canada, Charlottetown Research and Development Centre.

Ngouajio, M. et al. 2004. *Oilseed Radish: A New Cover Crop for Michigan*. Extension Bulletin E 2907, Michigan State University, Lansing, MI.

Nouri, A. et al. 2022. When do cover crops reduce nitrate leaching? A global meta-analysis. *Glob. Chang Biol.* 28(15): 4736–4749.

Pessereau, E. J. et al. 2025. Early spring-flowering winter cover crop (Camelina sativa L. Crantz) increases insect flower visits in Wisconsin (USA). *Agric. Ecosys. Environ.* 389: 109689.

Petersen, J. et al. 2001. Weed suppression by release of isothiocyanates from turnip-rape mulch. *Agron. J.* 93: 37–43.

Sassenrath, G. F. et al. 2019. *Controlling Soil-Borne Disease in Soybean with a Mustard Cover Crop*. Kandas Agricultural Experiment Research Report 5. Kansas Agricultural Experiment Station - Kansas State University, Manhattan, KS.

Snapp, S. S. et al. 2023. Cover crops for soil health. In Horwath, W. R. (ed.), *Improving Soil Health*. Burleigh Dodds Science Publishing, Cambridge, UK.

Sundermeier, A. 2008. *Oilseed Radish Cover Crop*. Ohio State University Extension Factsheet SAG 5-08. Ohio State University, Coumbus, OH.

Sustainable Agriculture Research and Education (SARE). 2007. Managing Cover Crops Profitably. 3rd ed., Handbook Series Book 9.

Taning, L. M. et al. 2023. Impact of cover crops on population density of a root-knot nematode. *Nematology* 25: 867–882.

Wallenhammar A. C. et al. 2024. Susceptibility of Oilseed Radish (*Raphanus sativus* subsp. *oleiferus)* Cultivars and Various Brassica Crops to *Plasmodiophora brassicae*. *Pathogens* 13(9): 739.

Weil, R. R. et al. 2006. *Forage Radish: New Multi-Purpose Cover Crop for the Mid-Atlantic*. University of Maryland Cooperative Extension FactSheet 824. University of Maryland, College Pary, MD.

White, E. A. et al. 2016. A review of the Benefits, Optimal Crop Management Practices and Knowledge Gaps associated with different Cover Crop Species. AHDB Cereals and Oilseeds. https://ahdb.org.uk/a-review-of-the-benefits-optimal-crop-management-practices-and-knowledge-gaps-associated-with-different-cover-crop-species.

Won, S. et al. 2024. Understanding the yield impacts of alternative cover crop families and mixtures: evidence from side-by-side plot-level panel data. *Agrosystems* 7: Article e70012.

Worthing, M. et al. 2025. Reduced canopy cover and development compromises weed suppression by buckwheat in cover crop mixtures. *Can. J. Plant Sci*. 105: 1–5.

Vieites-Álvarez, Y. et al. 2024. A decade of advances in the study of buckwheat for organic farming and agroecology (2013–2023). *Front Plant Sci*. 15: Article 1354672.

Part 4

The future of cover crops

Chapter 16

Strategies to increase cover crop adoption

Ken Staver, University of Maryland, USA

1 Introduction

Historically, cover crops were valued for reducing soil erosion, providing nitrogen for non-leguminous crops, conserving soil nutrients, and improving soil physical properties (Rogers and Giddens 1957). As chemical nitrogen became available after development of the Haber-Bosch process and national and global transportation systems developed for long-distance movement of other plant nutrients, the role of cover crops for maintaining soil nutrient stocks became less critical. In the last four decades, there has been renewed interest in cover crops as a partial solution to water quality degradation associated with intensifying agricultural production (e.g. Hargrove 1991), especially the leaching of nitrate (Meisinger et al. 1991; Staver and Brinsfield 1998; Kladivko et al. 2014). More recently, cover crops have moved to the forefront as a tool to build soil health and productivity, especially in areas with degraded soils or

http://dx.doi.org/10.19103/AS.2025.0162.20

limited resources (Blanco 2023). As the threat of climate change has become more apparent, cover crops are recognized as a possible tool for making production systems more resilient and as a two-pronged mitigation strategy that can help reduce agriculture's carbon footprint by increasing soil carbon storage and reducing fossil fuel–intensive nitrogen inputs in both temperate and tropical regions (Tellatin and Myers 2018; Delgado et al. 2021). Cover crop types and potential benefits are covered in depth in earlier chapters in this volume. Increasing the cumulative benefits of cover crops will depend on increasing cover crop implementation which will be the focus of this chapter.

The wide range in potential roles of cover crops creates an equally wide range of factors affecting rates of implementation. Although much of the interest in cover crops has emerged from efforts to reduce agricultural environmental impacts, producers' decisions whether to use cover crops tend to focus on justifying time and expense to plant and kill the cover crops and the potential for increased profitability. This requires an informed management approach that maximizes benefits while minimizing costs and potential negative effects on crop production. Numerous studies focused on the effect of cover crops on profitability have yielded a wide range of conclusions depending on the production system and the cover crop type (Bowman et al. 2025; Bergtold et al. 2017). Cover crop adoption rates, while increasing, remain relatively low despite extensive research and outreach, suggesting that farmers do not generally perceive cover crops as a tool to increase profitability without subsidies to offset some of the costs. In fact, national surveys on cover crop attitudes consistently find return on investment as one of the primary reasons for hesitancy regarding cover crop adoption (CTIC 2023). Even where subsidies and on-farm research are used to introduce farmers to cover crops, long-term adoption still hinges on return on investment (Pires et al. 2024; Sun et al. 2025). In a non-regulatory environment, developing cover crop management approaches that provide positive return on investment is likely the single most important factor for increasing cover crop implementation in the absence of subsidies. In the absence of cover crop options that are profitable at the farm level, subsidies will likely be needed to increase adoption levels, especially of cover crop types that provide maximum environmental benefits. Where funds for subsidies are not available, regulation may be required to achieve desired implementation levels.

2 What cover crops do?

Cover crop impacts start with their basic function of increasing total photosynthetic production in a cropping system, that is, increasing the total capture of solar energy and atmospheric carbon by extending the growing season beyond when harvested crops are growing. There is a wide range in added production capacity, depending on the climate and production system, which determine the available growing window (Cates et al. 2018). Cover crops can increase total annual photosynthetic production by 30% or more (Ketterings et al. 2015a; Ruis et al. 2019). This additional biomass production can provide a range of services such as (1) surface mulch to reduce raindrop impact, slow overland flow, moderate soil temperatures, and suppress weeds (2) roots that break up soil compaction, bind soil particles, uptake soluble nutrients that otherwise could be leached, and make soil nutrients more available, and (3) in the case of legumes, add nitrogen (N) to the soil for future use by harvested crops. In systems with livestock, cover crop production also can provide extra forage in addition to soil and water quality benefits. Finally, cover crop carbon (C) capture can offset overall agricultural greenhouse gas emissions in addition to the soil benefits resulting from the extra supply of organic material. Cover crop benefits generally increase with increasing biomass until competition with the following crop becomes an issue (Cates et al. 2018).

The wide range of benefits of cover crops can be divided into two main categories: those that potentially benefit farmers and those that address larger-scale and future issues but do not directly offset the added costs of implementing cover crops (Bergtold et al. 2017; Schipanski et al. 2014; Snapp et al. 2005). A few benefits, such as erosion control, fall into both categories, by protecting long-term productive capacity and reducing sediment pollution of surface waters. Within both categories, benefits can be further subdivided into short and long term. Producers will tend to first focus on short-term production benefits that produce immediate and apparent returns on investments, with secondary focus on production benefits that accrue in the long term and are less well researched and more difficult to quantify. Environmental benefits from cover crops typically are not the highest priority of producers. However, regardless of the intent, adding cover crops to the production system will have effects in both production and environmental categories. Without incentives or regulations, producers will make decisions about whether to use cover crops at all, what type to plant, and how to manage them based on achieving short- and long-term return on their investment in planting and managing cover crops. However, this focus may not result in cover crop practices that provide the suite of benefits that address larger-scale issues such as water quality and climate change that often are of interest beyond the agricultural sector. For example, a recent global analysis concluded that legume cover crops were likely to increase crop yields but would only minimally reduce greenhouse gas emissions, while grass cover crops would lead to the greatest reductions in greenhouse gas emissions but would negatively affect crop production (McClelland et al. 2025). Similarly, grass cover crops in the Midwestern US are most effective for reducing nitrate leaching (Kladivko et al 2014), a benefit that addresses major environmental and human health issues but provides little financial benefit to the farmer (Deines et al. 2023; Plastina et al. 2018). So, cover crops can do many things, but careful consideration of impacts on production and environmental outcomes will be needed to develop effective implementation strategies. The primary challenge is developing cover crop approaches that lead to high levels of implementation, but which also address larger-scale issues beyond increasing farm profitability.

3 Managing cover crops profitably: a continuing challenge

It is telling that one of the oldest and most used cover crop resources is titled *Managing Cover Crops Profitably*, third edition (Clark 2007; first edition published in 1993). Surveys of farmer attitudes consistently show that despite overall positive views about cover crops, uncertainty about their overall effect on profitability is one of the primary obstacles to implementation (CTIC 2023; Roesch-McNalley et al. 2017). Farmers' hesitancy in cash grain operations seems warranted as short-term effects on profitability have been reported to be highly variable and frequently negative (Schnitkey et al. 2024). Some studies report positive returns when cover crops are used for multiple years (Bergtold et al. 2017), but long-term studies that would give producers the high levels of confidence needed to wait multiple years for positive returns are limited (Bowman et al. 2025). A wide range of factors determine the balance between costs and benefits (Bergtold et al. 2017). The primary direct costs are for seed, planting expenses, and termination if extra herbicide or tillage passes are required. Labor and machinery availability are often at a premium during cover crop planting which often occurs during the harvest season for summer annual crops and the planting season for winter annual crops (Das et al. 2022).

While direct costs are straightforward, benefits and possible indirect costs are more complicated and variable depending on the production system, the following crop, and the management skills of the producer. Technical guidance on planting cover crops is plentiful (e.g. Clark 2007, Midwest Cover Crop Council, Northeast Cover Crop Council), but many benefits/indirect costs are related to how the cover crop is managed and unique to soil/climate setting and production system, and especially whether livestock are part of the system (Wallander et al. 2021). Although cover crops can help reduce weed control costs (Teasdale and Mohler 2000; Lu et al. 2000; Silva and Bagavathiannan 2023), most indirect costs and benefits are related to cover crop effects on crop yields that are the result of cover crop effects on soil water and nutrient status in the short term, or in the long term, soil health. Producer surveys often identify improved soil health as the primary cover crop goal (CTIC 2023), with the assumption that improving soil health will lead to yield increases sufficient to more than offset the cost of planting and managing a cover crop. The challenge that keeps emerging in this line of reasoning is that for the large row crop systems that dominate major production regions, it is difficult to achieve yield increases that make cover crops pay, especially if producers are not highly skilled in managing cover crops or do not have technical support from service providers (Crespo et al. 2024; Bowman et al. 2025). Positive yield effects in response to soil health building practices such as cover crops and reduced tillage are especially difficult to achieve in highly productive soils (Crespo et al. 2024). In some settings cover crops can reduce crop yields leading to significant negative returns (Lewis et al. 2018; Nielsen et al. 2016; Plastina et al. 2018, Reiter et al. 2008; Singh et al. 2020). Arbuckle and Roesch-McNally (2015) suggest that making producers aware of risks and increasing their risk management capacity are critical to increasing cover crop implementation.

4 Challenges in managing nitrogen availability with cover crops

The role cover crops play in regulating soil N levels is central to their value for reducing cropland N losses to the environment, but also critical to their effect on crop yields and profitability. The primary value of grass cover crops for reducing N losses from cropland results from nitrate uptake and lower soil nitrate concentrations during the non-growing season, making it unavailable for leaching which occurs predominately in late fall through early spring when transpiration levels are low (Meisinger et al. 1991; Staver and Brinsfield 1998; Thorup-Kristensen et al. 2002). But this function also tends to reduce N availability for the following crop, especially when grass cover crops are allowed to grow later into the spring to maximize biomass production (Wagger 1989). Increased biomass achieved through delayed termination is beneficial for weed suppression and increasing soil organic carbon (Teasdale and Mohler 2000; McClelland et al. 2021) but also results in depleted plant available soil N and higher cover crop carbon:nitrogen (C:N) ratios which can result in microbial immobilization of soil nitrate after cover crop termination (Thorup-Kristensen et al. 2002). Numerous studies have shown that N applications to non-leguminous crops need to be increased in the short term to prevent N limitation following grass cover crops, especially if cover crops are allowed to grow until C:N ratios increase above 20 (Clark et al. 1994; Finney et al. 2016; Qin et al. 2021). So, while grass cover crops offer the most effective solution to nitrate losses and greenhouse gas emissions, short-term reductions in soil plant available N can reduce crop yields resulting in negative economic outcomes for farmers, a major disincentive for implementation.

The role legume cover crops can play in supplying N to crops has a long history, given their central role as an N source before chemical N became widely available (Pieters and McKee 1938; Kladivko 2020). There was a resurgence in interest in legume cover crops as an N source in the late 1980s as concerns grew over water quality degradation and the impacts of fossil fuel use in the production of inorganic N fertilizer (Mitchell and Teel 1977; Power et al. 1983; Ebelhar et al. 1984; Hargrove 1986, Smith et al. 1987; Blevins et al. 1990; Ranells and Wagger 1992). Legume cover crops can supply large quantities of N, and where adequate growing windows are available can fully displace the need for chemical N (Sarrantonio and Scott 1988; Tonitto et al. 2006). More often, legume cover crops partially reduce the need for fertilizer N additions to produce maximum yields (Decker et al. 1994; Frye et al. 1985). Unfortunately, legume cover crops alone are much less effective than grass cover crops for rapidly producing biomass in the fall for soil protection during winter months and retaining N (Shipley et al. 1992). And while legume cover crops generally do not lead to negative impacts on crop yields, seed costs tend to be higher than for grass cover crops, and a major fraction of fertilizer N use has to be displaced to produce a positive return (Bergtold et al. 2017). Again, the fact that the historical use of legume green manures declined sharply after inorganic N fertilizer became widely available suggests farmers did not find legume cover crops as the most cost-effective way to supply N to crops. In organic systems where low-cost inorganic N is not an option, legume cover crops are one of the most cost-effective options for supplying N to non-leguminous crops despite their higher seed costs (Hayden et al. 2014; Wayman et al. 2016; Kaye et al. 2019). Manure, where available, often is the most cost-effective N source in organic production systems, but can lead to excessive soil P concentrations if used to meet all crop N needs (Nelson and Janke 2007).

This limitation of legume cover crops for addressing water quality problems and the tendency of grass cover crops to have negative impacts on yields has led to numerous studies in the last several decades looking at grass/legume cover crop mixtures. Mixtures may achieve the soil and water quality benefits of grass cover crops while minimizing yield reductions related to soil N immobilization by grass cover crops (Clark et al. 1997a; Ranells and Wagger 1997; Hayden et al. 2014; Poffenbarger et al. 2015; Finney et al. 2016; Abdalla et al. 2019; Thapa et al. 2018; Moreno-Cadena et al. 2024). Where longer growing windows are available, more complex cover crop mixes can be planted (Kaye et al. 2019), but in the large temperate zone maize and soybean production areas, late harvest of summer crops limits cover crop options to species that grow in cooler conditions (Ruis et al. 2019). Legume options in northern regions are limited to winter hardy species since most N fixation occurs in the spring (Holderbaum et al. 1990; Ranells and Wagger 1992; Clark 2007).

Adding legumes to grass cover crops has been shown to offset the N competition effects of grass cover crops alone (Abdalla et al. 2019; Lewis et al. 2018), but mixtures were not found to be universally superior to monoculture cover crops when considering general production benefits (Florence and McGuire 2020). Grass/legume mixtures can provide similar N leaching reduction as pure grass cover crops (Thapa et al. 2018). Not surprisingly, function of cover crop mixtures varies with the composition of the mixture and the availability of soil N (Hayden et al. 2014; Poffenbarger et al. 2015; Kaye et al. 2019; White et al. 2017), with high soil N favoring growth of the grass component and low soil N favoring legumes (Clark 1994; Bourgeois et al. 2022; He et al. 2025; Qin et al. 2023; Ma et al., 2025). While mixtures with legumes offer a solution to possible reduced N availability following grass cover crops, they add a direct input cost and complexity to cover crop management. Cover crop legume seed costs are generally higher than for many of the grasses and brassicas (Snapp et al. 2005; Hayden et al. 2014), so just neutralizing N-related yield decreases after monoculture grass cover crops may not increase cover crop cost-effectiveness. Since seed of some winter hardy legumes (e.g. crimson clover, *Trifolium incarnatum*) is much smaller than most grass cover crop seed, shallower planting depth is required than for winter cereals (e.g. rye, *Secale cereale*), and additional equipment or field passes may be required. In addition, legumes generally need to be planted earlier in the fall than winter cereals to ensure winter survival and need to be terminated later in the spring to allow time for significant N fixation and may require an additional herbicide where chemical termination is used (Clark 2007; Duiker and Curran 2005). Later termination can delay planting of the primary crop which may reduce yield potential (Chatterjee et al. 2020).

The cost savings from planting monoculture legume cover crops or mixing them with grasses come from reducing the need for inorganic N applications (Clark 2007). While not a direct benefit to farmers, reductions in inorganic N applications following legume cover crops also are a major component of cover crop effects on greenhouse gas emissions (Kaye and Quemada 2017), since fertilizer N inputs are the major source of carbon emissions from maize production (Holka and Bienkowski 2020). A wide range in reductions in inorganic N applications has been reported following legume/grass mixtures depending on the mixture, growing conditions, soil N status, and spring termination date and method (White et al. 2017). One of the key management challenges associated with legume/grass cover crop mixtures is accurately estimating the extent to which N applications can be reduced without limiting crop yield (Bergtold et al. 2017). Without adjustments to fertilizer N inputs following legume cover crops, it is unlikely that the added cost and management associated with legume cover crops will be offset. Standard soil nitrate tests can reliably estimate N needs of maize following incorporated legume cover crops but less reliably after spring termination in a no-till setting (Vyn et al. 2000). While most of the focus on legume cover crops has been for supplying N to maize and other nonlegume crops, there is evidence that soybean yields in high yield settings may be N limited (La Menza et al. 2017) creating an option for increasing the cost-effectiveness of legume cover crops by increasing soybean yields and therefore the likelihood of implementation.

5 Challenges in managing soil moisture with cover crops

Although not as intensely studied as N availability, the effect of cover crops on soil moisture is the other key issue that has been linked to crop yields and profitability which ultimately affects implementation decisions. Cover crops are credited with long-term improvements in soil moisture holding capacity by increasing soil organic matter and overall soil quality (Frye et al. 1985; Basche et al. 2016). Short-term effects can both increase and decrease soil moisture availability as a result of transpiration by a living cover crop and reduced evaporation from the soil surface by dead cover crop residue (Smith et al. 1987; Clark et al. 1997b). Cover crop impacts on soil water are especially critical in production areas with limited rainfall where total root zone water storage at the beginning of the growing season is critical (Nielsen et al. 2016; Nielsen et al. 2016; Mitchell et al. 2015; Unger and Vigil 1998) and can reduce yields even while adding N in the case of legume cover crops and protecting the soil from erosion. In a global meta-analysis Garba et al. (2022) identified a range of cover crop soil moisture and nitrogen impacts on crop yields that varied from positive to negative moving from wetter to drier climates and identified 700 mm annual precipitation as the level below which cover crop moisture competition was likely to limit crop yields.

While overall climate is critical in determining cover crop effects on total water availability during the growing season, short-term management also can play a critical role. Delaying cover crop termination to produce more biomass, which generally is beneficial, can increase soil moisture depletion during the critical cash crop establishment phase, even if later termination increases soil moisture availability later in the growing season by providing more soil cover (Qin et al. 2021). Terminating cover crops well in advance of crop planting minimizes yield impacts where growing season precipitation is adequate (Basche et al. 2016), but some N fixation by a legume cover crop will be sacrificed (Frye et al. 1985). Clark et al. (1997b) reported that summer water conservation by cover crop residues was more important than spring soil moisture depletion due to late terminated cover crops in determining corn yields in a region with more than 1000 mm of annual precipitation. Since the timing of termination affects both positive and negative cover crop services, the full range of effects of spring termination strategies needs to be considered to achieve the highest overall benefit, and particularly to avoid any negative impacts on crop yields which are known to strongly influence farmers' decisions to use cover crops.

6 Cover crops as forage: a win-win system

While a lack of positive return on investment has been broadly reported for cover crop use in cash grain production in the short term (Plastina et al. 2018; Bowman et al. 2025), more positive outcomes are generally reported where systems include livestock, whether by grazing cover crops or harvesting them for feed (CTIC 2023; Arbuckle and Roesch-McNally 2015; SARE 2019; Bowman et al. 2024; Ketterings et al. 2015b). Farm-scale studies of dairies in the Northeastern US have shown that winter cover crops can reduce erosion and N loss, and also increase total annual forage yields, reducing the need for off-farm feed and bringing farm nutrient budgets more into balance (Ketterings et al. 2015a). Similarly, winter cover crops in the Southeastern US were recommended as a viable forage option with grazing resulting in enhanced soil microbial activity (Franzluebbers and Stuedemann 2015). Since silage harvest, which tends to be associated with livestock production, occurs earlier than maize grain harvest, longer windows are available for fall cover crop growth which has been shown to be a strong predictor of cover crop biomass production (McClelland et al. 2021). In addition, since cropland associated with concentrated livestock operations often has enriched soil N status due to repeated manure applications (White et al. 2023), N limitations on crop yields due to cover crops are less likely. The combination of long cover crop growing windows and adequate plant available N creates the potential for increasing overall annual production by as much as 30% without the risk of reducing yields of the succeeding crop (Ruis et al. 2019; Baraibar et al 2020). The extra forage value of the cover crop is a short-term return on investment in cover crop planting (Bowman et al. 2024), and 76% of farmers who grazed their cover crops reported them as profitable (CTIC 2023). Soil health studies have found that integrating livestock into grain production, including grazing cover crops, has neutral to positive effects on soil health under proper management, but longer-term studies are needed (Franzluebbers and Stuedemann 2015: Anderson et al. 2022). Since livestock production often is associated with silage crops that have minimal residue cover following harvest and often receive manure applications, cover crops can help reduce the high levels of sediment and nutrient loss often associated with animal production (Hively et al. 2015; Hall et al. 2001). Using cover crops as forage is one of the most promising options for overcoming the implementation challenge of profitability while at the same time helping reduce water quality impacts and the overall carbon footprint of agriculture.

7 Overcoming the profitability challenge: the case for subsidies

Cover crops have long been shown to provide a wide range of ecosystem services related to soil and water quality, and more recently to help reduce agricultural greenhouse gas emissions. Cover crops also can improve farm profitability with proper management in the right setting. But despite intensified outreach and education programs, generally positive farmer attitudes, and evidence of positive long-term returns on planting and management costs, persistently low cover crop adoption rates in most regions suggest that subsidies may be needed when high levels of cover crop adoption are needed to help solve large-scale off-farm problems such as degraded water quality and greenhouse gas emissions (Wallander et al. 2021). The most reported exception is where cover crops can be utilized as forage, providing a short-term return on investment (SARE 2019; Bowman et al. 2024). This challenge is further complicated as grass cover crops generally provide the greatest overall environmental benefits but also are most likely to reduce crop yields, while legume cover crops generally have neutral to positive impacts on crop yields but are less effective for erosion control, reducing N losses and mitigating climate change (McClelland et al. 2025). So, while any cover crop provides some environmental benefit, for solving specific problems it may be necessary to encourage cover crop types and management approaches that would not be the most cost-effective for producers and therefore not likely to be implemented.

In the language of economists, cover crops provide both private and public goods, but the costs are private, that is, they are paid by the farmer (Wallander et al. 2020). Farmers are making rational decisions when focusing on the private goods, the on-farm or agronomic benefits provided by cover crops, when deciding whether or not to incur the costs to plant cover crops (Libby 1985). In contrast, much of the current interest in promoting cover crops is related to the public goods they provide such as improved water quality and reductions in greenhouse gas emissions (Kladivko et al. 2014; Kaye and Quemada 2017; Abdalla et al. 2019; Tellatin and Myers 2018; McClelland et al. 2021). Many studies about cover crop use have concluded that subsidies will likely be needed for cover crop implementation to reach levels needed to solve major environmental challenges (Snapp et al. 2005; Roesch-McNally et al. 2017: Qin et al. 2021; Schnitkey et al. 2024; Woodcock et al. 2025). Conservation practices that provide little direct benefit to farmers have been implemented almost exclusively using public support. For example, the USDA Conservation Reserve Program achieves conservation services on approximately 12 million ha through annual rental payments in exchange for converting cropland to natural vegetation. Almost a century ago, Federal direct payments were used during the Dust Bowl erosion crisis to convert nearly 12 million ha of US cropland to less erosion prone crops (Wells et al. 1936). Like cover crops, no-till methods are another conservation practice that has been widely promoted (e.g. USDA Conservation Stewardship Program, https://www.nrcs.usda.gov/programs-initiatives/conservation-stewardship-program; Maryland Income

Tax Subtraction for Conservation Equipment, https://mda.maryland.gov/resource conservation/documents/taxsubtraction.pdf) that also provides both on-farm benefits in addition to reducing water quality degradation due to soil erosion. Like cover crops, no-till and other reduced tillage methods also help reduce greenhouse emissions and improve soil health (McClelland et al. 2025). Gradually, with research support and technological advances, reduced tillage methods became widely viewed as providing enough on-farm benefits to be profitable, resulting in high rates of implementation without subsidies, a trend that has not developed to the same extent for cover crops (Myers and LaRose 2019).

Using subsidies to increase cover crop implementation is a model that has been used for water quality benefits in the Chesapeake Bay watershed (Figs. 1 and 2), and in combination with outreach efforts has led to higher implementation rates than throughout most of the US (Wallander et al. 2021). More recently, private schemes have emerged that offer payments for the carbon sequestration function of cover crops (Plastina et al. 2024). With current economic and policy conditions and the documented challenges of making cover crops profitable in the large commodity-producing regions, it appears likely that subsidies will be needed to tip the balance into a profitability range that will markedly increase implementation rates. This is especially the case for cover crop options that provide the most environmental benefits but that are not the most cost-effective options for farmers (McClelland et al. 2025). While subsidies may be needed to significantly increase cover crop implementation rates, continued efforts to develop cover crop approaches that are more

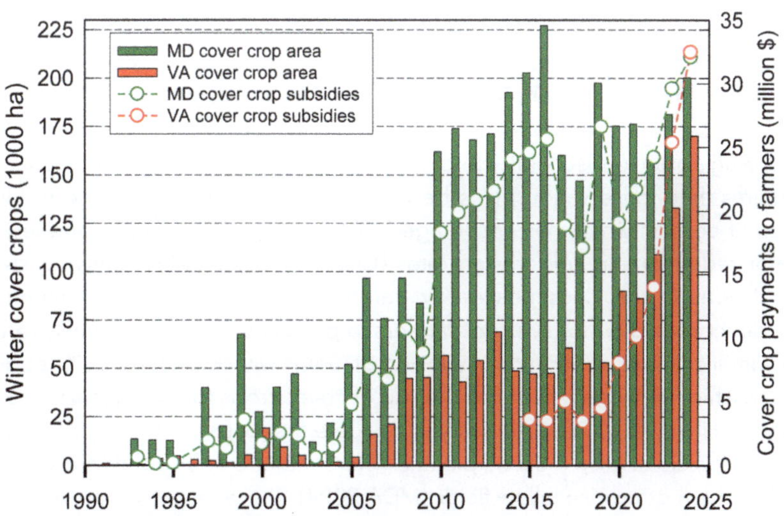

Figure 1 Winter cover crop area and total payments to farmers in Maryland and Virginia regions of the Chesapeake Bay watershed from 1990 to 2024.

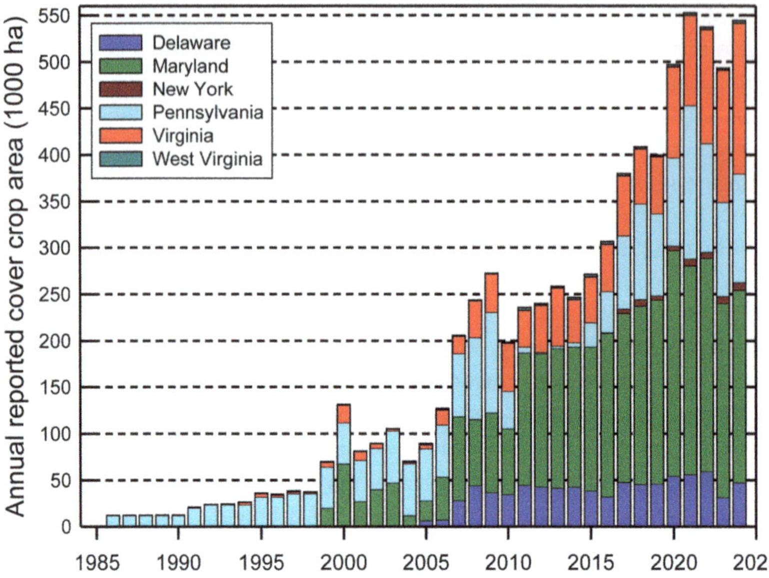

Figure 2 Winter cover crop area reported to the Chesapeake Bay Program for nutrient reduction credit for the states in the Chesapeake Bay watershed from 1985 to 2024.

cost-effective and systems to deliver that information to farmers will lessen the incentive levels needed to encourage implementation and the public funding needed to reach implementation goals.

While direct subsidies have been effective for increasing cover crops and other conservation practices and have been broadly identified as critical to further increase cover crop implementation rates, indirect financial incentives that require cover crops for eligibility for other benefits are another option for increasing implementation. Federal farm policy in the US requires conservation compliance to be eligible for participation in Federal support programs, including premium-subsidized crop insurance. Conservation compliance requires that conservation plans be put in place on cropland prone to high rates of erosion which has encouraged the use of reduced tillage and led to major reductions in soil erosion rates (Claassen et al. 2017). Similarly, state regulations in the Chesapeake Bay watershed provide incentives, although not direct subsidies, for livestock farmers to plant cover crops. In Pennsylvania, no winter manure application is allowed on fields without at least 25% residue cover, which essentially requires the use of cover crops following maize silage harvest (Hively et al. 2015). In Maryland, where the focus is on reducing nitrate leaching, a cover crop must be planted where manure or biosolids are applied in the fall on otherwise fallow fields (MDA 2024). These incentives are a hybrid of subsidy and regulatory approaches in that planting a cover crop is required to gain access to savings in manure storage and hauling costs. With the rise of nitrate as a contaminant of drinking water supplies and a driver of coastal water quality degradation (Staver 2020), including reduction of nitrate losses as part of Federal conservation compliance would be an option for increasing cover crop planting nationwide, similar to limited state-scale hybrid approaches. To date, few cases exist where pure regulatory approaches have been used to increase cover crop planting. Kathage et al. (2022) reported that policy is the strongest determinant of cover crop planting rates in the European Union where adoption rates also have remained low. They conclude that either stronger policy obligations or subsidies would be needed to increase adoption rates. Discussions regarding carrots (subsidies) and sticks (regulation) to achieve conservation goals have been a long-running discussion in the Chesapeake Bay restoration effort and elsewhere (Cook 1985), but with current limited requirements for cover crop planting, it appears that subsidies will be needed to achieve major increases in cover crop adoption where livestock are not integrated with crop production.

8 Case study: promoting cover crop use in the Chesapeake Bay watershed

In the 1980s an effort began to restore water quality in Chesapeake Bay, with the primary focus being on reducing elevated nutrient inputs which were driving excessive algae growth and formation of oxygen-depleted dead zones, as well as the dieback of rooted vascular plant communities (Kemp et al. 2005). Nitrogen in runoff from cropland was thought to be a primary source of N inputs in agriculturally dominated regions of the watershed, with nitrate-N loads transported via subsurface flow identified as the dominant route of delivery to surface waters (Bachman and Phillips 1996; Sanford and Pope 2013). Cover crops were identified as a potential solution, especially for subsurface nitrate-N loads.

The Bay watershed includes multiple states with varying dominant agricultural systems ranging from dairy in the northern part of the watershed to grain and concentrated poultry production in the Coastal Plain region of the watershed adjacent to the Bay. Maize and soybeans are the predominant crops, with maize silage being a major crop in livestock-producing areas. Cover crops are promoted throughout the watershed as part of the overall effort to reduce nutrient inputs to the Bay. The state of Maryland, which surrounds the northern half of the Bay, adopted cover crops as its primary agricultural N reduction practice in its first comprehensive strategy developed in the mid-1990s. But even modest goals proved very difficult to achieve for the next decade. Implementation resulted mostly from limited state subsidies. Notable is that most grain production in Maryland is not closely associated with livestock production, so options for utilization of cover crops as forage are limited. As frustration grew regarding a lack of progress on reaching N reduction goals, significant state funds were dedicated to increasing cover crop implementation in 2010. Farmers were quick to respond to incentives, and implementation grew over the next decade with additional public investment (Fig. 1). Maryland now has the highest percentage of cropland planted with cover crops in the US (Wallander et al. 2021). The focus of the program has been on cereal grain cover crops to reduce nitrate leaching with a tiered payment schedule to encourage early planting and the most effective species and planting methods (MDA 2025). The US Environmental Protection Agency (EPA) Chesapeake Bay watershed accounting system for tracking progress credits cover crops using a similar set of metrics related to effectiveness for reducing N losses (US EPA CPB 2017). In addition to incentive programs, Maryland also has regulations that require cover crops or cereal grains for harvest to be planted on any cropland where manure will be applied in the fall, which typically is a necessity for dairies and other livestock operations (MDA 2024). Virginia, which surrounds the lower half of Chesapeake Bay, went through a similar pattern of cover crop adoption with funding and adoption trending up sharply in the last five years (Fig. 1). Current payments to farmers in both states are approximately 150 USD/ha.

Dairy-producing regions in the Chesapeake Bay watershed have focused more on outreach and technical assistance to help operators utilize cover crops as a forage resource and conservation practice (Hively et al. 2015; Ketterings et al. 2015a). Since forage utilization often makes cover crops profitable (CTIC 2023; SARE 2019), adoption occurs more frequently in the absence of subsidies (Bowman 2024), making tracking of adoption somewhat more difficult than where farmers are enrolled in cover crop cost-share programs as in MD and VA. In PA, Hively et al. (2015) used satellite imagery to document increasing rates of adoption following a targeted outreach effort in settings where forage utilization is an option. Cover crop use in these settings is especially beneficial for meeting water quality goals since manure applications increase the potential for N losses. In both MD and PA there also are regulations requiring that cover crops be planted where manure is applied in the fall on cropland that otherwise would be fallow. The Chesapeake Bay restoration effort has used a multilayered approach that after more than three decades has resulted in cover crop adoption levels finally approaching target levels, by providing technical support to help farmers maximize on-farm benefits, direct subsidies to cover planting costs and minor yield losses, and policy incentives to encourage cover crop use where organic nutrient sources are applied (Fig. 2).

9 Future research needs

Studies of cover crop adoption often find that a lack of knowledge about how best to manage cover crops is a primary barrier to adoption (Snapp et al. 2005; Golden et al. 2023; Arbuckle and Roesch-McNally 2015; Roesch-McNally et al. 2017) and that farmers with management skills developed through experience and with input from crop advisors are most likely to achieve successful outcomes and continue to use cover crops (Dunn et al. 2016; CTIC 2025). This supports continuation of cover crop research to identify cost-effective cover crop management approaches that achieve both on-farm and public environmental objectives across the full spectrum of soil and climate conditions and production systems (Lal 2015; Daryanto et al. 2018; Kladivko 2020). Information presented to farmers developed from research in their unique environment is likely to be the most useful and accepted (Abdalla et al. 2019; Baumgart-Getz et al. 2012; Dunn et al. 2016; Reimer et al. 2012). Prokopy et al. (2019) suggested a need for research considering how farmers develop attitudes toward conservation practices, since attitudes toward environmental stewardship play a key role in farmers' willingness to implement conservation practices. Since many on-farm cover crop benefits and positive returns on investment take continued use to fully develop, long-term research that considers the full range of benefits is especially critical to give farmers confidence to take a longer-term perspective when they first adopt cover crops (Daryanto et al. 2018; DeVincentis et al. 2020; SARE 2019; Bowman et al. 2025).

A recurring conclusion of studies on cover crop effects on crop yields and profitability is that reduced soil N and water availability often lead to decreased yields and negative financial outcomes which is an often-cited concern in surveys of farmer attitudes about cover crops (CTIC 2023). Thorup-Kristensen et al. (2002) comprehensively reviewed cover crop effects on N in temperate zones and identified more accurate tools to predict fertilizer N applications following cover crops as a critical research need. With the recent intense interest in legume/grass cover crop mixtures as a way to overcome negative yield effects from grass cover crops (Abdalla et al. 2019), this topic remains a pressing research need, especially given the trend toward more diverse cover crop mixtures and later termination dates to maximize biomass production. Since reductions in N fertilizer applications following legume cover crops are critical to a positive return on investments in legumes, and also for reducing overall greenhouse gas emissions and nitrate leaching, having tools to accurately predict N application needs is critical to optimizing both on-farm and off-farm benefits. While not as complex as the situation for N, developing cover crop management approaches for optimizing soil water availability remains a critical research need, especially in regions with lower water availability, with termination timing a key management factor.

There is a long history of cover crop research, which until the last few decades, mostly addressed on-farm benefits. The impetus for much of the recent cover crop research is related to their potential to provide off-farm benefits such as improved water quality and an overall reduction in greenhouse gas emissions, benefits that provide little return on investment to farmers. The current challenge is developing approaches that achieve off-farm benefits but also provide on-farm benefits sufficient to encourage higher levels of adoption. The on-farm economics of cover crops emerge repeatedly as the critical factor determining adoption rates, but the off-farm economics, that is, the value of public benefits, also need to be well understood to develop policies and incentive programs that lead to cover crop adoption levels deemed necessary to meet objectives (Wallander et al. 2020). At present, despite focused research efforts to optimize on-farm benefits and considerable outreach and technical support efforts to encourage cover crop planting, adoption rates overall remain well below target levels. This suggests that public support will be needed if cover crops are to reach their full potential as a solution to larger environmental problems. An integrated research approach is needed that: first quantifies the ability of different cover crops to achieve off-farm benefits, then addresses how to optimize on-farm benefits from these types of cover crops and their effect on profitability, and finally considers the policy options (indirect incentives) and subsidies that will be necessary to develop strategies to achieve adoption goals. Current adoption rates suggest that focusing only on the on-farm benefits will lead to adoption of cover crop types that are profitable to farmers, but failure to consider on-farm economics when trying to use cover crops as a solution to off-farm problems will likely lead to insufficient adoption rates to achieve goals in the absence of policy incentives or subsidies to tip on-farm economics into the positive range.

10 Conclusion

The many potential benefits of cover crops have been studied extensively and covered thoroughly in earlier chapters. While much of the recent focus is on their potential role in solving large-scale environmental problems, farmers' willingness to adopt cover crops is determined primarily by on-farm benefits and effects on profitability. Technical assistance to help farmers maximize on-farm benefits from cover crops has expanded dramatically, but adoption rates – while increasing – have remained below target levels. Numerous studies have reported that the effects of reduced soil water and N availability on crop yields result in negative economic outcomes. Long-term locally relevant research is needed to address these challenges. Only when the on-farm profitability barrier is overcome will cover crop adoption reach levels needed to make major contributions to solving off-farm environmental problems. The four-decade-long effort to restore Chesapeake Bay has relied heavily on cover crops for reducing N inputs from agricultural activity in the watershed. The trajectory of that effort has borne out the general research findings that, despite the many ecosystem services provided by cover crops, their effects on profitability limit adoption. The exception to this is integrated crop-livestock operations, where the forage value of cover crops has led to increased adoption. After several decades of very slow progress toward cover crop implementation goals, expanded incentive programs overcame the on-farm economic barriers and resulted in rapid increases in cover crop adoption. Since favorable cover crop outcomes have been linked to producer management skills and experience, the incentive programs also support the development of cover crop management capacity of the entire agricultural system. Where livestock utilization of cover crops as forage is not an option and funding for subsidies is limited, regulatory approaches may be needed if cover crops are to yield their full potential of environmental benefits.

11 Where to look for further information

11.1 Cover crop selection and management

- Clark, A. SARE 2007.
- Midwest Cover Crops Council – https://www.midwestcovercrops.org/.
- Northeast Cover Crops Council – https://northeastcovercrops.com/.
- Center for Regenerative Agriculture, University of Missouri – https://cra.missouri.edu/.

11.2 Cover crop economic and policy barriers

- Bergtold, J.S., et al. (2017), A review of economic considerations for cover crops as a conservation practice, *Renewable Agriculture and Food Systems*, 34(1), 62–76.
- Roesch-McNally, G.E., et al. (2017), The trouble with cover crops: farmer' experience with overcoming barriers to adoption, *Renewable Agriculture and Food Systems*, 33(4), 322–333.
- Plastina, A., et al. (2018), Cover crops use in Midwestern US agriculture: perceived benefits and net returns. *Renewable Agriculture and Food Systems*, 35, 38–48.
- SARE, (2019), Cover crop economics: opportunities to improve your bottom line in row crops, SARE technical bulletin, USDA Sustainable Agriculture Research and Education, June 2019.
- Kathage, J., et al. (2022), How much is policy driving the adoption of cover crops? Evidence from four EU regions. *Land Use Policy*, 116, 106016.
- Schnitkey, G.D., et al. (2024), Cover crops, farm economics, and policy. *Applied Economic Perspectives and Policy*, 46, 595–608.

11.3 Cover crop soil nitrogen dynamics

- Thorup-Kristensen, K., et al. (2002), Catch crops and green manures as biological tools in nitrogen management in temperate zones, *Advances in Agronomy*, 79, 227–300.

11.4 Cover crop forage potential

- Bowman, M., et al. (2024), *Cover Crops on Livestock Operations: Potential for Expansion in the United States*, Report No. AP-120, U.S. Department of Agriculture, Economic Research Services.

12 References

Abdalla, M., et al. (2019), A critical review of the impacts of cover crops on nitrogen leaching, net greenhouse gas balance and crop productivity. *Global Change Biology*, 25, 2530–2543.

Anderson, L.K., et al. (2022), Cover crop grazing impacts on soil properties and crop yields under irrigated no-till corn-soybean management. *Soil Science Society of America Journal*, 86(1), 118–133.

Aiken, R.M., et al. (2013), Replacing fallow with continuous cropping reduces crop water productivity of semiarid wheat, *Agronomy Journal*, 105(1), 199–207.

Arbuckle Jr., J.G. and Roesch-McNally, G. (2015), Cover crop adoption in Iowa: The role of perceived practice characteristics, *Journal of Soil and Water Conservation*, 70(6), 418–429.

Bachman, L.J. and Phillips, P.J. (1996), Hydrologic landscapes on the Delmarva Peninsula Part 2: Estimates of base-flow nitrogen load in Chesapeake Bay, *Journal of the American Water Resources Association*, 32, 779–791.

Baraibar, B., et al. (2020), Cover crop mixture expression is influenced by nitrogen availability and growing degree days, *PLoS One*, 15(7) e0253868.

Basche, A., et al. (2016), Soil water improvements with the long-term use of a winter rye cover crop, *Agricultural Water Management*, 172, 40–50.

Baumgart-Getz, A., et al. (2012), Why farmers adopt best management practice in the United States: a meta-analysis of the adoption literature, *Journal of Environmental Management*, 96, 17–25.

Bergtold, J.S., et al. (2017), A review of economic considerations for cover crops as a conservation practice, *Renewable Agriculture and Food Systems*, 34(1), 62–76.

Blanco, H. (2023), *Cover Crops and Soil Ecosystem Services*, American Society of Agronomy, Crop Science Society of America, Soil Science of America, Hoboken, NJ.

Blevins, R.L., et al. (1990), Legume cover crops as a nitrogen source for no-till corn and grain sorghum, *Agronomy Journal*, 82(4), 769–772.

Bourgeois, B., et al. (2022), Interactive effects between cover crop management and the environment modulate benefits to cash crop yields: a meta-analysis, *Canadian Journal of Plant Science*, 102, 656–678.

Bowman, M., et al. (2024), *Cover Crops on Livestock Operations: Potential for Expansion in the United States*, Report No. AP-120, U.S. Department of Agriculture, Economic Research Services.

Bowman, M., et al. (2025), *Economics Outcomes of Soil Health and Conservation Practices on U.S. Cropland*, Report No. ERR-353, U.S. Department of Agriculture, Economic Research Services.

Cates, A.M., et al. (2018), What do we know about cover crops efficacy in the North Central United States? *Journal of Soil and Water Conservation*, 73(6), 153A–157A.

Chatterjee, N., et al. (2020), Simulating winter rye cover crop production under alternative management in a corn-soybean rotation, *Agronomy Journal*, 112, 4648–4665.

Claassen, R., et al. (2017), Conservation compliance: how farmer incentives are changing in the crop insurance era, ERR-234, Economic Research Service/USDA.

Clark, A. (2007), *Managing Cover Crops Profitably*, 3rd ed. Sustainable Agriculture Research and Education Program of CSREES, U.S. Department of Agriculture.

Clark, A.J., et al. (1994), Seeding rate and kill date effects on hairy vetch-cereal rye cover crop mixtures for corn production, *Agronomy Journal*, 86, 1065–1070.

Clark, A.J., et al. (1997a), Kill date of vetch, rye, and a vetch-rye mixture: I. cover crop and corn nitrogen, *Agronomy Journal*, 89, 427–434.

Clark, A.J., et al. (1997b), Kill date of vetch, rye, and a vetch-rye mixture: II. soil moisture and corn yield, *Agronomy Journal*, 89, 434–441.

Cook, K. (1985), Agricultural nonpoint pollution control: a time for sticks? *Journal of Soil and Water Conservation*, 40(1), 105–106.

Crespo, C., et al. (2024), Contrasting soil management systems had limited effects on soil health and crop yields in a North Central US Mollisol, *Journal of the Indian Society of Soil Science*, 88, 1723–1735.

CTIC (Conservation Technology Information Center) (2025), National Cover Crop Survey Report: Agricultural Advisors, Joint publication of the CTIC and Sustainable Agriculture Research & Education, American Seed Trade Association, https://www.ctic.org/data/Cover_Crops_Research_and_Demonstration_Cover_Crop_Survey.

CTIC (Conservation Technology Information Center) (2023), National Cover Crop Survey Report 2022–2023, Joint publication of the CTIC and Sustainable Agriculture Research & Education, American Seed Trade Association, https://www.ctic.org/data/Cover_Crops_Research_and_Demonstration_Cover_Crop_Survey.

Daryanto, S., et al. (2018), Quantitative synthesis on the ecosystem services of cover crops, *Earth-Science Reviews*, 185, 357–373.

Das, S., et al. (2022), Soil health, cover crop, and fertility management: Nebraska produces' perspectives on challenges and adoption, *Journal of Soil and Water Conservation*, 77(2), 126–134.

Decker, A.M., et al. (1994), Legume cover crop contributions to no-tillage corn production, *Agronomy Journal*, 86, 126–135.

Delgado, J.A., et al. (2021), Potential use of cover crops for soil and water conservation, nutrient management, and climate change adaptation across the tropics, *Advances in Agronomy*, 165, 175–247.

Deines, J.M., et al. (2023), Recent cover crop adoption is associated with small maize and soybean yield losses in the United States, *Global Change Biology*, 29, 794–807.

DeVincentis, A.J., et al. (2020), Using cost-benefit analysis to understand adoption of winter cover cropping in California's specialty crop systems, *Journal of Environmental Management*, 261, 110205.

Dou, Z., et al. (1994), Tillage effect on seasonal nitrogen availability in corn supplied with legume green manures, *Plant and Soil*, 162, 203–210.

Duiker, S.W. and Curran, W.S. (2005), Rye cover crop management for corn production in the Northern Mid-Atlantic Region, *Agronomy Journal*, 97, 1413–1418.

Dunn, M., et al. (2016), Perceptions and use of cover crops among early adopters: Findings from a national survey, *Journal of Soil and Water Conservation*, 71(1), 29–40.

Ebelhar, S.A., et al. (1984), Nitrogen from legume cover crops for no-tillage corn, *Agronomy Journal*, 76, 51–55.

Finney, D.M., et al. (2016), Biomass production and carbon/nitrogen ratio influence ecosystem from cover crop mixtures, *Agronomy Journal*, 108(1), 39–52.

Florence, A.M. and McGurie, A.M. (2020), Do diverse cover crop mixtures perform better than monocultures: a systematic review, *Agronomy Journal*, 112, 3513–3534.

Franzluebbers, A.J. and Stuedemann, J.A. (2015), Does grazing of cover crops impact biologically active soil carbon and nitrogen fractions under inversion or no tillage management? *Journal of Soil and Water Conservation*, 70(6), 365–373.

Frye, W.W., et al. (1985), Economics of winter cover crops as a source of nitrogen for no-till corn, *Journal of Soil and Water Conservation*, 40(2), 246–249.

Garba, I.I., et al. (2022), Cover crop legacy impacts on soil water and nitrogen dynamics, and on subsequent crop yields in drylands: a meta-analysis, *Agronomy for Sustainable Development*, 42, 34.

Golden, L.A., et al. (2023), Benefits, barriers, and use of cover crops in the western United States: regional survey results, *Journal of Soil and Water Conservation*, 78(3), 260–271.

Hall, M.D., et al. (2001), Regional nitrate leaching variability: What makes a difference in Northeastern Colorado, *Journal of the American Water Resources Association*, 37(1), 139–150.

Hargrove, W.L. (1991), *Cover Crops for Clean Water*, The proceedings of an international conference West Tennessee Experiment Station, Jackson, TN, Soil Water Conservation Society, Ankey, IA.

Hargrove, W.L. (1986), Winter legumes as a nitrogen source for no-till grain sorghum, *Agronomy Journal*, 78, 70–74.

Hayden, Z.D., et al. (2014), Rye-Vetch mixture proportion tradeoffs: cover crop productivity, nitrogen accumulation, and weed suppression, *Agronomy Journal*, 106(3), 904–914.

He, Q. et al. (2025), Optimizing cover crop application for sustainable crop prod. *Sustainable agriculture,* 3(1), 10.

Hively, W.D., et al. (2015), Remote sesing to monitor cover crop adoption in Southeastern Pennsylvania, *Journal of Soil and Water Conservation*, 70(6), 340–352.

Holderbaum, J.F., et al. (1990), Fall-seeded legume cover crops for no-tillage corn in the humid east, *Agronomy Journal*, 82, 117–124.

Holka, M. and Bieńkowski, J. (2020), Carbon footprint and life-cycle costs of maize production in conventional and non-inversion tillage systems, *Agronomy*, 10, 1877. https://doi.org/10.3390/agronomy10121877.

Kathage, J., et al. (2022), How much is policy driving the adoption of cover crops? Evidence from four EU regions, *Land Use Policy*, 116, 106016.

Kaye, J.P. and Quemada, M. (2017), Using cover crops to mitigate and adapt to climate change: a review, *Agronomy for Sustainable Development*, 37, 4. https://doi.org/10.1007/s13593-016-0410-x.

Kaye, J., et al. (2019), Managing nitrogen through cover crop species selection in the U.S. mid-Atlantic, *PLoS One*, 14(4), e0215448. https://doi.org/10.1371/journal.pone.0215448.

Kemp, W.M., et al. (2005), Eutrophication of Chesapeake Bay: historical trends and ecological interactions, *Marine Ecology Progress Series*, 303, 1–29.
Ketterings, Q.M., et al. (2015a), Integrating cover crops for nitrogen management in corn systems on the Northeastern U.S. dairies, *Agronomy Journal*, 107(4), 1365–1376.
Ketterings, Q.M., et al. (2015b), Winter cereals as double crops in corn rotations on New York dairy farms, *Journal of Agricultural Science*, 7(2), 18–25.
Kladivko, E.J., et al. (2014), Cover crops in the upper midwestern United States: potential adoption and reduction of nitrate leaching in the Mississippi River Basin, *Journal of Soil and Water Conservation*, 69(4), 279–291.
Kladivko, E.J. (2020), Cover crops: progress and outlook, In Delgado, J.A., et al. (eds.), *Soil and Water Conservation: A Celebration of 75 Years*, Ankeny, IA: Soil and Water Conservation Society, pp. 293–302.
La Menza, N.C., et al. (2017), Is soybean yield limited by nitrogen supply? *Field Crops Research*, 213, 204–212.
Lal, R. (2015), Soil carbon sequestration and aggregation by cover cropping, *Journal of Soil and Water Conservation*, 70(6), 329–339.
Lewis, K.L., et al. (2018), Soil benefits and yield limitations of cover crop use in Texas high plain cotton, *Agronomy Journal*, 110(4), 1616–1623.
Libby, L.W. (1985), Paying the nonpoint pollution control bill, *Journal of Soil and Water Conservation*, 40(1), 33–36.
Lu, Y., et al. (2000), Cover crops in sustainable food production, *Food Reviews International*, 16(2), 121–157.
Ma, J. et al. 2025. Legume-non-legume cover crop mixtures enhance soil nutrient availability and physical properties. *Agronomy*, 15(8), 1756.
MacLaren, C., et al. (2022), Long-term evidence for ecological intensification as a pathway to sustainable agriculture, *Nature Sustainability*, 5, 770–779.
Marcillo, G.S. and Miguez, F.E. (2017), Corn yield response to winter cover crops: an updated meta-analysis, *Journal of Soil and Water Conservation*, 72(3), 226–239.
McClelland, S.C., et al. (2021), Management of cover crops in temperate climates influences soil organic carbon stocks: a meta-analysis, *Ecological Applications*, 31(3), e02278.
McClelland, S.C., et al. (2025), Managing for climate and production goals on crop-lands, *Nature Climate Change*. https://doi.org/10.1038/s41558-025-02337-7.
MDA (2024), Maryland department of agriculture nutrient management regulations, https://mda.maryland.gov/resource_conservation/Documents/nm_manual/1-D1-1-1D1-6.pdf).
MDA (2025), Maryland Department of Agriculture cover crop cost-share program, https://mda.maryland.gov/resource_conservation/pages/cover_crop.aspx.
Meisinger, J.J., et al. (1991), In Hargrove, W.L. (ed.), *Cover Crops for Clean Water,* Soil and Water Conservation Society, The proceedings of an international conference, West Tennessee Experiment Station, Jackson, TN.
Mitchell, J.P., et al. (2015), Trade-offs between winter cover crop production and soil water depletion in the San Joaquin Valley, California, *Journal of Soil and Water Conservation*, 70(6), 430–440.
Mitchell, W.H. and Teel, M.R. (1977), Winter-annual cover crops for no-tillage corn production, *Agronomy Journal*, 69(4), 569–576.
Moreno-Cadena, P., et al. (2024), Productivity benefits of cereal-legume cover crop mixtures under variable soil nitrogen and termination times, *European Journal of Agronomy*, 155, 127114.

Myers, R. and LaRose, J. (2019), *Progress Report: Adoption of Soil Health Systems Based on Data from the 2017 U.S. Census of Agriculture*. Soil Health Institute, https://soilhealthinstitute.org/app/uploads/2022/01/Soil-Health-Adoption-Overview_2017-Census-Report_FINAL.pdf.

Nelson, N.O. and Janke, R.R. (2007), Phosphorus sources and management in organic production systems, *HorTechnology*, 17(4), 442–454.

Nielsen, D.C. and Vigil, M.F. (2005), Legume green fallow effect on soil water content at wheat planting and wheat yield, *Agronomy Journal*, 97, 684–689.

Nielsen, D.C., et al. (2016), Cover crop effects on subsequent wheat yield in the Central Great Plains, *Agronomy Journal*, 108(1), 243–256.

Pieters, A.J. and McKee, R. (1938), The use of cover and green-manure crops, In *Soils and Men, Yearbook of Agriculture 1938* USDA, 75th Congress, 2d Session, House Document No. 398, Washington, DC, pp. 431–444.

Pires, C.B., et al. (2024), Perspective of US farmers on collaborative on-farm agronomic research, *Agronomy Journal*, 116, 1590–1602.

Plastina, A., et al. (2018), Cover crops use in Midwestern US agriculture: perceived benefits and net returns, *Renewable Agriculture and Food Systems*, 35, 38–48.

Plastina, A., et al. (2024), The business case for carbon farming in the USA, *Carbon Balance and Management*, 19, 1–17.

Poffenbarger, H.J., et al. (2015), Biomass and nitrogen content of hairy vetch-cereal rye cover crop mixtures as influenced by species proportions, *Agronomy Journal*, 107, 2069–2082.

Power, J.F., et al. (1983), Legumes in conservation tillage systems: a research perspective, *Journal of Soil and Water Conservation*, 38(3), 217–218.

Prokopy, L.S., et al. (2019), Adoption of agricultural conservation practices in the United States: evidence form 35 years of quantitative literature, *Journal of Soil and Water Conservation*, 74(5), 520–534.

Qin, Z., et al. (2021), Assessing the impacts of cover crops on maize and soybean yield in the U.S. Midwestern agroecosystems, *Field Crops Research*, 273, 108264.

Qin, Z., et al. (2023), Assessing long-term impacts of cover crops on soil organic carbon in the central US Midwestern agroecosystems, *Global Change Biology*, 29, 2572–2590.

Ranells, N.N. and Wagger, M.G. (1997), Grass-legume bicultures as winter annual cover crops, *Agronomy Journal*, 89, 659–665.

Ranells, N.N., and Wagger, M.G. (1992), Nitrogen release from crimson clover in relation to plant growth stage and composition, *Agronomy Journal*, 84(3), 424–430.

Reimer, A.P., et al. (2012), The multi-dimensional nature of environmental attitudes among farmers in Indiana: implications for conservation adoption, *Agriculture and Human Values*, 29, 29–40.

Reiter, M.S., et al. (2008), Cotton nitrogen management in a high-residue conservation system: cover crop fertilization, *Soil Science Society of America Journal*, 72, 1321–1329.

Roesch-McNally, G.E., et al. (2017), The trouble with cover crops: Farmer' experience with overcoming barriers to adoption, *Renewable Agriculture and Food Systems*, 33(4), 322–333.

Rogers, T.H. and Giddens, J.E. (1957), Green manure and cover crops, In *Soil, the Yearbook of Agriculture 1957*, USDA, Washington, DC.

Ruis, S.J., et al. (2019), Cover crop biomass production in temperate agroecozones, *Agronomy Journal*, 111(4), 1535–1551.

Sanford, W.E. and Pope, J.P. (2013), Quantifying groundwater's role in delaying improvements to Chesapeake Bay water quality, *Environmental Science & Technology*, 47, 13330–13338.

Sare Technical Bulletin, (2019), *Cover Crop Economics: Opportunities to Improve Your Bottom Line in Row Crops*, USDA SARE June 2019.

Sarrantonio, M. and Scott, T.W. (1988), Tillage effects on availability of nitrogen to corn following a winter green manure crop, *Soil Science Society of America Journal*, 52, 1661–1668.

Schipanski, M.E., et al. (2014), A framework for evaluating ecosystem services provided by cover crops in agroecosystems, *Agricultural Systems*, 125, 12–22.

Schnitkey, G.D., et al. (2024), Cover crops, farm economics, and policy, *Applied Economic Perspectives and Policy*, 46, 595–608.

Shipley, P.R., et al. (1992), Conserving residual corn fertilizer nitrogen with winter cover crops, *Agronomy Journal*, 84(5), 869–876.

Silva, G.C. and Bagavathiannan, M. (2023), Mechanisms of weed suppression by cereal rye cover crop: a review, *Agronomy Journal*, 115, 1571–1585.

Singh, G., et al. (2020), Tillage and legume non-legume cover cropping effects on corn-soybean production, *Agronomy Journal*, 112, 2636–2648.

Snapp, S.S., et al. (2005), Evaluating cover crops for benefits, costs and performance within cropping system niches, *Agronomy Journal*, 97, 322–332.

Smith, M.S., et al. (1987), Legume winter cover crops, *Advances in Soil Science*, 7, 96–139.

Staver, K.W. (2020), The growing role of dissolved nutrients in soil and water conservation, In Delgado, J.A., et al. (eds.), *Soil and Water Conservation: A Celebration of 75 Years*, Ankeny, IA: Soil and Water Conservation Society, pp. 185–196.

Staver, K.W. and Brinsfield, R.B. (1998), Using cereal grain winter cover crops to reduce groundwater nitrate contamination in the mid-Atlantic coastal plain, *Journal of Soil and Water Conservation*, 53(3), 230–240.

Sun, H., et al. (2025), The flux of agricultural conservation: Understanding changes in Iowa farmers' adoption of cover crops and no-till over time, *Society and Natural Resources*. https://doi.org/10.1080/08941920.2025.2538175.

Teasdale, J.R. and Mohler, C.L. (2000), The quantitative relationship between weed emergence and the physical properties of mulches, *Weed Science*, 48, 385–392.

Tellatin, S. and Myers, R.L. (2018), Cover crop impact on US cropland carbon sequestration, *Journal of Soil and Water Conservation*, 73(5), 117A–121A.

Thapa, R., et al. (2018), Cover crops reduce nitrate leaching in Agroecosystems: a global meta-analysis, *Journal of Environmental Quality*, 47, 1400–1411.

Thorup-Kristensen, K., et al. (2002), Catch crops and green manures as biological tools in nitrogen management in temperate zones, *Advances in Agronomy*, 79, 227–300.

Tonitto, C., et al. (2006), Replacing bare fallows with cover crops in fertilizer-intensive cropping systems: a meta-analysis of crop yield and N dynamics, *Agriculture, Ecosystems and Environment*, 112, 58–72.

Unger, P.W. and Vigil, M.F. (1998), Cover crop effects on soil water relationships, *Journal of Soil and Water Conservation*, 53(3), 200–207.

US EPA CBP, (2017), US Environmental Protection Agency Chesapeake Bay Program 2017, Cover Crops Practices For Use in Phase 6.0 of the Chesapeake Bay Program Watershed Model, https://www.chesapeakebay.net/files/documents/Phase_6_CC_EP_Final_Report_12-16-2016-NEW_TEMPLATE_FINAL.pdf.

Vyn, T.J., et al. (2000), Cover crop effects on nitrogen availability to corn following wheat, *Agronomy Journal*, 92, 915–924.
Wagger, M.G. (1989), Cover crop management and nitrogen rate in relation to growth and yield of no-till corn, *Agronomy Journal*, 81, 533–538.
Wallander, S., et al. (2021), Cover crop trends, programs, and practices in the United States, EIB 222, U.S. Department of Agriculture, Economic Research Service.
Wallander, S., et al. (2020), A history of economic research on soil conservation incentives. In Delgado, J.A., et al. (eds.), *Soil and Water Conservation: A Celebration of 75 Years*, Ankeny, IA: Soil and Water Conservation Society, pp. 57–69.
Wayman, S. et al. (2016), Organic and conventional farmers differ in their perspectives on cover crop use. *Renewable Agriculture and Food Systems*, 32(4): 376–385.
Wells, O.V., et al. (1936), The remedies: direct aids to farmers, In *Soils and Men, Yearbook of Agriculture 1938* USDA, 75th Congress, 2nd Session, House Document No. 398, Washington, DC, pp. 279–288.
White, C.M., et al. (2017), Managing the tradeoff between nitrogen supply and retention with cover crop mixtures, *Agriculture, Ecosystems and Environment*, 237, 121–133.
White, C.M., et al. (2023), Recalibrating the Pennsylvania pre-side dress soil nitrate test recommendations for modern corn production, *Agronomy Journal*, 115, 2600–2613.
Woodcock, B.A., et al. (2025), Agroecological farming promotes yield and biodiversity but may require subsidy to be profitable. *Journal of Applied Ecology*, 62: 1902–1913.

Chapter 17

The future of cover crops: opportunities, challenges and R&D needs

Sjoerd W. Duiker, The Pennsylvania State University; and Andy Clark, formerly University of Maryland and USDA Sustainable Agriculture Research and Education (SARE), USA

1 Introduction

Cover crops are grown during ecological niches in crop rotations that would otherwise be fallow. Fallow periods make our cropping systems 'leaky'. With a loss of potentially captured sunlight through photosynthesis come decreases in soil health, losses of nutrients by runoff or leaching, lack of habitat for natural enemies, and loss of biodiversity. Cover crops are grown to fill these niches for nutrient management, water management, pest management, soil improvement, carbon sequestration, or the broader ecosystem services of water quality improvement, increased biodiversity, or climate resilience.

This volume highlights recent research across most facets of cover cropping with contributions from leaders in cover crop science. Roles of cover crops explored in Part 1 of this book (Chapters 1–6) focus on their effects on nutrient cycling, soil biological health, arthropod (insect, spiders, mites, etc.) diversity, crop pests, weed management, forage production, and crop yields. Part 2 expands on cover crop management (decision support tools, establishment, and termination), with Part 3 discussing common cover crop species. The book ends with a discussion of strategies to increase adoption. The authors present many new innovations in cover crop science and ways to fine-tune

http://dx.doi.org/10.19103/AS.2025.0162.21

management to address specific goals. They highlight the site-specific nature of cover cropping and how research results in better recommendations for cover crop species selection and management. We are particularly pleased to showcase methods to better fit cover crops into all sorts of cropping systems in locations once thought to be unsuitable.

2 Barriers to cover crop adoption

However, despite many reasons for using cover crops, their adoption remains limited (Chapter 16). Cover crops are only grown on 6–7% of U.S. acreage (22 million acres; https://carboncredits.com/cover-crops-carbon/; Fendrich et al., 2023), compared to 19% of European arable land (2016 data, Heller et al., 2024). Chapter 16 by Ken Staver identifies several barriers that prevent farmers from adopting cover crops, with cost being a major one. Because cover crops provide public goods such as improved water quality, he makes the case for financial incentives or subsidies to favor adoption, showing their effectiveness in the states of Maryland and Virginia where adoption rose dramatically after the institution of significant subsidies (averaging $150 per ha). Subsidies probably also explain high adoption in Europe, with new plans in the Common Agricultural Policy 2023–2027 to further incentivize the use of cover crops to improve soil structure and reduce nutrient leaching and erosion (Fendrich et al., 2023). Nonetheless, for poorly understood reasons, some farmers stop using cover crops with or without incentives (Purdue University, 2025; Sun et al., 2025). Besides cost, labor is one of the barriers cited most often by farmers in the U.S. – especially during the busy fall and spring when cash crop harvest and spring planting coincide with cover crop planting and termination, respectively (CTIC, 2023). Another barrier includes lack of information (CTIC, 2023).

Despite low overall adoption, cover crop adoption is high in some areas, even with limited subsidies. Climate plays a role, particularly water availability. Nowak et al. (2022) found that cover crop adoption in France was higher along the Atlantic coast, with wetter summers and warmer autumns than in the more continental parts of France, where dryer summers and cooler autumns prevail. Farmer surveys by Smit et al. (2019) found that the share of farmers growing cover crops was high in humid regions like Centre (France, 83.6% of farmers used cover crops), Overijssel (the Netherlands, 98.7%), although also high in semi-arid South Muntenia (Romania, 46.1%). Cover crop adopters in Castilla and Leon, Spain (12% adoption), usually had access to irrigation. As pointed out in Chapter 10, cover crop adoption in the United States is high where farmers can use cover crops as livestock feed (30% of dairy and feedlot operations reported adopting cover crops in the 2017 USDA Census of Agriculture; Bowman et al., 2024).

3 Key themes in cover crop research

Part 1 of this book discusses the latest research on selection and management of cover crop species to address specific farm or environmental goals. Whether growing cover crops for nutrient management, water management, pest management, soil improvement, carbon sequestration, or the broader ecosystem services of water quality improvement, increased biodiversity, or climate resilience, research-based information about the best species and their management can be found here. We have come so far in cover crop research that we now have the information needed to manage cover crops for beneficial arthropod populations, while in contrast, we are also learning how best to manage arthropod pests with cover crops. Better understanding of the benefits and drawbacks of cover crops will hopefully help researchers, service providers, policy makers, and farmers leading to more widespread adoption but also research and development to address remaining bottlenecks.

Parts 2 and 3 discuss species, decision-making tools, and management options to fit cover crops into the ecological niches in current cropping systems, but they also show ways to change cropping systems to allow adequate time for cover crop establishment and growth without sacrificing cash crop yield. Many innovations are described, including overseeding and interseeding of cover crops where the length of the growing season otherwise limits adoption, modifying the cropping system by using shorter-season hybrids, integrating livestock into cropping systems and harvesting cover crops for forage, or occasionally foregoing a cash crop in order to grow a soil-building or weed-suppressing cover crop.

4 Future research needs

We sincerely hope and have every reason to believe that cover crop adoption will continue to increase in the years to come. The research reviewed here has helped to advance our understanding of the many facets of cover crops in the landscape. But it has also left questions unanswered. Based on the contributions in this book, we identified the following priority needs:

1) Identify **new cover crop species** that can better fit in the ecological niches in common crop rotations. The authors in Part 3 discuss 32 cool and 9 warm season legume cover crops, 6 cool and 9 warm season grasses, and 8 non-legume broadleaves, most of which are often not able to produce sufficient biomass before needing to be terminated. The challenge we'd like to put before the research community is: can we screen the >300,000 known land plant species to find species that better fit in the ecological niches for cover crops in common crop rotations?

New tools such as Artificial Intelligence may enable us. Further, cover crop breeding research and development is also needed to fine-tune species' adaptation to different climates. Recent work at 12 sites across the U.S. evaluated cover crop species adaptation to diverse climates. Included were promising new varieties being developed specifically for adaptability to different climes (https://cra.missouri.edu/wp-content/uploads/2025/07/2024-2025-National-Cover-Crop-Variety-Trial.pdf). This research is part of the larger Cover Crop Breeding Network, which selects and breeds for winter hardiness, early vigor, high biomass, disease resistance, flowering time, seed yield, pod retention, and soft seed (https://www.covercropbreeding.com/publications).

2) There is an urgent need to identify **new legume cover crop species or varieties** that can produce sufficient biomass during the fallow period. Non-legume cover crops often lead to yield reductions in following non-legume economical crops while needing extra nitrogen fertilizer (Chapters 1, 7, and 13). This is likely why many farmers are still reluctant to adopt cover crops. Legume cover crops, on the other hand, typically have a neutral or positive effect on following non-legume cash crop yields, while enabling reduced nitrogen fertilizer applications. However, we lack affordable cool or warm season legumes that can produce substantial biomass (perhaps >2500 kg ha^{-1} is a good threshold) before needing to be terminated.
3) There is need for **new methods of termination** and management besides tillage or herbicides in Conservation Agriculture, considering the development of weed herbicide resistance and public concern about herbicides on top of the harm tillage does to soil. Examples are mowing, crushing, electrocuting, and grazing by livestock (Chapters 6 and 10).
4) Improve our understanding of how cover crops can be used to **reduce pests and diseases** by acting as non-hosts, biofumigants, trap crops, etc., as contrasted with negative cover crop effects (Chapters 3, 4, and 5).
5) A better understanding of the **soil health effects of cover crops** and how this translates into crop yield and environmental improvement under different management scenarios (Chapters 1, 2, and 7). Examples are root growth and architecture, root exudates, mycorrhizae, nutrient uptake, and mobilization dynamics.
6) Develop ways to **design cover crop mixes** for specific objectives (many chapters in this book). Is it possible to develop a location-specific cover crop mixture calculator to meet farmers' needs?
7) Quantify the **environmental trade-offs between non-harvested vs harvested (for forage) or grazed cover crops**. Adoption of cover crops

tends to increase dramatically if they have an immediate economic value to the farmer and might still fulfill most of the desired environmental functions (Chapters 6 and 16; Duiker and Richards, 2023). Cover crops could help stimulate livestock integration as part of regenerative agriculture systems.

8) A better understanding of how **cover crops can help mitigate the effects of climate change**. Erratic climate, especially but not limited to torrential rainfall events separated by drought periods and intense heat, continues to plague farmers worldwide. Cover crops reduce raindrop-induced soil compaction, increase infiltration, and reduce soil and wind erosion. Cover crop residue in Conservation Agriculture systems conserves soil moisture for the cash crop, while continuing to buffer the system against the impacts of excessive rainfall and protecting soil roots from excessive heat.
9) There is a need for **better cover crop decision support tools** (Chapter 8). Possible reductions in crop inputs in the following economic crop or environmental benefits depend on cover crop stand, composition, and biomass which can vary in time and space. Nitrogen applications, herbicide or tillage passes, and disease or pest control can be changed on the go with precision sensing and application equipment that needs to be developed.
10) **New ideas for government involvement** in assisting farmers to adopt cover crops (Chapter 16). Large sums are currently spent on farm subsidies, but governments might also help increase seed availability and affordability of a larger offering of cover crops, support technical assistance to farmers (e.g. crop consultants), and increase the knowledge and expertise about cover crops among agronomists advising farmers. This could help remove hurdles of cost, management, and time investment in cover crops that currently hinder adoption.

5 Conclusions

This compilation of the latest cover crop research reveals our increased understanding of the importance of cover crops for sustainable soil management, the ecosystem services they provide, and the potential they offer for yield improvement while reducing inputs. New cover crop management techniques and uses have been developed to increase benefits while avoiding drawbacks. However, to reap greater benefits from cover crops and increase adoption, much work still needs to be done, showing the need for increased funding for research, teaching, and development of cover crops. Cover crops have many environmental benefits that justify public investment. However, research is often limited to few species and few uses due to limited availability of funding. Greater investment in R&D of cover crops will lead to new discoveries of better cover crops and mixes, improved understanding of underlying principles, and more widespread adoption of cover crops.

6 Where to look for further information

Carbon farming references:

- University of Wisconsin-Madison (UW): An introduction to carbon markets, carbon credits (https://www.wibiomass.org/_files/ugd/f0664f_2971b766b50d44d895d52ef6331a5b9d.pdf).
- University of Nebraska-Lincoln (UNeb): Carbon Farming: A preliminary economic analysis (https://digitalcommons.unl.edu/cgi/viewcontent.cgi?article=2200&context=agecon_cornhusker).

7 References

Duiker, S.W. & Richards, S. (2023). Adoption of cover crops in Pennsylvania's Chesapeake Bay Watershed. *Journal of Soil and Water Conservation*, 78, 376–383. https://doi.org/10.2489/jswc.2023.00080.

Fendrich, A.N., et al. (2023). From regional to parcel scale: A high-resolution map of cover crops across Europe combining satellite data with statistical surveys. *Science of the Total Environment*, 873, Article 162306.

Heller, O., et al. (2024). Towards enhanced adoption of soil-improving management practices in Europe. *European Journal of Soil Science*, 75(2), e13483. https://doi.org/10.1111/ejss.13483.

Langemeier, M., et al. (2025). *Cover Crop Adoption: Results from 5 Years of Ag Economy Barometer Surveys*. Purdue University Center for Commercial Agriculture. https://ag.purdue.edu/commercialag/home/resource/2025/10/cover-crop-adoption/.

National Sustainable Agriculture Coalition (NSAC). (2025). Keeping Farmers on the Land. https://sustainableagriculture.net/blog/keeping-farmers-on-the-land/.

Nowak, B., et al. (2022). Soil-climate factors have a greater influence on the presence of winter cover crops than regulatory constraints in France. *Agronomy for Sustainable Development*, 42(2), 28. https://doi.org/10.1007/s13593-022-00770-y.

Smit, B., et al. (2019). *Adoption of Cover Crops for Climate Change Mitigation in the EU*. JRC Technical Reports JRC116730, pp. 1–76. Publications Office of the European Union. http://doi.org/10.2760/638382.

Sun, H., et al. (2025). The flux of agricultural conservation: understanding changes in Iowa farmers' adoption of cover crops and no-till over time. *Society & Natural Resources*, 1–20. https://doi.org/10.1080/08941920.2025.2538175.

Sutton, E., et al. (2025). *National Cover Crop Variety Testing Program 2024–2025 Trial Results*. Center for Regenerative Agriculture, University of Missouri; North Carolina State University-. https://cra.missouri.edu/wp-content/uploads/2025/07/2024-2025-National-Cover-Crop-Variety-Trial.pdf.

Index

www.ingramcontent.com/pod-product-compliance
Ingram Content Group UK Ltd.
Pitfield, Milton Keynes, MK11 3LW, UK
UKHW021008120926
8556IPUK00028B/140

9 781835 451205